DATA STRUCTURES AND ALGORITHM ANALYSIS IN C

MARK ALLEN WEISS
Florida International University

SECOND EDITION

Addison Wesley Longman

An imprint of Addison Wesley Longman, Inc.

Menlo Park, California • Reading, Massachusetts • Harlow, England
Berkeley, California • Don Mills, Ontario • Sydney • Bonn • Amsterdam • Tokyo • Mexico City

Original edition, entitled DATA STRUCTURES AND ALGORITHM ANALYSIS IN C, 2nd Edition, 0201498405 by WEISS, MARK ALLEN, published by Pearson Education, Inc, publishing as Addison-Wesley, Copyright © 1997.

ISBN 978-89-450-0617-2

DATA STRUCTURES AND ALGORITHM ANALYSIS IN C

PREFACE

Purpose/Goals

This book describes *data structures*, methods of organizing large amounts of data, and *algorithm analysis*, the estimation of the running time of algorithms. As computers become faster and faster, the need for programs that can handle large amounts of input becomes more acute. Paradoxically, this requires more careful attention to efficiency, since inefficiencies in programs become most obvious when input sizes are large. By analyzing an algorithm before it is actually coded, students can decide if a particular solution will be feasible. For example, in this text students look at specific problems and see how careful implementations can reduce the time constraint for large amounts of data from 16 years to less than a second. Therefore, no algorithm or data structure is presented without an explanation of its running time. In some cases, minute details that affect the running time of the implementation are explored.

Once a solution method is determined, a program must still be written. As computers have become more powerful, the problems they must solve have become larger and more complex, requiring development of more intricate programs. The goal of this text is to teach students good programming and algorithm analysis skills simultaneously so that they can develop such programs with the maximum amount of efficiency.

This book is suitable for either an advanced data structures (CS7) course or a first-year graduate course in algorithm analysis. Students should have some knowledge of intermediate programming, including such topics as pointers and recursion, and some background in discrete math.

Approach

I believe it is important for students to learn how to program for themselves, not how to copy programs from a book. On the other hand, it is virtually impossible to discuss realistic programming issues without including sample code. For this reason, the book usually provides about one-half to three-quarters of an implementation, and the student is encouraged to supply the rest. Chapter 12, which is new to this edition, discusses additional data structures with an emphasis on implementation details.

The algorithms in this book are presented in ANSI C, which, despite some flaws, is arguably the most popular systems programming language. The use of C instead of Pascal allows the use of dynamically allocated arrays (see, for instance, rehashing in Chapter 5). It also produces simplified code in several places, usually because the *and* (&&) operation is short-circuited.

Most criticisms of C center on the fact that it is easy to write code that is barely readable. Some of the more standard tricks, such as the simultaneous assignment and testing against 0 via

```
if (x=y)
```

are generally not used in the text, since the loss of clarity is compensated by only a few keystrokes and no increased speed. I believe that this book demonstrates that unreadable code can be avoided by exercising reasonable care.

Overview

Chapter 1 contains review material on discrete math and recursion. I believe the only way to be comfortable with recursion is to see good uses over and over. Therefore, recursion is prevalent in this text, with examples in every chapter except Chapter 5.

Chapter 2 deals with algorithm analysis. This chapter explains asymptotic analysis and its major weaknesses. Many examples are provided, including an in-depth explanation of logarithmic running time. Simple recursive programs are analyzed by intuitively converting them into iterative programs. More complicated divide-and-conquer programs are introduced, but some of the analysis (solving recurrence relations) is implicitly delayed until Chapter 7, where it is performed in detail.

Chapter 3 covers lists, stacks, and queues. The emphasis here is on coding these data structures using ADTs, fast implementation of these data structures, and an exposition of some of their uses. There are almost no programs (just routines), but the exercises contain plenty of ideas for programming assignments.

Chapter 4 covers trees, with an emphasis on search trees, including external search trees (B-trees). The UNIX file system and expression trees are used as examples. AVL trees and splay trees are introduced but not analyzed. Seventy-five percent of the code is written, leaving similar cases to be completed by the student. More careful treatment of search tree implementation details is found in Chapter 12. Additional coverage of trees, such as file compression and game trees, is deferred until Chapter 10. Data structures for an external medium are considered as the final topic in several chapters.

Chapter 5 is a relatively short chapter concerning hash tables. Some analysis is performed, and extendible hashing is covered at the end of the chapter.

Chapter 6 is about priority queues. Binary heaps are covered, and there is additional material on some of the theoretically interesting implementations of priority queues. The Fibonacci heap is discussed in Chapter 11, and the pairing heap is discussed in Chapter 12.

Chapter 7 covers sorting. It is very specific with respect to coding details and analysis. All the important general-purpose sorting algorithms are covered and compared. Four algorithms are analyzed in detail: insertion sort, Shellsort, heapsort, and quicksort. The analysis of the average-case running time of heapsort is new to this edition. External sorting is covered at the end of the chapter.

Chapter 8 discusses the disjoint set algorithm with proof of the running time. This is a short and specific chapter that can be skipped if Kruskal's algorithm is not discussed.

Chapter 9 covers graph algorithms. Algorithms on graphs are interesting, not only because they frequently occur in practice but also because their running time is so heavily dependent on the proper use of data structures. Virtually all of the standard algorithms are presented along with appropriate data structures, pseudocode, and analysis of running time. To place these problems in a proper context, a short discussion on complexity theory (including *NP*-completeness and undecidability) is provided.

Chapter 10 covers algorithm design by examining common problem-solving techniques. This chapter is heavily fortified with examples. Pseudocode is used in these later chapters so that the student's appreciation of an example algorithm is not obscured by implementation details.

Chapter 11 deals with amortized analysis. Three data structures from Chapters 4 and 6 and the Fibonacci heap, introduced in this chapter, are analyzed.

Chapter 12 is new to this edition. It covers search tree algorithms, the *k*-d tree, and the pairing heap. This chapter departs from the rest of the text by providing complete and careful implementations for the search trees and pairing heap. The material is structured so that the instructor can integrate sections into discussions from other chapters. For example, the top-down red black tree in Chapter 12 can be discussed under AVL trees (in Chapter 4).

Chapters 1–9 provide enough material for most one-semester data structures courses. If time permits, then Chapter 10 can be covered. A graduate course on algorithm analysis could cover Chapters 7–11. The advanced data structures analyzed in Chapter 11 can easily be referred to in the earlier chapters. The discussion of *NP*-completeness in Chapter 9 is far too brief to be used in such a course. Garey and Johnson's book on *NP*-completeness can be used to augment this text.

Exercises

Exercises, provided at the end of each chapter, match the order in which material is presented. The last exercises may address the chapter as a whole rather than a specific section. Difficult exercises are marked with an asterisk, and more challenging exercises have two asterisks.

A solutions manual containing solutions to almost all the exercises is available to instructors from the Addison-Wesley Publishing Company.

References

References are placed at the end of each chapter. Generally the references either are historical, representing the original source of the material, or they represent extensions and improvements to the results given in the text. Some references represent solutions to exercises.

Code Availability

The example program code in this book is available via anonymous ftp at `aw.com`. It is also accessible through the World Wide Web; the URL is `http://www.aw.com/cseng/` (follow the links from there). The exact location of this material may change.

Acknowledgments

Many, many people have helped me in the preparation of books in this series. Some are listed in other versions of the book; thanks to all.

For this edition, I would like to thank my editors at Addison-Wesley, Carter Shanklin and Susan Hartman. Teri Hyde did another wonderful job with the production, and Matthew Harris and his staff at Publication Services did their usual fine work putting the final pieces together.

M.A.W.

Miami, Florida
July, 1996

CONTENTS

CHAPTER 1

Introduction

In this chapter, we discuss the aims and goals of this text and briefly review programming concepts and discrete mathematics. We will

- See that how a program performs for reasonably large input is just as important as its performance on moderate amounts of input.
- Summarize the basic mathematical background needed for the rest of the book.
- Briefly review recursion.

1.1. What's the Book About?

Suppose you have a group of N numbers and would like to determine the kth largest. This is known as the *selection problem*. Most students who have had a programming course or two would have no difficulty writing a program to solve this problem. There are quite a few "obvious" solutions.

One way to solve this problem would be to read the N numbers into an array, sort the array in decreasing order by some simple algorithm such as bubblesort, and then return the element in position k.

A somewhat better algorithm might be to read the first k elements into an array and sort them (in decreasing order). Next, each remaining element is read one by one. As a new element arrives, it is ignored if it is smaller than the kth element in the array. Otherwise, it is placed in its correct spot in the array, bumping one element out of the array. When the algorithm ends, the element in the kth position is returned as the answer.

Both algorithms are simple to code, and you are encouraged to do so. The natural questions, then, are which algorithm is better and, more important, is either algorithm good enough? A simulation using a random file of 1 million elements and $k = 500{,}000$ will show that neither algorithm finishes in a reasonable amount of time; each requires several days of computer processing to terminate (albeit eventually with a correct answer). An alternative method, discussed in Chapter 7, gives a solution in about a second. Thus, although our proposed algorithms work, they cannot be considered good algorithms, because they are entirely impractical for input sizes that a third algorithm can handle in a reasonable amount of time.

A second problem is to solve a popular word puzzle. The input consists of a two-dimensional array of letters and a list of words. The object is to find the words in the puzzle. These words may be horizontal, vertical, or diagonal in any direction. As an example, the puzzle shown in Figure 1.1 contains the words *this, two, fat,* and *that.* The word *this* begins at row 1, column 1, or (1,1), and extends to (1,4); *two* goes from (1,1) to (3,1); *fat* goes from (4,1) to (2,3); and *that* goes from (4,4) to (1,1).

Again, there are at least two straightforward algorithms that solve the problem. For each word in the word list, we check each ordered triple (*row, column, orientation*) for the presence of the word. This amounts to lots of nested *for* loops but is basically straightforward.

Alternatively, for each ordered quadruple (*row, column, orientation, number of characters*) that doesn't run off an end of the puzzle, we can test whether the word indicated is in the word list. Again, this amounts to lots of nested *for* loops. It is possible to save some time if the maximum number of characters in any word is known.

It is relatively easy to code up either method of solution and solve many of the real-life puzzles commonly published in magazines. These typically have 16 rows, 16 columns, and 40 or so words. Suppose, however, we consider the variation where only the puzzle board is given and the word list is essentially an English dictionary. Both of the solutions proposed require considerable time to solve this problem and therefore are not acceptable. However, it is possible, even with a large word list, to solve the problem in a matter of seconds.

An important concept is that, in many problems, writing a working program is not good enough. If the program is to be run on a large data set, then the running time becomes an issue. Throughout this book we will see how to estimate the running time of a program for large inputs and, more important, how to compare the running times of two programs without actually coding them. We will see techniques for drastically improving the speed of a program and for determining program bottlenecks. These techniques will enable us to find the section of the code on which to concentrate our optimization efforts.

Figure 1.1 Sample word puzzle

	1	2	3	4
1	t	h	i	s
2	w	a	t	s
3	o	a	h	g
4	f	g	d	t

1.2. Mathematics Review

This section lists some of the basic formulas you need to memorize or be able to derive and reviews basic proof techniques.

1.2.1. *Exponents*

$$X^A X^B = X^{A+B}$$

$$\frac{X^A}{X^B} = X^{A-B}$$

$$(X^A)^B = X^{AB}$$

$$X^N + X^N = 2X^N \neq X^{2N}$$

$$2^N + 2^N = 2^{N+1}$$

1.2.2. *Logarithms*

In computer science, all logarithms are to the base 2 unless specified otherwise.

DEFINITION: $X^A = B$ if and only if $\log_X B = A$

Several convenient equalities follow from this definition.

THEOREM 1.1.

$$\log_A B = \frac{\log_C B}{\log_C A}; \quad C > 0$$

PROOF:
Let $X = \log_C B$, $Y = \log_C A$, and $Z = \log_A B$. Then, by the definition of logarithms, $C^X = B$, $C^Y = A$, and $A^Z = B$. Combining these three equalities yields $(C^Y)^Z = C^X = B$. Therefore, $X = YZ$, which implies $Z = X/Y$, proving the theorem.

THEOREM 1.2.

$$\log AB = \log A + \log B$$

PROOF:
Let $X = \log A$, $Y = \log B$, and $Z = \log AB$. Then, assuming the default base of 2, $2^X = A$, $2^Y = B$, and $2^Z = AB$. Combining the last three equalities yields $2^X 2^Y = 2^Z = AB$. Therefore, $X + Y = Z$, which proves the theorem.

Some other useful formulas, which can all be derived in a similar manner, follow.

$$\log A/B = \log A - \log B$$

$$\log(A^B) = B \log A$$

$$\log X < X \quad \text{for all } X > 0$$

$$\log 1 = 0, \qquad \log 2 = 1, \qquad \log 1{,}024 = 10, \qquad \log 1{,}048{,}576 = 20$$

1.2.3. Series

The easiest formulas to remember are

$$\sum_{i=0}^{N} 2^i = 2^{N+1} - 1$$

and the companion,

$$\sum_{i=0}^{N} A^i = \frac{A^{N+1} - 1}{A - 1}$$

In the latter formula, if $0 < A < 1$, then

$$\sum_{i=0}^{N} A^i \leq \frac{1}{1 - A}$$

and as N tends to ∞, the sum approaches $1/(1 - A)$. These are the "geometric series" formulas.

We can derive the last formula for $\sum_{i=0}^{\infty} A^i$ $(0 < A < 1)$ in the following manner. Let S be the sum. Then

$$S = 1 + A + A^2 + A^3 + A^4 + A^5 + \cdots$$

Then

$$AS = A + A^2 + A^3 + A^4 + A^5 + \cdots$$

If we subtract these two equations (which is permissible only for a convergent series), virtually all the terms on the right side cancel, leaving

$$S - AS = 1$$

which implies that

$$S = \frac{1}{1 - A}$$

We can use this same technique to compute $\sum_{i=1}^{\infty} i/2^i$, a sum that occurs frequently. We write

$$S = \frac{1}{2} + \frac{2}{2^2} + \frac{3}{2^3} + \frac{4}{2^4} + \frac{5}{2^5} + \cdots$$

and multiply by 2, obtaining

$$2S = 1 + \frac{2}{2} + \frac{3}{2^2} + \frac{4}{2^3} + \frac{5}{2^4} + \frac{6}{2^5} + \cdots$$

Subtracting these two equations yields

$$S = 1 + \frac{1}{2} + \frac{1}{2^2} + \frac{1}{2^3} + \frac{1}{2^4} + \frac{1}{2^5} + \cdots$$

Thus, $S = 2$.

Another type of common series in analysis is the arithmetic series. Any such series can be evaluated from the basic formula.

$$\sum_{i=1}^{N} i = \frac{N(N+1)}{2} \approx \frac{N^2}{2}$$

For instance, to find the sum $2 + 5 + 8 + \cdots + (3k - 1)$, rewrite it as $3(1 + 2 + 3 + \cdots + k) - (1 + 1 + 1 + \cdots + 1)$, which is clearly $3k(k + 1)/2 - k$. Another way to remember this is to add the first and last terms (total $3k + 1$), the second and next to last terms (total $3k + 1$), and so on. Since there are $k/2$ of these pairs, the total sum is $k(3k + 1)/2$, which is the same answer as before.

The next two formulas pop up now and then but are fairly uncommon.

$$\sum_{i=1}^{N} i^2 = \frac{N(N+1)(2N+1)}{6} \approx \frac{N^3}{3}$$

$$\sum_{i=1}^{N} i^k \approx \frac{N^{k+1}}{|k+1|} \qquad k \neq -1$$

When $k = -1$, the latter formula is not valid. We then need the following formula, which is used far more in computer science than in other mathematical disciplines. The numbers H_N are known as the harmonic numbers, and the sum is known as a harmonic sum. The error in the following approximation tends to $\gamma \approx 0.57721566$, which is known as *Euler's constant.*

$$H_N = \sum_{i=1}^{N} \frac{1}{i} \approx \log_e N$$

These two formulas are just general algebraic manipulations.

$$\sum_{i=1}^{N} f(N) = Nf(N)$$

$$\sum_{i=n_0}^{N} f(i) = \sum_{i=1}^{N} f(i) - \sum_{i=1}^{n_0-1} f(i)$$

1.2.4. Modular Arithmetic

We say that A is congruent to B modulo N, written $A \equiv B \pmod N$, if N divides $A - B$. Intuitively, this means that the remainder is the same when either A or B is divided by N. Thus, $81 \equiv 61 \equiv 1 \pmod{10}$. As with equality, if $A \equiv B \pmod N$, then $A + C \equiv B + C \pmod N$ and $AD \equiv BD \pmod N$.

There are many theorems that apply to modular arithmetic, and some of them require extraordinary proofs in number theory. We will use modular arithmetic sparingly, and the preceding theorems will suffice.

1.2.5. The P Word

The two most common ways of proving statements in data structure analysis are proof by induction and proof by contradiction (and occasionally proof by intimidation, used by professors only). The best way of proving that a theorem is false is by exhibiting a counterexample.

Proof by Induction

A proof by induction has two standard parts. The first step is proving a *base case,* that is, establishing that a theorem is true for some small (usually degenerate) value(s); this step is almost always trivial. Next, an *inductive hypothesis* is assumed. Generally this means that the theorem is assumed to be true for all cases up to some limit k. Using this assumption, the theorem is then shown to be true for the next value, which is typically $k + 1$. This proves the theorem (as long as k is finite).

As an example, we prove that the Fibonacci numbers, $F_0 = 1, F_1 = 1, F_2 = 2, F_3 = 3, F_4 = 5, \ldots, F_i = F_{i-1} + F_{i-2}$, satisfy $F_i < (5/3)^i$, for $i \geq 1$. (Some definitions have $F_0 = 0$, which shifts the series.) To do this, we first verify that the theorem is true for the trivial cases. It is easy to verify that $F_1 = 1 < 5/3$ and $F_2 = 2 < 25/9$; this proves the basis. We assume that the theorem is true for $i = 1, 2, \ldots, k$; this is the inductive hypothesis. To prove the theorem, we need to show that $F_{k+1} < (5/3)^{k+1}$. We have

$$F_{k+1} = F_k + F_{k-1}$$

by the definition, and we can use the inductive hypothesis on the right-hand side, obtaining

$$\begin{aligned} F_{k+1} &< (5/3)^k + (5/3)^{k-1} \\ &< (3/5)(5/3)^{k+1} + (3/5)^2(5/3)^{k+1} \\ &< (3/5)(5/3)^{k+1} + (9/25)(5/3)^{k+1} \end{aligned}$$

which simplifies to

$$\begin{aligned} F_{k+1} &< (3/5 + 9/25)(5/3)^{k+1} \\ &< (24/25)(5/3)^{k+1} \\ &< (5/3)^{k+1} \end{aligned}$$

proving the theorem.

As a second example, we establish the following theorem.

THEOREM 1.3.

If $N \geq 1$, *then* $\sum_{i=1}^{N} i^2 = \frac{N(N + 1)(2N + 1)}{6}$

PROOF:

The proof is by induction. For the basis, it is readily seen that the theorem is true when $N = 1$. For the inductive hypothesis, assume that the theorem is true for

$1 \le k \le N$. We will establish that, under this assumption, the theorem is true for $N + 1$. We have

$$\sum_{i=1}^{N+1} i^2 = \sum_{i=1}^{N} i^2 + (N+1)^2$$

Applying the inductive hypothesis, we obtain

$$\begin{aligned}
\sum_{i=1}^{N+1} i^2 &= \frac{N(N+1)(2N+1)}{6} + (N+1)^2 \\
&= (N+1)\left[\frac{N(2N+1)}{6} + (N+1)\right] \\
&= (N+1)\frac{2N^2 + 7N + 6}{6} \\
&= \frac{(N+1)(N+2)(2N+3)}{6}
\end{aligned}$$

Thus,

$$\sum_{i=1}^{N+1} = \frac{(N+1)[(N+1)+1][2(N+1)+1]}{6}$$

proving the theorem.

Proof by Counterexample

The statement $F_k \le k^2$ is false. The easiest way to prove this is to compute $F_{11} = 144 > 11^2$.

Proof by Contradiction

Proof by contradiction proceeds by assuming that the theorem is false and showing that this assumption implies that some known property is false, and hence the original assumption was erroneous. A classic example is the proof that there is an infinite number of primes. To prove this, we assume that the theorem is false, so that there is some largest prime P_k. Let $P_1, P_2, \ldots, P_k$ be all the primes in order and consider

$$N = P_1 P_2 P_3 \cdots P_k + 1$$

Clearly, N is larger than P_k, so by assumption N is not prime. However, none of $P_1, P_2, \ldots, P_k$ divides N exactly, because there will always be a remainder of 1. This is a contradiction, because every number is either prime or a product of primes. Hence, the original assumption, that P_k is the largest prime, is false, which implies that the theorem is true.

```
        int
        F( int X )
        {
/* 1*/      if( X == 0 )
/* 2*/          return 0;
            else
/* 3*/          return 2 * F( X - 1 ) + X * X;
        }
```

Figure 1.2 A recursive function

1.3. A Brief Introduction to Recursion

Most mathematical functions that we are familiar with are described by a simple formula. For instance, we can convert temperatures from Fahrenheit to Celsius by applying the formula

$$C = 5(F - 32)/9$$

Given this formula, it is trivial to write a C function; with declarations and braces removed, the one-line formula translates to one line of C.

Mathematical functions are sometimes defined in a less standard form. As an example, we can define a function F, valid on nonnegative integers, that satisfies $F(0) = 0$ and $F(X) = 2F(X - 1) + X^2$. From this definition we see that $F(1) = 1$, $F(2) = 6$, $F(3) = 21$, and $F(4) = 58$. A function that is defined in terms of itself is called *recursive*. C allows functions to be recursive.* It is important to remember that what C provides is merely an attempt to follow the recursive spirit. Not all mathematically recursive functions are efficiently (or correctly) implemented by C's simulation of recursion. The idea is that the recursive function F ought to be expressible in only a few lines, just like a nonrecursive function. Figure 1.2 shows the recursive implementation of F.

Lines 1 and 2 handle what is known as the *base case*, that is, the value for which the function is directly known without resorting to recursion. Just as declaring $F(X) = 2F(X - 1) + X^2$ is meaningless, mathematically, without including the fact that $F(0) = 0$, the recursive C function doesn't make sense without a base case. Line 3 makes the recursive call.

There are several important and possibly confusing points about recursion. A common question is: Isn't this just circular logic? The answer is that although we are defining a function in terms of itself, we are not defining a particular instance of the function in terms of itself. In other words, evaluating $F(5)$ by computing $F(5)$ would be circular. Evaluating $F(5)$ by computing $F(4)$ is not circular—unless, of course, $F(4)$ is evaluated by eventually computing $F(5)$. The two most important issues are

*Using recursion for numerical calculations is usually a bad idea. We have done so to illustrate the basic points.

probably the *how* and *why* questions. In Chapter 3, the *how* and *why* issues are formally resolved. We will give an incomplete description here.

It turns out that recursive calls are handled no differently from any others. If F is called with the value of 4, then line 3 requires the computation of $2 * F(3) + 4 * 4$. Thus, a call is made to compute $F(3)$. This requires the computation of $2 * F(2) + 3 * 3$. Therefore, another call is made to compute $F(2)$. This means that $2 * F(1) + 2 * 2$ must be evaluated. To do so, $F(1)$ is computed as $2 * F(0) + 1 * 1$. Now, $F(0)$ must be evaluated. Since this is a base case, we know a priori that $F(0) = 0$. This enables the completion of the calculation for $F(1)$, which is now seen to be 1. Then $F(2)$, $F(3)$, and finally $F(4)$ can be determined. All the bookkeeping needed to keep track of pending function calls (those started but waiting for a recursive call to complete), along with their variables, is done by the computer automatically. An important point, however, is that recursive calls will keep on being made until a base case is reached. For instance, an attempt to evaluate $F(-1)$ will result in calls to $F(-2)$, $F(-3)$, and so on. Since this will never get to a base case, the program won't be able to compute the answer (which is undefined anyway). Occasionally, a much more subtle error is made, which is exhibited in Figure 1.3. The error in the program in Figure 1.3 is that *Bad*(1) is defined, by line 3, to be *Bad*(1). Obviously, this doesn't give any clue as to what *Bad*(1) actually is. The computer will thus repeatedly make calls to *Bad*(1) in an attempt to resolve its values. Eventually, its bookkeeping system will run out of space, and the program will crash. Generally, we would say that this function doesn't work for one special case but is correct otherwise. This isn't true here, since *Bad*(2) calls *Bad*(1). Thus, *Bad*(2) cannot be evaluated either. Furthermore, *Bad*(3), *Bad*(4), and *Bad*(5) all make calls to *Bad*(2). Since *Bad*(2) is unevaluable, none of these values are either. In fact, this program doesn't work for any value of N, except 0. With recursive programs, there is no such thing as a "special case."

These considerations lead to the first two fundamental rules of recursion:

1. *Base cases*. You must always have some base cases, which can be solved without recursion.
2. *Making progress*. For the cases that are to be solved recursively, the recursive call must always be to a case that makes progress toward a base case.

Figure 1.3 A nonterminating recursive program

```
        int
        Bad( unsigned int N )
        {
/* 1*/      if( N == 0 )
/* 2*/          return 0;
            else
/* 3*/          return Bad( N / 3 + 1 ) + N - 1;
        }
```

Throughout this book, we will use recursion to solve problems. As an example of a nonmathematical use, consider a large dictionary. Words in dictionaries are defined in terms of other words. When we look up a word, we might not always understand the definition, so we might have to look up words in the definition. Likewise, we might not understand some of those, so we might have to continue this search for a while. Because the dictionary is finite, eventually either (1) we will come to a point where we understand all of the words in some definition (and thus understand that definition and retrace our path through the other definitions) or (2) we will find that the definitions are circular and we are stuck, or that some word we need to understand for a definition is not in the dictionary.

Our recursive strategy to understand words is as follows: If we know the meaning of a word, then we are done; otherwise, we look the word up in the dictionary. If we understand all the words in the definition, we are done; otherwise, we figure out what the definition means by ***recursively*** looking up the words we don't know. This procedure will terminate if the dictionary is well defined but can loop indefinitely if a word is either not defined or circularly defined.

Printing Out Numbers

Suppose we have a positive integer, N, that we wish to print out. Our routine will have the heading *PrintOut*(N). Assume that the only I/O routines available will take a single-digit number and output it to the terminal. We will call this routine *PrintDigit*; for example, *PrintDigit*(4) will output a 4 to the terminal.

Recursion provides a very clean solution to this problem. To print out 76234, we need to first print out 7623 and then print out 4. The second step is easily accomplished with the statement *PrintDigit*(N%10), but the first doesn't seem any simpler than the original problem. Indeed it is virtually the same problem, so we can solve it recursively with the statement *PrintOut*(N/10).

This tells us how to solve the general problem, but we still need to make sure that the program doesn't loop indefinitely. Since we haven't defined a base case yet, it is clear that we still have something to do. Our base case will be *PrintDigit*(N) if $0 \leq N < 10$. Now *PrintOut*(N) is defined for every positive number from 0 to 9, and larger numbers are defined in terms of a smaller positive number. Thus, there is no cycle. The entire procedure* is shown in Figure 1.4.

We have made no effort to do this efficiently. We could have avoided using the mod routine (which is very expensive) because $N \% 10 = N - \lfloor N/10 \rfloor * 10$.†

Recursion and Induction

Let us prove (somewhat) rigorously that the recursive number-printing program works. To do so, we'll use a proof by induction.

---

*The term *procedure* refers to a function that returns *void*.

†$\lfloor X \rfloor$ is the largest integer that is less than or equal to X.

```
void
PrintOut( unsigned int N )  /* Print nonnegative N */
{
    if( N >= 10 )
        PrintOut( N / 10 );
    PrintDigit( N % 10 );
}
```

Figure 1.4 Recursive routine to print an integer

THEOREM 1.4.

The recursive number-printing algorithm is correct for $N \geq 0$.

PROOF (*BY INDUCTION ON THE NUMBER OF DIGITS IN N*):

First, if N has one digit, then the program is trivially correct, since it merely makes a call to *PrintDigit*. Assume then that *PrintOut* works for all numbers of k or fewer digits. A number of $k + 1$ digits is expressed by its first k digits followed by its least significant digit. But the number formed by the first k digits is exactly $\lfloor N/10 \rfloor$, which, by the inductive hypothesis, is correctly printed, and the last digit is $N \bmod 10$, so the program prints out any $(k + 1)$-digit number correctly. Thus, by induction, all numbers are correctly printed.

This proof probably seems a little strange in that it is virtually identical to the algorithm description. It illustrates that in designing a recursive program, all smaller instances of the same problem (which are on the path to a base case) may be *assumed* to work correctly. The recursive program needs only to combine solutions to smaller problems, which are "magically" obtained by recursion, into a solution for the current problem. The mathematical justification for this is proof by induction. This gives the third rule of recursion:

3. *Design rule*. Assume that all the recursive calls work.

This rule is important because it means that when designing recursive programs, you generally don't need to know the details of the bookkeeping arrangements, and you don't have to try to trace through the myriad of recursive calls. Frequently, it is extremely difficult to track down the actual sequence of recursive calls. Of course, in many cases this is an indication of a good use of recursion, since the computer is being allowed to work out the complicated details.

The main problem with recursion is the hidden bookkeeping costs. Although these costs are almost always justifiable, because recursive programs not only simplify the algorithm design but also tend to give cleaner code, recursion should never be used as a substitute for a simple *for* loop. We'll discuss the overhead involved in recursion in more detail in Section 3.3.

When writing recursive routines, it is crucial to keep in mind the four basic rules of recursion:

1. *Base cases*. You must always have some base cases, which can be solved without recursion.

2. *Making progress.* For the cases that are to be solved recursively, the recursive call must always be to a case that makes progress toward a base case.
3. *Design rule.* Assume that all the recursive calls work.
4. *Compound interest rule.* Never duplicate work by solving the same instance of a problem in separate recursive calls.

The fourth rule, which will be justified (along with its nickname) in later sections, is the reason that it is generally a bad idea to use recursion to evaluate simple mathematical functions, such as the Fibonacci numbers. As long as you keep these rules in mind, recursive programming should be straightforward.

Summary

This chapter sets the stage for the rest of the book. The time taken by an algorithm confronted with large amounts of input will be an important criterion for deciding if it is a good algorithm. (Of course, correctness is most important.) Speed is relative. What is fast for one problem on one machine might be slow for another problem or a different machine. We will begin to address these issues in the next chapter and will use the mathematics discussed here to establish a formal model.

Exercises

1.1 Write a program to solve the selection problem. Let $k = N/2$. Draw a table showing the running time of your program for various values of N.

1.2 Write a program to solve the word puzzle problem.

1.3 Write a procedure to output an arbitrary real number (which might be negative) using only *PrintDigit* for I/O.

1.4 C allows statements of the form

```
#include filename
```

which reads *filename* and inserts its contents in place of the *include* statement. *Include* statements may be nested; in other words, the file *filename* may itself contain an *include* statement, but, obviously, a file can't include itself in any chain. Write a program that reads in a file and outputs the file as modified by the *include* statements.

1.5 Prove the following formulas:

a. $\log X < X$ for all $X > 0$

b. $\log(A^B) = B \log A$

1.6 Evaluate the following sums:

a. $\displaystyle\sum_{i=0}^{\infty} \frac{1}{4^i}$

b. $\displaystyle\sum_{i=0}^{\infty} \frac{i}{4^i}$

*c. $\sum_{i=0}^{\infty} \frac{i^2}{4^i}$

**d. $\sum_{i=0}^{\infty} \frac{i^N}{4^i}$

1.7 Estimate

$$\sum_{i=\lfloor N/2 \rfloor}^{N} \frac{1}{i}$$

*1.8 What is $2^{100} (\text{mod } 5)$?

1.9 Let F_i be the Fibonacci numbers as defined in Section 1.2. Prove the following:

a. $\sum_{i=1}^{N-2} F_i = F_N - 2$

b. $F_N < \phi^N$, with $\phi = (1 + \sqrt{5})/2$

**c. Give a precise closed-form expression for F_N.

1.10 Prove the following formulas:

a. $\sum_{i=1}^{N} (2i - 1) = N^2$

b. $\sum_{i=1}^{N} i^3 = \left(\sum_{i=1}^{N} i \right)^2$

References

There are many good textbooks covering the mathematics reviewed in this chapter. A small subset is [1], [2], [3], [9], [10], and [11]. Reference [9] is specifically geared toward the analysis of algorithms. It is the first volume of a three-volume series that will be cited throughout this text. More advanced material is covered in [5].

Throughout this book we will assume a knowledge of C [8]. Occasionally, we add a feature where necessary for clarity. We also assume familiarity with pointers and recursion (the recursion summary in this chapter is meant to be a quick review). We will attempt to provide hints on their use where appropriate throughout the textbook. Readers not familiar with these should consult [12] or any good intermediate programming textbook.

General programming style is discussed in several books. Some of the classics are [4], [6], and [7].

1. M. O. Albertson and J. P. Hutchinson, *Discrete Mathematics with Algorithms,* John Wiley & Sons, New York, 1988.
2. Z. Bavel, *Math Companion for Computer Science,* Reston Publishing Co., Reston, Va., 1982.
3. R. A. Brualdi, *Introductory Combinatorics,* North-Holland, New York, 1977.
4. E. W. Dijkstra, *A Discipline of Programming,* Prentice Hall, Englewood Cliffs, N.J., 1976.

5. R. L. Graham, D. E. Knuth, and O. Patashnik, *Concrete Mathematics,* Addison-Wesley, Reading, Mass., 1989.
6. D. Gries, *The Science of Programming,* Springer-Verlag, New York, 1981.
7. B. W. Kernighan and P. J. Plauger, *The Elements of Programming Style,* 2d ed., McGraw-Hill, New York, 1978.
8. B. W. Kernighan and D. M. Ritchie, *The C Programming Language,* 2d ed., Prentice Hall, Englewood Cliffs, N.J., 1988.
9. D. E. Knuth, *The Art of Computer Programming, Vol. 1: Fundamental Algorithms,* 2d ed., Addison-Wesley, Reading, Mass., 1973.
10. F. S. Roberts, *Applied Combinatorics,* Prentice Hall, Englewood Cliffs, N.J., 1984.
11. A. Tucker, *Applied Combinatorics,* 2d ed., John Wiley & Sons, New York, 1984.
12. M. A. Weiss, *Efficient C Programming: A Practical Approach,* Prentice Hall, Englewood Cliffs, N.J., 1995.

CHAPTER 2

Algorithm Analysis

An *algorithm* is a clearly specified set of simple instructions to be followed to solve a problem. Once an algorithm is given for a problem and decided (somehow) to be correct, an important step is to determine how much in the way of resources, such as time or space, the algorithm will require. An algorithm that solves a problem but requires a year is hardly of any use. Likewise, an algorithm that requires a gigabyte of main memory is not (currently) useful on most machines.

In this chapter, we shall discuss

- How to estimate the time required for a program.
- How to reduce the running time of a program from days or years to fractions of a second.
- The results of careless use of recursion.
- Very efficient algorithms to raise a number to a power and to compute the greatest common divisor of two numbers.

2.1. Mathematical Background

The analysis required to estimate the resource use of an algorithm is generally a theoretical issue, and therefore a formal framework is required. We begin with some mathematical definitions.

Throughout the book we will use the following four definitions:

DEFINITION: $T(N) = O(f(N))$ if there are positive *constants* c and n_0 such that $T(N) \leq cf(N)$ when $N \geq n_0$.

DEFINITION: $T(N) = \Omega(g(N))$ if there are positive *constants* c and n_0 such that $T(N) \geq cg(N)$ when $N \geq n_0$.

DEFINITION: $T(N) = \Theta(h(N))$ if and only if $T(N) = O(h(N))$ and $T(N) = \Omega(h(N))$.

DEFINITION: $T(N) = o(p(N))$ if $T(N) = O(p(N))$ and $T(N) \neq \Theta(p(N))$.

The idea of these definitions is to establish a relative order among functions. Given two functions, there are usually points where one function is smaller than the other function, so it does not make sense to claim, for instance, $f(N) < g(N)$. Thus, we compare their *relative rates of growth*. When we apply this to the analysis of algorithms, we shall see why this is the important measure.

Although $1{,}000N$ is larger than N^2 for small values of N, N^2 grows at a faster rate, and thus N^2 will eventually be the larger function. The turning point is $N = 1{,}000$ in this case. The first definition says that eventually there is some point n_0 past which $c \cdot f(N)$ is always at least as large as $T(N)$, so that if constant factors are ignored, $f(N)$ is at least as big as $T(N)$. In our case, we have $T(N) = 1{,}000N$, $f(N) = N^2$, $n_0 = 1{,}000$, and $c = 1$. We could also use $n_0 = 10$ and $c = 100$. Thus, we can say that $1{,}000N = O(N^2)$ (order N-squared). This notation is known as *Big-Oh* notation. Frequently, instead of saying "order...," one says "Big-Oh...."

If we use the traditional inequality operators to compare growth rates, then the first definition says that the growth rate of $T(N)$ is less than or equal to ($\le$) that of $f(N)$. The second definition, $T(N) = \Omega(g(N))$ (pronounced "omega"), says that the growth rate of $T(N)$ is greater than or equal to ($\ge$) that of $g(N)$. The third definition, $T(N) = \Theta(h(N))$ (pronounced "theta"), says that the growth rate of $T(N)$ equals ($=$) the growth rate of $h(N)$. The last definition, $T(N) = o(p(N))$ (pronounced "little-oh"), says that the growth rate of $T(N)$ is less than ($<$) the growth rate of $p(N)$. This is different from Big-Oh, because Big-Oh allows the possibility that the growth rates are the same.

To prove that some function $T(N) = O(f(N))$, we usually do not apply these definitions formally but instead use a repertoire of known results. In general, this means that a proof (or determination that the assumption is incorrect) is a very simple calculation and should not involve calculus, except in extraordinary circumstances (not likely to occur in an algorithm analysis).

When we say that $T(N) = O(f(N))$, we are guaranteeing that the function $T(N)$ grows at a rate no faster than $f(N)$; thus $f(N)$ is an ***upper bound*** on $T(N)$. Since this implies that $f(N) = \Omega(T(N))$, we say that $T(N)$ is a *lower bound* on $f(N)$.

As an example, N^3 grows faster than N^2, so we can say that $N^2 = O(N^3)$ or $N^3 = \Omega(N^2)$. $f(N) = N^2$ and $g(N) = 2N^2$ grow at the same rate, so both $f(N) = O(g(N))$ and $f(N) = \Omega(g(N))$ are true. When two functions grow at the same rate, then the decision of whether or not to signify this with $\Theta()$ can depend on the particular context. Intuitively, if $g(N) = 2N^2$, then $g(N) = O(N^4)$, $g(N) = O(N^3)$, and $g(N) = O(N^2)$ are all technically correct, but the last option is the best answer. Writing $g(N) = \Theta(N^2)$ says not only that $g(N) = O(N^2)$, but also that the result is as good (tight) as possible.

The important things to know are

RULE 1:

If $T_1(N) = O(f(N))$ *and* $T_2(N) = O(g(N))$, *then*

(*a*) $T_1(N) + T_2(N) = \max(O(f(N)), O(g(N)))$,

(*b*) $T_1(N) * T_2(N) = O(f(N) * g(N))$,

Function	Name
c	Constant
$\log N$	Logarithmic
$\log^2 N$	Log-squared
N	Linear
$N \log N$	
N^2	Quadratic
N^3	Cubic
2^N	Exponential

Figure 2.1 Typical growth rates

RULE 2:
If $T(N)$ is a polynomial of degree k, then $T(N) = \Theta(N^k)$.

RULE 3:
$\log^k N = O(N)$ *for any constant k. This tells us that logarithms grow very slowly.*

This information is sufficient to arrange most of the common functions by growth rate (see Figure 2.1).

Several points are in order. First, it is very bad style to include constants or low-order terms inside a Big-Oh. Do not say $T(N) = O(2N^2)$ or $T(N) = O(N^2 + N)$. In both cases, the correct form is $T(N) = O(N^2)$. This means that in any analysis that will require a Big-Oh answer, all sorts of shortcuts are possible. Lower-order terms can generally be ignored, and constants can be thrown away. Considerably less precision is required in these cases.

Second, we can always determine the relative growth rates of two functions $f(N)$ and $g(N)$ by computing $\lim_{N \to \infty} f(N)/g(N)$, using L'Hôpital's rule if necessary.* The limit can have four possible values:

- The limit is 0: This means that $f(N) = o(g(N))$.
- The limit is $c \neq 0$: This means that $f(N) = \Theta(g(N))$.
- The limit is ∞: This means that $g(N) = o(f(N))$.
- The limit oscillates: There is no relation (this will not happen in our context).

Using this method almost always amounts to overkill. Usually the relation between $f(N)$ and $g(N)$ can be derived by simple algebra. For instance, if $f(N) = N \log N$ and $g(N) = N^{1.5}$, then to decide which of $f(N)$ and $g(N)$ grows faster, one really needs to determine which of $\log N$ and $N^{0.5}$ grows faster. This is like determining

*L'Hôpital's rule states that if $\lim_{N \to \infty} f(N) = \infty$ and $\lim_{N \to \infty} g(N) = \infty$, then $\lim_{N \to \infty} f(N)/g(N) = \lim_{N \to \infty} f'(N)/g'(N)$, where $f'(N)$ and $g'(N)$ are the derivatives of $f(N)$ and $g(N)$, respectively.

which of $\log^2 N$ or N grows faster. This is a simple problem, because it is already known that N grows faster than any power of a log. Thus, $g(N)$ grows faster than $f(N)$.

One stylistic note: It is bad to say $f(N) \leq O(g(N))$, because the inequality is implied by the definition. It is wrong to write $f(N) \geq O(g(N))$, which does not make sense.

2.2. Model

In order to analyze algorithms in a formal framework, we need a model of computation. Our model is basically a normal computer, in which instructions are executed sequentially. Our model has the standard repertoire of simple instructions, such as addition, multiplication, comparison, and assignment, but, unlike the case with real computers, it takes exactly one time unit to do anything (simple). To be reasonable, we will assume that, like a modern computer, our model has fixed-size (say, 32-bit) integers and that there are no fancy operations, such as matrix inversion or sorting, that clearly cannot be done in one time unit. We also assume infinite memory.

This model clearly has some weaknesses. Obviously, in real life, not all operations take exactly the same time. In particular, in our model one disk read counts the same as an addition, even though the addition is typically several orders of magnitude faster. Also, by assuming infinite memory, we never worry about page faulting, which can be a real problem, especially for efficient algorithms.

2.3. What to Analyze

The most important resource to analyze is generally the running time. Several factors affect the running time of a program. Some, such as the compiler and computer used, are obviously beyond the scope of any theoretical model, so, although they are important, we cannot deal with them here. The other main factors are the algorithm used and the input to the algorithm.

Typically, the size of the input is the main consideration. We define two functions, $T_{avg}(N)$ and $T_{worst}(N)$, as the average and worst-case running time, respectively, used by an algorithm on input of size N. Clearly, $T_{avg}(N) \leq T_{worst}(N)$. If there is more than one input, these functions may have more than one argument.

Generally, the quantity required is the worst-case time, unless otherwise specified. One reason for this is that it provides a bound for all input, including particularly bad input, which an average-case analysis does not provide. The other reason is that average-case bounds are usually much more difficult to compute. In some instances, the definition of "average" can affect the result. (For instance, what is average input for the following problem?)

As an example, in the next section, we shall consider the following problem:

MAXIMUM SUBSEQUENCE SUM PROBLEM:
Given (possibly negative) integers $A_1, A_2, \ldots, A_N$, find the maximum value of $\sum_{k=i}^{j} A_k$. (For convenience, the maximum subsequence sum is 0 if all the integers are negative.)

Example:

For input $-2, 11, -4, 13, -5, -2$, *the answer is* 20 (A_2 through A_4).

This problem is interesting mainly because there are so many algorithms to solve it, and the performance of these algorithms varies drastically. We will discuss four algorithms to solve this problem. The running time on some computer (the exact computer is unimportant) for these algorithms is given in Figure 2.2.

There are several important things worth noting in this table. For a small amount of input, the algorithms all run in a blink of the eye, so if only a small amount of input is expected, it might be silly to expend a great deal of effort to design a clever algorithm. On the other hand, there is a large market these days for rewriting programs that were written five years ago based on a no-longer-valid assumption of small input size. These programs are now too slow, because they used poor algorithms. For large amounts of input, algorithm 4 is clearly the best choice (although algorithm 3 is still usable).

Second, the times given do not include the time required to read the input. For algorithm 4, the time merely to read in the input from a disk is likely to be an order of magnitude larger than the time required to solve the problem. This is typical of many efficient algorithms. Reading the data is generally the bottleneck; once the data are read, the problem can be solved quickly. For inefficient algorithms this is not true, and significant computer resources must be used. Thus it is important that, whenever possible, algorithms be efficient enough not to be the bottleneck of a problem.

Figure 2.3 shows the growth rates of the running times of the four algorithms. Even though this graph encompasses only values of N ranging from 10 to 100, the relative growth rates are still evident. Although the graph for algorithm 3 seems linear, it is easy to verify that it is not by using a straight-edge (or piece of paper). Figure 2.4 shows the performance for larger values. It dramatically illustrates how useless inefficient algorithms are for even moderately large amounts of input.

Figure 2.2 Running times of several algorithms for maximum subsequence sum (in seconds)

Algorithm		1	2	3	4
Time		$O(N^3)$	$O(N^2)$	$O(N \log N)$	$O(N)$
Input Size	$N = 10$	0.00103	0.00045	0.00066	0.00034
	$N = 100$	0.47015	0.01112	0.00486	0.00063
	$N = 1{,}000$	448.77	1.1233	0.05843	0.00333
	$N = 10{,}000$	NA	111.13	0.68631	0.03042
	$N = 100{,}000$	NA	NA	8.0113	0.29832

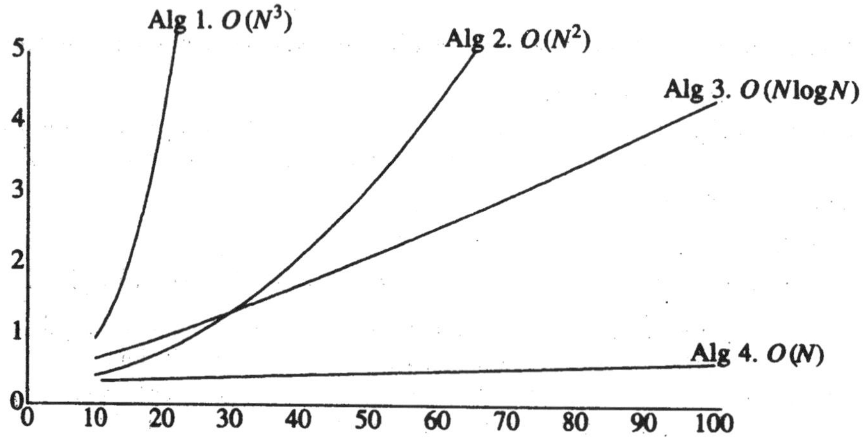

Figure 2.3 Plot (N vs. milliseconds) of various maximum subsequence sum algorithms

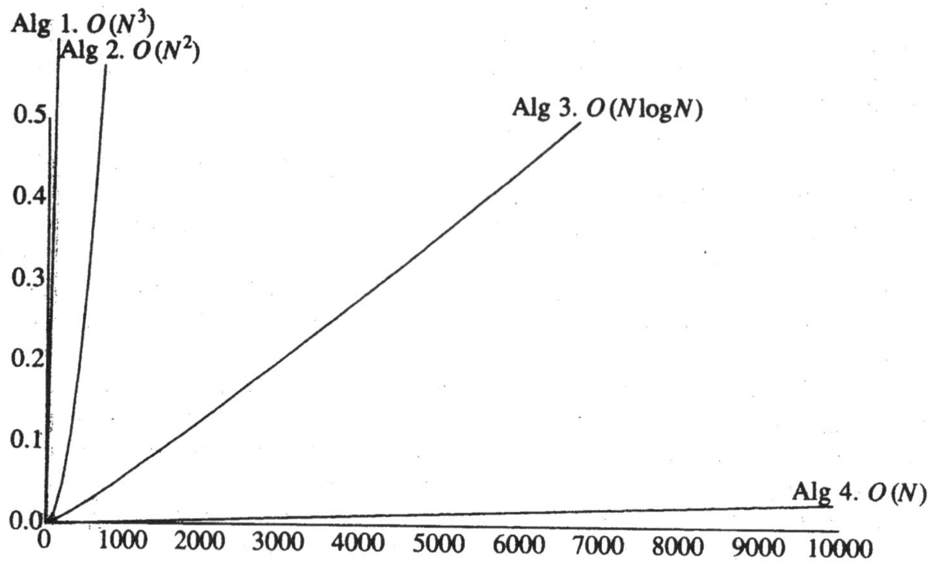

Figure 2.4 Plot (N vs. seconds) of various maximum subsequence sum algorithms

2.4. Running Time Calculations

There are several ways to estimate the running time of a program. The previous table was obtained empirically. If two programs are expected to take similar times, probably the best way to decide which is faster is to code them both up and run them!

Generally, there are several algorithmic ideas, and we would like to eliminate the bad ones early, so an analysis is usually required. Furthermore, the ability to do an analysis usually provides insight into designing efficient algorithms. The analysis also generally pinpoints the bottlenecks, which are worth coding carefully.

To simplify the analysis, we will adopt the convention that there are no particular units of time. Thus, we throw away leading constants. We will also throw away low-order terms, so what we are essentially doing is computing a Big-Oh running time. Since Big-Oh is an upper bound, we must be careful never to underestimate the running time of the program. In effect, the answer provided is a guarantee that the program will terminate within a certain time period. The program may stop earlier than this, but never later.

2.4.1. *A Simple Example*

Here is a simple program fragment to calculate $\sum_{i=1}^{N} i^3$:

```
        int
        Sum( int N )
        {
            int i, PartialSum;

/* 1*/      PartialSum = 0;
/* 2*/      for( i = 1; i <= N; i++ )
/* 3*/          PartialSum += i * i * i;
/* 4*/      return PartialSum;
        }
```

The analysis of this program is simple. The declarations count for no time. Lines 1 and 4 count for one unit each. Line 3 counts for four units per time executed (two multiplications, one addition, and one assignment) and is executed N times, for a total of $4N$ units. Line 2 has the hidden costs of initializing i, testing $i \leq N$, and incrementing i. The total cost of all these is 1 to initialize, $N + 1$ for all the tests, and N for all the increments, which is $2N + 2$. We ignore the costs of calling the function and returning, for a total of $6N + 4$. Thus, we say that this function is $O(N)$.

If we had to perform all this work every time we needed to analyze a program, the task would quickly become infeasible. Fortunately, since we are giving the answer in terms of Big-Oh, there are lots of shortcuts that can be taken without affecting the final answer. For instance, line 3 is obviously an $O(1)$ statement (per execution), so it is silly to count precisely whether it is two, three, or four units; it does not matter. Line 1 is obviously insignificant compared with the *for* loop, so it is silly to waste time here. This leads to several general rules.

2.4.2. *General Rules*

RULE 1—*FOR* LOOPS:

The running time of a for *loop is at most the running time of the statements inside the* for *loop (including tests) times the number of iterations.*

RULE 2—NESTED *FOR* LOOPS:

Analyze these inside out. The total running time of a statement inside a group of nested loops is the running time of the statement multiplied by the product of the sizes of all the for *loops.*

As an example, the following program fragment is $O(N^2)$:

```
for( i = 0; i < N; i++ )
    for( j = 0; j < N; j++ )
        k++;
```

RULE 3—CONSECUTIVE STATEMENTS:

These just add (which means that the maximum is the one that counts; see rule 1(a) on page 16).

As an example, the following program fragment, which has $O(N)$ work followed by $O(N^2)$ work, is also $O(N^2)$:

```
for( i = 0; i < N; i++ )
    A[ i ] = 0;
for( i = 0; i < N; i++ )
    for( j = 0; j < N; j++ )
        A[ i ] += A[ j ] + i + j;
```

RULE 4—*IF/ELSE*:

For the fragment

```
if( Condition )
    S1
else
    S2
```

the running time of an if/else *statement is never more than the running time of the test plus the larger of the running times of S1 and S2.*

Clearly, this can be an overestimate in some cases, but it is never an underestimate.

Other rules are obvious, but a basic strategy of analyzing from the inside (or deepest part) out works. If there are function calls, these must be analyzed first. If there are recursive procedures, there are several options. If the recursion is really just a thinly veiled *for* loop, the analysis is usually trivial. For instance, the following function is really just a simple loop and is $O(N)$:

```
long int
Factorial( int N )
{
    if( N <= 1 )
        return 1;
    else
        return N * Factorial( N - 1 );
}
```

This example is really a poor use of recursion. When recursion is properly used, it is difficult to convert the recursion into a simple loop structure. In this case, the analysis will involve a recurrence relation that needs to be solved. To see what might happen, consider the following program, which turns out to be a horrible use of recursion:

```
        long int
        Fib( int N )
        {
/* 1*/      if( N <= 1 )
/* 2*/          return 1;
            else
/* 3*/          return Fib( N - 1 ) + Fib( N - 2 );
        }
```

At first glance, this seems like a very clever use of recursion. However, if the program is coded up and run for values of N around 30, it becomes apparent that this program is terribly inefficient. The analysis is fairly simple. Let $T(N)$ be the running time for the function $Fib(N)$. If $N = 0$ or $N = 1$, then the running time is some constant value, which is the time to do the test at line 1 and return. We can say that $T(0) = T(1) = 1$ because constants do not matter. The running time for other values of N is then measured relative to the running time of the base case. For $N > 2$, the time to execute the function is the constant work at line 1 plus the work at line 3. Line 3 consists of an addition and two function calls. Since the function calls are not simple operations, they must be analyzed by themselves. The first function call is $Fib(N-1)$ and hence, by the definition of T, requires $T(N-1)$ units of time. A similar argument shows that the second function call requires $T(N-2)$ units of time. The total time required is then $T(N-1) + T(N-2) + 2$, where the 2 accounts for the work at line 1 plus the addition at line 3. Thus, for $N \geq 2$, we have the following formula for the running time of $Fib(N)$:

$$T(N) = T(N-1) + T(N-2) + 2$$

Since $Fib(N) = Fib(N-1) + Fib(N-2)$, it is easy to show by induction that $T(N) \geq Fib(N)$. In Section 1.2.5, we showed that $Fib(N) < (5/3)^N$. A similar calculation shows that (for $N > 4$) $Fib(N) \geq (3/2)^N$, and so the running time of this program grows *exponentially*. This is about as bad as possible. By keeping a simple array and using a *for* loop, the running time can be reduced substantially.

This program is slow because there is a huge amount of redundant work being performed, violating the fourth major rule of recursion (the compound interest rule), which was presented in Section 1.3. Notice that the first call on line 3, $Fib(N-1)$, actually computes $Fib(N-2)$ at some point. This information is thrown away and recomputed by the second call on line 3. The amount of information thrown away compounds recursively and results in the huge running time. This is perhaps the finest example of the maxim "Don't compute anything more than once" and should not scare you away from using recursion. Throughout this book, we shall see outstanding uses of recursion.

2.4.3. Solutions for the Maximum Subsequence Sum Problem

We will now present four algorithms to solve the maximum subsequence sum problem posed earlier. The first algorithm, which merely exhaustively tries all possibilities, is depicted in Figure 2.5. The indices in the *for* loop reflect the fact that in C, arrays begin at 0, instead of 1. Also, the algorithm does not compute the actual subsequences; additional code is required to do this.

Convince yourself that this algorithm works (this should not take much convincing). The running time is $O(N^3)$ and is entirely due to lines 5 and 6, which consist of an $O(1)$ statement buried inside three nested *for* loops. The loop at line 2 is of size N.

The second loop has size $N - i$ which could be small but could also be of size N. We must assume the worst, with the knowledge that this could make the final bound a bit high. The third loop has size $j - i + 1$, which, again, we must assume is of size N. The total is $O(1 \cdot N \cdot N \cdot N) = O(N^3)$. Statement 1 takes only $O(1)$ total, and statements 7 and 8 take only $O(N^2)$ total, since they are easy expressions inside only two loops.

It turns out that a more precise analysis, taking into account the actual size of these loops, shows that the answer is $\Theta(N^3)$ and that our estimate above was a factor of 6 too high (which is all right, because constants do not matter). This is generally true in these kinds of problems. The precise analysis is obtained from the sum $\sum_{i=0}^{N-1} \sum_{j=i}^{N-1} \sum_{k=i}^{j} 1$, which tells how many times line 6 is executed. The sum can be evaluated inside out, using formulas from Section 1.2.3. In particular, we will use the formulas for the sum of the first N integers and first N squares. First we have

$$\sum_{k=i}^{j} 1 = j - i + 1$$

Figure 2.5 Algorithm 1

```
        int
        MaxSubsequenceSum( const int A[ ], int N )
        {
            int ThisSum, MaxSum, i, j, k;

/* 1*/      MaxSum = 0;
/* 2*/      for( i = 0; i < N; i++ )
/* 3*/          for( j = i; j < N; j++ )
                {
/* 4*/              ThisSum = 0;
/* 5*/              for( k = i; k <= j; k++ )
/* 6*/                  ThisSum += A[ k ];

/* 7*/              if( ThisSum > MaxSum )
/* 8*/                  MaxSum = ThisSum;
                }
/* 9*/      return MaxSum;
        }
```

Next we evaluate

$$\sum_{j=i}^{N-1}(j-i+1) = \frac{(N-i+1)(N-i)}{2}$$

This sum is computed by observing that it is just the sum of the first $N - i$ integers. To complete the calculation, we evaluate

$$\begin{aligned}\sum_{i=0}^{N-1}\frac{(N-i+1)(N-i)}{2} &= \sum_{i=1}^{N}\frac{(N-i+1)(N-i+2)}{2}\\ &= \frac{1}{2}\sum_{i=1}^{N}i^2 - \left(N+\frac{3}{2}\right)\sum_{i=1}^{N}i + \frac{1}{2}(N^2+3N+2)\sum_{i=1}^{N}1\\ &= \frac{1}{2}\frac{N(N+1)(2N+1)}{6} - \left(N+\frac{3}{2}\right)\frac{N(N+1)}{2} + \frac{N^2+3N+2}{2}N\\ &= \frac{N^3+3N^2+2N}{6}\end{aligned}$$

We can avoid the cubic running time by removing a *for* loop. This is not always possible, but in this case there are an awful lot of unnecessary computations present in the algorithm. The inefficiency that the improved algorithm corrects can be seen by noticing that $\sum_{k=i}^{j} A_k = A_j + \sum_{k=i}^{j-1} A_k$, so the computation at lines 5 and 6 in algorithm 1 is unduly expensive. Figure 2.6 shows an improved algorithm. Algorithm 2 is clearly $O(N^2)$; the analysis is even simpler than before.

There is a recursive and relatively complicated $O(N \log N)$ solution to this problem, which we now describe. If there didn't happen to be an $O(N)$ (linear)

Figure 2.6 Algorithm 2

```
        int
        MaxSubSequenceSum( const int A[ ], int N )
        {
            int ThisSum, MaxSum, i, j;

/* 1*/      MaxSum = 0
/* 2*/      for( i = 0; i < N; i++ )
            {
/* 3*/          ThisSum = 0;
/* 4*/          for( j = i; j < N; j++ )
                {
/* 5*/              ThisSum += A[ j ];

/* 6*/              if( ThisSum > MaxSum )
/* 7*/                  MaxSum = ThisSum;
                }
            }
/* 8*/      return MaxSum;
        }
```

solution, this would be an excellent example of the power of recursion. The algorithm uses a "divide-and-conquer" strategy. The idea is to split the problem into two roughly equal subproblems, which are then solved recursively. This is the "divide" part. The "conquer" stage consists of patching together the two solutions of the subproblems, and possibly doing a small amount of additional work, to arrive at a solution for the whole problem.

In our case, the maximum subsequence sum can be in one of three places. Either it occurs entirely in the left half of the input, or entirely in the right half, or it crosses the middle and is in both halves. The first two cases can be solved recursively. The last case can be obtained by finding the largest sum in the first half that includes the last element in the first half, and the largest sum in the second half that includes the first element in the second half. These two sums can then be added together. As an example, consider the following input:

First Half				**Second Half**			
4	−3	5	−2	−1	2	6	−2

The maximum subsequence sum for the first half is 6 (elements A_1 through A_3) and for the second half is 8 (elements A_6 through A_7).

The maximum sum in the first half that includes the last element in the first half is 4 (elements A_1 through A_4), and the maximum sum in the second half that includes the first element in the second half is 7 (elements A_5 though A_7). Thus, the maximum sum that spans both halves and goes through the middle is $4 + 7 = 11$ (elements A_1 through A_7).

We see, then, that among the three ways to form a large maximum subsequence, for our example, the best way is to include elements from both halves. Thus, the answer is 11. Figure 2.7 shows an implementation of this strategy.

The code for algorithm 3 deserves some comment. The general form of the call for the recursive procedure is to pass the input array along with the left and right borders, which delimit the portion of the array that is operated upon. A one-line driver program sets this up by passing the borders 0 and $N - 1$ along with the array.

Lines 1 to 4 handle the base case. If *Left* == *Right*, there is one element, and it is the maximum subsequence if the element is nonnegative. The case *Left* > *Right* is not possible unless N is negative (although minor perturbations in the code could mess this up). Lines 6 and 7 perform the two recursive calls. We can see that the recursive calls are always on a smaller problem than the original, although minor perturbations in the code could destroy this property. Lines 8 to 12 and 13 to 17 calculate the two maximum sums that touch the center divider. The sum of these two values is the maximum sum that spans both halves. The pseudoroutine *Max3* returns the largest of the three possibilities.

Algorithm 3 clearly requires more effort to code than either of the two previous algorithms. However, shorter code does not always mean better code. As we have seen in the earlier table showing the running times of the algorithms, this algorithm is considerably faster than the other two for all but the smallest of input sizes.

```
static int
MaxSubSum( const int A[ ], int Left, int Right )
{
    int MaxLeftSum, MaxRightSum;
    int MaxLeftBorderSum, MaxRightBorderSum;
    int LeftBorderSum, RightBorderSum;
    int Center, i;

    if( Left == Right )  /* Base Case */
        if( A[ Left ] > 0 )
            return A[ Left ];
        else
            return 0;

    Center = ( Left + Right ) / 2;
    MaxLeftSum = MaxSubSum( A, Left, Center );
    MaxRightSum = MaxSubSum( A, Center + 1, Right );

    MaxLeftBorderSum = 0; LeftBorderSum = 0
    for( i = Center; i >= Left; i-- )
    {
        LeftBorderSum += A[ i ];
        if( LeftBorderSum > MaxLeftBorderSum )
            MaxLeftBorderSum = LeftBorderSum;
    }

    MaxRightBorderSum = 0; RightBorderSum = 0;
    for( i = Center + 1; i <= Right; i++ )
    {
        RightBorderSum += A[ i ];
        if( RightBorderSum > MaxRightBorderSum )
            MaxRightBorderSum = RightBorderSum;
    }

    return Max3( MaxLeftSum, MaxRightSum,
            MaxLeftBorderSum + MaxRightBorderSum );
}

int
MaxSubsequenceSum( const int A[ ], int N )
{
    return MaxSubSum( A, 0, N - 1 );
}
```

Figure 2.7 Algorithm 3

The running time is analyzed in much the same way as for the program that computes the Fibonacci numbers. Let $T(N)$ be the time it takes to solve a maximum subsequence sum problem of size N. If $N = 1$, then the program takes some constant amount of time to execute lines 1 to 4, which we shall call one unit. Thus, $T(1) = 1$. Otherwise, the program must perform two recursive calls, the two *for* loops between lines 9 and 17, and some small amount of bookkeeping, such as lines 5 and 18. The two *for* loops combine to touch every element from A_0 to A_{N-1}, and there is constant work inside the loops, so the time expended in lines 9 to 17 is $O(N)$. The code in lines 1 to 5, 8, 13, and 18 is all a constant amount of work and can thus be ignored compared with $O(N)$. The remainder of the work is performed in lines 6 and 7. These lines solve two subsequence problems of size $N/2$ (assuming N is even). Thus, these lines take $T(N/2)$ units of time each, for a total of $2T(N/2)$. The total time for the algorithm then is $2T(N/2) + O(N)$. This gives the equations

$$T(1) = 1$$

$$T(N) = 2T(N/2) + O(N)$$

To simplify the calculations, we can replace the $O(N)$ term in the equation above with N; since $T(N)$ will be expressed in Big-Oh notation anyway, this will not affect the answer. In Chapter 7, we shall see how to solve this equation rigorously. For now, if $T(N) = 2T(N/2)+N$, and $T(1) = 1$, then $T(2) = 4 = 2*2$, $T(4) = 12 = 4*3$, $T(8) = 32 = 8*4$, and $T(16) = 80 = 16*5$. The pattern that is evident, and can be derived, is that if $N = 2^k$, then $T(N) = N*(k+1) = N \log N + N = O(N \log N)$.

This analysis assumes N is even, since otherwise $N/2$ is not defined. By the recursive nature of the analysis, it is really valid only when N is a power of 2, since otherwise we eventually get a subproblem that is not an even size, and the equation is invalid. When N is not a power of 2, a somewhat more complicated analysis is required, but the Big-Oh result remains unchanged.

In future chapters, we will see several clever applications of recursion. Here, we present a fourth algorithm to find the maximum subsequence sum. This algorithm is simpler to implement than the recursive algorithm and also is more efficient. It is shown in Figure 2.8.

It should be clear why the time bound is correct, but it takes a little thought to see why the algorithm actually works; this is left to the reader. An extra advantage of this algorithm is that it makes only one pass through the data, and once $A[i]$ is read and processed, it does not need to be remembered. Thus, if the array is on a disk or tape, it can be read sequentially, and there is no need to store any part of it in main memory. Furthermore, at any point in time, the algorithm can correctly give an answer to the subsequence problem for the data it has already read (the other algorithms do not share this property). Algorithms that can do this are called *on-line algorithms*. An on-line algorithm that requires only constant space and runs in linear time is just about as good as possible.

2.4.4. *Logarithms in the Running Time*

The most confusing aspect of analyzing algorithms probably centers around the logarithm. We have already seen that some divide-and-conquer algorithms will

```
        int
        MaxSubsequenceSum( const int A[ ], int N )
        {
            int ThisSum, MaxSum, j;

/* 1*/      ThisSum = MaxSum = 0;
/* 2*/      for( j = 0; j < N; j++ )
            {
/* 3*/          ThisSum += A[ j ];

/* 4*/          if( ThisSum > MaxSum )
/* 5*/              MaxSum = ThisSum;
/* 6*/          else if( ThisSum < 0 )
/* 7*/              ThisSum = 0;
            }
/* 8*/      return MaxSum;
        }
```

Figure 2.8 Algorithm 4

run in $O(N \log N)$ time. Besides divide-and-conquer algorithms, the most frequent appearance of logarithms centers around the following general rule: *An algorithm is* $O(\log N)$ *if it takes constant* $(O(1))$ *time to cut the problem size by a fraction (which is usually* $\frac{1}{2}$*).* On the other hand, if constant time is required to merely reduce the problem by a constant *amount* (such as to make the problem smaller by 1), then the algorithm is $O(N)$.

It should be obvious that only special kinds of problems can be $O(\log N)$. For instance, if the input is a list of N numbers, an algorithm must take $\Omega(N)$ merely to read the input in. Thus, when we talk about $O(\log N)$ algorithms for these kinds of problems, we usually presume that the input is preread. We provide three examples of logarithmic behavior.

Binary Search

The first example is usually referred to as binary search.

BINARY SEARCH:
Given an integer X *and integers* $A_0, A_1, \ldots, A_{N-1}$, *which are presorted and already in memory, find* i *such that* $A_i = X$, *or return* $i = -1$ *if* X *is not in the input.*

The obvious solution consists of scanning through the list from left to right and runs in linear time. However, this algorithm does not take advantage of the fact that the list is sorted, and is thus not likely to be best. A better strategy is to check if X is the middle element. If so, the answer is at hand. If X is smaller than the middle element, we can apply the same strategy to the sorted subarray to the left of the

```
        int
        BinarySearch( const ElementType A[ ], ElementType X, int N )
        {
            int Low, Mid, High;

/* 1*/      Low = 0; High = N - 1;
/* 2*/      while( Low <= High )
            {
/* 3*/          Mid = ( Low + High ) / 2;
/* 4*/          if( A[ Mid ] < X )
/* 5*/              Low = Mid + 1;
                else
/* 6*/          if( A[ Mid ] > X )
/* 7*/              High = Mid - 1;
                else
/* 8*/              return Mid;  /* Found */
            }
/* 9*/      return NotFound;      /* NotFound is defined as -1 */
        }
```

Figure 2.9 Binary search

middle element; likewise, if X is larger than the middle element, we look to the right half. (There is also the case of when to stop.) Figure 2.9 shows the code for binary search (the answer is *Mid*). As usual, the code reflects C's convention that arrays begin with index 0.

Clearly, all the work done inside the loop takes $O(1)$ per iteration, so the analysis requires determining the number of times around the loop. The loop starts with $High - Low = N - 1$ and finishes with $High - Low \geq -1$. Every time through the loop the value $High - Low$ must be at least halved from its previous value; thus, the number of times around the loop is at most $\lceil \log(N - 1) \rceil + 2$. (As an example, if $High - Low = 128$, then the maximum values of $High - Low$ after each iteration are 64, 32, 16, 8, 4, 2, 1, 0, -1.) Thus, the running time is $O(\log N)$. Equivalently, we could write a recursive formula for the running time, but this kind of brute-force approach is usually unnecessary when you understand what is really going on and why.

Binary search can be viewed as our first data structure implementation. It supports the *Find* operation in $O(\log N)$ time, but all other operations (in particular *Insert*) require $O(N)$ time. In applications where the data are static (that is, insertions and deletions are not allowed), this could be very useful. The input would then need to be sorted once, but afterward accesses would be fast. An example is a program that needs to maintain information about the periodic table of elements (which arises in chemistry and physics). This table is relatively stable, as new elements are added infrequently. The element names could be kept sorted. Since there are only about 110 elements, at most eight accesses would be required to find an element. Performing a sequential search would require many more accesses.

Euclid's Algorithm

A second example is Euclid's algorithm for computing the greatest common divisor. The greatest common divisor (*Gcd*) of two integers is the largest integer that divides both. Thus, $Gcd(50, 15) = 5$. The algorithm in Figure 2.10 computes $Gcd(M, N)$, assuming $M \geq N$. (If $N > M$, the first iteration of the loop swaps them.)

The algorithm works by continually computing remainders until 0 is reached. The last nonzero remainder is the answer. Thus, if $M = 1{,}989$ and $N = 1{,}590$, then the sequence of remainders is 399, 393, 6, 3, 0. Therefore, $Gcd(1989, 1590) = 3$. As the example shows, this is a fast algorithm.

As before, estimating the entire running time of the algorithm depends on determining how long the sequence of remainders is. Although $\log N$ seems like a good answer, it is not at all obvious that the value of the remainder has to decrease by a constant factor, since we see that the remainder went from 399 to only 393 in the example. Indeed, the remainder *does not* decrease by a constant factor in one iteration. However, we can prove that after two iterations, the remainder is at most half of its original value. This would show that the number of iterations is at most $2 \log N = O(\log N)$ and establish the running time. This proof is easy, so we include it here. It follows directly from the following theorem.

THEOREM 2.1.

If $M > N$, *then* $M \bmod N < M/2$.

PROOF:

There are two cases. If $N \leq M/2$, then since the remainder is smaller than N, the theorem is true for this case. The other case is $N > M/2$. But then N goes into M once with a remainder $M - N < M/2$, proving the theorem.

One might wonder if this is the best bound possible, since $2 \log N$ is about 20 for our example, and only seven operations were performed. It turns out that the

Figure 2.10 Euclid's algorithm

```
        unsigned int
        Gcd( unsigned int M, unsigned int N )
        {
            unsigned int Rem;

/* 1*/      while( N > 0 )
            {
/* 2*/          Rem = M % N;
/* 3*/          M = N;
/* 4*/          N = Rem;
            }
/* 5*/      return M;
        }
```

constant can be improved slightly, to roughly $1.44 \log N$, in the worst case (which is achievable if M and N are consecutive Fibonacci numbers). The average-case performance of Euclid's algorithm requires pages and pages of highly sophisticated mathematical analysis, and it turns out that the average number of iterations is about $(12 \ln 2 \ln N)/\pi^2 + 1.47$.

Exponentiation

Our last example in this section deals with raising an integer to a power (which is also an integer). Numbers that result from exponentiation are generally quite large, so an analysis works only if we can assume that we have a machine that can store such large integers (or a compiler that can simulate this). We will count the number of multiplications as the measurement of running time.

The obvious algorithm to compute X^N uses $N-1$ multiplications. The recursive algorithm in Figure 2.11 does better. Lines 1 to 4 handle the base case of the recursion. Otherwise, if N is even, we have $X^N = X^{N/2} \cdot X^{N/2}$, and if N is odd, $X^N = X^{(N-1)/2} \cdot X^{(N-1)/2} \cdot X$.

For instance, to compute X^{62}, the algorithm does the following calculations, which involves only nine multiplications:

$$X^3 = (X^2)X, X^7 = (X^3)^2X, X^{15} = (X^7)^2X, X^{31} = (X^{15})^2X, X^{62} = (X^{31})^2$$

The number of multiplications required is clearly at most $2 \log N$, because at most two multiplications (if N is odd) are required to halve the problem. Again, a recurrence formula can be written and solved. Simple intuition obviates the need for a brute-force approach.

It is sometimes interesting to see how much the code can be tweaked without affecting correctness. In Figure 2.11, lines 3 to 4 are actually unnecessary, because if N is 1, then line 7 does the right thing. Line 7 can also be rewritten as

```
/* 7*/          return Pow( X, N - 1 ) * X;
```

without affecting the correctness of the program. Indeed, the program will still run in $O(\log N)$, because the sequence of multiplications is the same as before. However,

Figure 2.11 Efficient exponentiation

```
            long int
            Pow( long int X, unsigned in N )
            {
/* 1*/          if( N == 0 )
/* 2*/              return 1;
/* 3*/          if( N == 1 )
/* 4*/              return X;
/* 5*/          if( IsEven( N ) )
/* 6*/              return Pow( X * X, N / 2 );
                else
/* 7*/              return Pow( X * X, N / 2 ) * X;
            }
```

all of the following alternatives for line 6 are bad, even though they look correct:

```
/* 6a*/        return Pow( Pow( X, 2 ), N / 2 );
/* 6b*/        return Pow( Pow( X, N / 2 ), 2 );
/* 6c*/        return Pow( X, N / 2 ) * Pow( X, N / 2 );
```

Both lines 6a and 6b are incorrect because when N is 2, one of the recursive calls to *Pow* has 2 as the second argument. Thus no progress is made, and an infinite loop results (in an eventual crash).

Using line 6c affects the efficiency, because there are now two recursive calls of size $N/2$ instead of only one. An analysis will show that the running time is no longer $O(\log N)$. We leave it as an exercise to the reader to determine the new running time.

2.4.5. *Checking Your Analysis*

Once an analysis has been performed, it is desirable to see if the answer is correct and as good as possible. One way to do this is to code up the program and see if the empirically observed running time matches the running time predicted by the analysis. When N doubles, the running time goes up by a factor of 2 for linear programs, 4 for quadratic programs, and 8 for cubic programs. Programs that run in logarithmic time take only an additive constant longer when N doubles, and programs that run in $O(N \log N)$ take slightly more than twice as long to run under the same circumstances. These increases can be hard to spot if the lower-order terms have relatively large coefficients and N is not large enough. An example is the jump from $N = 10$ to $N = 100$ in the running time for the various implementations of the maximum subsequence sum problem. It also can be very difficult to differentiate linear programs from $O(N \log N)$ programs purely on empirical evidence.

Another commonly used trick to verify that some program is $O(f(N))$ is to compute the values $T(N)/f(N)$ for a range of N (usually spaced out by factors of 2), where $T(N)$ is the empirically observed running time. If $f(N)$ is a tight answer for the running time, then the computed values converge to a positive constant. If $f(N)$ is an overestimate, the values converge to zero. If $f(N)$ is an underestimate and hence wrong, the values diverge.

As an example, the program fragment in Figure 2.12 computes the probability that two distinct positive integers, less than or equal to N and chosen randomly, are relatively prime. (As N gets large, the answer approaches $6/\pi^2$.)

You should be able to do the analysis for this program instantaneously. Figure 2.13 shows the actual observed running time for this routine on a real computer. The table shows that the last column is most likely, and thus the analysis that you should have gotten is probably correct. Notice that there is not a great deal of difference between $O(N^2)$ and $O(N^2 \log N)$, since logarithms grow so slowly.

2.4.6. *A Grain of Salt*

Sometimes the analysis is shown empirically to be an overestimate. If this is the case, then either the analysis needs to be tightened (usually by a clever observation), or it may be that the *average* running time is significantly less than the worst-case

```
Rel = 0; Tot = 0;
for( i = 1; i <= N; i++ )
    for( j = i + 1; j <= N; j++ )
    {
        Tot++;
        if( Gcd( i, j ) == 1 )
            Rel++;
    }
printf( "Percentage of relatively prime pairs is %f\n",
            ( double ) Rel / Tot );
```

Figure 2.12 Estimate the probability that two random numbers are relatively prime

N	CPU time (T)	T/N^2	T/N^3	$T/N^2 \log N$
100	022	.002200	.000022000	.0004777
200	056	.001400	.000007000	.0002642
300	118	.001311	.000004370	.0002299
400	207	.001294	.000003234	.0002159
500	318	.001272	.000002544	.0002047
600	466	.001294	.000002157	.0002024
700	644	.001314	.000001877	.0002006
800	846	.001322	.000001652	.0001977
900	1,086	.001341	.000001490	.0001971
1,000	1,362	.001362	.000001362	.0001972
1,500	3,240	.001440	.000000960	.0001969
2,000	5,949	.001482	.000000740	.0001947
4,000	25,720	.001608	.000000402	.0001938

Figure 2.13 Empirical running times for the previous routine

running time and no improvement in the bound is possible. For many complicated algorithms the worst-case bound is achievable by some bad input but is usually an overestimate in practice. Unfortunately, for most of these problems, an average-case analysis is extremely complex (in many cases still unsolved), and a worst-case bound, even though overly pessimistic, is the best analytical result known.

Summary

This chapter gives some hints on how to analyze the complexity of programs. Unfortunately, it is not a complete guide. Simple programs usually have simple analyses, but this is not always the case. As an example, later in the text we shall see a sorting algorithm (Shellsort, Chapter 7) and an algorithm for maintaining disjoint sets (Chapter 8), each of which requires about 20 lines of code. The analysis of Shellsort is still not complete, and the disjoint set algorithm has an analysis that is extremely difficult and requires pages and pages of intricate calculations. Most of the analyses that we will encounter here will be simple and involve counting through loops.

An interesting kind of analysis, which we have not touched upon, is lower-bound analysis. We will see an example of this in Chapter 7, where it is proved that any algorithm that sorts by using only comparisons requires $\Omega(N \log N)$ comparisons in the worst case. Lower-bound proofs are generally the most difficult, because they apply not to an algorithm but to a class of algorithms that solve a problem.

We close by mentioning that some of the algorithms described here have real-life application. The *Gcd* algorithm and the exponentiation algorithm are both used in cryptography. Specifically, a 200-digit number is raised to a large power (usually another 200-digit number), with only the low 200 or so digits retained after each multiplication. Since the calculations require dealing with 200-digit numbers, efficiency is obviously important. The straightforward algorithm for exponentiation would require about 10^{200} multiplications, whereas the algorithm presented requires only about 1,200.

Exercises

2.1 Order the following functions by growth rate: N, $\sqrt{N}$, $N^{1.5}$, N^2, $N \log N$, $N \log\log N$, $N \log^2 N$, $N \log(N^2)$, $2/N$, 2^N, $2^{N/2}$, 37, $N^2 \log N$, N^3. Indicate which functions grow at the same rate.

2.2 Suppose $T_1(N) = O(f(N))$ and $T_2(N) = O(f(N))$. Which of the following are true?

a. $T_1(N) + T_2(N) = O(f(N))$

b. $T_1(N) - T_2(N) = o(f(N))$

c. $\dfrac{T_1(N)}{T_2(N)} = O(1)$

d. $T_1(N) = O(T_2(N))$

2.3 Which function grows faster: $N \log N$ or $N^{1+\epsilon/\sqrt{\log N}}$, $\epsilon > 0$?

2.4 Prove that for any constant, k, $\log^k N = o(N)$.

2.5 Find two functions $f(N)$ and $g(N)$ such that neither $f(N) = O(g(N))$ nor $g(N) = O(f(N))$.

2.6 For each of the following six program fragments:

a. Give an analysis of the running time (Big-Oh will do).

b. Implement the code in the language of your choice, and give the running time for several values of N.

c. Compare your analysis with the actual running times.

```
(1) Sum = 0;
    for( i = 0; i < N; i++ )
            Sum++;

(2) Sum = 0;
    for( i = 0; i < N; i++ )
        for( j = 0; j < N; j++ )
            Sum++;

(3) Sum = 0;
    for( i = 0; i < N; i++ )
        for( j = 0; j < N * N; j++ )
            Sum++;

(4) Sum = 0;
    for( i = 0; i < N; i++ )
        for( j = 0; j < i; j++ )
            Sum++;

(5) Sum = 0;
    for( i = 0; i < N; i++ )
        for( j = 0; j < i * i; j++ )
            for( k = 0; k < j; k++ )
                Sum++;

(6) Sum = 0;
    for( i = 1; i < N; i++ )
        for( j = 1; j < i * i; j++ )
            if( j % i == 0 )
                for( k = 0; k < j; k++ )
                    Sum++;
```

2.7 Suppose you need to generate a *random* permutation of the first N integers. For example, {4, 3, 1, 5, 2} and {3, 1, 4, 2, 5} are legal permutations, but {5, 4, 1, 2, 1} is not, because one number (1) is duplicated and another (3) is missing. This routine is often used in simulation of algorithms. We assume the existence of a random number generator, $RandInt(i, j)$, which generates integers between i and j with equal probability. Here are three algorithms:

1. Fill the array A from $A[0]$ to $A[N-1]$ as follows: To fill $A[i]$, generate random numbers until you get one that is not already in $A[0], A[1], \ldots, A[i-1]$.
2. Same as algorithm (1), but keep an extra array called the *Used* array. When a random number, *Ran*, is first put in the array A, set $Used[Ran] = 1$. This means that when filling $A[i]$ with a random number, you can test in one step to see whether the random number has been used, instead of the (possibly) i steps in the first algorithm.

3. Fill the array such that $A[i] = i + 1$. Then

```
for( i = 1; i < N; i++ )
    Swap( &A[ i ], &A[ RandInt( 0, i ) ] );
```

a. Prove that all three algorithms generate only legal permutations and that all permutations are equally likely.

b. Give as accurate (Big-Oh) an analysis as you can of the *expected* running time of each algorithm.

c. Write (separate) programs to execute each algorithm 10 times, to get a good average. Run program (1) for $N = 250, 500, 1{,}000, 2{,}000$; program (2) for $N = 2{,}500, 5{,}000, 10{,}000, 20{,}000, 40{,}000, 80{,}000$; and program (3) for $N = 10{,}000, 20{,}000, 40{,}000, 80{,}000, 160{,}000, 320{,}000, 640{,}000$.

d. Compare your analysis with the actual running times.

e. What is the worst-case running time of each algorithm?

2.8 Complete the table in Figure 2.2 with estimates for the running times that were too long to simulate. Interpolate the running times for these algorithms and estimate the time required to compute the maximum subsequence sum of 1 million numbers. What assumptions have you made?

2.9 How much time is required to compute $F(X) = \sum_{i=0}^{N} A_i X^i$:

a. Using a simple routine to perform exponentiation?

b. Using the routine in Section 2.4.4?

2.10 Consider the following algorithm (known as *Horner's rule*) to evaluate $F(X) = \sum_{i=0}^{N} A_i X^i$:

```
Poly = 0;
for( i = N; i >= 0; i-- )
    Poly = X * Poly + A[i];
```

a. Show how the steps are performed by this algorithm for $X = 3$, $F(X) = 4X^4 + 8X^3 + X + 2$.

b. Explain why this algorithm works.

c. What is the running time of this algorithm?

2.11 Give an efficient algorithm to determine if there exists an integer i such that $A_i = i$ in an array of integers $A_1 < A_2 < A_3 < \cdots < A_N$. What is the running time of your algorithm?

2.12 Give efficient algorithms (along with running time analyses) to:

a. Find the minimum subsequence sum.

*b. Find the minimum *positive* subsequence sum.

*c. Find the maximum subsequence *product*.

2.13 a. Write a program to determine if a positive integer, N, is prime.

b. In terms of N, what is the worst-case running time of your program? (You should be able to do this in $O(\sqrt{N})$.)

c. Let B equal the number of bits in the binary representation of N. What is the value of B?

d. In terms of B, what is the worst-case running time of your program?

e. Compare the running times to determine if a 20-bit number and a 40-bit number are prime.

f. Is it more reasonable to give the running time in terms of N or B? Why?

*2.14 The Sieve of Erastothenes is a method used to compute all primes less than N. We begin by making a table of integers 2 to N. We find the smallest integer, i, that is not crossed out, print i, and cross out $i, 2i, 3i, \ldots$. When $i > \sqrt{N}$, the algorithm terminates. What is the running time of this algorithm?

2.15 Show that X^{62} can be computed with only eight multiplications.

2.16 Write the fast exponentiation routine without recursion.

2.17 Give a precise count on the number of multiplications used by the fast exponentiation routine. (*Hint:* Consider the binary representation of N.)

2.18 Programs A and B are analyzed and found to have worst-case running times no greater than $150N \log_2 N$ and N^2, respectively. Answer the following questions, if possible:

a. Which program has the better guarantee on the running time, for large values of N ($N > 10{,}000$)?

b. Which program has the better guarantee on the running time, for small values of N ($N < 100$)?

c. Which program will run faster *on average* for $N = 1{,}000$?

d. Is it possible that program B will run faster than program A on *all* possible inputs?

2.19 A majority element in an array, A, of size N is an element that appears more than $N/2$ times (thus, there is at most one). For example, the array

3, 3, 4, 2, 4, 4, 2, 4, 4

has a majority element (4), whereas the array

3, 3, 4, 2, 4, 4, 2, 4

does not. If there is no majority element, your program should indicate this. Here is a sketch of an algorithm to solve the problem:

> First, a *candidate* majority element is found (this is the harder part). This candidate is the only element that could possibly be the majority element. The second step determines if this candidate is actually the majority. This is just a sequential search through the array. To find a candidate in the array, A, form a second array, B. Then compare A_1 and A_2. If they are equal, add one of these to B; otherwise do nothing. Then compare A_3 and A_4. Again if they are equal, add one of these to B; otherwise do nothing. Continue in this fashion until the entire array is read. Then recursively find a candidate for B; this is the candidate for A (why?).

a. How does the recursion terminate?

*b. How is the case where N is odd handled?

*c. What is the running time of the algorithm?

d. How can we avoid using an extra array B?

*e. Write a program to compute the majority element.

*2.20 Why is it important to assume that integers in our computer model have a fixed size?

2.21 Consider the word puzzle problem described in Chapter 1. Suppose we fix the size of the longest word to be 10 characters.

a. In terms of R and C, which are the number of rows and columns in the puzzle, and W, which is the number of words, what are the running times of the algorithms described in Chapter 1?

b. Suppose the word list is presorted. Show how to use binary search to obtain an algorithm with significantly better running time.

2.22 Suppose that line 5 in the binary search routine had the statement $Low = Mid$ instead of $Low = Mid + 1$. Would the routine still work?

2.23 Implement the binary search so that only one two-way comparison is performed in each iteration.

2.24 Suppose that lines 6 and 7 in algorithm 3 (Fig. 2.7) are replaced by

```
/* 6*/    MaxLeftSum  = MaxSubSum( A, Left, Center - 1 );
/* 7*/    MaxRightSum = MaxSubSum( A, Center, Right );
```

Would the routine still work?

*2.25 The inner loop of the cubic maximum subsequence sum algorithm performs $N(N + 1)(N + 2)/6$ iterations of the innermost code. The quadratic version performs $N(N + 1)/2$ iterations. The linear version performs N iterations. What pattern is evident? Can you give a combinatoric explanation of this phenomenon?

References

Analysis of the running time of algorithms was first made popular by Knuth in the three-part series [5], [6], and [7]. Analysis of the *Gcd* algorithm appears in [6]. Another early text on the subject is [1].

Big-Oh, big-omega, big-theta, and little-oh notation were advocated by Knuth in [8]. There is still not uniform agreement on the matter, especially when it comes to using $\Theta()$. Many people prefer to use $O()$, even though it is less expressive. Additionally, $O()$ is still used in some corners to express a lower bound, when $\Omega()$ is called for.

The maximum subsequence sum problem is from [3]. The series of books [2], [3], and [4] show how to optimize programs for speed.

1. A. V. Aho, J. E. Hopcroft, and J. D. Ullman, *The Design and Analysis of Computer Algorithms,* Addison-Wesley, Reading, Mass., 1974.
2. J. L. Bentley, *Writing Efficient Programs,* Prentice Hall, Englewood Cliffs, N.J., 1982.
3. J. L. Bentley, *Programming Pearls,* Addison-Wesley, Reading, Mass., 1986.
4. J. L. Bentley, *More Programming Pearls,* Addison-Wesley, Reading, Mass., 1988.
5. D. E. Knuth, *The Art of Computer Programming, Vol 1: Fundamental Algorithms,* 2d ed., Addison-Wesley, Reading, Mass., 1973.
6. D. E. Knuth, *The Art of Computer Programming, Vol 2: Seminumerical Algorithms,* 2d ed., Addison-Wesley, Reading, Mass., 1981.
7. D. E. Knuth, *The Art of Computer Programming, Vol 3: Sorting and Searching,* Addison-Wesley, Reading, Mass., 1973.
8. D. E. Knuth, "Big Omicron and Big Omega and Big Theta," *ACM SIGACT News,* 8 (1976), 18–23.

CHAPTER 3

Lists, Stacks, and Queues

This chapter discusses three of the most simple and basic data structures. Virtually every significant program will use at least one of these structures explicitly, and a stack is always implicitly used in a program, whether or not you declare one. Among the highlights of this chapter, we will

- Introduce the concept of Abstract Data Types (ADTS).
- Show how to efficiently perform operations on lists.
- Introduce the stack ADT and its use in implementing recursion.
- Introduce the queue ADT and its use in operating systems and algorithm design.

Because these data structures are so important, one might expect that they are hard to implement. In fact, they are extremely easy to code up; the main difficulty is maintaining enough discipline to write good general-purpose code for routines that are generally only a few lines long.

3.1. Abstract Data Types (ADTS)

One of the basic rules concerning programming is that no routine should ever exceed a page. This is accomplished by breaking the program down into *modules*. Each module is a logical unit and does a specific job. Its size is kept small by calling other modules. Modularity has several advantages. First, it is much easier to debug small routines than large routines. Second, it is easier for several people to work on a modular program simultaneously. Third, a well-written modular program places certain dependencies in only one routine, making changes easier. For instance, if output needs to be written in a certain format, it is certainly important to have one routine to do this. If printing statements are scattered throughout the program, it will take considerably longer to make modifications. The idea that global variables and side effects are bad is directly attributable to the idea that modularity is good.

An *abstract data type* (ADT) is a set of operations. Abstract data types are mathematical abstractions; nowhere in an ADT's definition is there any mention of *how* the set of operations is implemented. This can be viewed as an extension of modular design.

Objects such as lists, sets, and graphs, along with their operations, can be viewed as abstract data types, just as integers, reals, and booleans are data types. Integers, reals, and booleans have operations associated with them, and so do abstract data types. For the set ADT, we might have such operations as *union, intersection, size,* and *complement.* Alternatively, we might only want the two operations *union* and *find,* which would define a different ADT on the set.

The basic idea is that the implementation of these operations is written once in the program, and any other part of the program that needs to perform an operation on the ADT can do so by calling the appropriate function. If for some reason implementation details need to be changed, it should be easy to do so by merely changing the routines that perform the ADT operations. This change, in a perfect world, would be completely transparent to the rest of the program.

There is no rule telling us which operations must be supported for each ADT; this is a design decision. Error handling and tie breaking (where appropriate) are also generally up to the program designer. The three data structures that we will study in this chapter are primary examples of ADTs. We will see how each can be implemented in several ways, but if they are done correctly, the programs that use them will not need to know which implementation was used.

3.2. The List ADT

We will deal with a general list of the form $A_1, A_2, A_3, \ldots, A_N$. We say that the size of this list is N. We will call the special list of size 0 an *empty list.*

For any list except the empty list, we say that A_{i+1} follows (or succeeds) A_i ($i < N$) and that A_{i-1} precedes A_i ($i > 1$). The first element of the list is A_1, and the last element is A_N. We will not define the predecessor of A_1 or the successor of A_N. The *position* of element A_i in a list is i. Throughout this discussion, we will assume, to simplify matters, that the elements in the list are integers, but in general, arbitrarily complex elements are allowed.

Associated with these "definitions" is a set of operations that we would like to perform on the list ADT. Some popular operations are *PrintList* and *MakeEmpty,* which do the obvious things; *Find,* which returns the position of the first occurrence of a key; *Insert* and *Delete,* which generally insert and delete some key from some position in the list; and *FindKth,* which returns the element in some position (specified as an argument). If the list is 34, 12, 52, 16, 12, then *Find*(52) might return 3; *Insert*$(X, 3)$ might make the list into 34, 12, 52, X, 16, 12 (if we insert after the position given); and *Delete*(52) might turn that list into 34, 12, X, 16, 12.

Of course, the interpretation of what is appropriate for a function is entirely up to the programmer, as is the handling of special cases (for example, what does *Find*(1) return above?). We could also add operations such as *Next* and *Previous,* which would take a position as argument and return the position of the successor and predecessor, respectively.

3.2.1. Simple Array Implementation of Lists

All of these instructions can be implemented just by using an array. Even if the array is dynamically allocated, an estimate of the maximum size of the list is required. Usually this requires a high overestimate, which wastes considerable space. This could be a serious limitation, especially if there are many lists of unknown size.

An array implementation allows *PrintList* and *Find* to be carried out in linear time, which is as good as can be expected, and the *FindKth* operation takes constant time. However, insertion and deletion are expensive. For example, inserting at position 0 (which amounts to making a new first element) requires first pushing the entire array down one spot to make room, whereas deleting the first element requires shifting all the elements in the list up one, so the worst case of these operations is $O(N)$. On average, half of the list needs to be moved for either operation, so linear time is still required. Merely building a list by N successive inserts would require quadratic time.

Because the running time for insertions and deletions is so slow and the list size must be known in advance, simple arrays are generally not used to implement lists.

3.2.2. Linked Lists

In order to avoid the linear cost of insertion and deletion, we need to ensure that the list is not stored contiguously, since otherwise entire parts of the list will need to be moved. Figure 3.1 shows the general idea of a *linked list.*

The linked list consists of a series of structures, which are not necessarily adjacent in memory. Each structure contains the element and a pointer to a structure containing its successor. We call this the *Next* pointer. The last cell's *Next* pointer points to *NULL;* this value is defined by C and cannot be confused with another pointer. ANSI C specifies that *NULL* is zero.

Recall that a pointer variable is just a variable that contains the address where some other data are stored. Thus, if P is declared to be a pointer to a structure, then the value stored in P is interpreted as the location, in main memory, where a structure can be found. A field of that structure can be accessed by *P->FieldName,* where *FieldName* is the name of the field we wish to examine. Figure 3.2 shows the actual representation of the list in Figure 3.1. The list contains five structures, which happen

Figure 3.1 A linked list

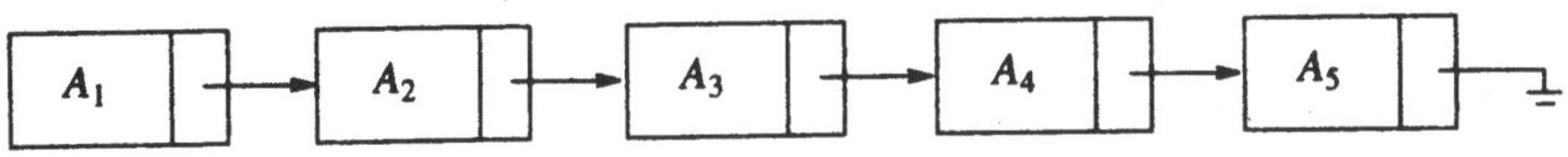

Figure 3.2 Linked list with actual pointer values

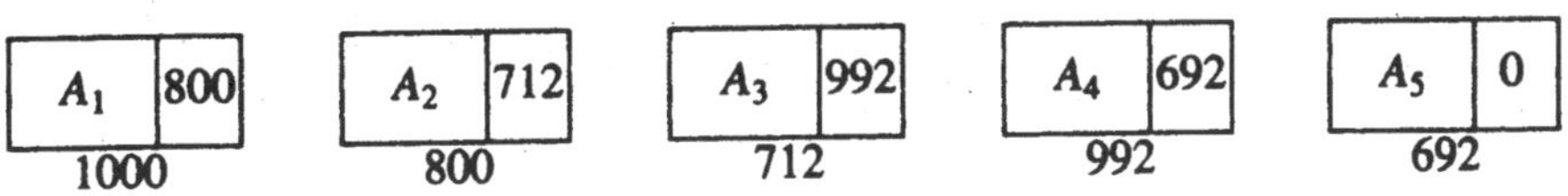

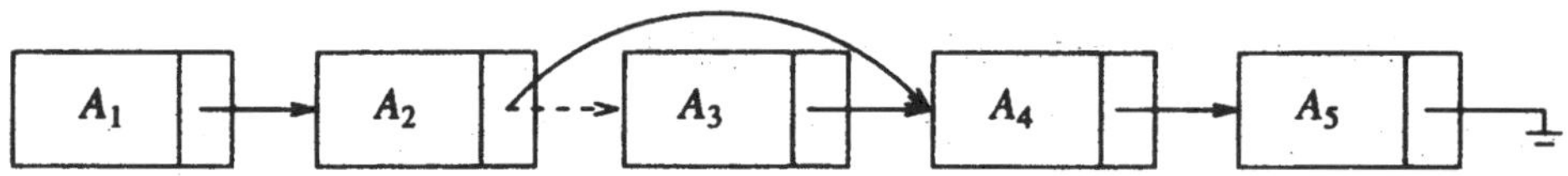

Figure 3.3 Deletion from a linked list

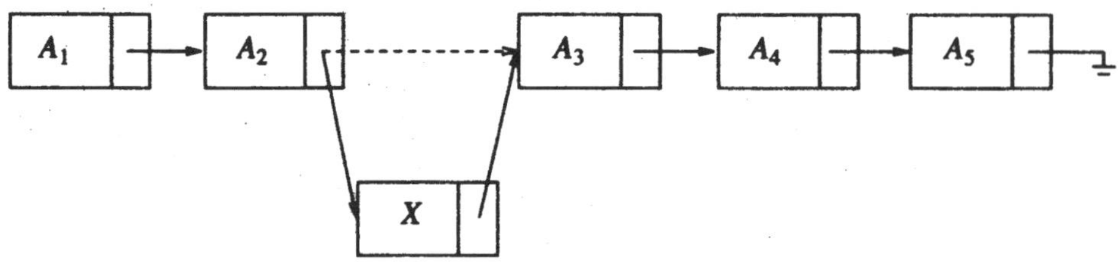

Figure 3.4 Insertion into a linked list

to reside in memory locations 1000, 800, 712, 992, and 692, respectively. The *Next* pointer in the first structure has the value 800, which provides the indication of where the second structure is. The other structures each have a pointer that serves a similar purpose. Of course, in order to access this list, we need to know where the first cell can be found. A pointer variable can be used for this purpose. It is important to remember that a pointer is just a number. For the rest of this chapter, we will draw pointers with arrows, because they are more illustrative.

To execute *PrintList(L)* or *Find(L,Key)*, we merely pass a pointer to the first element in the list and then traverse the list by following the *Next* pointers. This operation is clearly linear-time, although the constant is likely to be larger than if an array implementation were used. The *FindKth* operation is no longer quite as efficient as an array implementation; *FindKth(L,i)* takes $O(i)$ time and works by traversing down the list in the obvious manner. In practice, this bound is pessimistic, because frequently the calls to *FindKth* are in sorted order (by i). As an example, *FindKth(L,2)*, *FindKth(L,3)*, *FindKth(L,4)*, and *FindKth(L,6)* can all be executed in one scan down the list.

The *Delete* command can be executed in one pointer change. Figure 3.3 shows the result of deleting the third element in the original list.

The *Insert* command requires obtaining a new cell from the system by using a *malloc* call (more on this later) and then executing two pointer maneuvers. The general idea is shown in Figure 3.4. The dashed line represents the old pointer.

3.2.3. Programming Details

The description above is actually enough to get everything working, but there are several places where you are likely to go wrong. First of all, there is no really obvious way to insert at the front of the list from the definitions given. Second, deleting from the front of the list is a special case, because it changes the start of the list; careless coding will lose the list. A third problem concerns deletion in general. Although the

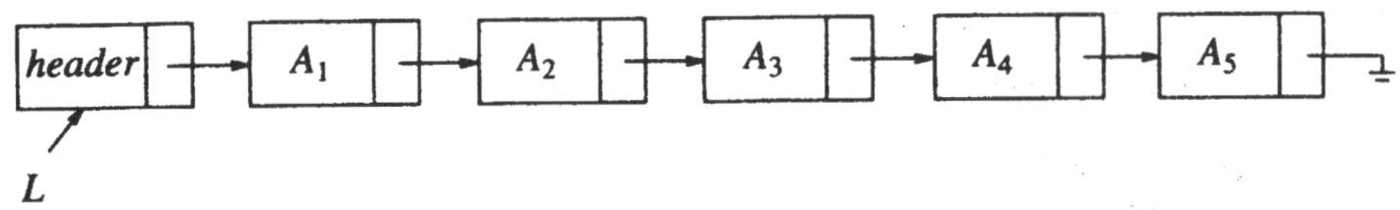

Figure 3.5 Linked list with a header

pointer moves above are simple, the deletion algorithm requires us to keep track of the cell *before* the one that we want to delete.

It turns out that one simple change solves all three problems. We will keep a sentinel node, which is sometimes referred to as a *header* or *dummy* node. This is a common practice, which we will see several times in the future. Our convention will be that the header is in position 0. Figure 3.5 shows a linked list with a header representing the list $A_1, A_2, \ldots, A_5$.

To avoid the problems associated with deletions, we need to write a routine *FindPrevious,* which will return the position of the predecessor of the cell we wish to delete. If we use a header, then if we wish to delete the first element in the list, *FindPrevious* will return the position of the header. The use of a header node is somewhat controversial. Some people argue that avoiding special cases is not sufficient justification for adding fictitious cells; they view the use of header nodes as little more than old-style hacking. Even so, we will use them here, precisely because they allow us to show the basic pointer manipulations without obscuring the code with special cases. Otherwise, whether or not a header should be used is a matter of personal preference.

As examples, we will write about half of the list ADT routines. First, we need our declarations, which are given in Figure 3.6. Following C conventions, the types *List* and *Position,* as well as the function prototypes, are listed in what is to be a *.h* file. The actual *Node* declaration is in a *.c* file.

The first function that we will write tests for an empty list. When we write code for any data structure that involves pointers, it is always best to draw a picture first. Figure 3.7 shows an empty list; from the figure it is easy to write the function in Figure 3.8.

The next function, which is shown in Figure 3.9, tests whether the current element, which by assumption exists, is the last of the list.

The next routine we will write is *Find. Find,* shown in Figure 3.10, returns the position in the list of some element. Line 2 takes advantage of the fact that the *and* (&&) operation is *short-circuited:* if the first half of the *and* is false, the result is automatically false and the second half is not executed.

Some programmers find it tempting to code the *Find* routine recursively, possibly because it avoids the sloppy termination condition. We shall see later that this is a very bad idea and should be avoided at all costs.

Our fourth routine will delete some element X in list L. We need to decide what to do if X occurs more than once or not at all. Our routine deletes the first occurrence of X and does nothing if X is not in the list. To do this, we find P, which is the cell prior to the one containing X, via a call to *FindPrevious*. The code to

```
#ifndef _List_H

struct Node;
typedef struct Node *PtrToNode;
typedef PtrToNode List;
typedef PtrToNode Position;

List MakeEmpty( List L );
int IsEmpty( List L );
int IsLast( Position P, List L );
Position Find( ElementType X, List L );
void Delete( ElementType X, List L );
Position FindPrevious( ElementType X, List L );
void Insert( ElementType X, List L, Position P );
void DeleteList( List L );
Position Header( List L );
Position First( List L );
Position Advance( Position P );
ElementType Retrieve( Position P );

#endif    /* _List_H */


/* Place in the implementation file */
struct Node
{
    ElementType Element;
    Position    Next;
};
```

Figure 3.6 Type declarations for linked lists

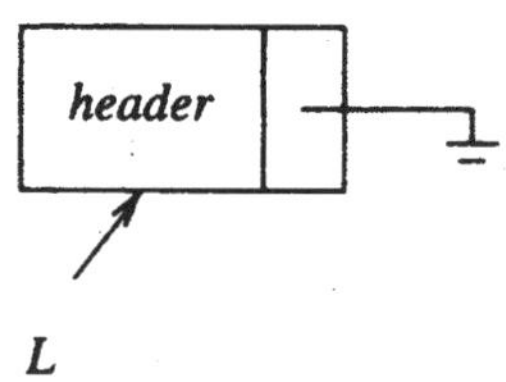

Figure 3.7 Empty list with header

```
/* Return true if L is empty */

int
IsEmpty( List L )
{
    return L->Next == NULL;
}
```

Figure 3.8 Function to test whether a linked list is empty

```
/* Return true if P is the last position in list L */
/* Parameter L is unused in this implementation */

int
IsLast( Position P, List L )
{
    return P->Next == NULL;
}
```

Figure 3.9 Function to test whether current position is the last in a linked list

```
/* Return Position of X in L; NULL if not found */

Position
Find( ElementType X, List L )
{
    Position P;

    P = L->Next;
    while( P != NULL && P->Element != X )
        P = P->Next;

    return P;
}
```

Figure 3.10 *Find* routine

implement this is shown in Figure 3.11. The *FindPrevious* routine is similar to *Find* and is shown in Figure 3.12.

The last routine we will write is an insertion routine. We will pass an element to be inserted along with the list L and a position P. Our particular insertion routine

```
        /* Delete first occurrence of X from a list */
        /* Assume use of a header node *

        void
        Delete( ElementType X, List L )
        {
            Position P, TmpCell;

            P = FindPrevious( X, L );

            if( !IsLast( P, L ) )  /* Assumption of header use */
            {                      /* X is found; delete it */
                TmpCell = P->Next;
                P->Next = TmpCell->Next;  /* Bypass deleted cell */
                free( TmpCell );
            }
        }
```

Figure 3.11 Deletion routine for linked lists

```
        /* If X is not found, then Next field of returned */
        /* Position is NULL */
        /* Assumes a header */

        Position
        FindPrevious( ElementType X, List L )
        {
            Position P;

/* 1*/      P = L;
/* 2*/      while( P->Next != NULL && P->Next->Element != X )
/* 3*/          P = P->Next;

/* 4*/      return P;
        }
```

Figure 3.12 *FindPrevious*—the *Find* routine for use with *Delete*

will insert an element *after* the position implied by *P*. This decision is arbitrary and is meant to show that there are no set rules for what insertion does. It is quite possible to insert the new element into position *P* (which means before the element currently in position *P*), but doing this requires knowledge of the element before position *P*. This could be obtained by a call to *FindPrevious*. It is thus important to comment what you are doing. This has been done in Figure 3.13.

```
        /* Insert (after legal position P) */
        /* Header implementation assumed */
        /* Parameter L is unused in this implementation */

        void
        Insert( ElementType X, List L, Position P )
        {
            Position TmpCell;

/* 1*/      TmpCell = malloc( sizeof( struct Node ) );
/* 2*/      if( TmpCell == NULL )
/* 3*/          FatalError( "Out of space!!!" );

/* 4*/      TmpCell->Element = X;
/* 5*/      TmpCell->Next = P->Next;
/* 6*/      P->Next = TmpCell;
        }
```

Figure 3.13 Insertion routine for linked lists

Notice that we have passed the list to the *Insert* and *IsLast* routines, even though it was never used. We did this because another implementation might need this information, and so not passing the list would defeat the idea of using ADTs.*

With the exception of the *Find* and *FindPrevious* routines (and *Delete*, which calls *FindPrevious*), all of the operations we have coded take $O(1)$ time. This is because in all cases only a fixed number of instructions are performed, no matter how large the list is. For the *Find* and *FindPrevious* routines, the running time is $O(N)$ in the worst case, because the entire list might need to be traversed if the element either is not found or is last in the list. On average, the running time is $O(N)$, because on average, half the list must be traversed.

Additional routines, listed in Figure 3.6, are fairly straightforward. We could also write a routine to implement *Previous*. We leave these as exercises.

3.2.4. Common Errors

The most common error you will encounter is that your program will crash with a nasty error message from the system, such as "memory access violation" or "segmentation violation." This message usually means that a pointer variable contains a bogus address. One common reason is failure to initialize the variable. For instance, if line 1 in Figure 3.14 is omitted, then *P* is undefined and is not likely to be pointing at a valid part of memory. Another typical error concerns line 6 in Figure 3.13. If *P* is *NULL*, then the indirection is illegal. This function knows that *P* is not *NULL*, so the routine is OK. Of course, you should comment this so that

*This is legal, but some compilers will issue a warning.

```
/* Incorrect DeleteList algorithm */

void
DeleteList( List L )
{
    Position P;

    P = L->Next;  /* Header assumed */
    L->Next = NULL;
    while( P != NULL )
    {
        free( P );
        P = P->Next;
    }
}
```

Figure 3.14 Incorrect way to delete a list

the routine that calls *Insert* will ensure this. *Whenever you do an indirection, you must make sure that the pointer is not* NULL. Some C compilers will implicitly do this check for you, but this is not part of the C standard. When you port a program from one compiler to another, you may find that it no longer works. This is one of the common reasons why.

The second common mistake concerns when and when not to use *malloc* to get a new cell. You must remember that declaring a pointer to a structure does not create the structure but only gives enough space to hold the address where some structure might be. The only way to create a record that is not already declared is to use the *malloc* library routine. *malloc*(*HowManyBytes*) has the system create, magically, a new structure and return a pointer to it. If, on the other hand, if you want to use a pointer variable to run down a list, there is no need to create a new structure; in that case the *malloc* command is inappropriate. A type cast is needed on very old compilers to make both sides of the assignment operator compatible. The C library provides other variations of *malloc* such as *calloc*. Both of these routines require the inclusion of *stdlib.h*.

When things are no longer needed, you can issue a *free* command to inform the system that it may reclaim the space. A consequence of the *free*(*P*) command is that the address that *P* is pointing to is unchanged, but the data that reside at that address are now undefined.

If you never delete from a linked list, the number of calls to *malloc* should equal the size of the list, plus 1 if a header is used. Any fewer, and you cannot possibly have a working program. Any more, and you are wasting space and probably time. Occasionally, if your program uses a lot of space, the system may be unable to satisfy your request for a new cell. In this case a *NULL* pointer is returned.

After a deletion in a linked list, it is usually a good idea to free the cell, especially if there are lots of insertions and deletions intermingled and memory might become a problem. You need to keep a temporary variable set to the cell to be disposed of,

```
/* Correct DeleteList algorithm */

void
DeleteList( List L )
{
    Position P, Tmp;

    P = L->Next;  /* Header assumed */
    L->Next = NULL;
    while( P != NULL )
    {
        Tmp = P->Next;
        free( P );
        P = Tmp;
    }
}
```

Figure 3.15 Correct way to delete a list

because after the pointer moves are finished, you will not have a reference to it. As an example, the code in Figure 3.14 is not the correct way to delete an entire list (although it may work on some systems).

Figure 3.15 shows the correct way to do this. Disposal is not necessarily a fast thing, so you might want to check to see if the disposal routine is causing any slow performance, and comment it out if this is the case. This author has written a program (see the exercises) that was made 25 times faster by commenting out the disposal (of 10,000 nodes). It turned out that the cells were freed in a rather peculiar order and apparently caused an otherwise linear program to spend $O(N \log N)$ time to dispose of N cells.

One last warning: *malloc*(*sizeof*(*PtrToNode*)) is legal, but it doesn't allocate enough space for a structure. It allocates space only for a pointer.

3.2.5. *Doubly Linked Lists*

Sometimes it is convenient to traverse lists backwards. The standard implementation does not help here, but the solution is simple. Merely add an extra field to the data structure, containing a pointer to the previous cell. The cost of this is an extra link, which adds to the space requirement and also doubles the cost of insertions and deletions because there are more pointers to fix. On the other hand, it simplifies deletion, because you no longer have to refer to a key by using a pointer to the previous cell; this information is now at hand. Figure 3.16 shows a doubly linked list.

Figure 3.16 A doubly linked list

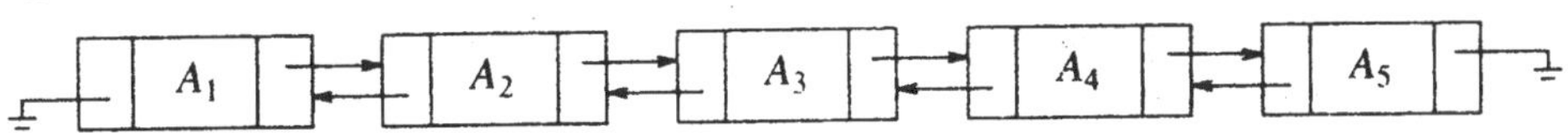

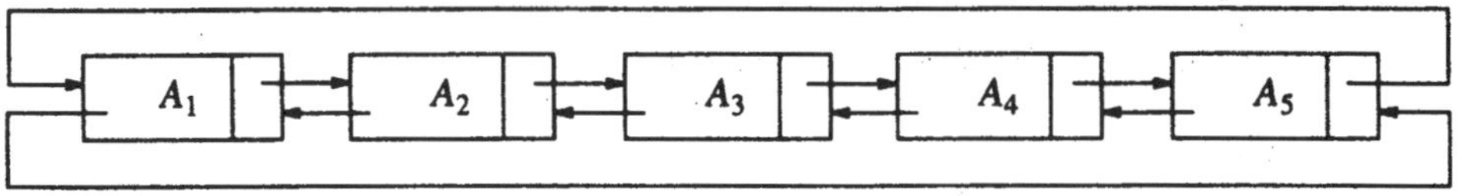

Figure 3.17 A double circularly linked list

3.2.6. *Circularly Linked Lists*

A popular convention is to have the last cell keep a pointer back to the first. This can be done with or without a header (if the header is present, the last cell points to it) and can also be done with doubly linked lists (the first cell's previous pointer points to the last cell). This clearly affects some of the tests, but the structure is popular in some applications. Figure 3.17 shows a double circularly linked list with no header.

3.2.7. *Examples*

We provide three examples that use linked lists. The first is a simple way to represent single-variable polynomials. The second is a method to sort in linear time, for some special cases. Finally, we show a complicated example of how linked lists might be used to keep track of course registration at a university.

The Polynomial ADT

We can define an abstract data type for single-variable polynomials (with nonnegative exponents) by using a list. Let $F(X) = \sum_{i=0}^{N} A_i X^i$. If most of the coefficients A_i are nonzero, we can use a simple array to store the coefficients. We could then write routines to perform addition, subtraction, multiplication, differentiation, and other operations on these polynomials. In this case, we might use the type declarations given in Figure 3.18. We could then write routines to perform various operations. Two possibilities are addition and multiplication; these are shown in Figures 3.19 to 3.21. Ignoring the time to initialize the output polynomials to zero, the running time of the multiplication routine is proportional to the product of the degree of the two input polynomials. This is adequate for dense polynomials, where most of the terms are present, but if $P_1(X) = 10X^{1000} + 5X^{14} + 1$ and $P_2(X) = 3X^{1990} - 2X^{1492} + 11X + 5$, then the running time is likely to be unacceptable. One can see that

Figure 3.18 Type declarations for array implementation of the polynomial ADT

```
typedef struct
{
    int CoeffArray[ MaxDegree + 1 ];
    int HighPower;
} * Polynomial;
```

```
void
ZeroPolynomial( Polynomial Poly )
{
    int i;

    for( i = 0; i <= MaxDegree; i++ )
        Poly->CoeffArray[ i ] = 0;
    Poly->HighPower = 0;
}
```

Figure 3.19 Procedure to initialize a polynomial to zero

```
void
AddPolynomial( const Polynomial Poly1,
            const Polynomial Poly2, Polynomial PolySum )
{
    int i;

    ZeroPolynomial( PolySum );
    PolySum->HighPower = Max( Poly1->HighPower,
                              Poly2->HighPower );

    for( i = PolySum->HighPower; i >= 0; i-- )
        PolySum->CoeffArray[ i ] = Poly1->CoeffArray[ i ]
                                 + Poly2->CoeffArray[ i ];
}
```

Figure 3.20 Procedure to add two polynomials

```
void
MultPolynomial( const Polynomial Poly1,
            const Polynomial Poly2, Polynomial PolyProd )
{
    int i, j;

    ZeroPolynomial( PolyProd );
    PolyProd->HighPower = Poly1->HighPower + Poly2->HighPower;

    if( PolyProd->HighPower > MaxDegree )
        Error( "Exceeded array size" );
    else
        for( i = 0; i <= Poly1->HighPower; i++ )
            for( j = 0; j <= Poly2->HighPower; j++ )
                PolyProd->CoeffArray[ i + j ] +=
                        Poly1->CoeffArray[ i ] *
                        Poly2->CoeffArray[ j ];
}
```

Figure 3.21 Procedure to multiply two polynomials

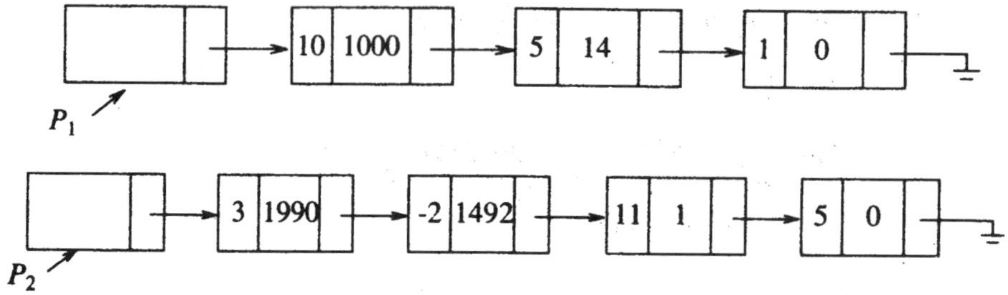

Figure 3.22 Linked list representations of two polynomials

```
typedef struct Node *PtrToNode;

struct Node
{
    int Coefficient;
    int Exponent;
    PtrToNode Next;
};

typedef PtrToNode Polynomial;   /* Nodes sorted by exponent */
```

Figure 3.23 Type declaration for linked list implementation of the Polynomial ADT

most of the time is spent multiplying zeros and stepping through what amounts to nonexistent parts of the input polynomials. This is always undesirable.

An alternative is to use a singly linked list. Each term in the polynomial is contained in one cell, and the cells are sorted in decreasing order of exponents. For instance, the linked lists in Figure 3.22 represent $P_1(X)$ and $P_2(X)$. We could then use the declarations in Figure 3.23.

The operations would then be straightforward to implement. The only potential difficulty is that when two polynomials are multiplied, the resultant polynomial will have to have like terms combined. There are several ways to do this, but we leave this as an exercise.

Radix Sort

A second example where linked lists are used is called *radix sort.* Radix sort is sometimes known as *card sort,* because it was used, until the advent of modern computers, to sort old-style punch cards.

If we have N integers in the range 1 to M (or 0 to $M - 1$), we can use this information to obtain a fast sort known as *bucket sort.* We keep an array called *Count,* of size M, which is initialized to zero. Thus, *Count* has M cells (or buckets), which are initially empty. When A_i is read, increment (by 1) $Count[A_i]$. After all the input is read, scan the *Count* array, printing out a representation of the sorted list. This algorithm takes $O(M + N)$; the proof is left as an exercise. If $M = \Theta(N)$, then bucket sort is $O(N)$.

Radix sort is a generalization of this. The easiest way to see what happens is by example. Suppose we have 10 numbers, in the range 0 to 999, that we would like to sort. In general, this is N numbers in the range 0 to $N^P - 1$ for some constant P. Obviously, we cannot use bucket sort; there would be too many buckets. The trick is to use several passes of bucket sort. The natural algorithm would be to bucket-sort by the most significant "digit" (digit is taken to base N), then the next most significant, and so on. That algorithm does not work, but if we perform bucket sorts by the least significant "digit" first, then the algorithm works. Of course, more than one number could fall into the same bucket and, unlike the original bucket sort, these numbers could be different, so we keep them in a list. Notice that all the numbers could have some digit in common, so if a simple array were used for the lists, each array would have to be of size N, for a total space requirement of $\Theta(N^2)$.

The following example shows the action of radix sort on 10 numbers. The input is 64, 8, 216, 512, 27, 729, 0, 1, 343, 125 (the first 10 cubes, arranged randomly). The first step bucket-sorts by the least significant digit. In this case the math is in base 10 (to make things simple), but do not assume this in general. The buckets are as shown in Figure 3.24, so the list, sorted by least significant digit, is 0, 1, 512, 343, 64, 125, 216, 27, 8, 729. These are now sorted by the next least significant digit (the tens digit here) (see Fig. 3.25). Pass 2 gives output 0, 1, 8, 512, 216, 125, 27, 729, 343, 64. This list is now sorted with respect to the two least significant digits. The final pass, shown in Figure 3.26, bucket-sorts by the most significant digit. The final list is 0, 1, 8, 27, 64, 125, 216, 343, 512, 729.

To see that the algorithm works, notice that the only possible failure would occur if two numbers came out of the same bucket in the wrong order. But the previous passes ensure that when several numbers enter a bucket, they enter in sorted order. The running time is $O(P(N + B))$ where P is the number of passes, N is the number of elements to sort, and B is the number of buckets. In our case, $B = N$.

Figure 3.24 Buckets after first step of radix sort

0	1	512	343	64	125	216	27	8	729
0	1	2	3	4	5	6	7	8	9

Figure 3.25 Buckets after the second pass of radix sort

8 1 0	216 512	729 27 125		343		64			
0	1	2	3	4	5	6	7	8	9

64 27 8 1 0	125	216	343		512		729		
0	1	2	3	4	5	6	7	8	9

Figure 3.26 Buckets after the last pass of radix sort

As an example, we could sort all integers that are representable on a computer (32 bits) by radix sort, if we did three passes over a bucket size of 2^{11}. This algorithm would always be $O(N)$ on this computer, but probably still not as efficient as some of the algorithms we shall see in Chapter 7, because of the high constant involved. (Remember that a factor of $\log N$ is not all that high, and this algorithm would have the overhead of maintaining linked lists.)

Multilists

Our last example shows a more complicated use of linked lists. A university with 40,000 students and 2,500 courses needs to be able to generate two types of reports. The first report lists the registration for each class, and the second report lists, by student, the classes that each student is registered for.

The obvious implementation might be to use a two-dimensional array. Such an array would have 100 million entries. The average student registers for about three courses, so only 120,000 of these entries, or roughly 0.1 percent, would actually have meaningful data.

What is needed is a list for each class containing the students in the class. We also need a list for each student containing the classes the student is registered for. Figure 3.27 shows our implementation.

As the figure shows, we have combined two lists into one. All lists use a header and are circular. To list all of the students in class C3, we start at C3 and traverse its list (by going right). The first cell belongs to student S1. Although there is no explicit information to this effect, this can be determined by following the student's linked list until the header is reached. Once this is done, we return to C3's list (we stored the position we were at in the course list before we traversed the student's list) and find another cell, which can be determined to belong to S3. We can continue and find that S4 and S5 are also in this class. In a similar manner, we can determine, for any student, all of the classes in which the student is registered.

Using a circular list saves space but does so at the expense of time. In the worst case, if the first student was registered for every course, then every entry would need to be examined to determine all the course names for that student. Because in this application there are relatively few courses per student and few students per course, this is not likely to happen. If it were suspected that this could cause a problem, then each of the (nonheader) cells could have pointers directly back to the student and class header. This would double the space requirement but would simplify and speed the implementation.

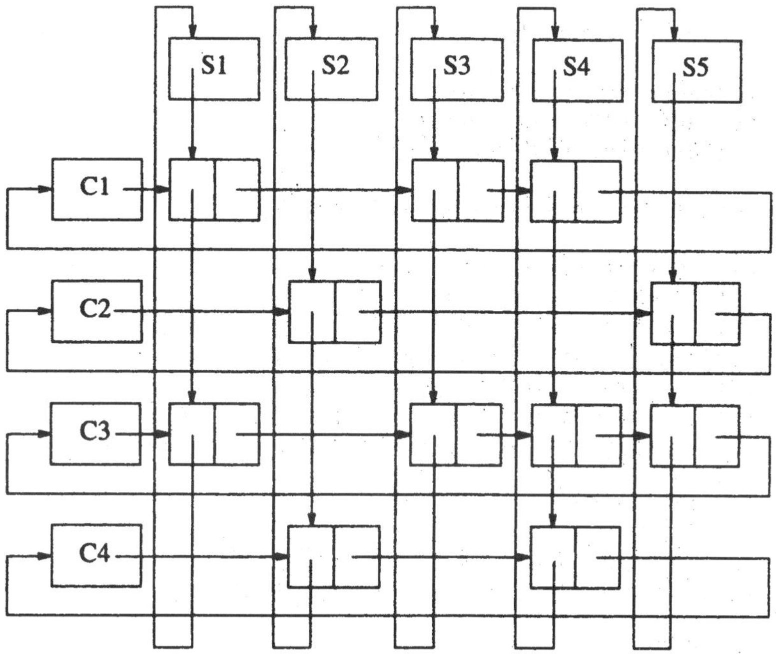

Figure 3.27 Multilist implementation for registration problem

3.2.8. Cursor Implementation of Linked Lists

Many languages, such as BASIC and FORTRAN, do not support pointers. If linked lists are required and pointers are not available, then an alternative implementation must be used. The method we will describe is known as a *cursor* implementation.

The two important features present in a pointer implementation of linked lists are as follows:

1. The data are stored in a collection of structures. Each structure contains data and a pointer to the next structure.
2. A new structure can be obtained from the system's global memory by a call to *malloc* and released by a call to *free.*

Our cursor implementation must be able to simulate this. The logical way to satisfy condition 1 is to have a global array of structures. For any cell in the array, its array index can be used in place of an address. Figure 3.28 gives the declarations for a cursor implementation of linked lists.

We must now simulate condition 2 by allowing the equivalent of *malloc* and *free* for cells in the *CursorSpace* array. To do this, we will keep a list (the *freelist*) of cells that are not in any list. The list will use cell 0 as a header. The initial configuration is shown in Figure 3.29.

A value of 0 for *Next* is the equivalent of a *NULL* pointer. The initialization of *CursorSpace* is a straightforward loop, which we leave as an exercise. To perform a *malloc*, the first element (after the header) is removed from the freelist. To perform

```
#ifndef _Cursor_H

typedef int PtrToNode;
typedef PtrToNode List;
typedef PtrToNode Position;

void InitializeCursorSpace( void );

List MakeEmpty( List L );
int IsEmpty( const List L );
int IsLast( const Position P, const List L );
Position Find( ElementType X, const List L );
void Delete( ElementType X, List L );
Position FindPrevious( ElementType X, const List L );
void Insert( ElementType X, List L, Position P );
void DeleteList( List L );
Position Header( const List L );
Position First( const List L );
Position Advance( const Position P );
ElementType Retrieve( const Position P );

#endif    /* _Cursor_H */


/* Place in the implementation file */
struct Node
{
    ElementType Element;
    Position    Next;
};

struct Node CursorSpace[ SpaceSize ];
```

Figure 3.28 Declarations for cursor implementation of linked lists

a *free,* we place the cell at the front of the freelist. Figure 3.30 shows the cursor implementation of *malloc* and *free.* Notice that if there is no space available, our routine does the correct thing by setting $P = 0$. This indicates that there are no more cells left, and also makes the second line of *CursorAlloc* a nonoperation (no-op).

Given this, the cursor implementation of linked lists is straightforward. For consistency, we will implement our lists with a header node. As an example, in Figure 3.31, if the value of L is 5 and the value of M is 3, then L represents the list $a, b, e,$ and M represents the list c, d, f.

To write the functions for a cursor implementation of linked lists, we must pass and return the identical parameters as the pointer implementation. The routines are straightforward. Figure 3.32 implements a function to test whether a list is empty. Figure 3.33 implements the test of whether the current position is the

Slot	Element	Next
0		1
1		2
2		3
3		4
4		5
5		6
6		7
7		8
8		9
9		10
10		0

Figure 3.29 An initialized *CursorSpace*

```
static Position
CursorAlloc( void )
{
    Position P;

    P = CursorSpace[ 0 ].Next;
    CursorSpace[ 0 ].Next = CursorSpace[ P ].Next;

    return P;
}

static void
CursorFree( Position P )
{
    CursorSpace[ P ].Next = CursorSpace[ 0 ].Next;
    CursorSpace[ 0 ].Next = P;
}
```

Figure 3.30 Routines: *CursorAlloc* and *CursorFree*

last in a linked list. The function *Find* in Figure 3.34 returns the position of X in list L. The code to implement deletion is shown in Figure 3.35. Again, the interface for the cursor implementation is identical to the pointer implementation. Finally, Figure 3.36 shows a cursor implementation of *Insert*.

The rest of the routines are similarly coded. The crucial point is that these routines follow the ADT specification. They take specific arguments and perform specific operations. The implementation is transparent to the user. The cursor implementation could be used instead of the linked list implementation, with virtually no change required in the rest of the code. If relatively few *Finds* are

Slot	Element	Next
0	–	6
1	b	9
2	f	0
3	header	7
4	–	0
5	header	10
6	–	4
7	c	8
8	d	2
9	e	0
10	a	1

Figure 3.31 Example of a cursor implementation of linked lists

```
/* Return true if L is empty */

int
IsEmpty( List L )
{
    return CursorSpace[ L ].Next == 0;
}
```

Figure 3.32 Function to test whether a linked list is empty—cursor implementation

```
/* Return true if P is the last position in list L */
/* Parameter L is unused in this implementation */

int
IsLast( Position P, List L )
{
    return CursorSpace[ P ].Next == 0;
}
```

Figure 3.33 Function to test whether *P* is last in a linked list—cursor implementation

```
/* Return Position of X in L; 0 if not found */
/* Uses a header node */

Position
Find( ElementType X, List L )
{
    Position P;

    P = CursorSpace[ L ].Next;
    while( P && CursorSpace[ P ].Element != X )
        P = CursorSpace[ P ].Next;

    return P;
}
```

Figure 3.34 *Find* routine—cursor implementation

```
/* Delete first occurrence of X from a list */
/* Assume use of a header node */

void
Delete( ElementType X, List L )
{
    Position P, TmpCell;

    P = FindPrevious( X, L );

    if( !IsLast( P, L ) )   /* Assumption of header use */
    {                       /* X is found; delete it */
        TmpCell = CursorSpace[ P ].Next;
        CursorSpace[ P ].Next = CursorSpace[ TmpCell ].Next;
        CursorFree( TmpCell );
    }
}
```

Figure 3.35 Deletion routine for linked lists—cursor implementation

performed, the cursor implementation could be significantly faster because of the lack of memory management routines.

The freelist represents an interesting data structure in its own right. The cell that is removed from the freelist is the one that was most recently placed there by

```
/* Insert (after legal position P) */
/* Header implementation assumed */
/* Parameter L is unused in this implementation */

void
Insert( ElementType X, List L, Position P )
{
    Position TmpCell;

    TmpCell = CursorAlloc( );
    if( TmpCell == 0 )
        FatalError( "Out of space!!!" );

    CursorSpace[ TmpCell ].Element = X;
    CursorSpace[ TmpCell ].Next = CursorSpace[ P ].Next;
    CursorSpace[ P ].Next = TmpCell;
}
```

Figure 3.36 Insertion routine for linked lists—cursor implementation

virtue of *free*. Thus, the last cell placed on the freelist is the first cell taken off. The data structure that also has this property is known as a *stack*, and is the topic of the next section.

3.3. The Stack ADT

3.3.1. Stack Model

A *stack* is a list with the restriction that insertions and deletions can be performed in only one position, namely, the end of the list, called the *top*. The fundamental operations on a stack are *Push*, which is equivalent to an insert, and *Pop*, which deletes the most recently inserted element. The most recently inserted element can be examined prior to performing a *Pop* by use of the *Top* routine. A *Pop* or *Top* on an empty stack is generally considered an error in the stack ADT. On the other hand, running out of space when performing a *Push* is an implementation error but not an ADT error.

Stacks are sometimes known as LIFO (last in, first out) lists. The model depicted in Figure 3.37 signifies only that *Push*es are input operations and *Pop*s and *Top*s are output. The usual operations to make empty stacks and test for emptiness are part of the repertoire, but essentially all that you can do to a stack is *Push* and *Pop*.

Figure 3.38 shows an abstract stack after several operations. The general model is that there is some element that is at the top of the stack, and it is the only element that is visible.

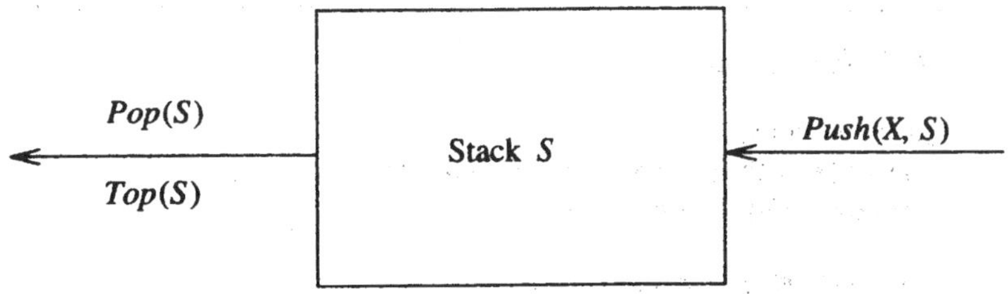

Figure 3.37 Stack model: input to a stack is by *Push*, output is by *Pop*

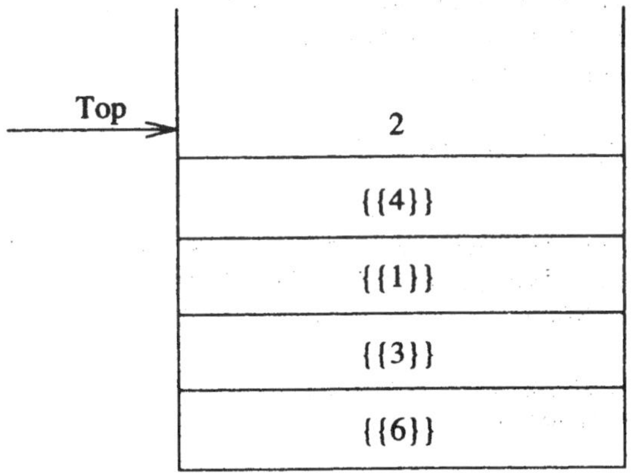

Figure 3.38 Stack model: only the top element is accessible

3.3.2. Implementation of Stacks

Since a stack is a list, any list implementation will do. We will give two popular implementations. One uses pointers and the other uses an array, but, as we saw in the previous section, if we use good programming principles, the calling routines do not need to know which method is being used.

Linked List Implementation of Stacks

The first implementation of a stack uses a singly linked list. We perform a *Push* by inserting at the front of the list. We perform a *Pop* by deleting the element at the front of the list. A *Top* operation merely examines the element at the front of the list, returning its value. Sometimes the *Pop* and *Top* operations are combined into one. We could use calls to the linked list routines of the previous section, but we will rewrite the stack routines from scratch for the sake of clarity.

First, we give the definitions in Figure 3.39. We implement the stack using a header. Figure 3.40 shows that an empty stack is tested for in the same manner as an empty list.

Creating an empty stack is also simple. We merely create a header node; *MakeEmpty* sets the *Next* pointer to *NULL* (see Fig. 3.41). The *Push* is implemented

```
#ifndef _Stack_h

struct Node;
typedef struct Node *PtrToNode;
typedef PtrToNode Stack;

int IsEmpty( Stack S );
Stack CreateStack( void );
void DisposeStack( Stack S );
void MakeEmpty( Stack S );
void Push( ElementType X, Stack S );
ElementType Top( Stack S );
void Pop( Stack S );

#endif  /* _Stack_h */


/* Place in implementation file */
/* Stack implementation is a linked list with a header */
struct Node
{
    ElementType Element;
    PtrToNode   Next;
};
```

Figure 3.39 Type declaration for linked list implementation of the stack ADT

```
int
IsEmpty( Stack S )
{
    return S->Next == NULL;
}
```

Figure 3.40 Routine to test whether a stack is empty—linked list implementation

as an insertion into the front of a linked list, where the front of the list serves as the top of the stack (see Fig. 3.42). The *Top* is performed by examining the element in the first position of the list (see Fig. 3.43). Finally, we implement *Pop* as a deletion from the front of the list (see Fig. 3.44).

It should be clear that all the operations take constant time, because nowhere in any of the routines is there even a reference to the size of the stack (except for emptiness), much less a loop that depends on this size. The drawback of this implementation is that the calls to *malloc* and *free* are expensive, especially in

```
Stack
CreateStack( void )
{
    Stack S;

    S = malloc( sizeof( struct Node ) );
    if( S == NULL )
        FatalError( "Out of space!!!" );
    MakeEmpty( S );
    return S;
}

void
MakeEmpty( Stack S )
{
    if( S == NULL )
        Error( "Must use CreateStack first" );
    else
        while( !IsEmpty( S ) )
            Pop( S );
}
```

Figure 3.41 Routine to create an empty stack—linked list implementation

```
void
Push( ElementType X, Stack S )
{
    PtrToNode TmpCell;

    TmpCell = malloc( sizeof( struct Node ) );
    if( TmpCell == NULL )
        FatalError( "Out of space!!!" );
    else
    {
        TmpCell->Element = X;
        TmpCell->Next = S->Next;
        S->Next = TmpCell;
    }
}
```

Figure 3.42 Routine to push onto a stack—linked list implementation

```
ElementType
Top( Stack S )
{
    if( !IsEmpty( S ) )
        return S->Next->Element;
    Error( "Empty stack" );
    return 0;  /* Return value used to avoid warning */
}
```

Figure 3.43 Routine to return top element in a stack—linked list implementation

```
void
Pop( Stack S )
{
    PtrToNode FirstCell;

    if( IsEmpty( S ) )
        Error( "Empty stack" );
    else
    {
        FirstCell = S->Next;
        S->Next = S->Next->Next;
        free( FirstCell );
    }
}
```

Figure 3.44 Routine to pop from a stack—linked list implementation

comparison to the pointer manipulation routines. Some of this can be avoided by using a second stack, which is initially empty. When a cell is to be dropped from the first stack, it is merely placed on the second stack. Then, when new cells are needed for the first stack, the second stack is checked first.

Array Implementation of Stacks

An alternative implementation avoids pointers and is probably the more popular solution. The only potential hazard with this strategy is that we need to declare an array size ahead of time. Generally this is not a problem, because in typical applications, even if there are quite a few stack operations, the actual number of elements in the stack at any time never gets too large. It is usually easy to declare the array to be large enough without wasting too much space. If this is not possible, then a safe course would be to use a linked list implementation.

If we use an array implementation, the implementation is trivial. Associated with each stack is *TopOfStack*, which is -1 for an empty stack (this is how an empty stack is initialized). To push some element X onto the stack, we increment *TopOfStack* and then set *Stack*[*TopOfStack*] = X, where *Stack* is the array representing the actual stack. To pop, we set the return value to *Stack*[*TopOfStack*] and then decrement *TopOfStack*. Of course, since there are potentially several stacks, the *Stack* array and *TopOfStack* are part of one structure representing a stack. It is almost always a bad idea to use global variables and fixed names to represent this (or any) data structure, because in most real-life situations there will be more than one stack. When writing your actual code, you should attempt to follow the model as closely as possible, so that no part of your code, except for the stack routines, can attempt to access the array or top-of-stack variable implied by each stack. This is true for *all* ADT operations. Modern languages such as Ada and C++ can actually enforce this rule.

Notice that these operations are performed in not only constant time, but very fast constant time. On some machines, *Push*es and *Pop*s (of integers) can be written in one machine instruction, operating on a register with auto-increment and auto-decrement addressing. The fact that most modern machines have stack operations as part of the instruction set enforces the idea that the stack is probably the most fundamental data structure in computer science, after the array.

One problem that affects the efficiency of implementing stacks is error testing. Our linked list implementation carefully checked for errors. As described above, a *Pop* on an empty stack or a *Push* on a full stack will overflow the array bounds and cause a crash. This is obviously undesirable, but if checks for these conditions were put in the array implementation, they would likely take as much time as the actual stack manipulation. For this reason, it has become a common practice to skimp on error checking in the stack routines, except where error handling is crucial (as in operating systems). Although you can probably get away with this in most cases by declaring the stack to be large enough not to overflow and ensuring that routines that use *Pop* never attempt to *Pop* an empty stack, this can lead to code that barely works at best, especially when programs are large and are written by more than one person or at more than one time. Because stack operations take such fast constant time, it is rare that a significant part of the running time of a program is spent in these routines. This means that it is generally not justifiable to omit error checks. You should always write the error checks; if they are redundant, you can always comment them out if they really cost too much time. Having said all this, we can now write routines to implement a general stack using arrays.

A *Stack* is defined in Figure 3.45 as a pointer to a structure. The structure contains the *TopOfStack* and *Capacity* fields. Once the maximum size is known, the stack array can be dynamically allocated. Figure 3.46 creates a stack of a given maximum size. Lines 3–5 allocate the stack structure, and lines 6–8 allocate the stack array. Lines 9 and 10 initialize the *TopOfStack* and *Capacity* fields. The stack array does not need to be initialized. The stack is returned at line 11.

The routine *DisposeStack* should be written to free the stack structure. This routine first frees the stack array and then the stack structure (see Fig. 3.47). Since *CreateStack* requires an argument in the array implementation, but not in the linked list implementation, the routine that uses a stack will need to know which implementation is being used unless a dummy parameter is added for the

```
#ifndef _Stack_h

struct StackRecord;
typedef struct StackRecord *Stack;

int IsEmpty( Stack S );
int IsFull( Stack S );
Stack CreateStack( int MaxElements );
void DisposeStack( Stack S );
void MakeEmpty( Stack S );
void Push( ElementType X, Stack S );
ElementType Top( Stack S );
void Pop( Stack S );
ElementType TopAndPop( Stack S );

#endif  /* _Stack_h */


/* Place in implementatioin file */
/* Stack implementation is a dynamically allocated array */
#define EmptyTOS ( -1 )
#define MinStackSize ( 5 )

struct StackRecord
{
    int Capacity;
    int TopOfStack;
    ElementType *Array;
};
```

Figure 3.45 *Stack* declarations—array implementation

later implementation. Unfortunately, efficiency and software idealism often create conflicts.

We have assumed that all stacks deal with the same type of element. In many languages, if there are different types of stacks, then we need to rewrite a new version of the stack routines for each different type, giving each version a different name. A cleaner alternative is provided in C++, which allows one to write a set of generic stack routines that work for any type. C++ also allows stacks of several different types to retain the same procedure and function names (such as *Push* and *Pop*): The compiler decides which routines are implied by checking the type of the calling routine.

Having said all this, we will now rewrite the four stack routines. In true ADT spirit, we will make the function and procedure heading look identical to the linked list implementation. The routines themselves are very simple and follow the written description exactly (see Figs. 3.48 to 3.52).

```
Stack
CreateStack( int MaxElements )
{
    Stack S;

    if( MaxElements < MinStackSize )
        Error( "Stack size is too small" );

    S = malloc( sizeof( struct StackRecord ) );
    if( S == NULL )
        FatalError( "Out of space!!!" );

    S->Array = malloc( sizeof( ElementType ) * MaxElements );
    if( S->Array == NULL )
        FatalError( "Out of space!!!" );
    S->Capacity = MaxElements;
    MakeEmpty( S );

    return S;
}
```

Figure 3.46 Stack creation—array implementation

```
void
DisposeStack( Stack S )
{
    if( S != NULL )
    {
        free( S->Array );
        free( S );
    }
}
```

Figure 3.47 Routine for freeing stack—array implementation

```
int
IsEmpty( Stack S )
{
    return S->TopOfStack == EmptyTOS;
}
```

Figure 3.48 Routine to test whether a stack is empty—array implementation

```
void
MakeEmpty( Stack S )
{
    S->TopOfStack = EmptyTOS;
}
```

Figure 3.49 Routine to create an empty stack—array implementation

```
void
Push( ElementType X, Stack S )
{
    if( IsFull( S ) )
        Error( "Full stack" );
    else
        S->Array[ ++S->TopOfStack ] = X;
}
```

Figure 3.50 Routine to push onto a stack—array implementation

```
ElementType
Top( Stack S )
{
    if( !IsEmpty( S ) )
        return S->Array[ S->TopOfStack ];
    Error( "Empty stack" );
    return 0;  /* Return value used to avoid warning */
}
```

Figure 3.51 Routine to return top of stack—array implementation

```
void
Pop( Stack S )
{
    if( IsEmpty( S ) )
        Error( "Empty stack" );
    else
        S->TopOfStack--;
}
```

Figure 3.52 Routine to pop from a stack—array implementation

```
ElementType
TopAndPop( Stack S )
{
    if( !IsEmpty( S ) )
        return S->Array[ S->TopOfStack-- ];
    Error( "Empty stack" );
    return 0;  /* Return value used to avoid warning */
}
```

Figure 3.53 Routine to give top element and pop a stack—array implementation

Pop is occasionally written as a function that returns the popped element (and alters the stack). Although current thinking suggests that functions should not change their input variables, Figure 3.53 illustrates that this is the most convenient method in C.

3.3.3. Applications

It should come as no surprise that if we restrict the operations allowed on a list, those operations can be performed very quickly. The big surprise, however, is that the small number of operations left are so powerful and important. We give three of the many applications of stacks. The third application gives a deep insight into how programs are organized.

Balancing Symbols

Compilers check your programs for syntax errors, but frequently a lack of one symbol (such as a missing brace or comment starter) will cause the compiler to spill out a hundred lines of diagnostics without identifying the real error.

A useful tool in this situation is a program that checks whether everything is balanced. Thus, every right brace, bracket, and parenthesis must correspond to its left counterpart. The sequence [()] is legal, but [(]) is wrong. Obviously, it is not worthwhile writing a huge program for this, but it turns out that it is easy to check these things. For simplicity, we will just check for balancing of parentheses, brackets, and braces and ignore any other character that appears.

The simple algorithm uses a stack and is as follows:

> Make an empty stack. Read characters until end of file. If the character is an opening symbol, push it onto the stack. If it is a closing symbol, then if the stack is empty report an error. Otherwise, pop the stack. If the symbol popped is not the corresponding opening symbol, then report an error. At end of file, if the stack is not empty report an error.

You should be able to convince yourself that this algorithm works. It is clearly linear and actually makes only one pass through the input. It is thus on-line and quite fast. Extra work can be done to attempt to decide what to do when an error is reported—such as identifying the likely cause.

Postfix Expressions

Suppose we have a pocket calculator and would like to compute the cost of a shopping trip. To do so, we add a list of numbers and multiply the result by 1.06; this computes the purchase price of some items with local sales tax added. If the items are 4.99, 5.99, and 6.99, then a natural way to enter this would be the sequence

4.99 + 5.99 + 6.99 * 1.06 =

Depending on the calculator, this produces either the intended answer, 19.05, or the scientific answer, 18.39. Most simple four-function calculators will give the first answer, but many advanced calculators know that multiplication has higher precedence than addition.

On the other hand, some items are taxable and some are not, so if only the first and last items were actually taxable, then the sequence

4.99 * 1.06 + 5.99 + 6.99 * 1.06 =

would give the correct answer (18.69) on a scientific calculator and the wrong answer (19.37) on a simple calculator. A scientific calculator generally comes with parentheses, so we can always get the right answer by parenthesizing, but with a simple calculator we need to remember intermediate results.

A typical evaluation sequence for this example might be to multiply 4.99 and 1.06, saving this answer as A_1. We then add 5.99 and A_1, saving the result in A_1. We multiply 6.99 and 1.06, saving the answer in A_2, and finish by adding A_1 and A_2, leaving the final answer in A_1. We can write this sequence of operations as follows:

4.99 1.06 * 5.99 + 6.99 1.06 * +

This notation is known as *postfix* or *reverse Polish* notation and is evaluated exactly as we have described above. The easiest way to do this is to use a stack. When a number is seen, it is pushed onto the stack; when an operator is seen, the operator is applied to the two numbers (symbols) that are popped from the stack, and the result is pushed onto the stack. For instance, the postfix expression

6 5 2 3 + 8 * +3 + *

is evaluated as follows: The first four symbols are placed on the stack. The resulting stack is

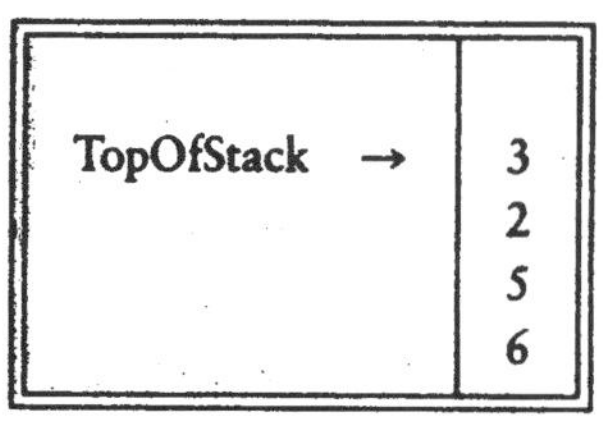

Next a '+' is read, so 3 and 2 are popped from the stack and their sum, 5, is pushed.

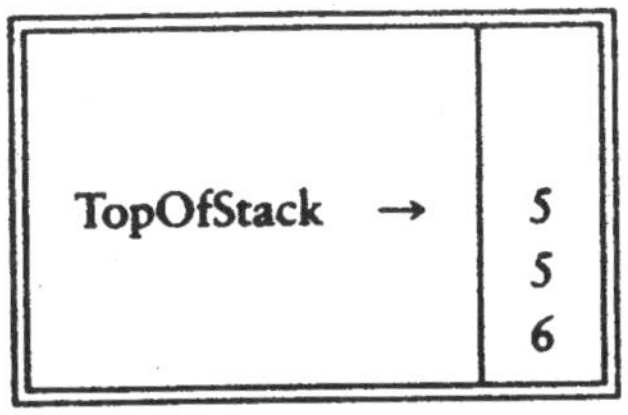

Next 8 is pushed.

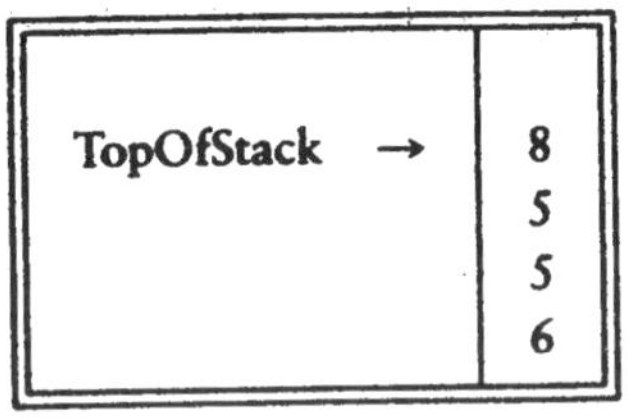

Now a '*' is seen, so 8 and 5 are popped and $5 * 8 = 40$ is pushed.

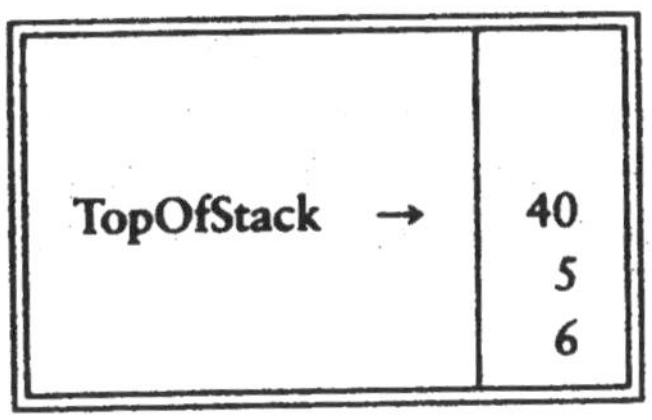

Next a '+' is seen, so 40 and 5 are popped and $5 + 40 = 45$ is pushed.

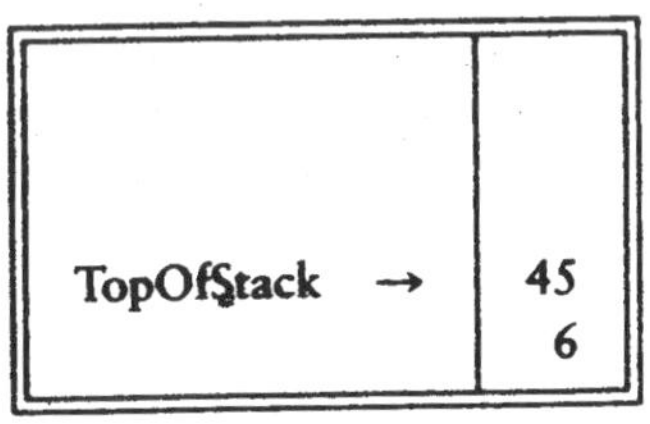

Now, 3 is pushed.

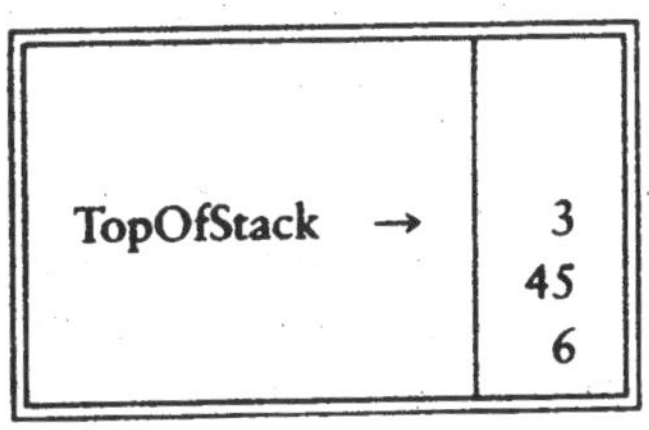

Next '+' pops 3 and 45 and pushes 45 + 3 = 48.

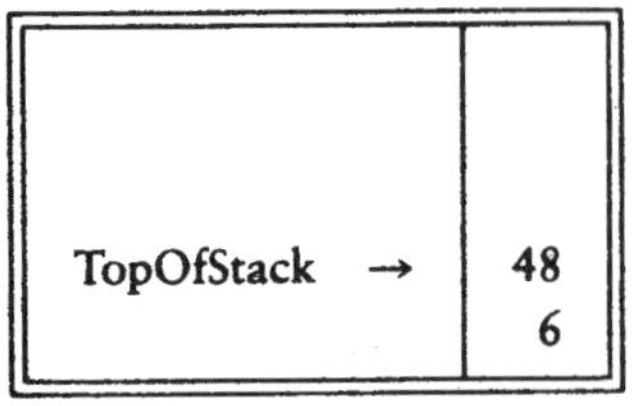

Finally, a '*' is seen and 48 and 6 are popped; the result, 6 * 48 = 288, is pushed.

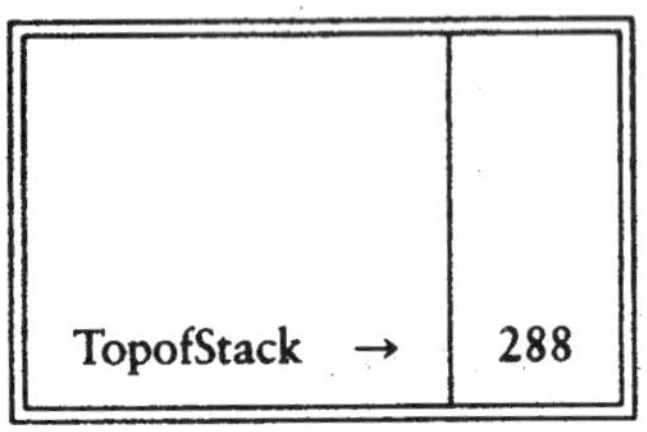

The time to evaluate a postfix expression is $O(N)$, because processing each element in the input consists of stack operations and thus takes constant time. The algorithm to do so is very simple. Notice that when an expression is given in postfix notation, there is no need to know any precedence rules; this is an obvious advantage.

Infix to Postfix Conversion

Not only can a stack be used to evaluate a postfix expression, but we can also use a stack to convert an expression in standard form (otherwise known as *infix*) into postfix. We will concentrate on a small version of the general problem by allowing only the operators +, *, (,), and insisting on the usual precedence rules. We will further assume that the expression is legal. Suppose we want to convert the infix expression

a + b * c + (d * e + f) * g

into postfix. A correct answer is a b c * + d e * f + g * +.

When an operand is read, it is immediately placed onto the output. Operators are not immediately output, so they must be saved somewhere. The correct thing to do is to place operators that have been seen, but not placed on the output, onto the stack. We will also stack left parentheses when they are encountered. We start with an initially empty stack.

If we see a right parenthesis, then we pop the stack, writing symbols until we encounter a (corresponding) left parenthesis, which is popped but not output.

If we see any other symbol ('+', '*', '('), then we pop entries from the stack until we find an entry of lower priority. One exception is that we never remove a

'(' from the stack except when processing a ')'. For the purposes of this operation, '+' has lowest priority and '(' highest. When the popping is done, we push the operator onto the stack.

Finally, if we read the end of input, we pop the stack until it is empty, writing symbols onto the output.

To see how this algorithm performs, we will convert the infix expression above into its postfix form. First, the symbol *a* is read, so it is passed through to the output. Then '+' is read and pushed onto the stack. Next *b* is read and passed through to the output. The state of affairs at this juncture is as follows:

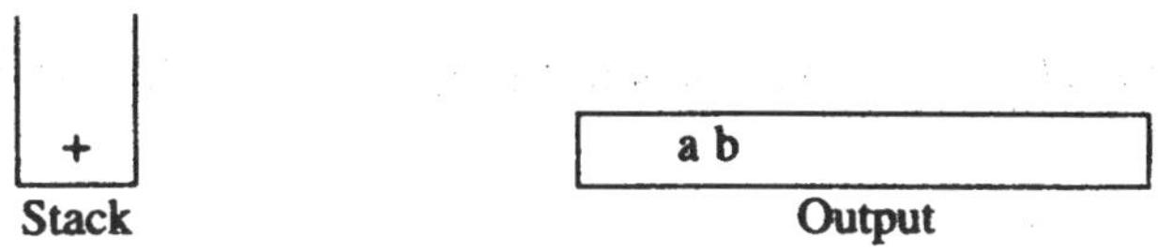

Next a '*' is read. The top entry on the operator stack has lower precedence than '*', so nothing is output and '*' is put on the stack. Next, *c* is read and output. Thus far, we have

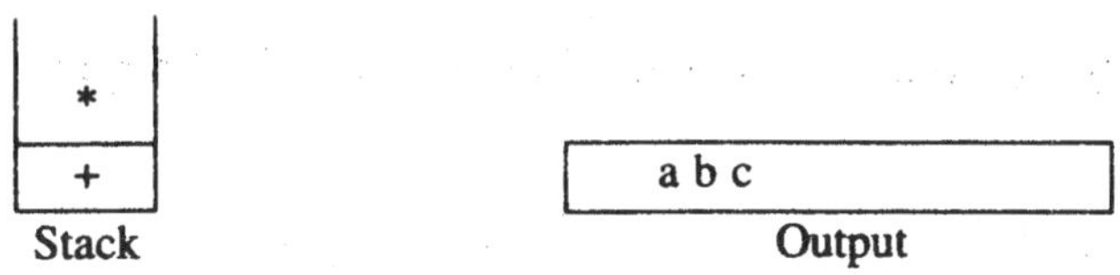

The next symbol is a '+'. Checking the stack, we find that we will pop a '*' and place it on the output; pop the other '+', which is not of *lower* but equal priority, on the stack; and then push the '+'.

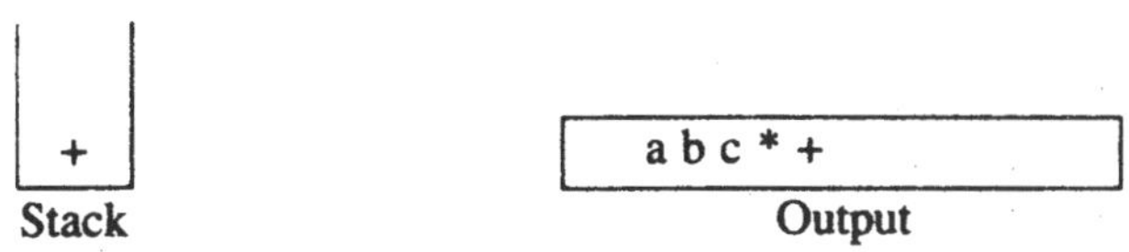

The next symbol read is a '(', which, being of highest precedence, is placed on the stack. Then *d* is read and output.

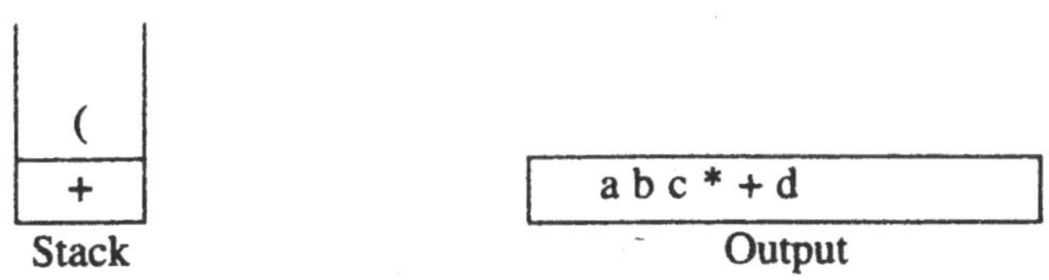

We continue by reading a '*'. Since open parentheses do not get removed except when a closed parenthesis is being processed, there is no output. Next, *e* is read and output.

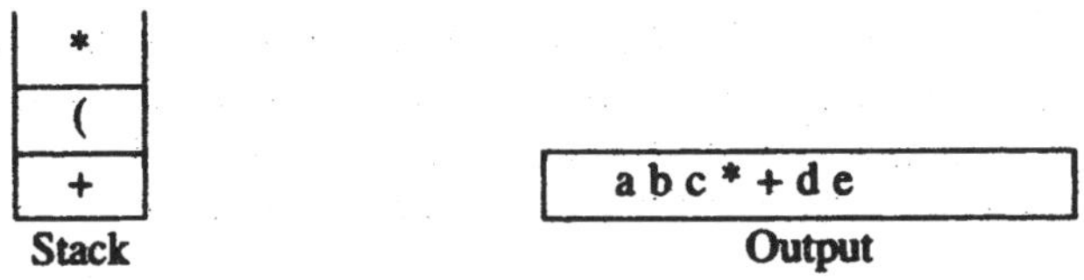

The next symbol read is a '+'. We pop and output '*' and then push '+'. Then we read and output *f*.

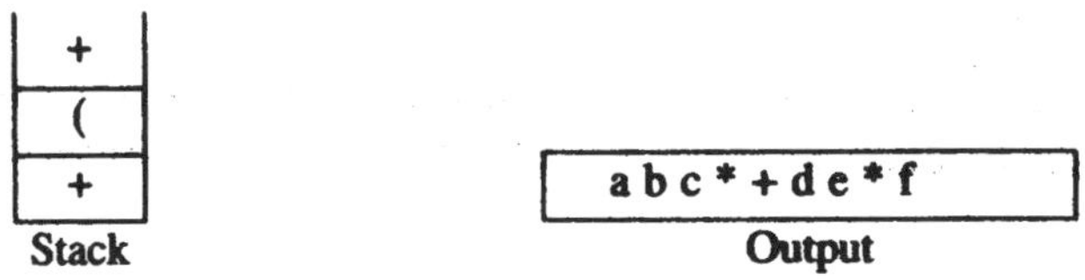

Now we read a ')', so the stack is emptied back to the '('. We output a '+'.

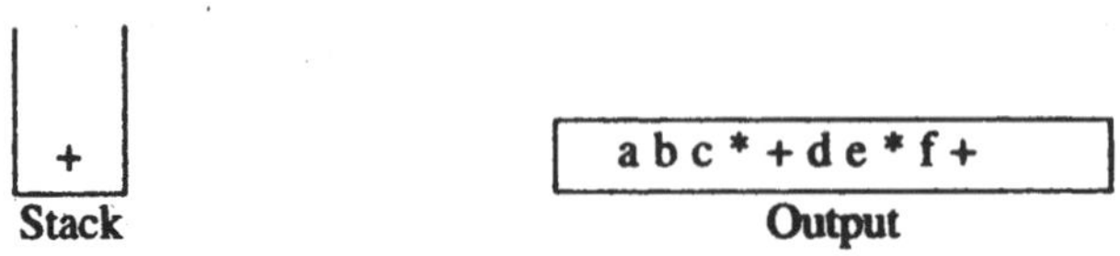

We read a '*' next; it is pushed onto the stack. Then *g* is read and output.

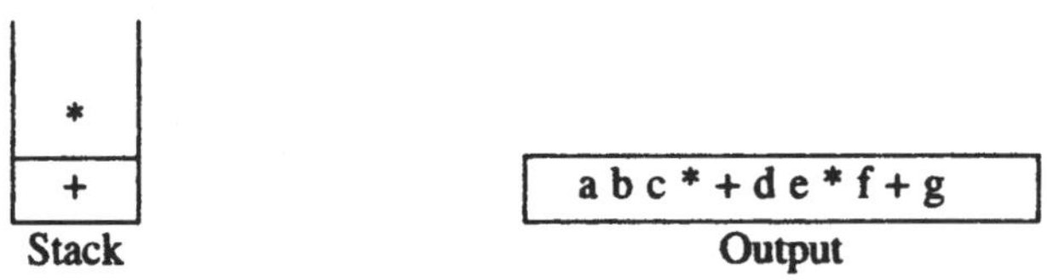

The input is now empty, so we pop and output symbols from the stack until it is empty.

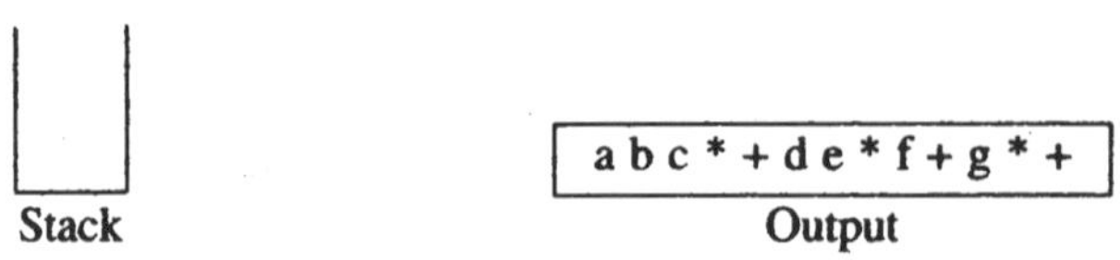

As before, this conversion requires only $O(N)$ time and works in one pass through the input. We can add subtraction and division to this repertoire by assigning subtraction and addition equal priority and multiplication and division equal priority. A subtle point is that the expression $a - b - c$ will be converted to $ab - c-$ and not $abc - -$. Our algorithm does the right thing, because these operators associate from left to right. This is not necessarily the case in general, since exponentiation associates right to left: $2^{2^3} = 2^8 = 256$, not $4^3 = 64$. We leave as an exercise the problem of adding exponentiation to the repertoire of assignments.

Function Calls

The algorithm to check balanced symbols suggests a way to implement function calls. The problem here is that when a call is made to a new function, all the variables local to the calling routine need to be saved by the system, since otherwise the new function will overwrite the calling routine's variables. Furthermore, the current location in the routine must be saved so that the new function knows where to go after it is done. The variables have generally been assigned by the compiler to machine registers, and there are certain to be conflicts (usually all procedures get some variables assigned to register #1), especially if recursion is involved. The reason that this problem is similar to balancing symbols is that a function call and function return are essentially the same as an open parenthesis and closed parenthesis, so the same ideas should work.

When there is a function call, all the important information that needs to be saved, such as register values (corresponding to variable names) and the return address (which can be obtained from the program counter, which is typically in a register), is saved "on a piece of paper" in an abstract way and put at the top of a pile. Then the control is transferred to the new function, which is free to replace the registers with its values. If it makes other function calls, it follows the same procedure. When the function wants to return, it looks at the "paper" at the top of the pile and restores all the registers. It then makes the return jump.

Clearly, all of this work can be done using a stack, and that is exactly what happens in virtually every programming language that implements recursion. The information saved is called either an *activation record* or *stack frame*. Typically, a slight adjustment is made: The current environment is represented at the top of the stack. Thus, a return gives the previous environment (without copying). The stack in a real computer frequently grows from the high end of your memory partition downward, and on many systems there is no checking for overflow. There is always the possibility that you will run out of stack space by having too many simultaneously active functions. Needless to say, running out of stack space is always a fatal error.

In languages and systems that do not check for stack overflow, your program will crash without an explicit explanation. On these systems, strange things may happen when your stack gets too big, because your stack will run into part of your program. It could be the main program, or it could be part of your data, especially if you have a big array. If it runs into your program, your program will be corrupted; you will have nonsense instructions and will crash as soon as they are executed. If the stack runs into your data, what is likely to happen is that when you write

```
        /* Bad use of recursion: Printing a linked list */
        /* No header */

        void
        PrintList( List L )
        {
/* 1*/      if( L != NULL )
            {
/* 2*/          PrintElement( L->Element );
/* 3*/          PrintList( L->Next);
            }
        }
```

Figure 3.54 A bad use of recursion: printing a linked list

something into your data, it will destroy stack information—probably the return address—and your program will attempt to return to some weird address and crash.

In normal events, you should not run out of stack space; doing so is usually an indication of runaway recursion (forgetting a base case). On the other hand, some perfectly legal and seemingly innocuous program can cause you to run out of stack space. The routine in Figure 3.54, which prints out a linked list, is perfectly legal and actually correct. It properly handles the base case of an empty list, and the recursion is fine. This program can be *proven* correct. Unfortunately, if the list contains 20,000 elements, there will be a stack of 20,000 activation records representing the nested calls of line 3. Activation records are typically large because of all the information they contain, so this program is likely to run out of stack space. (If 20,000 elements are not enough to make the program crash, replace the number with a larger one.)

This program is an example of an extremely bad use of recursion known as *tail recursion*. Tail recursion refers to a recursive call at the last line. Tail recursion can be mechanically eliminated by changing the recursive call to a *goto* preceded by one assignment per function argument. This simulates the recursive call because nothing needs to be saved; after the recursive call finishes, there is really no need to know the saved values. Because of this, we can just go to the top of the function with the values that would have been used in a recursive call. The program in Figure 3.55 shows the improved version. Keep in mind that *you* should use the more natural *while* loop construction. The *goto* is used here to show how a compiler might automatically remove the recursion.

Removal of tail recursion is so simple that some compilers do it automatically. Even so, it is best not to find out that yours does not.

Recursion can always be completely removed (the compiler does so in converting to assembly language), but doing so can be quite tedious. The general strategy requires using a stack and is worthwhile only if you can manage to put the bare minimum on the stack. We will not dwell on this further, except to point out that although nonrecursive programs are certainly generally faster than equivalent recursive programs, the speed advantage rarely justifies the lack of clarity that results from removing the recursion.

```
/* Printing a linked list non-recursively */
/* Uses a mechanical translation */
/* No header */

void
PrintList( List L )
{
  top:
    if( L != NULL )
    {
        PrintElement( L->Element );
        L = L->Next;
        goto top;
    }
}
```

Figure 3.55 Printing a list without recursion; a compiler might do this (you should not)

3.4. The Queue ADT

Like stacks, *queues* are lists. With a queue, however, insertion is done at one end, whereas deletion is performed at the other end.

3.4.1. Queue Model

The basic operations on a queue are *Enqueue,* which inserts an element at the end of the list (called the rear), and *Dequeue,* which deletes (and returns) the element at the start of the list (known as the front). Figure 3.56 shows the abstract model of a queue.

3.4.2. Array Implementation of Queues

As with stacks, any list implementation is legal for queues. Like stacks, both the linked list and array implementations give fast O(1) running times for every operation. The linked list implementation is straightforward and left as an exercise. We will now discuss an array implementation of queues.

Figure 3.56 Model of a queue

For each queue data structure, we keep an array, *Queue*[], and the positions *Front* and *Rear,* which represent the ends of the queue. We also keep track of the number of elements that are actually in the queue, *Size.* All this information is part of one structure, and as usual, except for the queue routines themselves, no routine should ever access these directly. The following figure shows a queue in some intermediate state. By the way, the cells that are blanks have undefined values in them. In particular, the first three cells have elements that used to be in the queue.

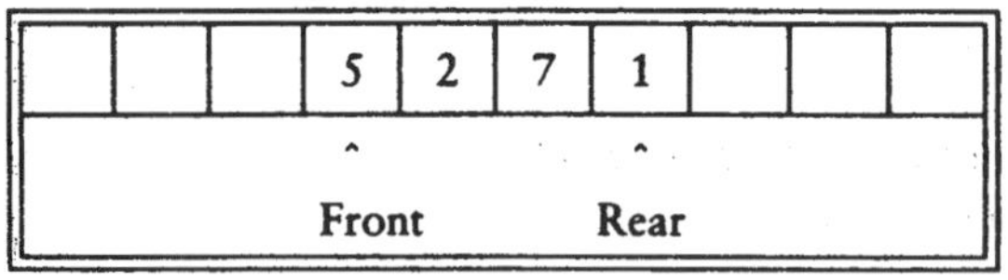

The operations should be clear. To *Enqueue* an element X, we increment *Size* and *Rear,* then set *Queue*[*Rear*] = X. To *Dequeue* an element, we set the return value to *Queue*[*Front*], decrement *Size,* and then increment *Front.* Other strategies are possible (this is discussed later). We will comment on checking for errors presently.

There is one potential problem with this implementation. After 10 *Enqueues*, the queue appears to be full, since *Rear* is now 10, and the next *Enqueue* would be in a nonexistent position. However, there might only be a few elements in the queue, because several elements may have already been dequeued. Queues, like stacks, frequently stay small even in the presence of a lot of operations.

The simple solution is that whenever *Front* or *Rear* gets to the end of the array, it is wrapped around to the beginning. The following figure shows the queue during some operations. This is known as a *circular array* implementation.

The extra code required to implement the wraparound is minimal (although it probably doubles the running time). If incrementing either *Rear* or *Front* causes it to go past the array, the value is reset to the first position in the array.

Initial State

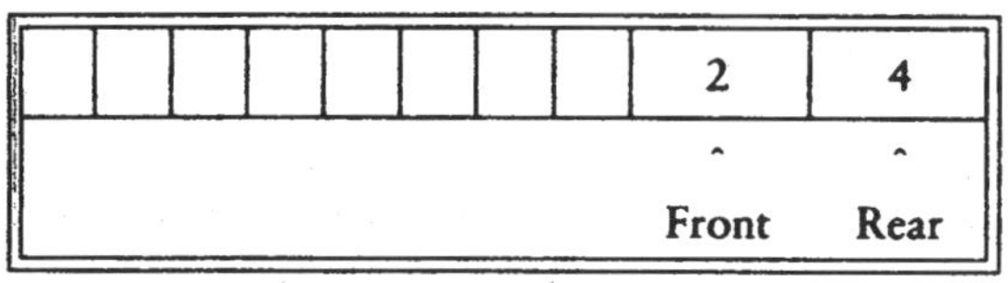

After *Enqueue*(1)

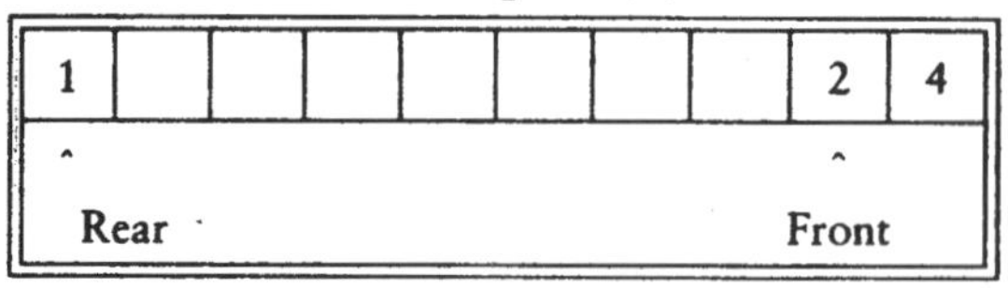

After *Enqueue*(3)

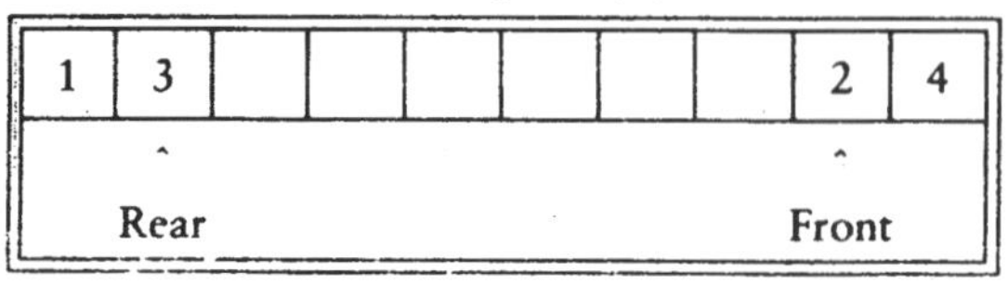

After *Dequeue,* Which Returns 2

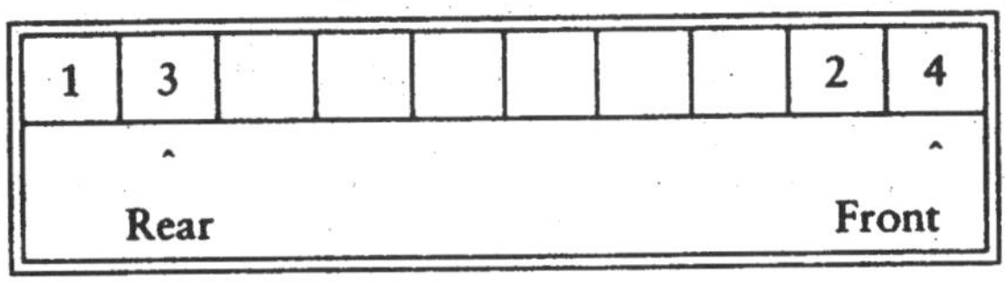

After *Dequeue,* Which Returns 4

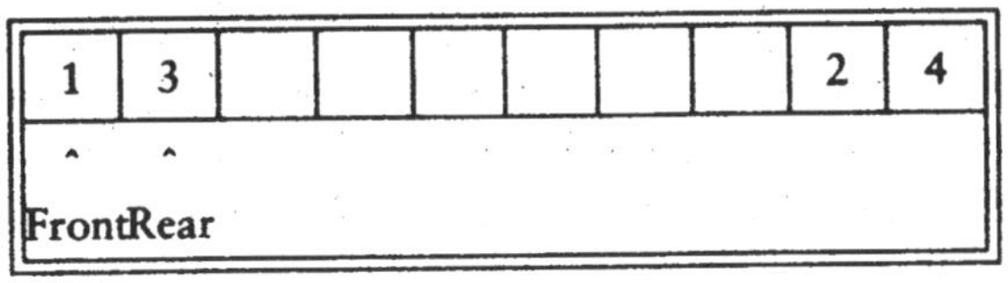

After *Dequeue,* Which Returns 1

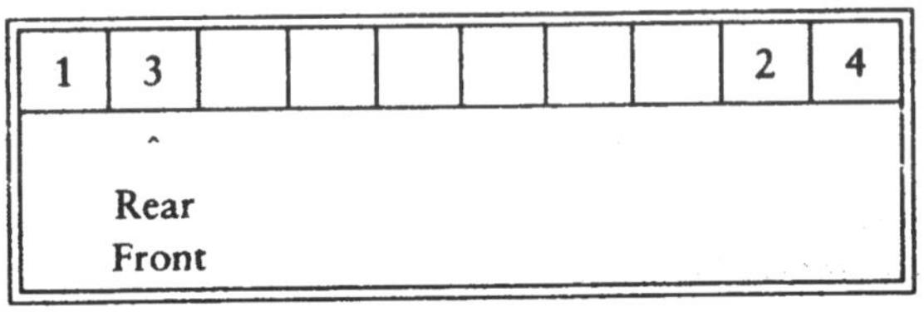

After *Dequeue,* Which Returns 3
and Makes the Queue Empty

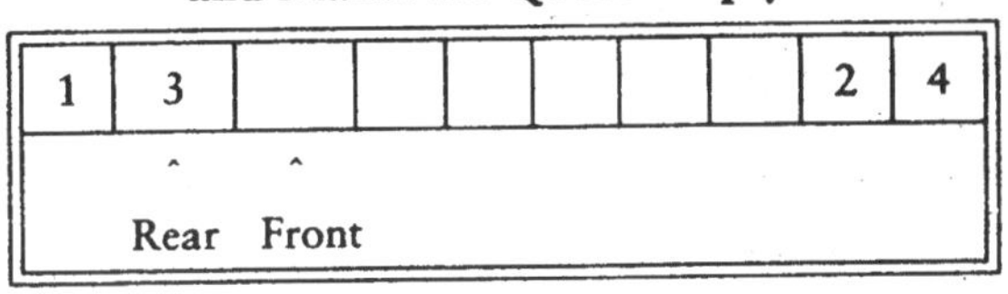

There are two warnings about the circular array implementation of queues. First, it is important to check the queue for emptiness, because a *Dequeue* when the queue is empty will return an undefined value, silently.

Second, some programmers use different ways of representing the front and rear of a queue. For instance, some do not use an entry to keep track of the size, because they rely on the base case that when the queue is empty, $Rear = Front - 1$. The size is computed implicitly by comparing *Rear* and *Front*. This is a very tricky way to go, because there are some special cases, so be very careful if you need to modify code written this way. If the size is not part of the structure, then if the array size is *ASize*, the queue is full when there are $ASize - 1$ elements, since only *ASize* different sizes can be differentiated, and one of these is 0. Pick any style you like and make sure that all your routines are consistent. Since there are a few options for implementation, it is probably worth a comment or two in the code, if you don't use the size field.

In applications where you are sure that the number of *Enqueues* is not larger than the size of the queue, the wraparound is not necessary. As with stacks, *Dequeues* are rarely performed unless the calling routines are certain that the queue is not empty. Thus error calls are frequently skipped for this operation, except in critical code. This is generally not justifiable, because the time savings that you are likely to achieve are minimal.

We finish this section by writing some of the queue routines. We leave the others as an exercise. First, we give the queue declarations in Figure 3.57. We add a maximum size field, as was done for the array implementation of the stack; *CreateQueue* and *DisposeQueue* routines also need to be provided We also provide routines to test whether a queue is empty and to make an empty queue (Figs. 3.58 and 3.59). The reader can write the function *IsFull,* which performs the test

Figure 3.57 Type declarations for queue—array implementation

```
#ifndef _Queue_h

struct QueueRecord;
typedef struct QueueRecord *Queue;

int IsEmpty( Queue Q );
int IsFull( Queue Q );
Queue CreateQueue( int MaxElements );
void DisposeQueue( Queue Q );
void MakeEmpty( Queue Q );
void Enqueue( ElementType X, Queue Q );
ElementType Front( Queue Q );
void Dequeue( Queue Q );
ElementType FrontAndDequeue( Queue Q );

#endif  /* _Queue_h */


/* Place in implementation file */
/* Queue implementation is a dynamically allocated array */
#define MinQueueSize ( 5 )

struct QueueRecord
{
    int Capacity;
    int Front;
    int Rear;
    int Size;
    ElementType *Array;
};
```

```
int
IsEmpty( Queue Q )
{
    return Q->Size == 0;
}
```

Figure 3.58 Routine to test whether a queue is empty—array implementation

```
void
MakeEmpty( Queue Q )
{
    Q->Size = 0;
    Q->Front = 1;
    Q->Rear = 0;
}
```

Figure 3.59 Routine to make an empty queue—array implementation

```
static int
Succ( int Value, Queue Q )
{
    if( ++Value == Q->Capacity )
        Value = 0;
    return Value;
}

void
Enqueue( ElementType X, Queue Q )
{
    if( IsFull( Q ) )
        Error( "Full queue" );
    else
    {
        Q->Size++;
        Q->Rear = Succ( Q->Rear, Q );
        Q->Array[ Q->Rear ] = X;
    }
}
```

Figure 3.60 Routines to enqueue—array implementation

implied by its name. Notice that *Rear* is preinitialized to 1 before *Front*. The final operation we will write is the *Enqueue* routine. Following the exact description above, we arrive at the implementation in Figure 3.60.

3.4.3. Applications of Queues

There are several algorithms that use queues to give efficient running times. Several of these are found in graph theory, and we will discuss them in Chapter 9. For now, we will give some simple examples of queue usage.

When jobs are submitted to a printer, they are arranged in order of arrival. Thus, essentially, jobs sent to a line printer are placed on a queue.*

Virtually every real-life line is (supposed to be) a queue. For instance, lines at ticket counters are queues, because service is first-come first-served.

Another example concerns computer networks. There are many network setups of personal computers in which the disk is attached to one machine, known as the *file server*. Users on other machines are given access to files on a first-come first-served basis, so the data structure is a queue.

Further examples include the following:

- Calls to large companies are generally placed on a queue when all operators are busy.
- In large universities, where resources are limited, students must sign a waiting list if all terminals are occupied. The student who has been at a terminal the longest is forced off first, and the student who has been waiting the longest is the next user to be allowed on.

A whole branch of mathematics, known as *queueing theory,* deals with computing, probabilistically, how long users expect to wait on a line, how long the line gets, and other such questions. The answer depends on how frequently users arrive to the line and how long it takes to process a user once the user is served. Both of these parameters are given as probability distribution functions. In simple cases, an answer can be computed analytically. An example of an easy case would be a phone line with one operator. If the operator is busy, callers are placed on a waiting line (up to some maximum limit). This problem is important for businesses, because studies have shown that people are quick to hang up the phone.

If there are k operators, then this problem is much more difficult to solve. Problems that are difficult to solve analytically are often solved by a simulation. In our case, we would need to use a queue to perform the simulation. If k is large, we also need other data structures to do this efficiently. We shall see how to do this simulation in Chapter 6. We could then run the simulation for several values of k and choose the minimum k that gives a reasonable waiting time.

Additional uses for queues abound, and as with stacks, it is staggering that such a simple data structure can be so important.

---

*We say *essentially* because jobs can be killed. This amounts to a deletion from the middle of the queue, which is a violation of the strict definition.

Summary

This chapter describes the concept of ADTs and illustrates the concept with three of the most common abstract data types. The primary objective is to separate the implementation of the abstract data types from their function. The program must know what the operations do, but it is actually better off not knowing how it is done.

Lists, stacks, and queues are perhaps the three fundamental data structures in all of computer science, and their use is documented through a host of examples. In particular, we saw how stacks are used to keep track of procedure and function calls and how recursion is actually implemented. This is important to understand, not just because it makes procedural languages possible, but because knowing how recursion is implemented removes a good deal of the mystery that surrounds its use. Although recursion is very powerful, it is not an entirely free operation; misuse and abuse of recursion can result in programs crashing.

Exercises

3.1 Write a program to print out the elements of a singly linked list.

3.2 You are given a linked list, L, and another linked list, P, containing integers sorted in ascending order. The operation *PrintLots(L,P)* will print the elements in L that are in positions specified by P. For instance, if $P = 1, 3, 4, 6$, the first, third, fourth, and sixth elements in L are printed. Write the procedure *PrintLots(L,P)*. You should use only the basic list operations. What is the running time of your procedure?

3.3 Swap two adjacent elements by adjusting only the pointers (and not the data) using:

a. Singly linked lists.

b. Doubly linked lists.

3.4 Given two sorted lists, L_1 and L_2, write a procedure to compute $L_1 \cap L_2$ using only the basic list operations.

3.5 Given two sorted lists, L_1 and L_2, write a procedure to compute $L_1 \cup L_2$ using only the basic list operations.

3.6 Write a function to add two polynomials. Do not destroy the input. Use a linked list implementation. If the polynomials have M and N terms, respectively, what is the time complexity of your program?

3.7 Write a function to multiply two polynomials, using a linked list implementation. You must make sure that the output polynomial is sorted by exponent and has at most one term of any power.

a. Give an algorithm to solve this problem in $O(M^2N^2)$ time.

*b. Write a program to perform the multiplication in $O(M^2N)$ time, where M is the number of terms in the polynomial of fewer terms.

*c. Write a program to perform the multiplication in $O(MN \log(MN))$ time.

d. Which time bound above is the best?

3.8 Write a program that takes a polynomial, $F(X)$, and computes $(F(X))^P$. What is the complexity of your program? Propose at least one alternative solution that could be competitive for some plausible choices of $F(X)$ and P.

3.9 Write an arbitrary-precision integer arithmetic package. You should use a strategy similar to polynomial arithmetic. Compute the distribution of the digits 0 to 9 in $2^{4,000}$.

3.10 The *Josephus problem* is the following game: N people, numbered 1 to N, are sitting in a circle. Starting at person 1, a hot potato is passed. After M passes, the person holding the hot potato is eliminated, the circle closes ranks, and the game continues with the person who was sitting after the eliminated person picking up the hot potato. The last remaining person wins. Thus, if $M = 0$ and $N = 5$, players are eliminated in order, and player 5 wins. If $M = 1$ and $N = 5$, the order of elimination is 2, 4, 1, 5.

a. Write a program to solve the Josephus problem for general values of M and N. Try to make your program as efficient as possible. Make sure you dispose of cells.

b. What is the running time of your program?

c. If $M = 1$, what is the running time of your program? How is the actual speed affected by the *free* routine for large values of N ($N > 10{,}000$)?

3.11 Write a program to find a particular element in a singly linked list. Do this both recursively and nonrecursively, and compare the running times. How big does the list have to be before the recursive version crashes?

3.12 a. Write a nonrecursive procedure to reverse a singly linked list in $O(N)$ time.

*b. Write a procedure to reverse a singly linked list in $O(N)$ time using constant extra space.

3.13 You have to sort an array of student records by social security number. Write a program to do this, using radix sort with 1,000 buckets and three passes.

3.14 Write a program to read a graph into adjacency lists using:

a. Linked lists.

b. Cursors.

3.15 a. Write an array implementation of self-adjusting lists. A *self-adjusting* list is like a regular list, except that all insertions are performed at the front, and when an element is accessed by a *Find*, it is moved to the front of the list without changing the relative order of the other items.

b. Write a linked list implementation of self-adjusting lists.

*c. Suppose each element has a fixed probability, p_i, of being accessed. Show that the elements with highest access probability are expected to be close to the front.

3.16 Suppose we have an array-based list $A[0..N-1]$ and we want to delete all duplicates. *LastPosition* is initially $N - 1$, but gets smaller as elements

```
for( i = 0; i < LastPosition; i++ )
{
    j = i + 1;
    while( j < LastPosition )
        if( A[ i ] == A[ j ] )
            Delete( j );
        else
            j++;
}
```

Figure 3.61 Routine to remove duplicates from a list—array implementation

are deleted. Consider the pseudocode program fragment in Figure 3.61. The procedure *Delete* deletes the element in position *j* and collapses the list.

a. Explain how this procedure works.

b. Rewrite this procedure using general list operations.

*c. Using a standard array implementation, what is the running time of this procedure?

d. What is the running time using a linked list implementation?

*e. Give an algorithm to solve this problem in $O(N \log N)$ time.

**f. Prove that any algorithm to solve this problem requires $\Omega(N \log N)$ comparisons if only comparisons are used. (*Hint:* Look to Chapter 7.)

*g. Prove that if we allow operations besides comparisons, and the keys are real numbers, then we can solve the problem without using comparisons between elements.

3.17 An alternative to the deletion strategy we have given is to use *lazy deletion.* To delete an element, we merely mark it deleted (using an extra bit field). The number of deleted and nondeleted elements in the list is kept as part of the data structure. If there are as many deleted elements as nondeleted elements, we traverse the entire list, performing the standard deletion algorithm on all marked nodes.

a. List the advantages and disadvantages of lazy deletion.

b. Write routines to implement the standard linked list operations using lazy deletion.

3.18 Write a program to check for balancing symbols in the following languages:

a. Pascal (*begin/end,* ( ), [ ], { }).

b. C (/* */, ( ), [ ], { }).

*c. Explain how to print out an error message that is likely to reflect the probable cause.

3.19 Write a program to evaluate a postfix expression.

3.20 a. Write a program to convert an infix expression which includes '(', ')', '+', '−', '*', and '/' to postfix.

b. Add the exponentiation operator to your repertoire.

c. Write a program to convert a postfix expression to infix.

3.21 Write routines to implement two stacks using only one array. Your stack routines should not declare an overflow unless every slot in the array is used.

3.22*a. Propose a data structure that supports the stack *Push* and *Pop* operations and a third operation *FindMin,* which returns the smallest element in the data structure, all in $O(1)$ worst case time.

*b. Prove that if we add the fourth operation *DeleteMin* which finds and removes the smallest element, then at least one of the operations must take $\Omega(\log N)$ time. (This requires reading Chapter 7.)

3.23 * Show how to implement three stacks in one array.

3.24 If the recursive routine in Section 2.4 used to compute Fibonacci numbers is run for $N = 50$, is stack space likely to run out? Why or why not?

3.25 Write the routines to implement queues using:

a. Linked lists

b. Arrays

3.26 A *deque* is a data structure consisting of a list of items, on which the following operations are possible:

Push(X,D): Insert item X on the front end of deque D.

Pop(D): Remove the front item from deque D and return it.

Inject(X,D): Insert item X on the rear end of deque D.

Eject(D): Remove the rear item from deque D and return it.

Write routines to support the deque that take $O(1)$ time per operation.

CHAPTER 4

Trees

For large amounts of input, the linear access time of linked lists is prohibitive. In this chapter we look at a simple data structure for which the running time of most operations is $O(\log N)$ on average. We also sketch a conceptually simple modification to this data structure that guarantees the above time bound in the worst case and discuss a second modification that essentially gives an $O(\log N)$ running time per operation for a long sequence of instructions.

The data structure that we are referring to is known as a *binary search tree*. *Trees* in general are very useful abstractions in computer science, so we will discuss their use in other, more general applications. In this chapter, we will

- See how trees are used to implement the file system of several popular operating systems.
- See how trees can be used to evaluate arithmetic expressions.
- Show how to use trees to support searching operations in $O(\log N)$ average time, and how to refine these ideas to obtain $O(\log N)$ worst-case bounds. We will also see how to implement these operations when the data are stored on a disk.

4.1. Preliminaries

A *tree* can be defined in several ways. One natural way to define a tree is recursively. A tree is a collection of nodes. The collection can be empty; otherwise, a tree consists of a distinguished node r, called the *root*, and zero or more nonempty (sub)trees T_1, T_2, ..., T_k, each of whose roots are connected by a directed *edge* from r.

The root of each subtree is said to be a *child* of r, and r is the *parent* of each subtree root. Figure 4.1 shows a typical tree using the recursive definition.

From the recursive definition, we find that a tree is a collection of N nodes, one of which is the root, and $N - 1$ edges. That there are $N - 1$ edges follows from the fact that each edge connects some node to its parent, and every node except the root has one parent (see Fig. 4.2).

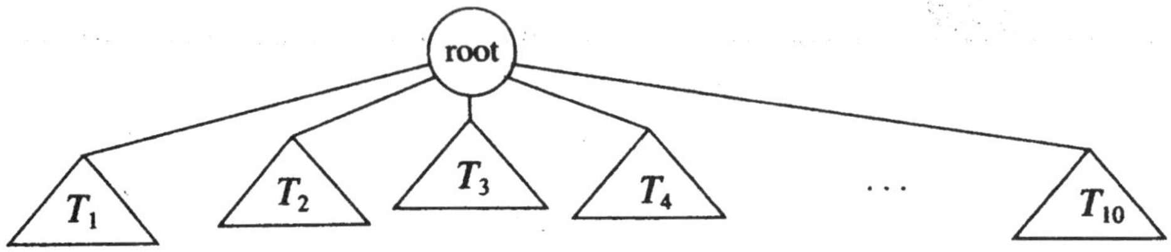

Figure 4.1 Generic tree

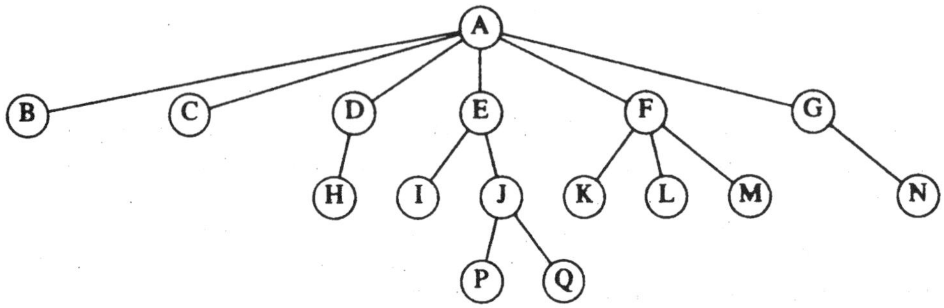

Figure 4.2 A tree

In the tree of Figure 4.2, the root is A. Node F has A as a parent and K, L, and M as children. Each node may have an arbitrary number of children, possibly zero. Nodes with no children are known as *leaves;* the leaves in the tree above are $B, C, H, I, P, Q, K, L, M$, and N. Nodes with the same parent are *siblings;* thus K, L, and M are all siblings. *Grandparent* and *grandchild* relations can be defined in a similar manner.

A *path* from node n_1 to n_k is defined as a sequence of nodes $n_1, n_2, \ldots, n_k$ such that n_i is the parent of n_{i+1} for $1 \leq i < k$. The *length* of this path is the number of edges on the path, namely $k - 1$. There is a path of length zero from every node to itself. Notice that in a tree there is exactly one path from the root to each node.

For any node n_i, the *depth* of n_i is the length of the unique path from the root to n_i. Thus, the root is at depth 0. The *height* of n_i is the length of the longest path from n_i to a leaf. Thus all leaves are at height 0. The height of a tree is equal to the height of the root. For the tree in Figure 4.2, E is at depth 1 and height 2; F is at depth 1 and height 1; the height of the tree is 3. The depth of a tree is equal to the depth of the deepest leaf; this is always equal to the height of the tree.

If there is a path from n_1 to n_2, then n_1 is an *ancestor* of n_2 and n_2 is a *descendant* of n_1. If $n_1 \neq n_2$, then n_1 is a *proper ancestor* of n_2 and n_2 is a *proper descendant* of n_1.

4.1.1. Implementation of Trees

One way to implement a tree would be to have in each node, besides its data, a pointer to each child of the node. However, since the number of children per node can vary so greatly and is not known in advance, it might be infeasible to make the

```
typedef struct TreeNode *PtrToNode;

struct TreeNode
{
    ElementType Element;
    PtrToNode   FirstChild;
    PtrToNode   NextSibling;
}
```

Figure 4.3 Node declarations for trees

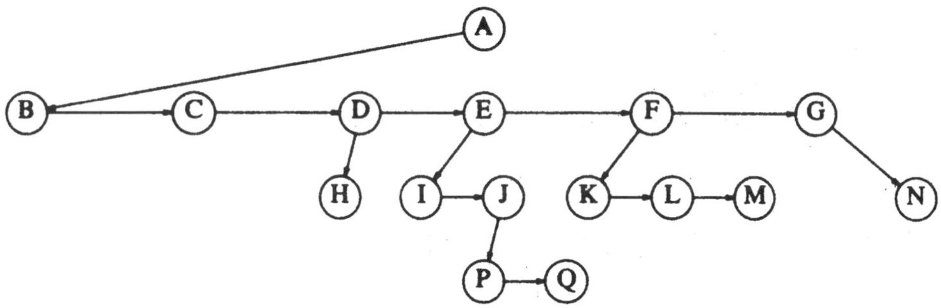

Figure 4.4 First child/next sibling representation of the tree shown in Figure 4.2

children direct links in the data structure, because there would be too much wasted space. The solution is simple: Keep the children of each node in a linked list of tree nodes. The declaration in Figure 4.3 is typical.

Figure 4.4 shows how a tree might be represented in this implementation. Arrows that point downward are *FirstChild* pointers. Arrows that go left to right are *NextSibling* pointers. Null pointers are not drawn, because there are too many.

In the tree of Figure 4.4, node E has both a pointer to a sibling (F) and a pointer to a child (I), while some nodes have neither.

4.1.2. Tree Traversals with an Application

There are many applications for trees. One of the popular uses is the directory structure in many common operating systems, including UNIX, VAX/VMS, and DOS. Figure 4.5 is a typical directory in the UNIX file system.

The root of this directory is */usr*. (The asterisk next to the name indicates that */usr* is itself a directory.) */usr* has three children, *mark*, *alex*, and *bill*, which are themselves directories. Thus, */usr* contains three directories and no regular files. The filename */usr/mark/book/ch1.r* is obtained by following the leftmost child three times. Each / after the first indicates an edge; the result is the full *pathname*. This hierarchical file system is very popular, because it allows users to organize their data logically. Furthermore, two files in different directories can share the same name, because they must have different paths from the root and thus have different

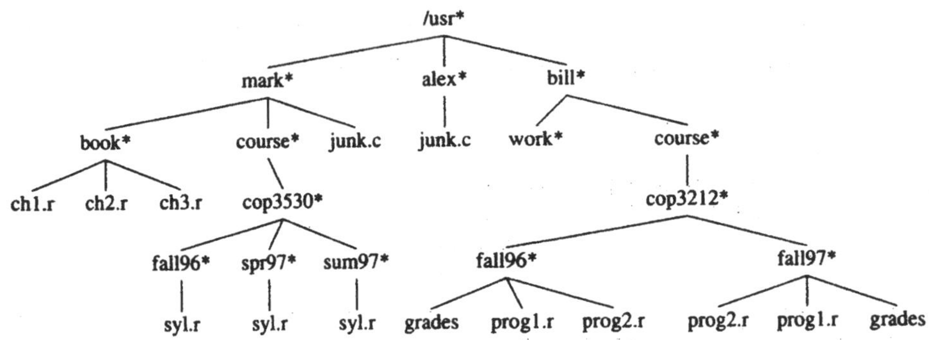

Figure 4.5 Unix directory

```
static void
ListDir( DirectoryOrFile D, int Depth )
{
    if( D is a legitimate entry )
    {
        PrintName( D, Depth );
        if( D is a directory )
            for each child, C, of D
                ListDir( C, Depth + 1 );
    }
}

void
ListDirectory( DirectoryOrFile D )
{
    ListDir( D, 0 );
}
```

Figure 4.6 Routine to list a directory in a hierarchical file system

pathnames. A directory in the UNIX file system is just a file with a list of all its children, so the directories are structured almost exactly in accordance with the type declaration above.* Indeed, if the normal command to print a file is applied to a directory, then the names of the files in the directory can be seen in the output (along with other non-ASCII information).

Suppose we would like to list the names of all of the files in the directory. Our output format will be that files that are depth d_i will have their names indented by d_i tabs. Our algorithm is given in Figure 4.6.

The heart of the algorithm is the recursive procedure *ListDir*. This routine needs to be started with the directory name and a depth of 0, to signify no indenting for

*Each directory in the UNIX file system also has one entry that points to itself and another entry that points to the parent of the directory. Thus, technically, the UNIX file system is not a tree, but is treelike.

the root. This depth is an internal bookkeeping variable, and is hardly a parameter that a calling routine should be expected to know about. Thus the driver routine *ListDirectory* is used to interface the recursive routine to the outside world.

The logic of the algorithm is simple to follow. The argument to *ListDir* is some sort of reference into the tree. As long as the reference is valid, the name implied by the reference is printed out with the appropriate number of tabs. If the entry is a directory, then we process all children recursively, one by one. These children are one level deeper, and thus need to be indented an extra space. The output is in Figure 4.7.

This traversal strategy is known as a *preorder* traversal. In a preorder traversal, work at a node is performed before (*pre*) its children are processed. When this program is run, it is clear that line 2 is executed exactly once per node, since each name is output once. Since line 2 is executed at most once per node, line 3 must also be executed once per node. Furthermore, line 5 can be executed at most once for each child of each node. But the number of children is exactly one less than the

Figure 4.7 The (preorder) directory listing

```
/usr
mark
    book
        chr1.c
        chr2.c
        chr3.c
    course
        cop3530
            fall96
                syl.r
            spr97
                syl.r
            sum97
                syl.r
    junk.c
alex
    junk.c
bill
    work
    course
        cop3212
            fall96
                grades
                prog1.r
                prog2.r
            fall97
                prog2.r
                prog1.r
                grades
```

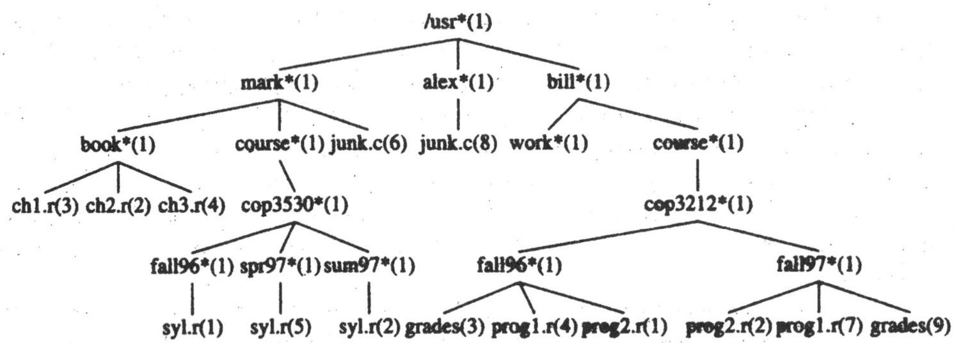

Figure 4.8 Unix directory with file sizes obtained via postorder traversal

number of nodes. Finally, the *for* loop iterates once per execution of line 5, plus once each time the loop ends. Each *for* loop terminates on a *NULL* pointer, but there is at most one of those per node. Thus, the total amount of work is constant per node. If there are N file names to be output, then the running time is $O(N)$.

Another common method of traversing a tree is the *postorder* traversal. In a postorder traversal, the work at a node is performed after (*post*) its children are evaluated. As an example, Figure 4.8 represents the same directory structure as before, with the numbers in parentheses representing the number of disk blocks taken up by each file.

Since the directories are themselves files, they have sizes too. Suppose we would like to calculate the total number of blocks used by all the files in the tree. The most natural way to do this would be to find the number of blocks contained in the subdirectories */usr/mark* (30), */usr/alex* (9), and */usr/bill* (32). The total number of blocks is then the total in the subdirectories (71) plus the one block used by */usr*, for a total of 72. The function *SizeDirectory* in Figure 4.9 implements this strategy.

Figure 4.9 Routine to calculate the size of a directory

```
        static void
        SizeDirectory( DirectoryOrFile D )
        {
            int TotalSize;

/* 1*/      TotalSize = 0;
/* 2*/      if( D is a legitimate entry )
            {
/* 3*/          TotalSize = FileSize( D );
/* 4*/          if( D is a directory )
/* 5*/              for each child, C, of D
/* 6*/                  TotalSize += SizeDirectory( C );
            }
/* 7*/      return TotalSize;
        }
```

```
          ch1.r            3
          ch2.r            2
          ch3.r            4
      book                10
                  syl.r    1
              fall96       2
                  syl.r    5
              spr97        6
                  syl.r    2
              sum97        3
          cop3530         12
      course              13
      junk.c               6
  mark                    30
      junk.c               8
  alex                     9
      work                 1
                  grades   3
                  prog1.r  4
                  prog2.r  1
              fall96       9
                  prog2.r  2
                  prog1.r  7
                  grades   9
              fall97      19
          cop3212         29
      course              30
  bill                    32
  /usr                    72
```

Figure 4.10 Trace of the *SizeDirectory* function

If D is not a directory, then *SizeDirectory* merely returns the number of blocks used by D. Otherwise, the number of blocks used by D is added to the number of blocks (recursively) found in all of the children. To see the difference between the postorder traversal strategy and the preorder traversal strategy, Figure 4.10 shows how the size of each directory or file is produced by the algorithm.

4.2. Binary Trees

A binary tree is a tree in which no node can have more than two children.

Figure 4.11 shows that a binary tree consists of a root and two subtrees, T_L and T_R, both of which could possibly be empty.

A property of a binary tree that is sometimes important is that the depth of an average binary tree is considerably smaller than N. An analysis shows that the

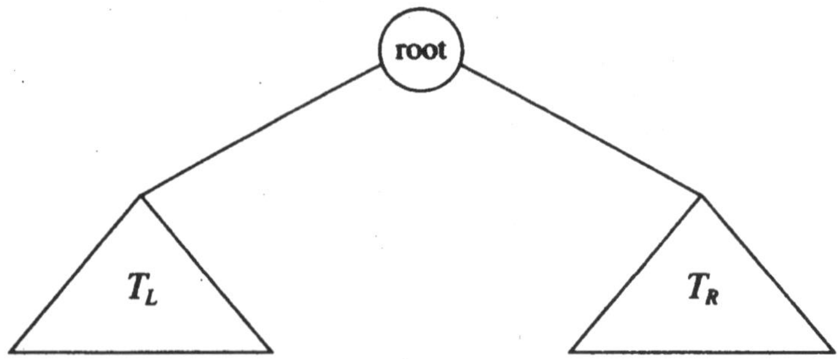

Figure 4.11 Generic binary tree

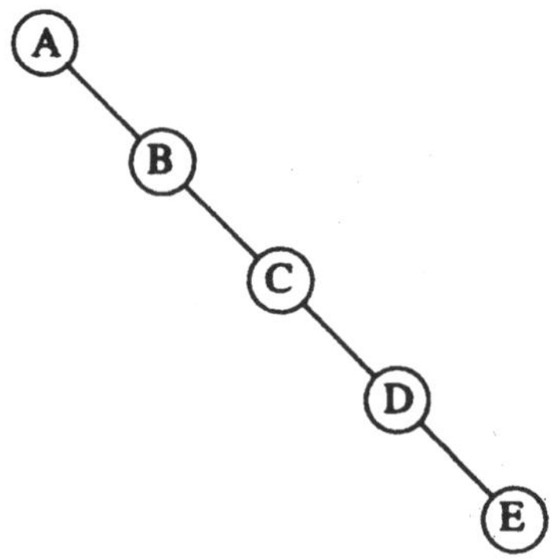

Figure 4.12 Worst-case binary tree

average depth is $O(\sqrt{N})$, and that for a special type of binary tree, namely the *binary search tree,* the average value of the depth is $O(\log N)$. Unfortunately, the depth can be as large as $N - 1$, as the example in Figure 4.12 shows.

4.2.1. *Implementation*

Because a binary tree has at most two children, we can keep direct pointers to them. The declaration of tree nodes is similar in structure to that for doubly linked lists, in that a node is a structure consisting of the *Key* information plus two pointers (*Left* and *Right*) to other nodes (see Fig. 4.13).

Many of the rules that apply to linked lists will apply to trees as well. In particular, when an insertion is performed, a node will have to be created by a call to *malloc*. Nodes can be freed after deletion by calling *free*.

We could draw the binary trees using the rectangular boxes that are customary for linked lists, but trees are generally drawn as circles connected by lines, because they are actually graphs. We also do not explicitly draw *NULL* pointers when

```
typedef struct TreeNode *PtrToNode;
typedef struct PtrToNode Tree;

struct TreeNode
{
    ElementType Element;
    Tree        Left;
    Tree        Right;
};
```

Figure 4.13 Binary tree node declarations

referring to trees, because every binary tree with N nodes would require $N + 1$ *NULL* pointers.

Binary trees have many important uses not associated with searching. One of the principal uses of binary trees is in the area of compiler design, which we will now explore.

4.2.2. Expression Trees

Figure 4.14 shows an example of an *expression tree*. The leaves of an expression tree are *operands*, such as constants or variable names, and the other nodes contain *operators*. This particular tree happens to be binary, because all of the operations are binary, and although this is the simplest case, it is possible for nodes to have more than two children. It is also possible for a node to have only one child, as is the case with the ***unary minus*** operator. We can evaluate an expression tree, T, by applying the operator at the root to the values obtained by recursively evaluating the left and right subtrees. In our example, the left subtree evaluates to $a + (b * c)$ and the right subtree evaluates to $((d * e) + f) * g$. The entire tree therefore represents $(a + (b * c)) + (((d * e) + f) * g)$.

We can produce an (overly parenthesized) infix expression by recursively producing a parenthesized left expression, then printing out the operator at the

Figure 4.14 Expression tree for $(a + b * c) + ((d * e + f) * g)$

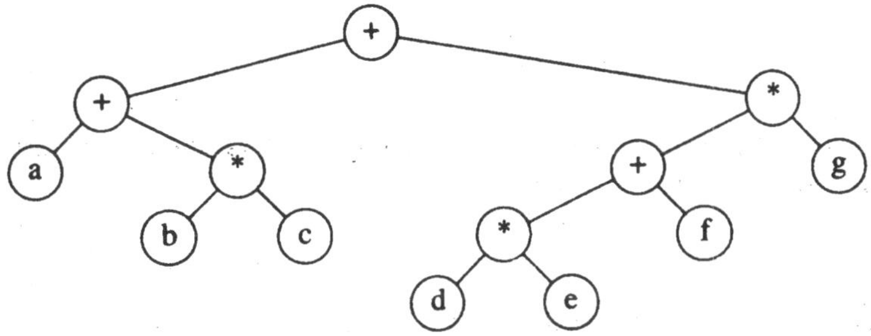

root, and finally recursively producing a parenthesized right expression. This general strategy (left, node, right) is known as an *inorder* traversal; it is easy to remember because of the type of expression it produces.

An alternate traversal strategy is to recursively print out the left subtree, the right subtree, and then the operator. If we apply this strategy to our tree above, the output is a b c * + d e * f + g * +, which is easily seen to be the postfix representation of Section 3.3.3. This traversal strategy is generally known as a *postorder* traversal. We have seen this traversal strategy earlier in Section 4.1.

A third traversal strategy is to print out the operator first and then recursively print out the left and right subtrees. The resulting expression, + + a * b c * + * d e f g, is the less useful *prefix* notation and the traversal strategy is a *preorder* traversal, which we have also seen earlier in Section 4.1. We will return to these traversal strategies later in the chapter.

Constructing an Expression Tree

We now give an algorithm to convert a postfix expression into an expression tree. Since we already have an algorithm to convert infix to postfix, we can generate expression trees from the two common types of input. The method we describe strongly resembles the postfix evaluation algorithm of Section 3.2.3. We read our expression one symbol at a time. If the symbol is an operand, we create a one-node tree and push a pointer to it onto a stack. If the symbol is an operator, we pop pointers to two trees T_1 and T_2 from the stack (T_1 is popped first) and form a new tree whose root is the operator and whose left and right children point to T_2 and T_1, respectively. A pointer to this new tree is then pushed onto the stack.

As an example, suppose the input is

a b + c d e + * *

The first two symbols are operands, so we create one-node trees and push pointers to them onto a stack.*

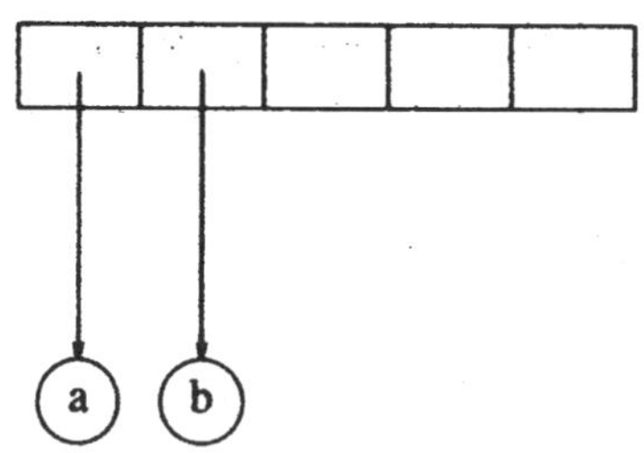

Next, a '+' is read, so two pointers to trees are popped, a new tree is formed, and a pointer to it is pushed onto the stack.

*For convenience, we will have the stack grow from left to right in the diagrams.

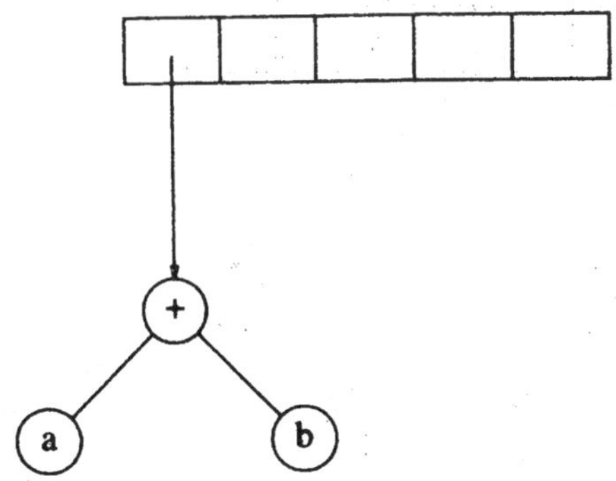

Next, c, d, and e are read, and for each a one-node tree is created and a pointer to the corresponding tree is pushed onto the stack.

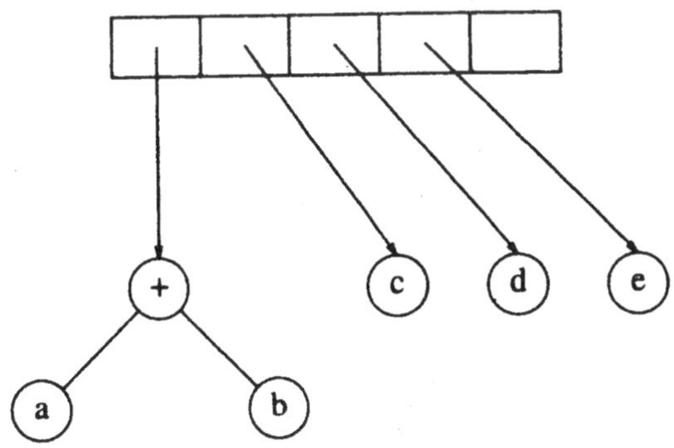

Now a '+' is read, so two trees are merged.

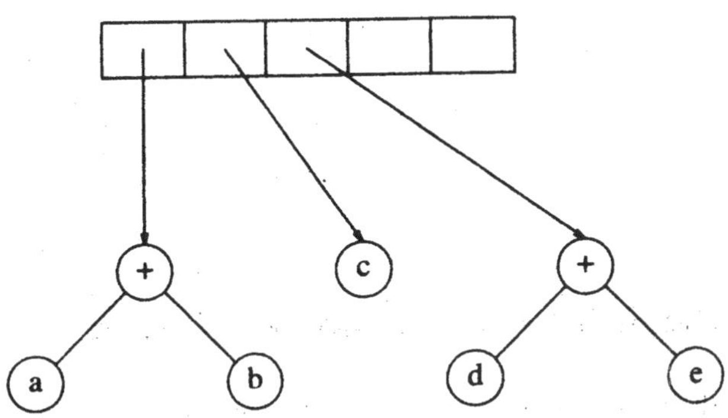

Continuing, a '*' is read, so we pop two tree pointers and form a new tree with a '*' as root.

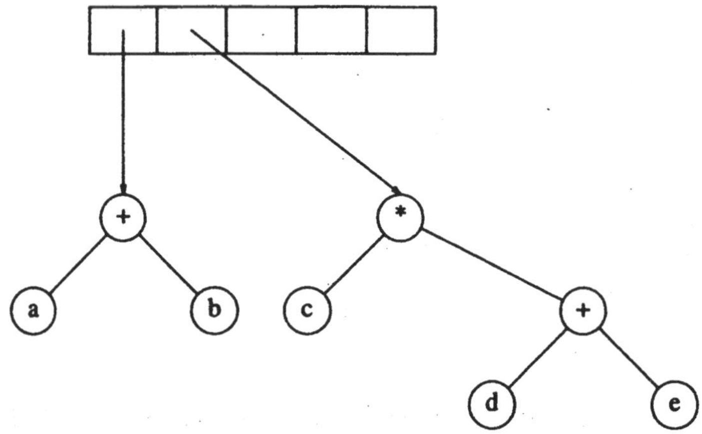

Finally, the last symbol is read, two trees are merged, and a pointer to the final tree is left on the stack.

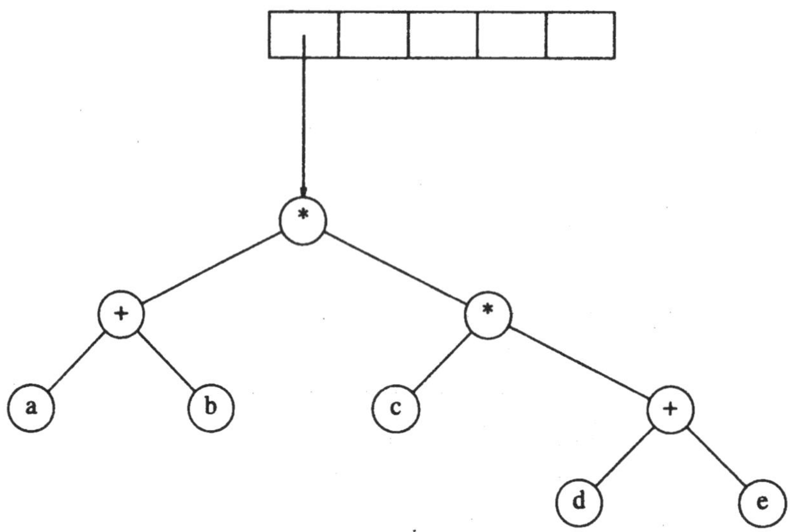

4.3. The Search Tree ADT—Binary Search Trees

An important application of binary trees is their use in searching. Let us assume that each node in the tree is assigned a key value. In our examples, we will assume for simplicity that these are integers, although arbitrarily complex keys are allowed. We will also assume that all the keys are distinct, and deal with duplicates later.

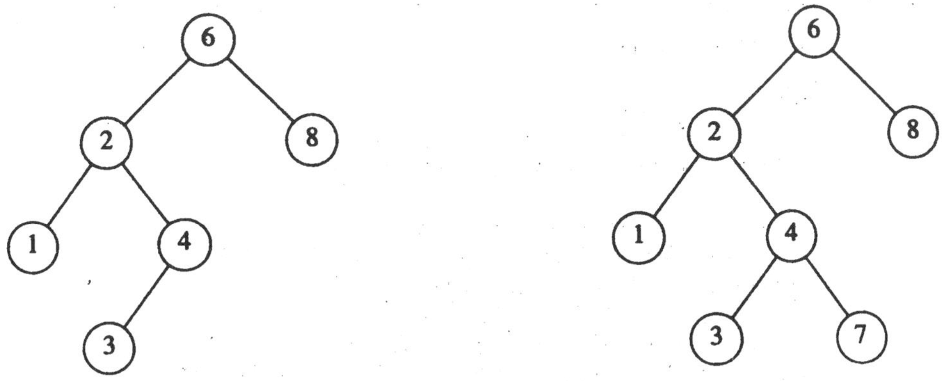

Figure 4.15 Two binary trees (only the left tree is a search tree)

The property that makes a binary tree into a binary search tree is that for every node, X, in the tree, the values of all the keys in its left subtree are smaller than the key value in X, and the values of all the keys in its right subtree are larger than the key value in X. Notice that this implies that all the elements in the tree can be ordered in some consistent manner. In Figure 4.15, the tree on the left is a binary search tree, but the tree on the right is not. The tree on the right has a node with key 7 in the left subtree of a node with key 6 (which happens to be the root).

We now give brief descriptions of the operations that are usually performed on binary search trees. Note that because of the recursive definition of trees, it is common to write these routines recursively. Because the average depth of a binary search tree is $O(\log N)$, we generally do not need to worry about running out of stack space. We repeat our type definition in Figure 4.16 and list the function prototypes. Since all the elements can be ordered, we will assume that the operators $<$, $>$, and $=$ can be applied to them, even if this might be syntactically erroneous for some types.

4.3.1. MakeEmpty

This operation is mainly for initialization. Some programmers prefer to initialize the first element as a one-node tree, but our implementation follows the recursive definition of trees more closely. It is also a simple routine, as evidenced by Figure 4.17.

4.3.2. Find

This operation generally requires returning a pointer to the node in tree T that has key X, or *NULL* if there is no such node. The structure of the tree makes this simple. If T is *NULL*, then we can just return *NULL*. Otherwise, if the key stored at T is X, we can return T. Otherwise, we make a recursive call on a subtree of T, either left or right, depending on the relationship of X to the key stored in T. The code in Figure 4.18 is an implementation of this strategy.

```
#ifndef _Tree_H

struct TreeNode;
typedef struct TreeNode *Position;
typedef struct TreeNode *SearchTree;

SearchTree MakeEmpty( SearchTree T );
Position Find( ElementType X, SearchTree T );
Position FindMin( SearchTree T );
Position FindMax( SearchTree T );
SearchTree Insert( ElementType X, SearchTree T );
SearchTree Delete( ElementType X, SearchTree T );
ElementType Retrieve( Position P );

#endif /* _Tree_H */


/* Place in the implementation file */
struct TreeNode
{
    ElementType Element;
    SearchTree  Left;
    SearchTree  Right;
};
```

Figure 4.16 Binary search tree declarations

```
SearchTree
MakeEmpty( SearchTree T )
{
    if( T != NULL )
    {
        MakeEmpty( T->Left );
        MakeEmpty( T->Right );
        free( T );
    }
    return NULL;
}
```

Figure 4.17 Routine to make an empty tree

```
Position
Find( ElementType X, SearchTree T )
{
    if( T == NULL )
        return NULL;
    if( X < T->Element )
        return Find( X, T->Left );
    else
    if( X > T->Element )
        return Find( X, T->Right );
    else
        return T;
}
```

Figure 4.18 *Find* operation for binary search trees

Notice the order of the tests. It is crucial that the test for an empty tree be performed first, since otherwise the indirections would be on a *NULL* pointer. The remaining tests are arranged with the least likely case last. Also note that both recursive calls are actually tail recursions and can be easily removed with an assignment and a *goto*. The use of tail recursion is justifiable here because the simplicity of algorithmic expression compensates for the decrease in speed, and the amount of stack space used is expected to be only $O(\log N)$.

4.3.3. FindMin and FindMax

These routines return the position of the smallest and largest elements in the tree, respectively. Although returning the exact values of these elements might seem more reasonable, this would be inconsistent with the *Find* operation. It is important that similar-looking operations do similar things. To perform a *FindMin*, start at the root and go left as long as there is a left child. The stopping point is the smallest element. The *FindMax* routine is the same, except that branching is to the right child.

Figure 4.19 Recursive implementation of *FindMin* for binary search trees

```
Position
FindMin( SearchTree T )
{
    if( T == NULL )
        return NULL;
    else
    if( T->Left == NULL )
        return T;
    else
        return FindMin( T->Left );
}
```

```
Position
FindMax( SearchTree T )
{
    if( T != NULL )
        while( T->Right != NULL )
            T = T->Right;

    return T;
}
```

Figure 4.20 Nonrecursive implementation of *FindMax* for binary search trees

This is so easy that many programmers do not bother using recursion. We will code the routines both ways by doing *FindMin* recursively and *FindMax* nonrecursively (see Figs. 4.19 and 4.20).

Notice how we carefully handle the degenerate case of an empty tree. Although this is always important to do, it is especially crucial in recursive programs. Also notice that it is safe to change *T* in *FindMax*, since we are only working with a copy. Always be extremely careful, however, because a statement such as *T−>Right* = *T−>Right−>Right* will make changes.

4.3.4. Insert

The insertion routine is conceptually simple. To insert *X* into tree *T*, proceed down the tree as you would with a *Find*. If *X* is found, do nothing (or "update" something). Otherwise, insert *X* at the last spot on the path traversed. Figure 4.21 shows what happens. To insert 5, we traverse the tree as though a *Find* were occurring. At the node with key 4, we need to go right, but there is no subtree, so 5 is not in the tree, and this is the correct spot.

Duplicates can be handled by keeping an extra field in the node record indicating the frequency of occurrence. This adds some extra space to the entire tree, but is

Figure 4.21 Binary search trees before and after inserting 5

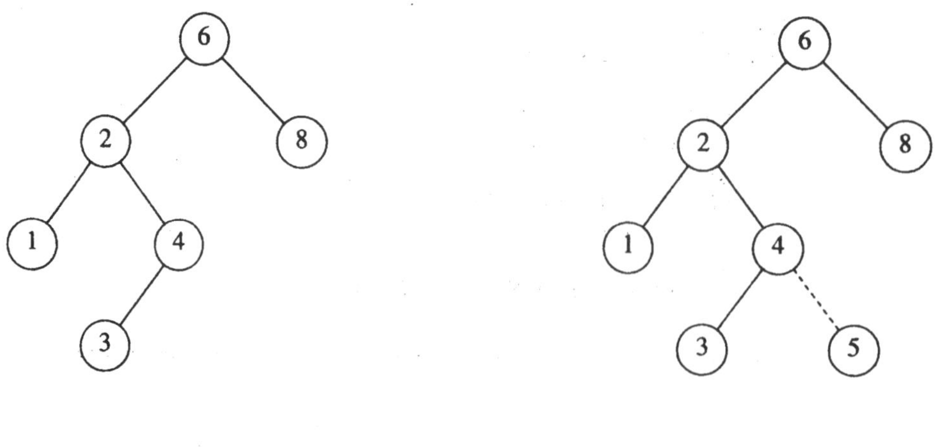

```
        SearchTree
        Insert( ElementType X, SearchTree T )
        {
/* 1*/      if( T == NULL )
            {
                /* Create and return a one-node tree */
/* 2*/          T = malloc( sizeof( struct TreeNode ) );
/* 3*/          if( T == NULL )
/* 4*/              FatalError( "Out of space!!!" );
                else
                {
/* 5*/              T->Element = X;
/* 6*/              T->Left = T->Right = NULL;
                }
            }
            else
/* 7*/      if( X < T->Element )
/* 8*/          T->Left = Insert( X, T->Left );
            else
/* 9*/      if( X > T->Element )
/*10*/          T->Right = Insert( X, T->Right );
            /* Else X is in the tree already; we'll do nothing */

/*11*/      return T;  /* Do not forget this line!! */
        }
```

Figure 4.22 Insertion into a binary search tree

better than putting duplicates in the tree (which tends to make the tree very deep). Of course this strategy does not work if the key is only part of a larger structure. If that is the case, then we can keep all of the structures that have the same key in an auxiliary data structure, such as a list or another search tree.

Figure 4.22 shows the code for the insertion routine. Since T points to the root of the tree, and the root changes on the first insertion, *Insert* is written as a function that returns a pointer to the root of the new tree. Lines 8 and 10 recursively insert and attach X into the appropriate subtree.

4.3.5. *Delete*

As is common with many data structures, the hardest operation is deletion. Once we have found the node to be deleted, we need to consider several possibilities.

If the node is a leaf, it can be deleted immediately. If the node has one child, the node can be deleted after its parent adjusts a pointer to bypass the node (we will draw the pointer directions explicitly for clarity). See Figure 4.23. Notice that the deleted node is now unreferenced and can be disposed of only if a pointer to it has been saved.

The complicated case deals with a node with two children. The general strategy is to replace the data of this node with the smallest data of the right subtree (which

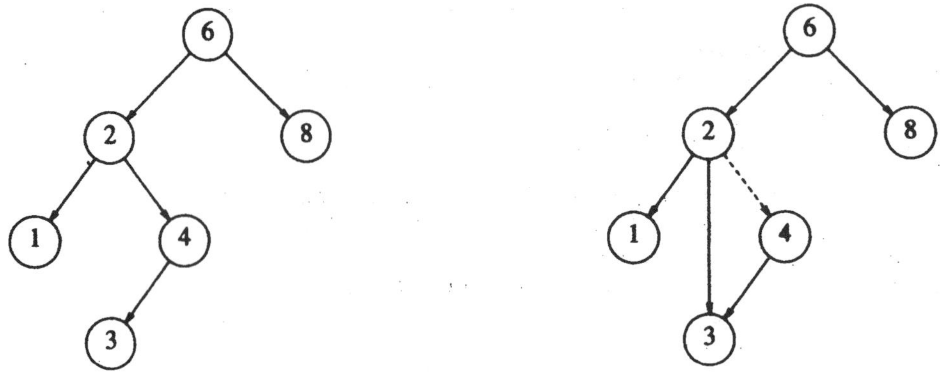

Figure 4.23 Deletion of a node (4) with one child, before and after

is easily found) and recursively delete that node (which is now empty). Because the smallest node in the right subtree cannot have a left child, the second *Delete* is an easy one. Figure 4.24 shows an initial tree and the result of a deletion. The node to be deleted is the left child of the root; the key value is 2. It is replaced with the smallest data in its right subtree (3), and then that node is deleted as before.

The code in Figure 4.25 performs deletion. It is inefficient, because it makes two passes down the tree to find and delete the smallest node in the right subtree when this is appropriate. It is easy to remove this inefficiency, by writing a special *DeleteMin* function, and we have left it in only for simplicity.

If the number of deletions is expected to be small, then a popular strategy to use is *lazy deletion:* When an element is to be deleted, it is left in the tree and merely *marked* as being deleted. This is especially popular if duplicate keys are

Figure 4.24 Deletion of a node (2) with two children, before and after

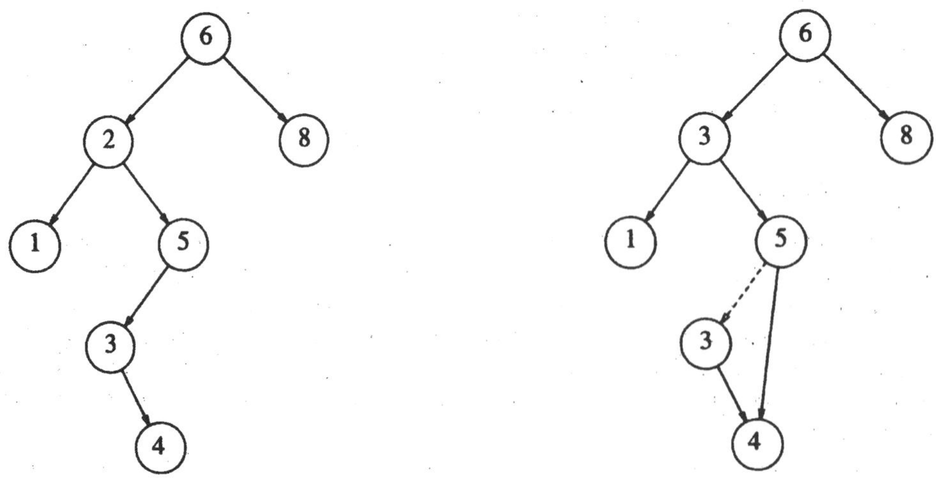

```
SearchTree
Delete( ElementType X, SearchTree T )
{
    Position TmpCell;

    if( T == NULL )
        Error( "Element not found" );
    else
    if( X < T->Element )  /* Go left */
        T->Left = Delete( X, T->Left );
    else
    if( X > T->Element )  /* Go right */
        T->Right = Delete( X, T->Left );
    else  /* Found element to be deleted */
    if( T->Left && T->Right )  /* Two children */
    {
        /* Replace with smallest in right subtree */
        TmpCell = FindMin( T->Right );
        T->Element = TmpCell->Element;
        T->Right = Delete( T->Element, T->Right );
    }
    else  /* One or zero children */
    {
        TmpCell = T;
        if( T->Left == NULL ) /* Also handles 0 children */
            T = T->Right;
        else if( T->Right == NULL )
            T = T->Left;
        free( TmpCell );
    }

    return T;
}
```

Figure 4.25 Deletion routine for binary search trees

present, because then the field that keeps count of the frequency of appearance can be decremented. If the number of real nodes in the tree is the same as the number of "deleted" nodes, then the depth of the tree is only expected to go up by a small constant (why?), so there is a very small time penalty associated with lazy deletion. Also, if a deleted key is reinserted, the overhead of allocating a new cell is avoided.

4.3.6. *Average-Case Analysis*

Intuitively, we expect that all of the operations of the previous section, except *MakeEmpty*, should take $O(\log N)$ time, because in constant time we descend a level in the tree, thus operating on a tree that is now roughly half as large. Indeed, the running time of all the operations, except *MakeEmpty*, is $O(d)$, where d is the depth of the node containing the accessed key.

We prove in this section that the average depth over all nodes in a tree is $O(\log N)$ on the assumption that all trees are equally likely.

The sum of the depths of all nodes in a tree is known as the *internal path length*. We will now calculate the average internal path length of a binary search tree, where the average is taken over all possible insertion sequences into binary search trees.

Let $D(N)$ be the internal path length for some tree T of N nodes. $D(1) = 0$. An N-node tree consists of an i-node left subtree and an $(N - i - 1)$-node right subtree, plus a root at depth zero for $0 \le i < N$. $D(i)$ is the internal path length of the left subtree with respect to its root. In the main tree, all these nodes are one level deeper. The same holds for the right subtree. Thus, we get the recurrence

$$D(N) = D(i) + D(N - i - 1) + N - 1$$

If all subtree sizes are equally likely, which is true for binary search trees (since the subtree size depends only on the relative rank of the first element inserted into the tree), but not binary trees, then the average value of both $D(i)$ and $D(N - i - 1)$ is $(1/N)\sum_{j=0}^{N-1} D(j)$. This yields

$$D(N) = \frac{2}{N}\left[\sum_{j=0}^{N-1} D(j)\right] + N - 1$$

This recurrence will be encountered and solved in Chapter 7, obtaining an average value of $D(N) = O(N \log N)$. Thus, the expected depth of any node is $O(\log N)$. As an example, the randomly generated 500-node tree shown in Figure 4.26 has nodes at expected depth 9.98.

Figure 4.26 A randomly generated binary search tree

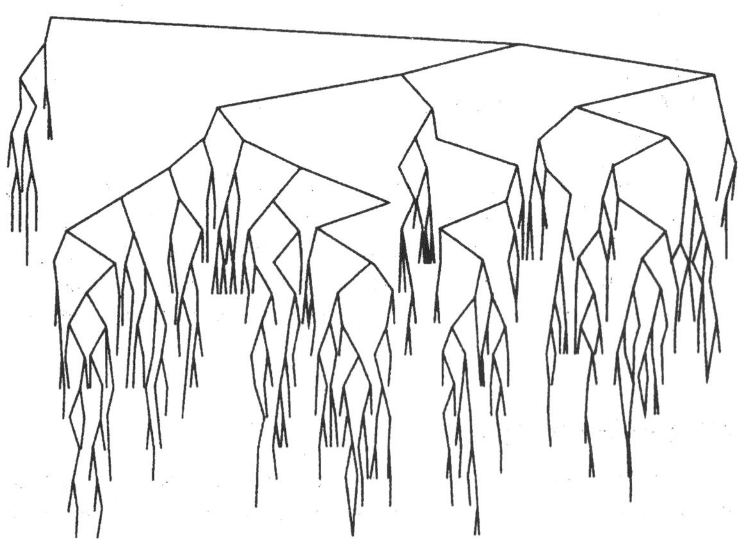

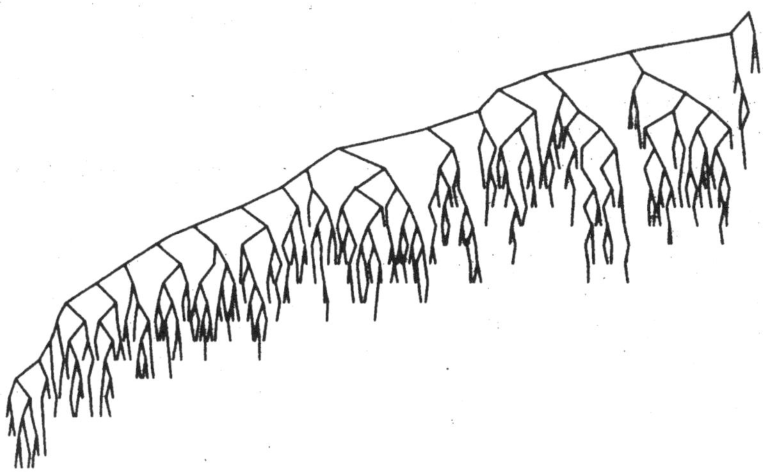

Figure 4.27 Binary search tree after $\Theta(N^2)$ *Insert/Delete* pairs

It is tempting to say immediately that this result implies that the average running time of all the operations discussed in the previous section is $O(\log N)$, but this is not entirely true. The reason for this is that because of deletions, it is not clear that all binary search trees are equally likely. In particular, the deletion algorithm described above favors making the left subtrees deeper than the right, because we are always replacing a deleted node with a node from the right subtree. The exact effect of this strategy is still unknown, but it seems only to be a theoretical novelty. It has been shown that if we alternate insertions and deletions $\Theta(N^2)$ times, then the trees will have an expected depth of $\Theta(\sqrt{N})$. After a quarter-million random *Insert/Delete* pairs, the tree that was somewhat right-heavy in Figure 4.26 looks decidedly unbalanced (average depth = 12.51). See Figure 4.27.

We could try to eliminate the problem by randomly choosing between the smallest element in the right subtree and the largest in the left when replacing the deleted element. This apparently eliminates the bias and should keep the trees balanced, but nobody has actually proved this. In any event, this phenomenon appears to be mostly a theoretical novelty, because the effect does not show up at all for small trees, and stranger still, if $o(N^2)$ *Insert/Delete* pairs are used, then the tree seems to gain balance!

The main point of this discussion is that deciding what "average" means is generally extremely difficult and can require assumptions that may or may not be valid. In the absence of deletions, or when lazy deletion is used, it can be shown that all binary search trees are equally likely and we can conclude that the average running times of the operations above are $O(\log N)$. Except for strange cases like the one discussed above, this result is very consistent with observed behavior.

If the input comes into a tree presorted, then a series of *Insert*s will take quadratic time and give a very expensive implementation of a linked list, since the tree will consist only of nodes with no left children. One solution to the problem is to insist on an extra structural condition called *balance:* no node is allowed to get too deep.

There are quite a few general algorithms to implement balanced trees. Most are quite a bit more complicated than a standard binary search tree, and all take longer on average for updates. They do, however, provide protection against the embarrassingly simple cases. Below, we will sketch one of the oldest forms of balanced search trees, the AVL tree.

A second, newer method is to forego the balance condition and allow the tree to be arbitrarily deep, but after every operation, a restructuring rule is applied that tends to make future operations efficient. These types of data structures are generally classified as *self-adjusting*. In the case of a binary search tree, we can no longer guarantee an $O(\log N)$ bound on any single operation, but can show that any *sequence* of M operations takes total time $O(M \log N)$ in the worst case. This is generally sufficient protection against a bad worst case. The data structure we will discuss is known as a *splay tree;* its analysis is fairly intricate and is discussed in Chapter 11.

4.4. AVL Trees

An AVL (Adelson-Velskii and Landis) tree is a binary search tree with a *balance* condition. The balance condition must be easy to maintain, and it ensures that the depth of the tree is $O(\log N)$. The simplest idea is to require that the left and right subtrees have the same height. As Figure 4.28 shows, this idea does not force the tree to be shallow.

Another balance condition would insist that every node must have left and right subtrees of the same height. If the height of an empty subtree is defined to be -1 (as is usual), then only perfectly balanced trees of $2^k - 1$ nodes would satisfy this criterion. Thus, although this guarantees trees of small depth, the balance condition is too rigid to be useful and needs to be relaxed.

An AVL tree is identical to a binary search tree, except that for every node in the tree, the height of the left and right subtrees can differ by at most 1. (The height of an empty tree is defined to be -1.) In Figure 4.29 the tree on the left is an AVL tree, but the tree on the right is not. Height information is kept for each node (in the node structure). It can be shown that the height of an AVL tree is at most roughly

Figure 4.28 A bad binary tree. Requiring balance at the root is not enough.

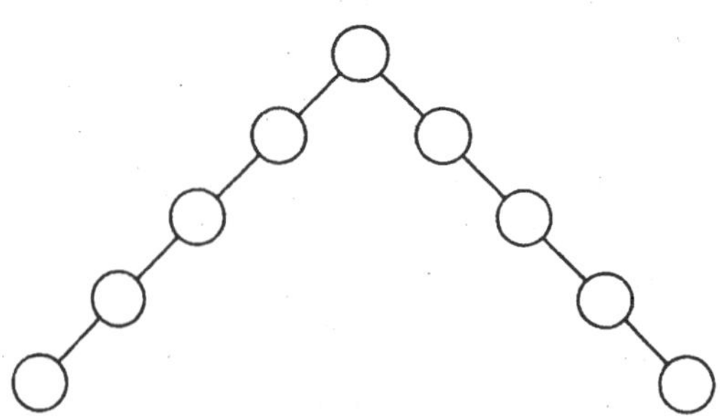

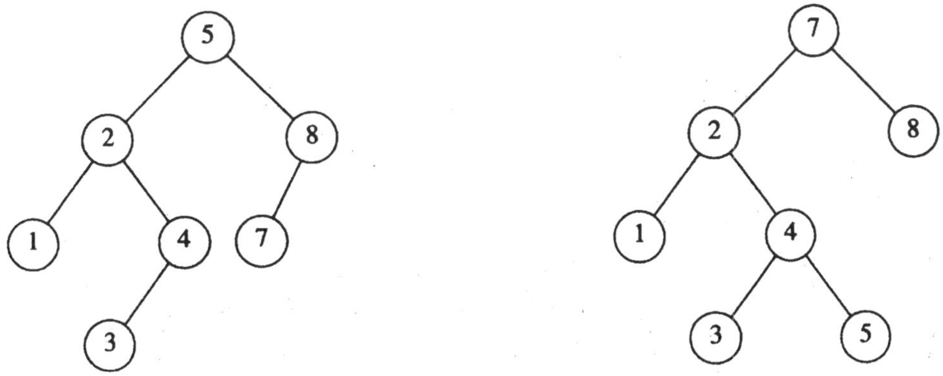

Figure 4.29 Two binary search trees. Only the left tree is AVL.

$1.44 \log(N + 2) - .328$, but in practice it is only slightly more than $\log N$. As an example, the AVL tree of height 9 with the fewest nodes (143) is shown in Figure 4.30. This tree has as a left subtree an AVL tree of height 7 of minimum size. The right subtree is an AVL tree of height 8 of minimum size. This tells us that the minimum number of nodes, $S(h)$, in an AVL tree of height h is given by $S(h) = S(h-1) + S(h-2) + 1$. For $h = 0$, $S(h) = 1$. For $h = 1$, $S(h) = 2$. The function $S(h)$ is closely related to the Fibonacci numbers, from which the bound claimed above on the height of an AVL tree follows.

Thus, all the tree operations can be performed in $O(\log N)$ time, except possibly insertion (we will assume lazy deletion). When we do an insertion, we need to update all the balancing information for the nodes on the path back to the root, but the reason that insertion is potentially difficult is that inserting a node could violate the AVL tree property. (For instance, inserting 6 into the AVL tree in Figure 4.29 would destroy the balance condition at the node with key 8.) If this is the case, then the property has to be restored before the insertion step is considered over. It turns out that this can always be done with a simple modification to the tree, known as a *rotation*.

After an insertion, only nodes that are on the path from the insertion point to the root might have their balance altered because only those nodes have their subtrees altered. As we follow the path up to the root and update the balancing information, we may find a node whose new balance violates the AVL condition. We will show how to rebalance the tree at the first (i.e., deepest) such node, and we will prove that this rebalancing guarantees that the entire tree satisfies the AVL property.

Let us call the node that must be rebalanced α. Since any node has at most two children, and a height imbalance requires that α's two subtrees' height differ by two, it is easy to see that a violation might occur in four cases:

1. An insertion into the left subtree of the left child of α.
2. An insertion into the right subtree of the left child of α.
3. An insertion into the left subtree of the right child of α.
4. An insertion into the right subtree of the right child of α.

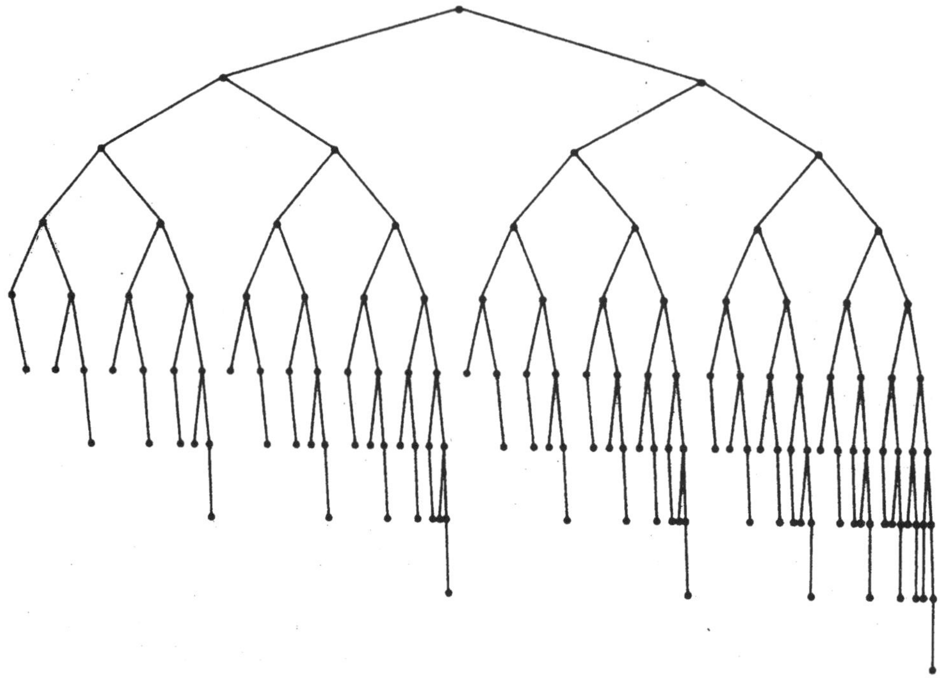

Figure 4.30 Smallest AVL tree of height 9

Cases 1 and 4 are mirror image symmetries with respect to α, as are cases 2 and 3. Consequently, as a matter of theory, there are two basic cases. From a programming perspective, of course, there are still four cases.

The first case, in which the insertion occurs on the "outside" (i.e., left–left or right–right), is fixed by a *single rotation* of the tree. The second case, in which the insertion occurs on the "inside" (i.e., left–right or right–left) is handled by the slightly more complex *double rotation*. These are fundamental operations on the tree that we'll see used several times in balanced-tree algorithms. The remainder of this section describes these rotations, proves that they suffice to maintain balance, and gives a casual implementation of the AVL tree. Chapter 12 describes other balanced-tree methods with an eye toward a more careful implementation.

4.4.1. *Single Rotation*

Figure 4.31 shows the *single rotation* that fixes case 1. The before picture is on the left, and the after is on the right. Let us analyze carefully what is going on. Node k_2 violates the AVL balance property because its left subtree is two levels deeper than its right subtree (the dashed lines in the middle of the diagram mark the levels). The situation depicted is the only possible case 1 scenario that allows k_2 to satisfy the AVL property before an insertion but violate it afterwards. Subtree X has grown to an extra level, causing it to be exactly two levels deeper than Z. Y cannot be at the same level as the new X because then k_2 would have been out of balance *before* the insertion, and Y cannot be at the same level as Z because then k_1 would be the first node on the path toward the root that was in violation of the AVL balancing condition.

To ideally rebalance the tree, we would like to move X up a level and Z down a level. Note that this is actually more than the AVL property would require. To do this, we rearrange nodes into an equivalent tree as shown in the second part of Figure 4.31. Here is an abstract scenario: visualize the tree as being flexible, grab the child node k_1, close your eyes, and shake it, letting gravity take hold. The result is that k_1 will be the new root. The binary search tree property tells us that in the original tree $k_2 > k_1$, so k_2 becomes the right child of k_1 in the new tree. X and Z remain as the left child of k_1 and right child of k_2, respectively. Subtree Y, which holds items that are between k_1 and k_2 in the original tree, can be placed as k_2's left child in the new tree and satisfy all the ordering requirements.

As a result of this work, which requires only a few pointer changes, we have another binary search tree that is an AVL tree. This happens because X moves up one level, Y stays at the same level, and Z moves down one level. k_2 and k_1 not only satisfy the AVL requirements, but they also have subtrees that are exactly the same height. Furthermore, the new height of the entire subtree is *exactly the same* as the height of the original subtree prior to the insertion that caused X to grow. Thus no further updating of heights on the path to the root is needed, and consequently *no further rotations are needed.* Figure 4.32 shows that after the insertion of 6 into

Figure 4.31 Single rotation to fix case 1

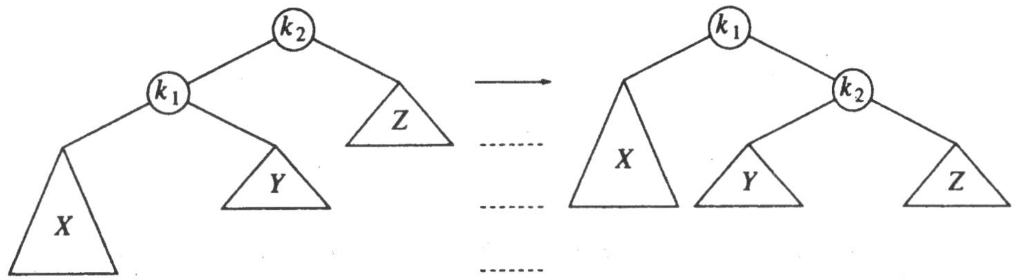

Figure 4.32 AVL property destroyed by insertion of 6, then fixed by a single rotation

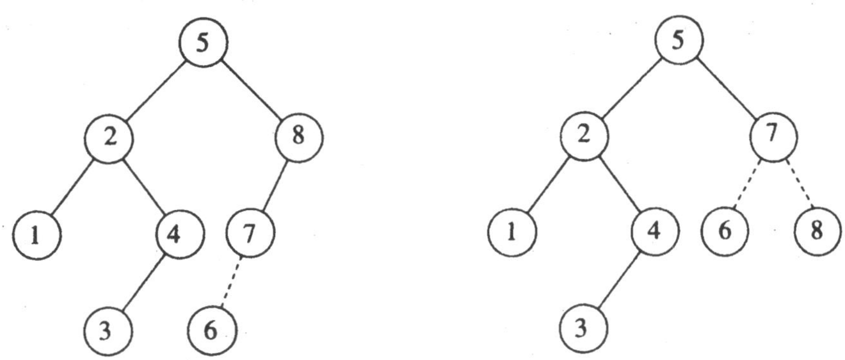

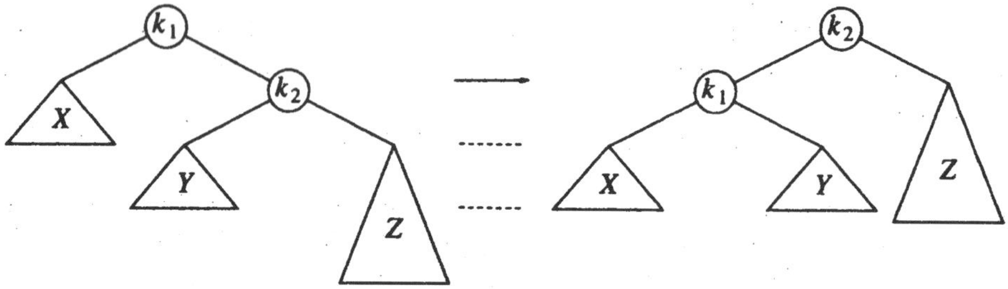

Figure 4.33 Single rotation fixes case 4

the original AVL tree on the left, node 8 becomes unbalanced. Thus, we do a single rotation between 7 and 8, obtaining the tree on the right.

As we mentioned earlier, case 4 represents a symmetric case. Figure 4.33 shows how a single rotation is applied. Let us work through a rather long example. Suppose we start with an initially empty AVL tree and insert the keys 3, 2, 1, and then 4 through 7 in sequential order. The first problem occurs when it is time to insert key 1 because the AVL property is violated at the root. We perform a single rotation between the root and its left child to fix the problem. Here are the before and after trees:

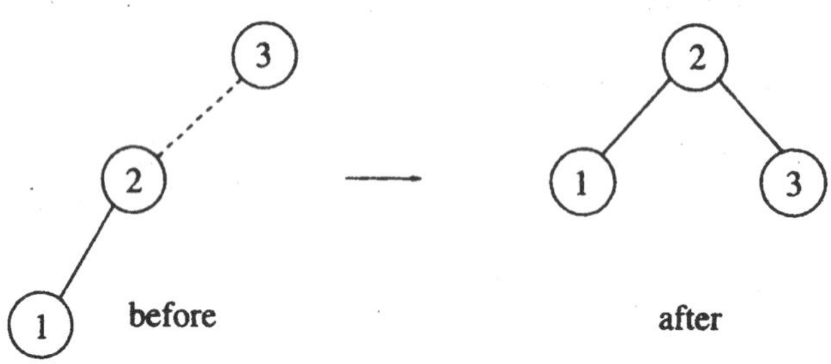

A dashed line joins the two nodes that are the subject of the rotation. Next we insert the key 4, which causes no problems, but the insertion of 5 creates a violation at node 3 that is fixed by a single rotation. Besides the local change caused by the rotation, the programmer must remember that the rest of the tree has to be informed of this change. Here this means that 2's right child must be reset to point to 4 instead of 3. Forgetting to do so is easy and would destroy the tree (4 would be inaccessible).

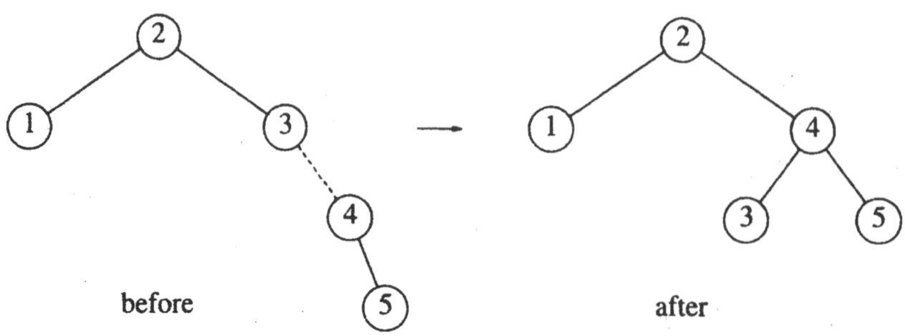

Next we insert 6. This causes a balance problem at the root, since its left subtree is of height 0 and its right subtree would be height 2. Therefore, we perform a single rotation at the root between 2 and 4.

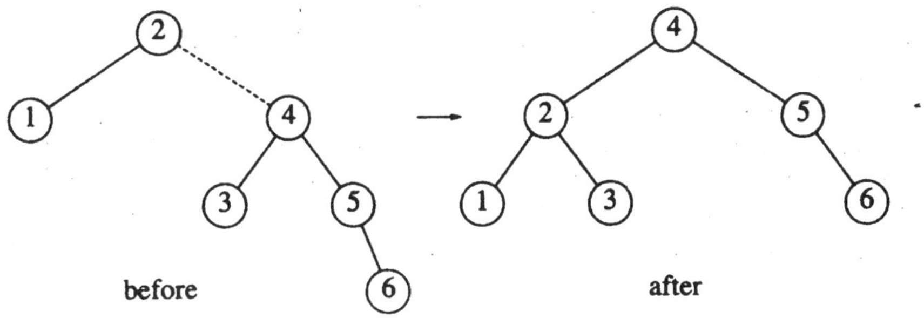

The rotation is performed by making 2 a child of 4 and 4's original left subtree the new right subtree of 2. Every key in this subtree must lie between 2 and 4, so this transformation makes sense. The next key we insert is 7, which causes another rotation:

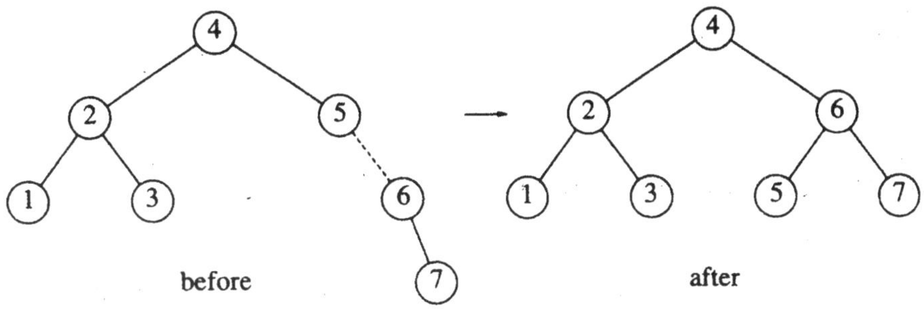

4.4.2. *Double Rotation*

The algorithm described above has one problem: as Figure 4.34 shows, it does not work for cases 2 or 3. The problem is that subtree Y is too deep, and a single rotation does not make it any less deep. The *double rotation* that solves the problem is shown in Figure 4.35.

The fact that subtree Y in Figure 4.34 has had an item inserted into it guarantees that it is nonempty. Thus, we may assume that it has a root and two subtrees. Consequently, the tree may be viewed as four subtrees connected by three nodes. As the diagram suggests, exactly one of tree B or C is two levels deeper than D (unless all are empty), but we cannot be sure which one. It turns out not to matter; in Figure 4.35, both B and C are drawn at $1\frac{1}{2}$ levels below D.

To rebalance, we see that we cannot leave k_3 as the root, and a rotation between k_3 and k_1 was shown in Figure 4.34 to not work, so the only alternative is to place k_2 as the new root. This forces k_1 to be k_2's left child and k_3 to be its right child, and it also completely determines the resulting locations of the four subtrees. It is easy to see that the resulting tree satisfies the AVL tree property, and as was the case with the single rotation, it restores the height to what it was before the insertion, thus guaranteeing that all rebalancing and height updating is complete. Figure 4.36 shows that the symmetric case 3 can also be fixed by a double rotation. In both cases the effect is the same as rotating between α's child and grandchild, and then between α and its new child.

We will continue our previous example by inserting the keys 10 through 16 in reverse order, followed by 8 and then 9. Inserting 16 is easy, since it does not destroy the balance property, but inserting 15 causes a height imbalance at node 7. This is case 3, which is solved by a right–left double rotation. In our example, the right–left double rotation will involve 7, 16, and 15. In this case, k_1 is the node with key 7, k_3

Figure 4.34 Single rotation fails to fix case 2

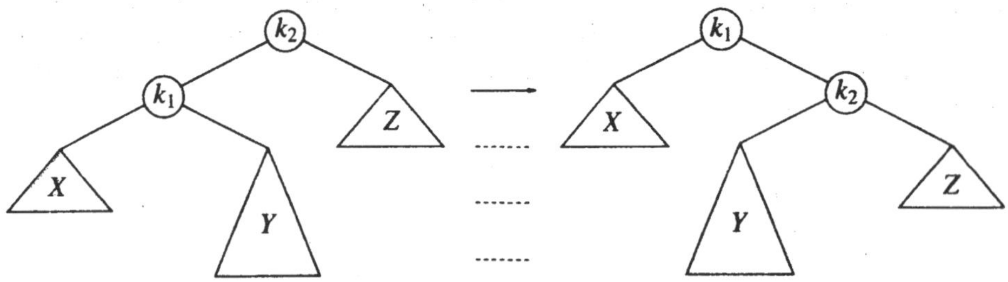

Figure 4.35 Left–right double rotation to fix case 2

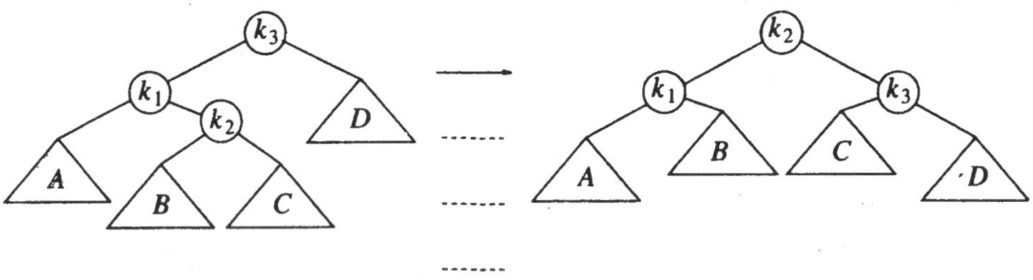

Figure 4.36 Right–left double rotation to fix case 3

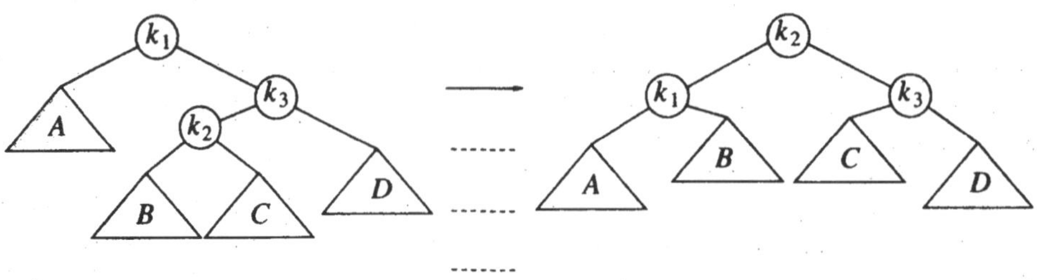

is the node with key 16, and k_2 is the node with key 15. Subtrees A, B, C, and D are empty.

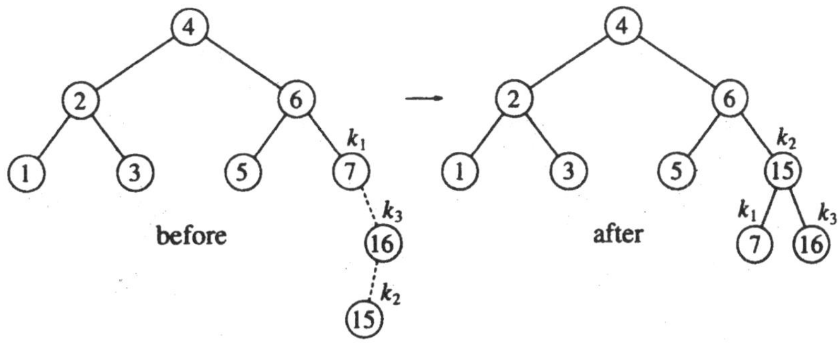

Next we insert 14, which also requires a double rotation. Here the double rotation that will restore the tree is again a right–left double rotation that will involve 6, 15, and 7. In this case, k_1 is the node with key 6, k_2 is the node with key 7, and k_3 is the node with key 15. Subtree A is the tree rooted at the node with key 5; subtree B is the empty subtree that was originally the left child of the node with key 7, subtree C is the tree rooted at the node with key 14, and finally, subtree D is the tree rooted at the node with key 16.

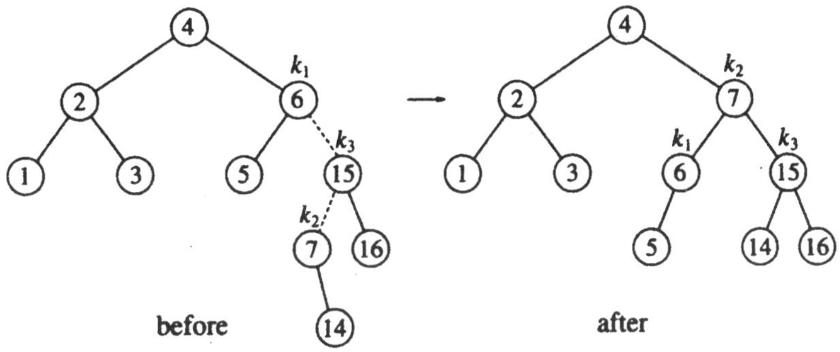

If 13 is now inserted, there is an imbalance at the root. Since 13 is not between 4 and 7, we know that the single rotation will work.

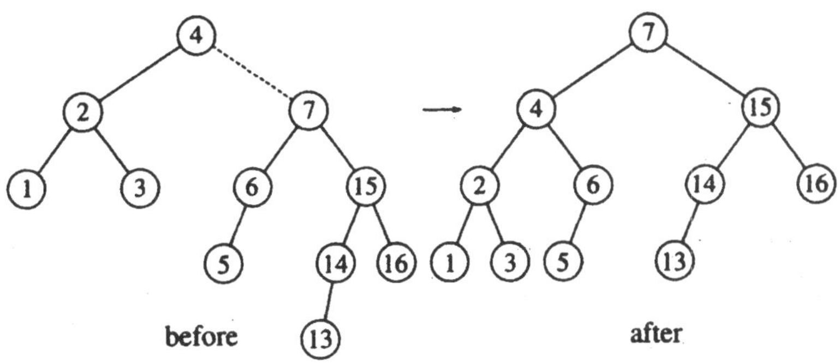

Insertion of 12 will also require a single rotation:

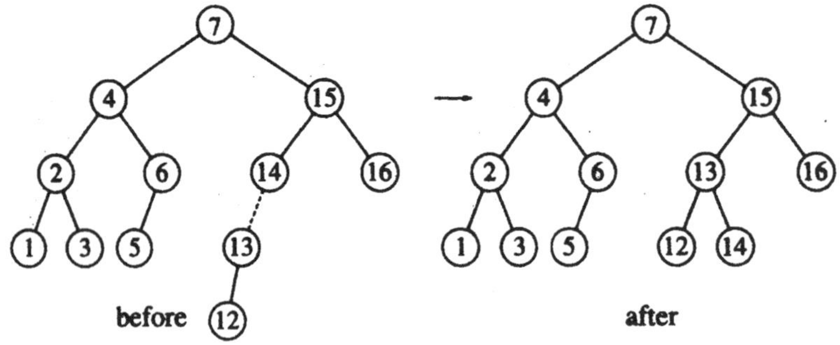

To insert 11, a single rotation needs to be performed, and the same is true for the subsequent insertion of 10. We insert 8 without a rotation creating an almost perfectly balanced tree:

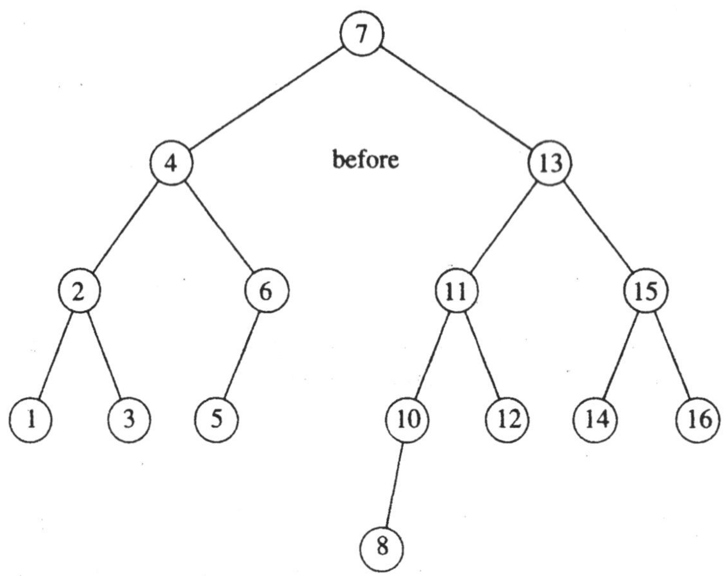

Finally, we will insert 9 to show the symmetric case of the double rotation. Notice that 9 causes the node containing 10 to become unbalanced. Since 9 is between 10 and 8 (which is 10's child on the path to 9), a double rotation needs to be performed, yielding the following tree:

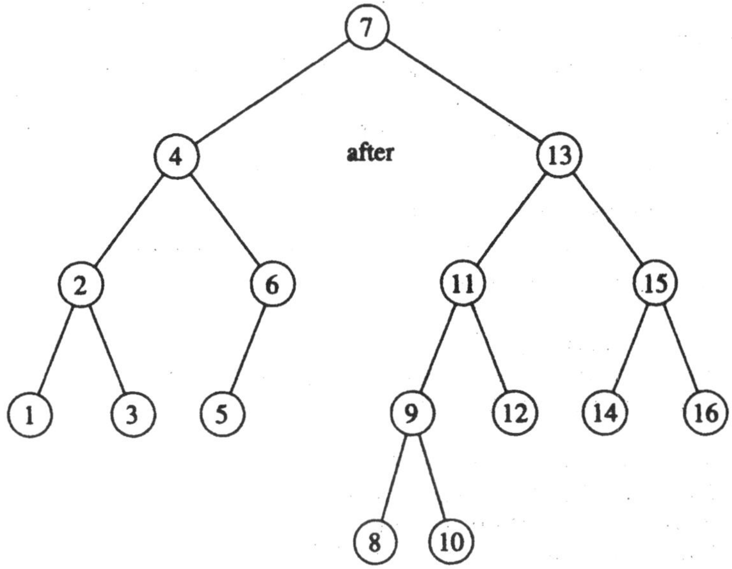

Let us summarize what happens. The programming details are fairly straightforward except that there are several cases. To insert a new node with key X into an AVL tree T, we recursively insert X into the appropriate subtree of T (let us call this T_{LR}). If the height of T_{LR} does not change, then we are done. Otherwise, if a height imbalance appears in T, we do the appropriate single or double rotation depending on X and the keys in T and T_{LR}, update the heights (making the connection from the rest of the tree above), and are done. Since one rotation always suffices, a carefully coded nonrecursive version generally turns out to be significantly faster than the recursive version. However, nonrecursive versions are quite difficult to code correctly, so many programmers implement AVL trees recursively.

Another efficiency issue concerns storage of the height information. Since all that is really required is the difference in height, which is guaranteed to be small, we could get by with two bits (to represent $+1$, 0, -1) if we really try. Doing so will avoid repetitive calculation of balance factors but results in some loss of clarity. The resulting code is somewhat more complicated than if the height were stored at each node. If a recursive routine is written, then speed is probably not the main consideration. In this case, the slight speed advantage obtained by storing balance factors hardly seems worth the loss of clarity and relative simplicity. Furthermore, since most machines will align this to at least an 8-bit boundary anyway, there is not likely to be any difference in the amount of space used. Eight bits will allow us to store absolute heights of up to 255. Since the tree is balanced, it is inconceivable that this would be insufficient (see the exercises).

With all this, we are ready to write the AVL routines. We will do only a partial job and leave the rest as an exercise. First, we need the declarations. These are given in Figure 4.37. We also need a quick function to return the height of a node. This function is necessary to handle the annoying case of a *NULL* pointer. This is shown in Figure 4.38. The basic insertion routine is easy to write, since it consists mostly of function calls (see Fig. 4.39).

For the trees in Figure 4.40, *SingleRotateWithLeft* converts the tree on the left to the tree on the right, returning a pointer to the new root. *SingleRotateWithRight* is symmetric. The code is shown in Figure 4.41.

Figure 4.37 Node declaration for AVL trees

```
#ifndef _AvlTree_H

struct AvlNode;
typedef struct AvlNode *Position;
typedef struct AvlNode *AvlTree;

AvlTree MakeEmpty( AvlTree T );
Position Find( ElementType X, AvlTree T );
Position FindMin( AvlTree T );
Position FindMax( AvlTree T );
AvlTree Insert( ElementType X, AvlTree T );
AvlTree Delete( ElementType X, AvlTree T );
ElementType Retrieve( Position P );

#endif  /* _AvlTree_H */


/* Place in the implementation file */
struct AvlNode
{
    ElementType Element;
    AvlTree  Left;
    AvlTree  Right;
    int      Height;
};
```

Figure 4.38 Function to compute height of an AVL node

```
static int
Height( Position P )
{
    if( P == NULL )
        return -1;
    else
        return P->Height;
}
```

```
AvlTree
Insert( ElementType X, AvlTree T )
{
    if( T == NULL )
    {
        /* Create and return a one-node tree */
        T = malloc( sizeof( struct AvlNode ) );
        if( T == NULL )
            FatalError( "Out of space!!!" );
        else
        {
            T->Element = X; T->Height = 0;
            T->Left = T->Right = NULL;
        }
    }
    else
    if( X < T->Element )
    {
        T->Left = Insert( X, T->Left );
        if( Height( T->Left ) - Height( T->Right ) == 2 )
            if( X < T->Left->Element )
                T = SingleRotateWithLeft( T );
            else
                T = DoubleRotateWithLeft( T );
    }
    else
    if( X > T->Element )
    {
        T->Right = Insert( X, T->Right );
        if( Height( T->Right ) - Height( T->Left ) == 2 )
            if( X > T->Right->Element )
                T = SingleRotateWithRight( T );
            else
                T = DoubleRotateWithRight( T );
    }
    /* Else X is in the tree already; we'll do nothing */

    T->Height = Max( Height( T->Left ), Height( T->Right ) ) + 1;
    return T;
}
```

Figure 4.39 Insertion into an AVL tree

The last function we will write will perform the double rotation pictured in Figure 4.42, for which the code is shown in Figure 4.43.

Deletion in AVL trees is somewhat more complicated than insertion. Lazy deletion is probably the best strategy if deletions are relatively infrequent.

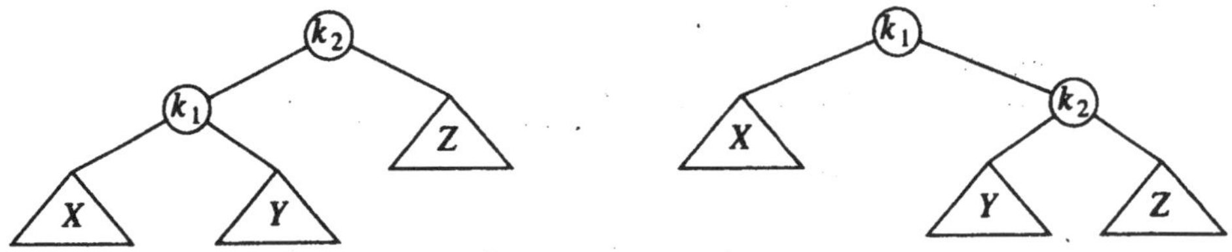

Figure 4.40 Single Rotation

```
/* This function can be called only if K2 has a left child */
/* Perform a rotate between a node (K2) and its left child */
/* Update heights, then return new root */

static Position
SingleRotateWithLeft( Position K2 )
{
    Position K1;

    K1 = K2->Left;
    K2->Left = K1->Right;
    K1->Right = K2;

    K2->Height = Max( Height( K2->Left ),
                      Height( K2->Right ) ) + 1;
    K1->Height = Max( Height( K1->Left ), K2->Height ) + 1;

    return K1;  /* New root */
}
```

Figure 4.41 Routine to perform single rotation

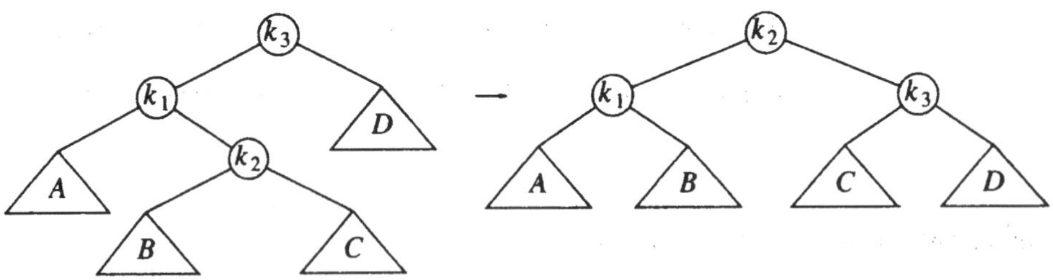

Figure 4.42 Double rotation

```
/* This function can be called only if K3 has a left */
/* child and K3's left child has a right child */
/* Do the left-right double rotation */
/* Update heights, then return new root */

static Position
DoubleRotateWithLeft( Position K3 )
{
    /* Rotate between K1 and K2 */
    K3->Left = SingleRotateWithRight( K3->Left );

    /* Rotate between K3 and K2 */
    return SingleRotateWithLeft( K3 );
}
```

Figure 4.43 Routine to perform double rotation

4.5. Splay Trees

We now describe a relatively simple data structure, known as a *splay tree*, that guarantees that any M consecutive tree operations starting from an empty tree take at most $O(M \log N)$ time. Although this guarantee does not preclude the possibility that any *single* operation might take $O(N)$ time, and thus the bound is not as strong as an $O(\log N)$ worst-case bound per operation, the net effect is the same: There are no bad input sequences. Generally, when a sequence of M operations has total worst-case running time of $O(M F(N))$, we say that the *amortized* running time is $O(F(N))$. Thus, a splay tree has an $O(\log N)$ amortized cost per operation. Over a long sequence of operations, some may take more, some less.

Splay trees are based on the fact that the $O(N)$ worst-case time per operation for binary search trees is not bad, as long at it occurs relatively infrequently. Any one access, even if it takes $O(N)$, is still likely to be extremely fast. The problem with binary search trees is that it is possible, and not uncommon, for a whole sequence of bad accesses to take place. The cumulative running time then becomes noticeable. A search tree data structure with $O(N)$ worst-case time, but a *guarantee* of at most $O(M \log N)$ for any M consecutive operations, is certainly satisfactory, because there are no bad sequences.

If any particular operation is allowed to have an $O(N)$ worst-case time bound, and we still want an $O(\log N)$ amortized time bound, then it is clear that whenever a node is accessed, it must be moved. Otherwise, once we find a deep node, we could keep performing *Find*s on it. If the node does not change location, and each access costs $O(N)$, then a sequence of M accesses will cost $O(M \cdot N)$.

The basic idea of the splay tree is that after a node is accessed, it is pushed to the root by a series of AVL tree rotations. Notice that if a node is deep, there are many nodes on the path that are also relatively deep, and by restructuring we can make future accesses cheaper on all these nodes. Thus, if the node is unduly deep, then we want this restructuring to have the side effect of balancing the tree (to some extent). Besides giving a good time bound in theory, this method is likely to have practical utility, because in many applications when a node is accessed, it is likely to be accessed again in the near future. Studies have shown that this happens much more often than one would expect. Splay trees also do not require the maintenance of height or balance information, thus saving space and simplifying the code to some extent (especially when careful implementations are written).

4.5.1. A Simple Idea (That Does Not Work)

One way of performing the restructuring described above is to perform single rotations, bottom up. This means that we rotate every node on the access path with its parent. As an example, consider what happens after an access (a *Find*) on k_1 in the following tree.

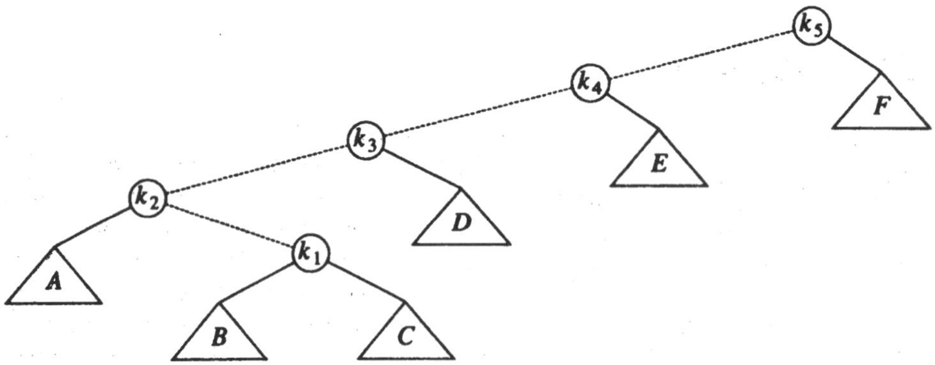

The access path is dashed. First, we would perform a single rotation between k_1 and its parent, obtaining the following tree.

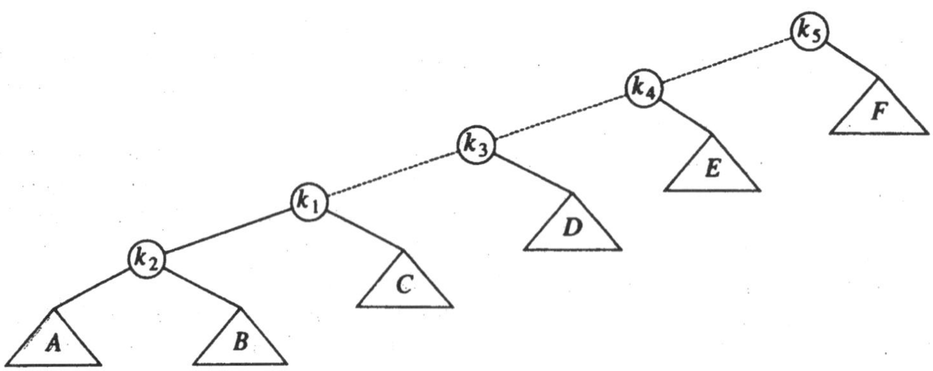

Then, we rotate between k_1 and k_3, obtaining the next tree.

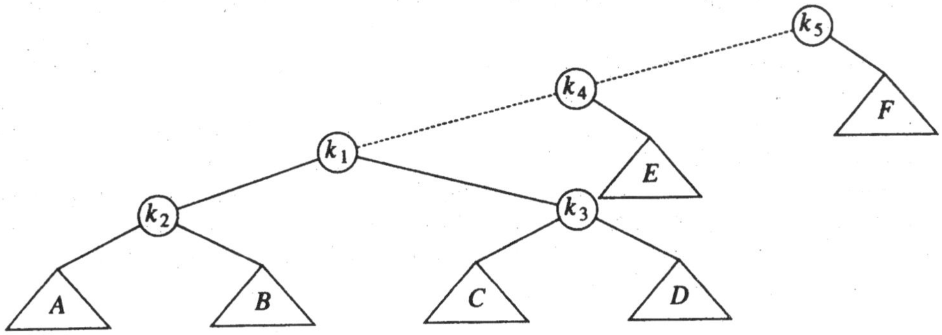

Then two more rotations are performed until we reach the root.

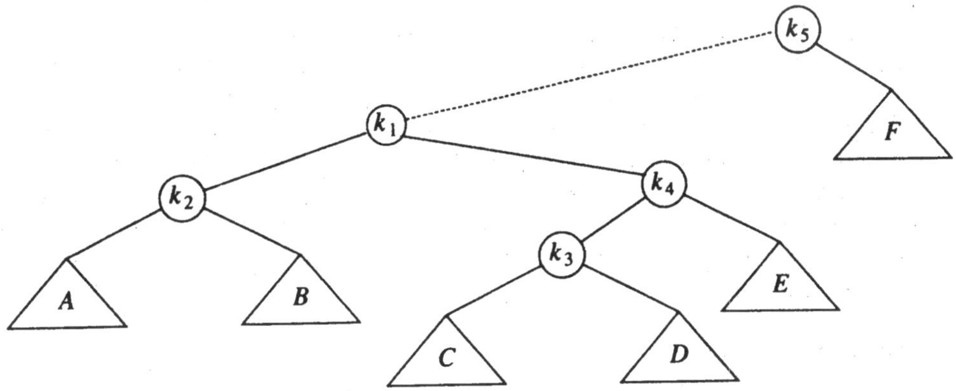

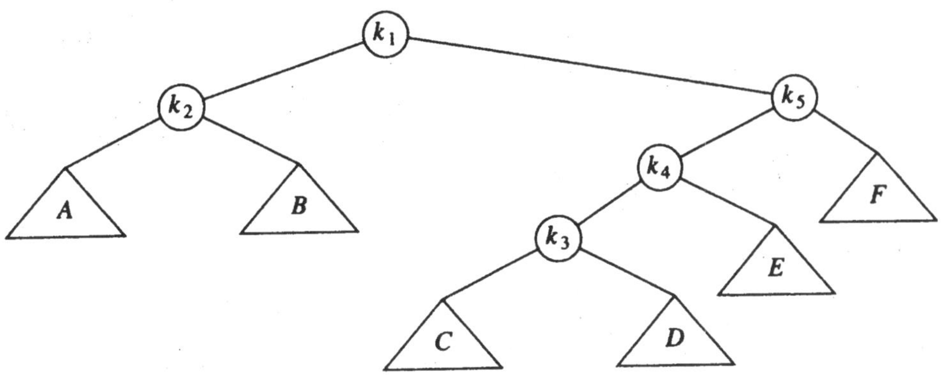

These rotations have the effect of pushing k_1 all the way to the root, so that future accesses on k_1 are easy (for a while). Unfortunately, it has pushed another node (k_3) almost as deep as k_1 used to be. An access on that node will then push another node deep, and so on. Although this strategy makes future accesses of k_1 cheaper, it has not significantly improved the situation for the other nodes on the (original) access path. It turns out that it is possible to prove that using this strategy, there is a sequence of M operations requiring $\Omega(M \cdot N)$ time, so this idea is not quite good enough. The simplest way to show this is to consider the tree formed by inserting keys $1, 2, 3, \ldots, N$ into an initially empty tree (work this example out). This gives a tree consisting of only left children. This is not necessarily bad, though, since the time to build this tree is $O(N)$ total. The bad part is that accessing the node with key 1 takes $N - 1$ units of time. After the rotations are complete, an access of the node with key 2 takes $N - 2$ units of time. The total for accessing all the keys in order is $\sum_{i=1}^{N-1} i = \Omega(N^2)$. After they are accessed, the tree reverts to its original state, and we can repeat the sequence.

4.5.2. *Splaying*

The splaying strategy is similar to the rotation idea above, except that we are a little more selective about how rotations are performed. We will still rotate bottom up along the access path. Let X be a (nonroot) node on the access path at which we are rotating. If the parent of X is the root of the tree, we merely rotate X and the root. This is the last rotation along the access path. Otherwise, X has both a parent (P) and a grandparent (G), and there are two cases, plus symmetries, to consider. The first case is the *zig-zag* case (see Fig. 4.44). Here X is a right child and P is a left child (or vice versa). If this is the case, we perform a double rotation, exactly like an AVL double rotation. Otherwise, we have a *zig-zig* case: X and P are either both left children or both right children. In that case, we transform the tree on the left of Figure 4.45 to the tree on the right.

Figure 4.44 Zig-zag

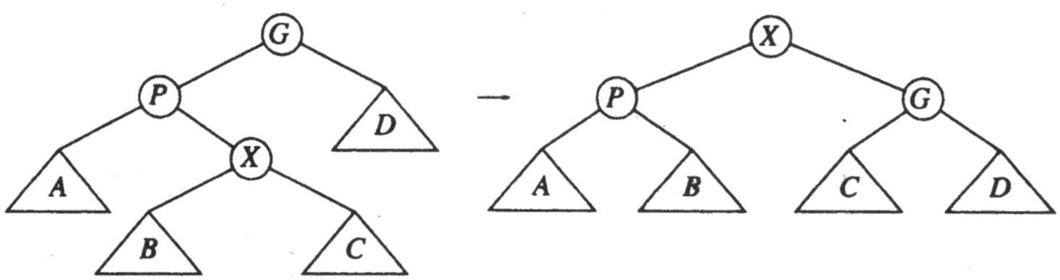

Figure 4.45 Zig-zig

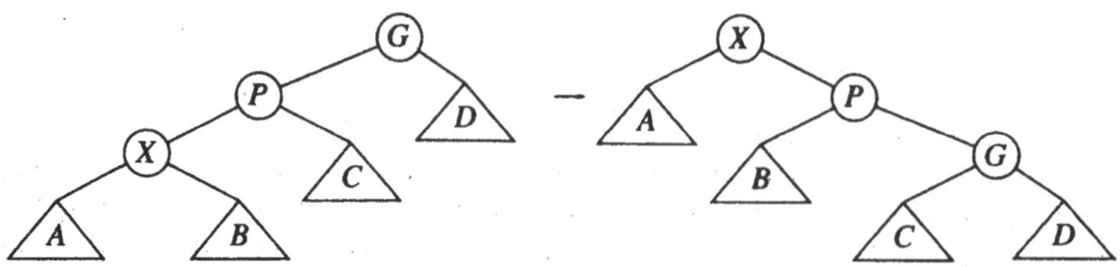

As an example, consider the tree from the last example, with a *Find* on k_1:

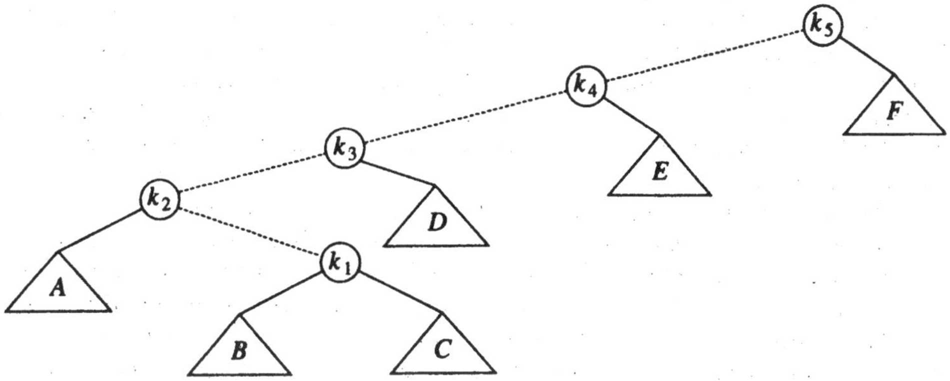

The first splay step is at k_1, and is clearly a *zig-zag*, so we perform a standard AVL double rotation using k_1, k_2, and k_3. The resulting tree follows.

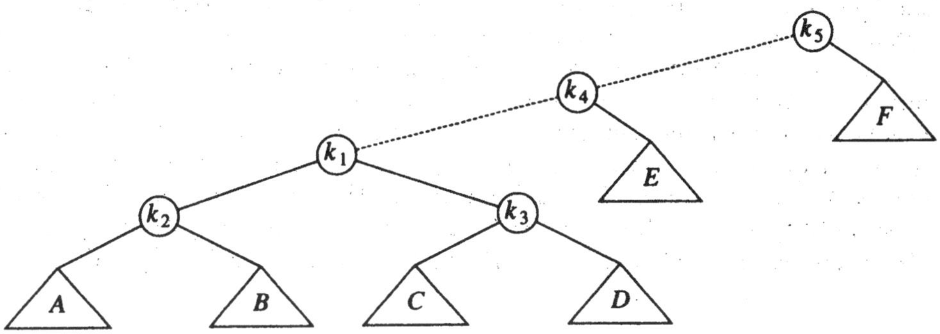

The next splay step at k_1 is a *zig-zig*, so we do the *zig-zig* rotation with k_1, k_4, and k_5, obtaining the final tree.

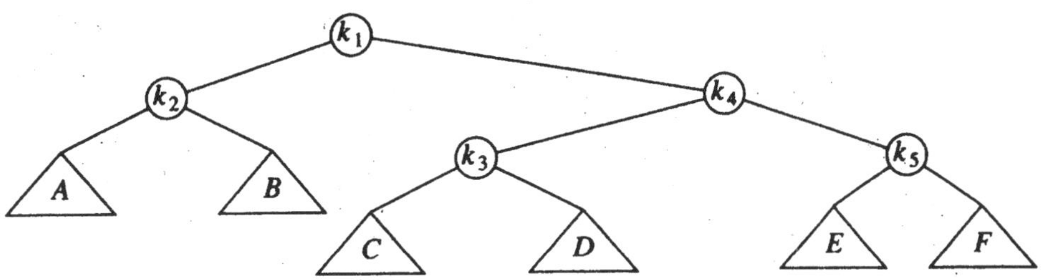

Although it is hard to see from small examples, splaying not only moves the accessed node to the root, but also has the effect of roughly halving the depth of most nodes on the access path (some shallow nodes are pushed down at most two levels).

To see the difference that splaying makes over simple rotation, consider again the effect of inserting keys 1, 2, 3, ..., N into an initially empty tree. This takes a total of $O(N)$, as before, and yields the same tree as simple rotations. Figure 4.46 shows the result of splaying at the node with key 1. The difference is that after an access of the node with key 1, which takes $N - 1$ units, the access on the node with key 2 will only take about $N/2$ units instead of $N - 2$ units; there are no nodes quite as deep as before.

An access on the node with key 2 will bring nodes to within $N/4$ of the root, and this is repeated until the depth becomes roughly $\log N$ (an example with $N = 7$ is too small to see the effect well). Figures 4.47 to 4.55 show the result of accessing keys 1 through 9 in a 32-node tree that originally contains only left children. Thus we do not get the same bad behavior from splay trees that is prevalent in the simple rotation strategy. (Actually, this turns out to be a very good case. A rather complicated proof shows that for this example, the N accesses take a total of $O(N)$ time.)

These figures highlight the fundamental and crucial property of splay trees. When access paths are long, thus leading to a longer-than-normal search time, the rotations tend to be good for future operations. When accesses are cheap, the rotations are not as good and can be bad. The extreme case is the initial tree formed by the insertions. All the insertions were constant-time operations leading to a bad initial tree. At that point in time, we had a very bad tree, but we were running ahead of schedule and had the compensation of less total running time. Then a couple of really horrible accesses left a nearly balanced tree, but the cost was that we had to give back some of the time that had been saved. The main theorem, which we will prove in Chapter 11, is that we never fall behind a pace of $O(\log N)$ per operation: We are always on schedule, even though there are occasionally bad operations.

Figure 4.46 Result of splaying at node 1

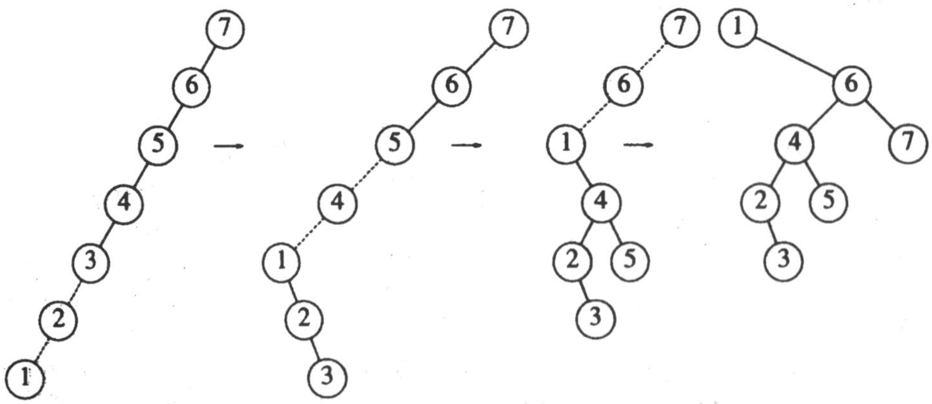

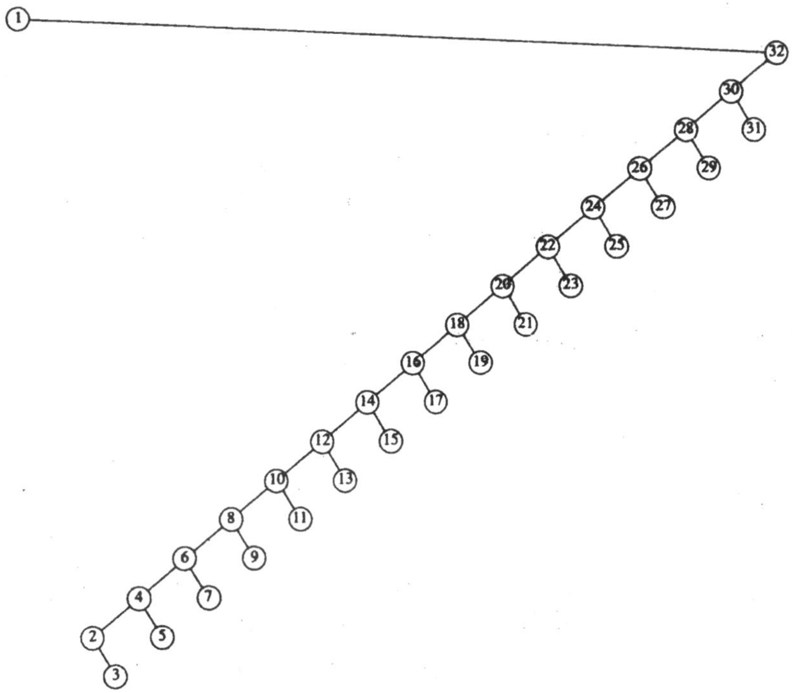

Figure 4.47 Result of splaying at node 1 a tree of all left children

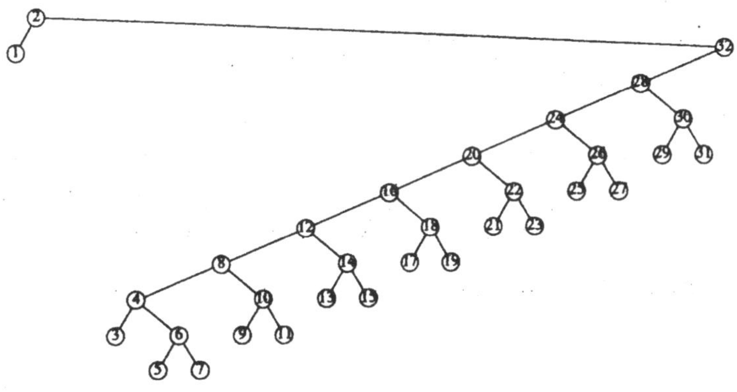

Figure 4.48 Result of splaying previous tree at node 2

We can perform deletion by accessing the node to be deleted. This puts the node at the root. If it is deleted, we get two subtrees T_L and T_R (left and right). If we find the largest element in T_L (which is easy), then this element is rotated to the root of T_L, and T_L will now have a root with no right child. We can finish the deletion by making T_R the right child.

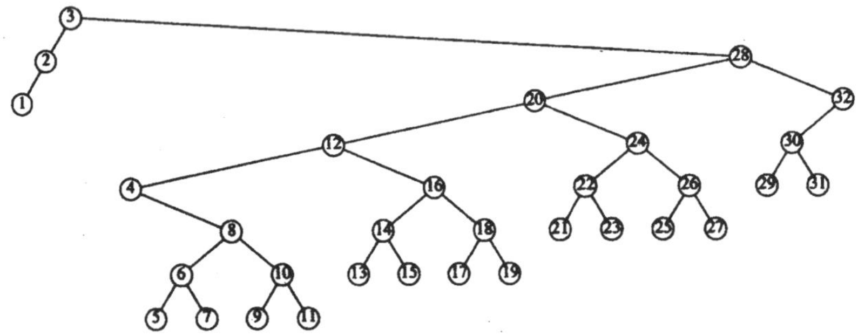

Figure 4.49 Result of splaying previous tree at node 3

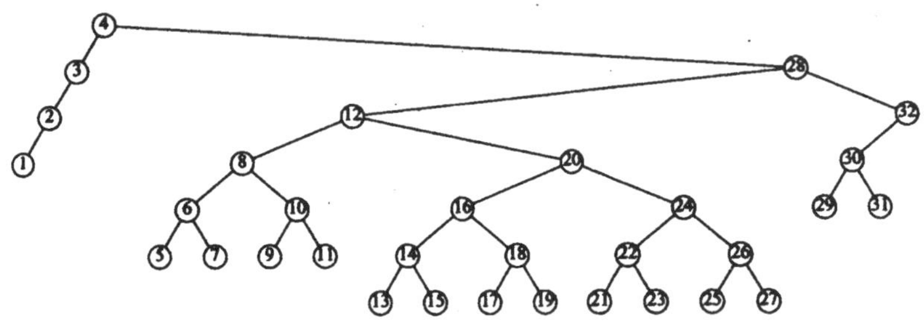

Figure 4.50 Result of splaying previous tree at node 4

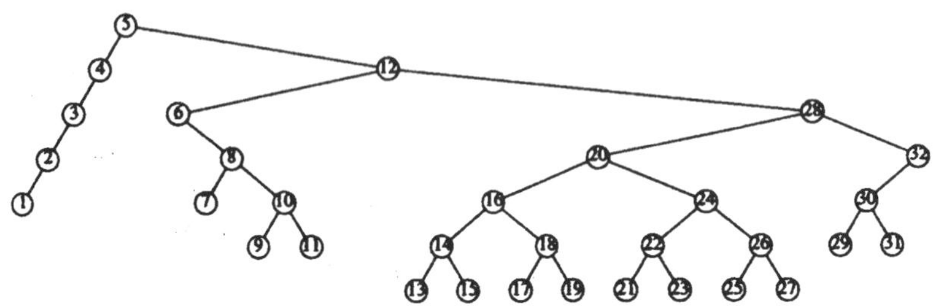

Figure 4.51 Result of splaying previous tree at node 5

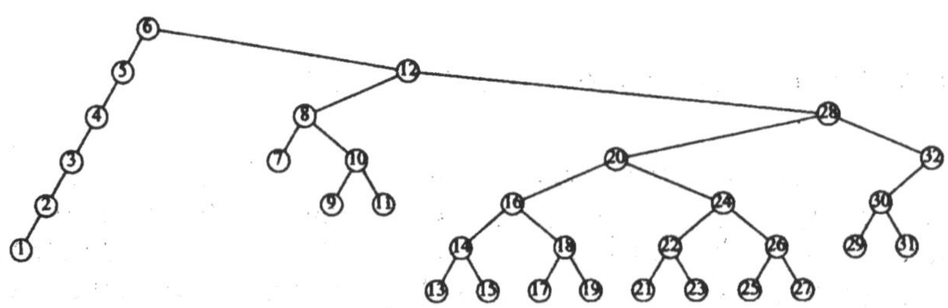

Figure 4.52 Result of splaying previous tree at node 6

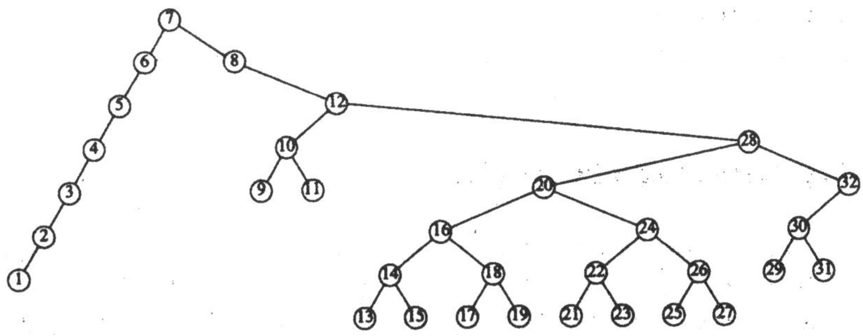

Figure 4.53 Result of splaying previous tree at node 7

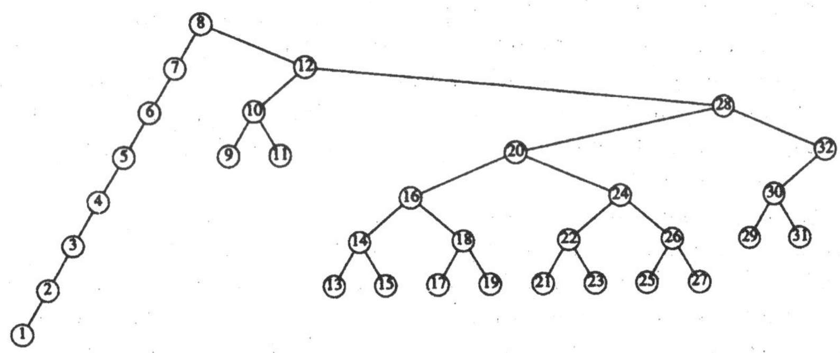

Figure 4.54 Result of splaying previous tree at node 8

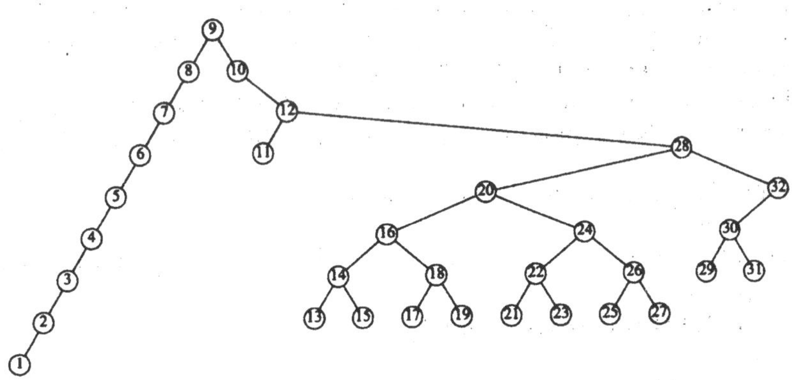

Figure 4.55 Result of splaying previous tree at node 9

The analysis of splay trees is difficult, because it must take into account the ever-changing structure of the tree. On the other hand, splay trees are much simpler to program than AVL trees, since there are fewer cases to consider and no balance information to maintain. Some empirical evidence suggests that this translates into faster code in practice, although the case for this is far from complete. Finally, we point out that there are several variations of splay trees that can perform even better in practice. One variation is completely coded in Chapter 12.

4.6. Tree Traversals (Revisited)

Because of the ordering information in a binary search tree, it is simple to list all the keys in sorted order. The recursive procedure in Figure 4.56 does this.

Convince yourself that this procedure works. As we have seen before, this kind of routine when applied to trees is known as an *inorder* traversal (which makes sense, since it lists the keys in order). The general strategy of an inorder traversal is to process the left subtree first, then perform processing at the current node, and finally process the right subtree. The interesting part about this algorithm, aside from its simplicity, is that the total running time is $O(N)$. This is because there is constant work being performed at every node in the tree. Each node is visited once, and the work performed at each node is testing against *NULL*, setting up two procedure calls, and doing a *PrintElement*. Since there is constant work per node and N nodes, the running time is $O(N)$.

Sometimes we need to process both subtrees first before we can process a node. For instance, to compute the height of a node, we need to know the height of the subtrees first. The code in Figure 4.57 computes this. Since it is always a good idea to check the special cases—and crucial when recursion is involved—notice that the routine will declare the height of a leaf to be zero, which is correct. This general order of traversal, which we have also seen before, is known as a *postorder* traversal. Again, the total running time is $O(N)$, because constant work is performed at each node.

The third popular traversal scheme that we have seen is *preorder* traversal. Here, the node is processed before the children. This could be useful, for example, if you wanted to label each node with its depth.

The common idea in all of these routines is that you handle the *NULL* case first, and then the rest. Notice the lack of extraneous variables. These routines pass only

Figure 4.56 Routine to print a binary search tree in order

```
void
PrintTree( SearchTree T )
{
    if( T != NULL )
    {
        PrintTree( T->Left );
        PrintElement( T->Element );
        PrintTree( T->Right );
    }
}
```

```
int
Height( Tree T )
{
    if( T == NULL )
        return -1;
    else
        return 1 + Max( Height( T->Left ),
                        Height( T->Right ) );
}
```

Figure 4.57 Routine to compute the height of a tree using a postorder traversal

the tree, and do not declare or pass any extra variables. The more compact the code, the less likely that a silly bug will turn up. A fourth, less often used, traversal (which we have not seen yet) is *level-order* traversal. In a level-order traversal, all nodes at depth D are processed before any node at depth $D + 1$. Level-order traversal differs from the other traversals in that it is not done recursively; a queue is used, instead of the implied stack of recursion.

4.7. B-Trees

Although all of the search trees we have seen so far are binary, there is a popular search tree that is not binary. This tree is known as a *B-tree.*

A B-tree of order M is a tree with the following structural properties:

- The root is either a leaf or has between 2 and M children.
- All nonleaf nodes (except the root) have between $\lceil M/2 \rceil$ and M children.
- All leaves are at the same depth.

All data are stored at the leaves. Contained in each interior node are pointers $P_1, P_2, \ldots, P_M$ to the children, and values $k_1, k_2, \ldots, k_{M-1}$, representing the smallest key found in the subtrees $P_2, P_3, \ldots, P_M$, respectively. Of course, some of these pointers might be *NULL,* and the corresponding k_i would then be undefined. For every node, all the keys in subtree P_1 are smaller than the keys in subtree P_2, and so on. The leaves contain all the actual data, which are either the keys themselves or pointers to records containing the keys. We will assume the former to keep our examples simple. There are various definitions of B-trees that change this structure in mostly minor ways, but this definition is one of the popular forms. (A popular alternative structure allows the actual data to be stored in both leaves and internal nodes, as is done in binary search trees.) We will also insist (for now) that the number of keys in a (nonroot) leaf is also between $\lceil M/2 \rceil$ and M.

The tree in Figure 4.58 is an example of a B-tree of order 4.

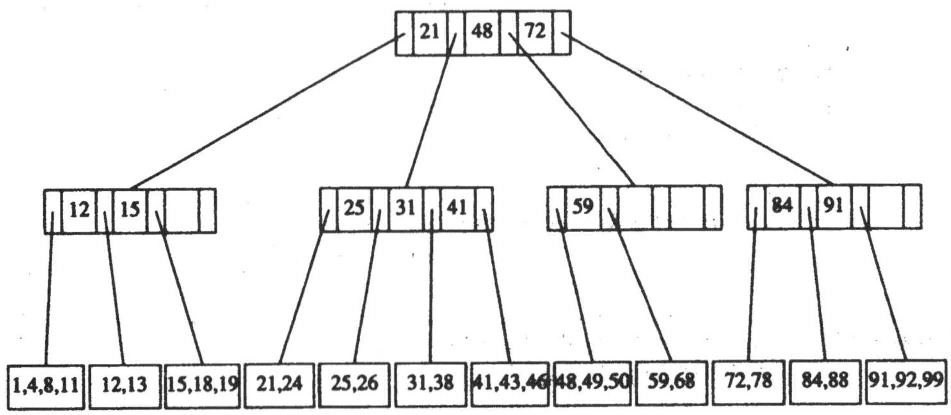

Figure 4.58 B-tree of order 4

A B-tree of order 4 is more popularly known as a 2–3–4 tree, and a B-tree of order 3 is known as a 2–3 tree. We will describe the operation of B-trees by using the special case of 2–3 trees. Our starting point is the 2–3 tree that follows.

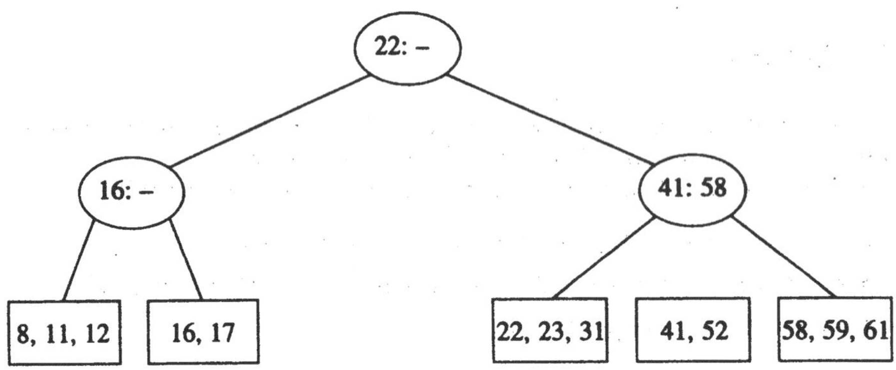

We have drawn interior nodes (nonleaves) in ellipses, which contain the two pieces of data for each node. A dash line as a second piece of information in an interior node indicates that the node has only two children. Leaves are drawn in boxes, which contain the keys. The keys in the leaves are ordered. To perform a *Find,* we start at the root and branch in one of (at most) three directions, depending on the relation of the key we are looking for to the two (possibly one) values stored at the node.

To perform an *Insert* on a previously unseen key, X, we follow the path as though we were performing a *Find.* When we get to a leaf node, we have found the correct place to put X. Thus, to insert a node with key 18, we can just add it to a leaf without causing any violations of the 2–3 tree properties. The result is shown in the following figure.

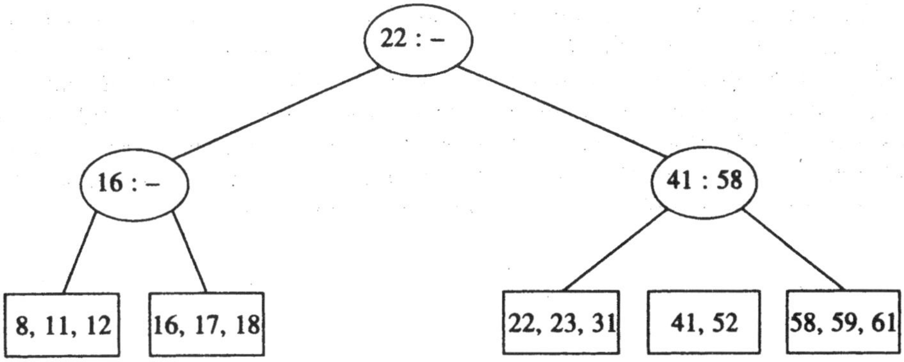

Unfortunately, since a leaf can hold only two or three keys, this might not always be possible. If we now try to insert 1 into the tree, we find that the node where it belongs is already full. Placing our new key into this node would give it a fourth element, which is not allowed. This can be solved by making two nodes of two keys each and adjusting the information in the parent.

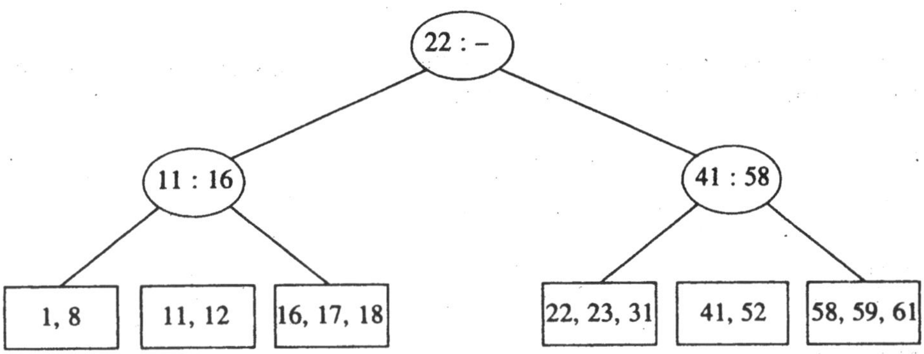

Unfortunately, this idea does not always work, as can be seen by an attempt to insert 19 into the current tree. If we make two nodes of two keys each, we obtain the following tree.

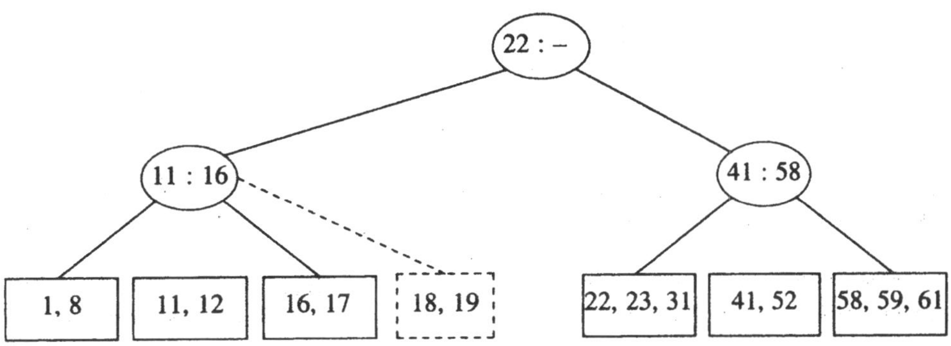

This tree has an internal node with four children, but we only allow three per node. The solution is simple. We merely split this node into two nodes with two children. Of course, this node might be one of three children itself, and thus splitting it would create a problem for its parent (which would now have four children), but we can keep on splitting nodes on the way up to the root until we either get to the root or find a node with only two children. In our case, we can get by with splitting only the first internal node we see, obtaining the following tree.

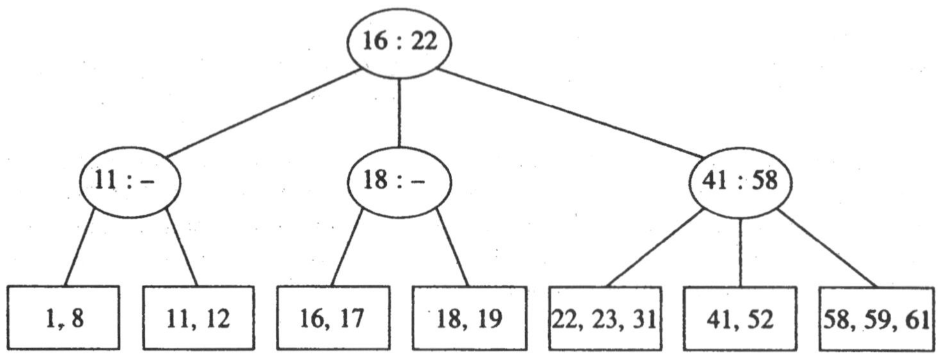

If we now insert an element with key 28, we create a leaf with four children, which is split into two leaves of two children:

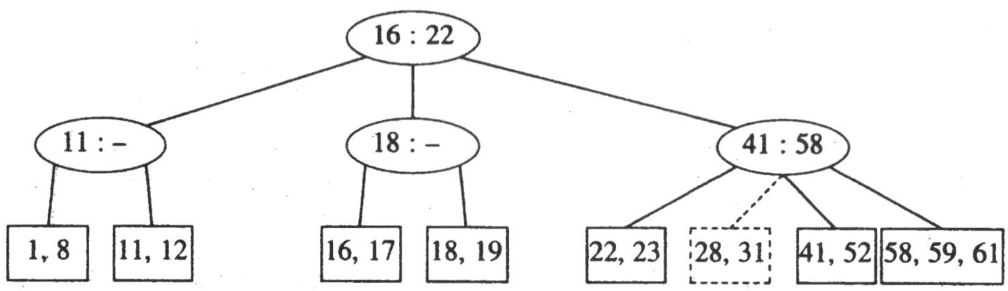

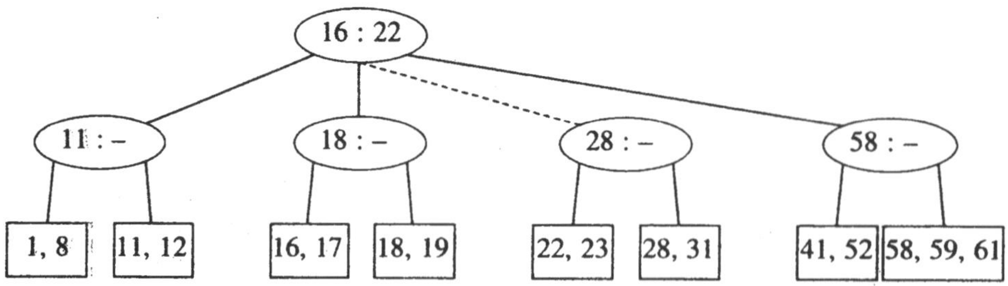

This creates an internal node with four children, which is then split into two children. What we have done here is split the root into two nodes. When we do this, we have a special case, which we finish by creating a new root. This is how (the only way) a 2–3 tree gains height.

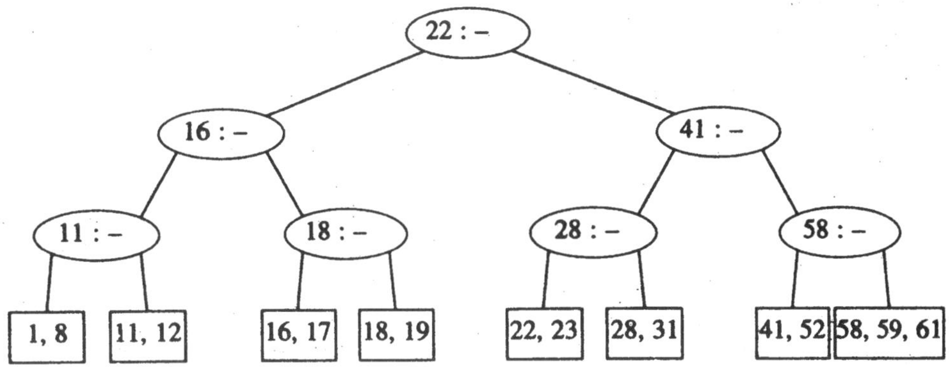

Notice also that when a key is inserted, the only changes to internal nodes occur on the access path. These changes can be made in time proportional to the length of this path; but be forewarned that there are quite a few cases to handle, and it is easy to do this wrong.

There are other ways to handle the case where a node becomes overloaded with children, but the method we have described is probably the simplest. When attempting to add a fourth key to a leaf, instead of splitting the node into two we can first attempt to find a sibling with only two keys. For instance, to insert 70 into the tree above, we could move 58 to the leaf containing 41 and 52, place 70 with 59 and 61, and adjust the entries in the internal nodes. This strategy can also be applied to internal nodes and tends to keep more nodes full. The cost of this is slightly more complicated routines, but less space tends to be wasted.

We can perform deletion by finding the key to be deleted and removing it. If this key was one of only two keys in a node, then its removal leaves only one key. We can fix this by combining this node with a sibling. If the sibling has three keys, we can steal one and have both nodes with two keys. If the sibling has only two keys, we combine the two nodes into a single node with three keys. The parent of this node now loses a child, so we might have to percolate this strategy all the way to the top. If the root loses its second child, then the root is also deleted and the tree becomes one level shallower. As we combine nodes, we must remember to update the information kept at the internal nodes.

With general B-trees of order M, when a key is inserted, the only difficulty arises when the node that is to accept the key already has M keys. This key gives the node $M + 1$ keys, which we can split into two nodes with $\lceil (M + 1)/2 \rceil$ and $\lfloor (M + 1)/2 \rfloor$ keys, respectively. As this gives the parent an extra node, we have to check whether this node can be accepted by the parent and split the parent if it already has M children. We repeat this until we find a parent with less than M children. If we split the root, we create a new root with two children.

The depth of a B-tree is at most $\lceil \log_{\lceil M/2 \rceil} N \rceil$. At each node on the path, we perform $O(\log M)$ work to determine which branch to take (using a binary search), but an *Insert* or *Delete* could require $O(M)$ work to fix up all the information at the node. The worst-case running time for each of the *Insert* and *Delete* operations is thus $O(M \log_M N) = O((M/\log M)\log N)$, but a *Find* takes only $O(\log N)$. The best (legal) choice of M for running time considerations has been shown empirically to be either $M = 3$ or $M = 4$; this agrees with the bounds above, which show that as M gets larger, the insertion and deletion times increase. If we are only concerned with main memory speed, higher order B-trees, such as 5–9 trees, are not an advantage.

The real use of B-trees lies in database systems, where the tree is kept on a physical disk instead of main memory. Accessing a disk is typically several orders of magnitude slower than any main memory operation. If we use a B-tree of order M, then the number of *disk accesses* is $O(\log_M N)$. Although each disk access carries the overhead of $O(\log M)$ to determine the direction to branch, the time to perform this computation is typically much smaller than the time to read a block of memory and can thus be considered inconsequential (as long as M is chosen reasonably). Even if updates are performed and $O(M)$ computing time is required at each node, these too are generally not significant. The value of M is then chosen to be the largest value that still allows an interior node to fit into one disk block, and is typically in the range $32 \leq M \leq 256$. The maximum number of elements that are stored in a leaf is chosen so that if the leaf is full, it fits in one block. This means that a record can always be found in very few disk accesses, since a typical B-tree will have a depth of only 2 or 3, and the root (and possibly the first level) can be kept in main memory.

Analysis suggests that a B-tree will be $\ln 2 = 69$ percent full. Better space utilization can be obtained if, instead of always splitting a node when the tree obtains its $(M + 1)$th entry, the routine searches for a sibling that can take the extra child. The details can be found in the references.

Summary

We have seen uses of trees in operating systems, compiler design, and searching. Expression trees are a small example of a more general structure known as a *parse tree*, which is a central data structure in compiler design. Parse trees are not binary, but are relatively simple extensions of expression trees (although the algorithms to build them are not quite so simple).

Search trees are of great importance in algorithm design. They support almost all the useful operations, and the logarithmic average cost is very small. Nonrecursive implementations of search trees are somewhat faster, but the recursive versions are sleeker, more elegant, and easier to understand and debug. The problem with search trees is that their performance depends heavily on the input being random. If this is not the case, the running time increases significantly, to the point where search trees become expensive linked lists.

We saw several ways to deal with this problem. AVL trees work by insisting that all nodes' left and right subtrees differ in heights by at most one. This ensures that the tree cannot get too deep. The operations that do not change the tree, as insertion does, can all use the standard binary search tree code. Operations that change the tree must restore the tree. This can be somewhat complicated, especially in the case of deletion. We showed how to restore the tree after insertions in $O(\log N)$ time.

We also examined the splay tree. Nodes in splay trees can get arbitrarily deep, but after every access the tree is adjusted in a somewhat mysterious manner. The net effect is that any sequence of M operations takes $O(M \log N)$ time, which is the same as a balanced tree would take.

B-trees are balanced M-way (as opposed to 2-way or binary) trees, which are well suited for disks; a special case is the 2–3 tree, which is another common method of implementing balanced search trees.

In practice, the running time of all the balanced tree schemes is worse (by a constant factor) than the simple binary search tree, but this is generally acceptable in view of the protection being given against easily obtained worst-case input. Chapter 12 discusses some additional search tree data structures and provides detailed implementations.

A final note: By inserting elements into a search tree and then performing an inorder traversal, we obtain the elements in sorted order. This gives an $O(N \log N)$ algorithm to sort, which is a worst-case bound if any sophisticated search tree is used. We shall see better ways in Chapter 7, but none that have a lower time bound.

Exercises

Questions 4.1 to 4.3 refer to the tree in Figure 4.59.

4.1 For the tree in Figure 4.59:

a. Which node is the root?

b. Which nodes are leaves?

4.2 For each node in the tree of Figure 4.59:

a. Name the parent node.

b. List the children.

c. List the siblings.

d. Compute the depth.

e. Compute the height.

4.3 What is the depth of the tree in Figure 4.59?

4.4 Show that in a binary tree of N nodes, there are $N + 1$ *NULL* pointers representing children.

4.5 Show that the maximum number of nodes in a binary tree of height H is $2^{H+1} - 1$.

4.6 A *full node* is a node with two children. Prove that the number of full nodes plus one is equal to the number of leaves in a nonempty binary tree.

4.7 Suppose a binary tree has leaves $l_1, l_2, \ldots, l_M$ at depths $d_1, d_2, \ldots, d_M$, respectively. Prove that $\sum_{i=1}^{M} 2^{-d_i} \le 1$ and determine when the equality is true.

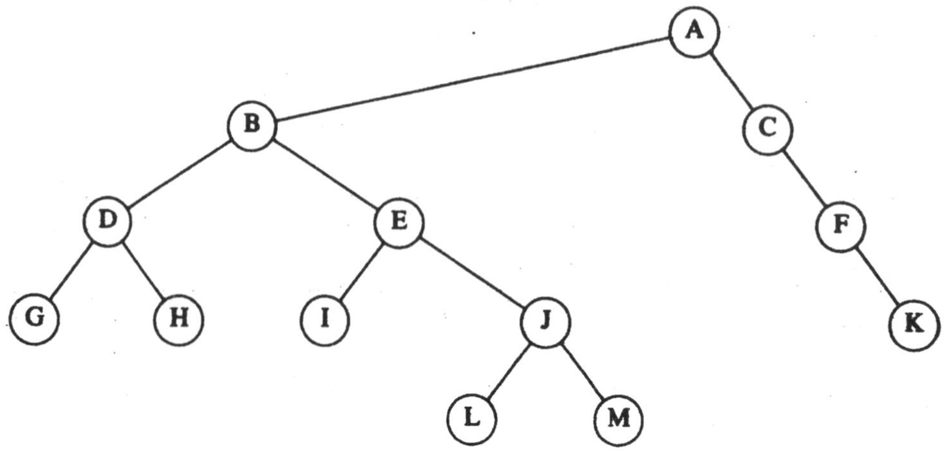

Figure 4.59

4.8 Give the prefix, infix, and postfix expressions corresponding to the tree in Figure 4.60.

4.9 a. Show the result of inserting 3, 1, 4, 6, 9, 2, 5, 7 into an initially empty binary search tree.

b. Show the result of deleting the root.

4.10 Write routines to implement the basic binary search tree operations.

4.11 Binary search trees can be implemented with cursors, using a strategy similar to a cursor linked list implementation. Write the basic binary search tree routines using a cursor implementation.

4.12 Suppose you want to perform an experiment to verify the problems that can be caused by random *Insert/Delete* pairs. Here is a strategy that is not perfectly random, but close enough. You build a tree with N elements by inserting N

Figure 4.60 Tree for Exercise 4.8

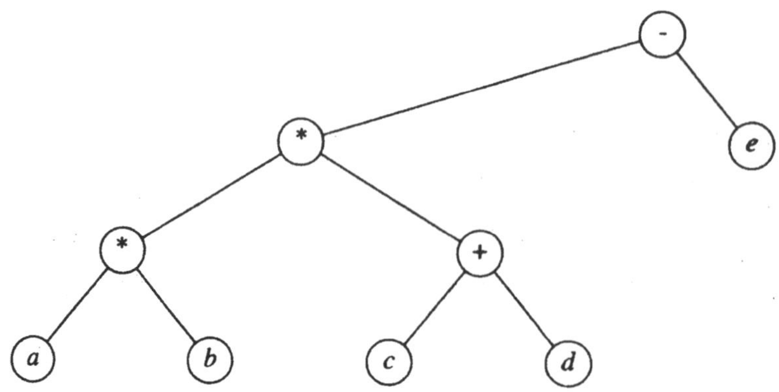

elements chosen at random from the range 1 to $M = \alpha N$. You then perform N^2 pairs of insertions followed by deletions. Assume the existence of a routine, *RandomInteger*(A, B), which returns a uniform random integer between A and B inclusive.

a. Explain how to generate a random integer between 1 and M that is not already in the tree (so a random insert can be performed). In terms of N and α, what is the running time of this operation?

b. Explain how to generate a random integer between 1 and M that is already in the tree (so a random delete can be performed). What is the running time of this operation?

c. What is a good choice of α? Why?

4.13 Write a program to evaluate empirically the following strategies for deleting nodes with two children:

a. Replace with the largest node, X, in T_L and recursively delete X.

b. Alternately replace with the largest node in T_L and the smallest node in T_R, and recursively delete the appropriate node.

c. Replace with either the largest node in T_L or the smallest node in T_R (recursively deleting the appropriate node), making the choice randomly.

Which strategy seems to give the most balance? Which takes the least CPU time to process the entire sequence?

4.14 ** Prove that the depth of a random binary search tree (depth of the deepest node) is $O(\log N)$, on average.

4.15 *a. Give a precise expression for the minimum number of nodes in an AVL tree of height H.

b. What is the minimum number of nodes in an AVL tree of height 15?

4.16 Show the result of inserting 2, 1, 4, 5, 9, 3, 6, 7 into an initially empty AVL tree.

4.17 * Keys $1, 2, \ldots, 2^k - 1$ are inserted in order into an initially empty AVL tree. Prove that the resulting tree is perfectly balanced.

4.18 Write the remaining procedures to implement AVL single and double rotations.

4.19 Write a nonrecursive function to insert into an AVL tree.

4.20 * How can you implement (nonlazy) deletion in AVL trees?

4.21 a. How many bits are required per node to store the height of a node in an N-node AVL tree?

b. What is the smallest AVL tree that overflows an 8-bit height counter?

4.22 Write the functions to perform the double rotation without the inefficiency of doing two single rotations.

4.23 Show the result of accessing the keys 3, 9, 1, 5 in order in the splay tree in Figure 4.61.

4.24 Show the result of deleting the element with key 6 in the resulting splay tree for the previous exercise.

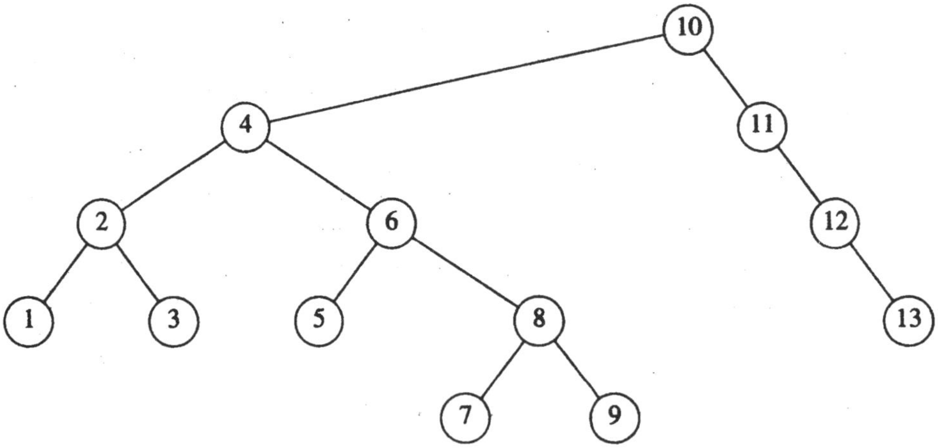

Figure 4.61

4.25 Nodes 1 through $N = 1024$ form a splay tree of left children.

a. What is the internal path length of the tree (exactly)?

*b. Calculate the internal path length after each of *Find*(1), *Find*(2), *Find*(3), *Find*(4), *Find*(5), *Find*(6).

*c. If the sequence of successive *Find*s is continued, when is the internal path length minimized?

4.26 a. Show that if all nodes in a splay tree are accessed in sequential order, the resulting tree consists of a chain of left children.

**b. Show that if all nodes in a splay tree are accessed in sequential order, then the total access time is $O(N)$, regardless of the initial tree.

4.27 Write a program to perform random operations on splay trees. Count the total number of rotations performed over the sequence. How does the running time compare to AVL trees and unbalanced binary search trees?

4.28 Write efficient functions that take only a pointer to the root of a binary tree, T, and compute:

a. The number of nodes in T.

b. The number of leaves in T.

c. The number of full nodes in T.

What is the running time of your routines?

4.29 Write a function to generate an N-node random binary search tree with distinct keys 1 through N. What is the running time of your routine?

4.30 Write a function to generate the AVL tree of height H with fewest nodes. What is the running time of your function?

4.31 Write a function to generate a perfectly balanced binary search tree of height H with keys 1 through $2^{H+1} - 1$. What is the running time of your function?

4.32 Write a function that takes as input a binary search tree, T, and two keys k_1 and k_2, which are ordered so that $k_1 \leq k_2$, and prints all elements X in the tree such that $k_1 \leq Key(X) \leq k_2$. Do not assume any information about the type of keys except that they can be ordered (consistently). Your program should run in $O(K + \log N)$ average time, where K is the number of keys printed. Bound the running time of your algorithm.

4.33 The larger binary trees in this chapter were generated automatically by a program. This was done by assigning an (x, y) coordinate to each tree node, drawing a circle around each coordinate (this is hard to see in some pictures), and connecting each node to its parent. Assume you have a binary search tree stored in memory (perhaps generated by one of the routines above) and that each node has two extra fields to store the coordinates.

a. The x coordinate can be computed by assigning the inorder traversal number. Write a routine to do this for each node in the tree.

b. The y coordinate can be computed by using the negative of the depth of the node. Write a routine to do this for each node in the tree.

c. In terms of some imaginary unit, what will the dimensions of the picture be? How can you adjust the units so that the tree is always roughly two-thirds as high as it is wide?

d. Prove that using this system no lines cross, and that for any node, X, all elements in X's left subtree appear to the left of X and all elements in X's right subtree appear to the right of X.

4.34 Write a general-purpose tree-drawing program that will convert a tree into the following graph-assembler instructions:

a. $Circle(X, Y)$

b. $DrawLine(i, j)$

The first instruction draws a circle at (X, Y), and the second instruction connects the ith circle to the jth circle (circles are numbered in the order drawn). You should either make this a program and define some sort of input language or make this a function that can be called from any program. What is the running time of your routine?

4.35 Write a routine to list out the nodes of a binary tree in *level-order*. List the root, then nodes at depth 1, followed by nodes at depth 2, and so on. You must do this in linear time. Prove your time bound.

4.36 a. Show the result of inserting the following keys into an initially empty 2–3 tree: 3, 1, 4, 5, 9, 2, 6, 8, 7, 0.

b. Show the result of deleting 0 and then 9 from the 2–3 tree created in part (a).

4.37 *a. Write a routine to perform insertion into a B-tree.

*b. Write a routine to perform deletion from a B-tree. When a key is deleted, is it necessary to update information in the internal nodes?

*c. Modify your insertion routine so that if an attempt is made to add into a node that already has M entries, a search is performed for a sibling with less than M children before the node is split.

4.38 A *B^*-tree* of order M is a B-tree in which each interior node has between $2M/3$ and M children. Describe a method to perform insertion into a B^*-tree.

4.39 Show how the tree in Figure 4.62 is represented using a child/sibling pointer implementation.

4.40 Write a procedure to traverse a tree stored with child/sibling links.

4.41 Two binary trees are similar if they are both empty or both nonempty and have similar left and right subtrees. Write a function to decide whether two binary trees are similar. What is the running time of your program?

4.42 Two trees, T_1 and T_2, are *isomorphic* if T_1 can be transformed into T_2 by swapping left and right children of (some of the) nodes in T_1. For instance, the two trees in Figure 4.63 are isomorphic because they are the same if the children of A, B, and G, but not the other nodes, are swapped.

a. Give a polynomial time algorithm to decide if two trees are isomorphic.

*b. What is the running time of your program (there is a linear solution)?

4.43 *a. Show that via AVL single rotations, any binary search tree T_1 can be transformed into another search tree T_2 (with the same keys).

*b. Give an algorithm to perform this transformation using $O(N \log N)$ rotations on average.

**c. Show that this transformation can be done with $O(N)$ rotations, worst-case.

Figure 4.62 Tree for Exercise 4.39

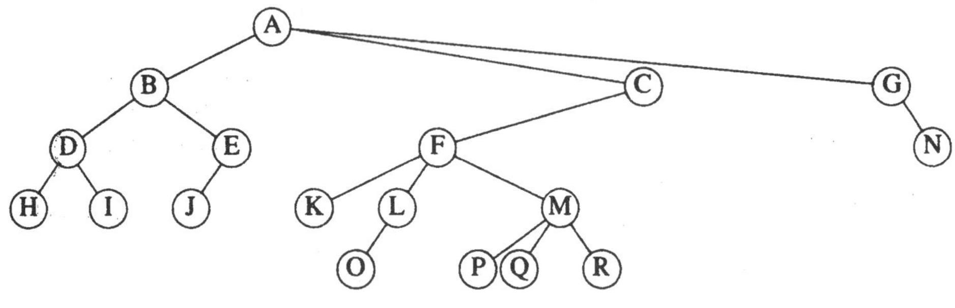

Figure 4.63 Two isomorphic trees

4.44 Suppose we want to add the operation *FindKth* to our repertoire. The operation *FindKth*(T, i) returns the element in tree T with ith smallest key. Assume all elements have distinct keys. Explain how to modify the binary search tree to support this operation in $O(\log N)$ average time, without sacrificing the time bounds of any other operation.

4.45 Since a binary search tree with N nodes has $N + 1$ *NULL* pointers, half the space allocated in a binary search tree for pointer information is wasted. Suppose that if a node has a *NULL* left child, we make its left child point to its inorder predecessor, and if a node has a *NULL* right child, we make its right child point to its inorder successor. This is known as a ***threaded tree*** and the extra pointers are called ***threads***.

a. How can we distinguish threads from real children pointers?

b. Write routines to perform insertion and deletion into a tree threaded in the manner described above.

c. What is the advantage of using threaded trees?

4.46 A binary search tree presupposes that searching is based on only one key per record. Suppose we would like to be able to perform searching based on either of two keys, Key_1 or Key_2.

a. One method is to build two separate binary search trees. How many extra pointers does this require?

b. An alternative method is a 2-d *tree*. A 2-d tree is similar to a binary search tree, except that branching at even levels is done with respect to Key_1, and branching at odd levels is done with Key_2. Figure 4.64 shows a 2-d tree, with the first and last names as keys, for post–WWII presidents. The presidents' names were inserted chronologically (Truman, Eisenhower, Kennedy, Johnson, Nixon, Ford, Carter, Reagan, Bush, Clinton). Write a routine to perform insertion into a 2-d tree.

c. Write an efficient procedure that prints all records in the tree that simultaneously satisfy the constraints $Low_1 \le Key_1 \le High_1$ and $Low_2 \le Key_2 \le High_2$.

Figure 4.64 A 2-d tree

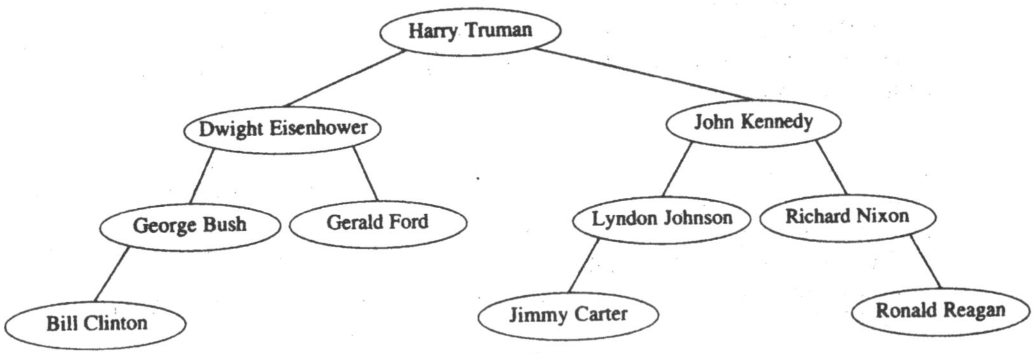

d. Show how to extend the 2-d tree to handle more than two search keys. The resulting strategy is known as a *k*-d tree.

References

More information on binary search trees, and in particular the mathematical properties of trees, can be found in the two books by Knuth, [23] and [24].

Several papers deal with the lack of balance caused by biased deletion algorithms in binary search trees. Hibbard's paper [20] proposed the original deletion algorithm and established that one deletion preserves the randomness of the trees. A complete analysis has been performed only for trees with three nodes [21] and four nodes [5]. Eppinger's paper [15] provided early empirical evidence of nonrandomness, and the papers by Culberson and Munro, [11], [12] provided some analytical evidence (but not a complete proof for the general case of intermixed insertions and deletions).

AVL trees were proposed by Adelson-Velskii and Landis [1]. Simulation results for AVL trees, and variants in which the height imbalance is allowed to be at most *k* for various values of *k*, are presented in [22]. A deletion algorithm for AVL trees can be found in [24]. Analysis of the average search cost in AVL trees is incomplete, but some results are contained in [25].

[3] and [9] considered self-adjusting trees like the type in Section 4.5.1. Splay trees are described in [29].

B-trees first appeared in [6]. The implementation described in the original paper allows data to be stored in internal nodes as well as leaves. The data structure we have described is sometimes known as a B^+-tree. A survey of the different types of B-trees is presented in [10]. Empirical results of the various schemes are reported in [18]. Analysis of 2–3 trees and B-trees can be found in [4], [14], and [33].

Exercise 4.14 is deceptively difficult. A solution can be found in [16]. Exercise 4.26 is from [32]. Information on B^*-trees, described in Exercise 4.38, can be found in [13]. Exercise 4.42 is from [2]. A solution to Exercise 4.43 using $2N - 6$ rotations is given in [30]. Using threads, à la Exercise 4.45, was first proposed in [28]. *k*-d trees were first proposed in [7]. Their major drawback is that both deletion and balancing are difficult. [8] discusses *k*-d trees and other methods used for multidimensional searching; a short discussion is also provided in Chapter 12.

Other popular balanced search trees are red black trees [19] and weight-balanced trees [27]. More balanced tree schemes can be found in Chapter 12, as well as in the books [17], [26], and [31].

1. G. M. Adelson-Velskii and E. M. Landis, "An Algorithm for the Organization of Information," *Soviet. Mat. Doklady,* 3 (1962), 1259–1263.
2. A. V. Aho, J. E. Hopcroft, and J. D. Ullman, *The Design and Analysis of Computer Algorithms,* Addison-Wesley, Reading, Mass., 1974.
3. B. Allen and J. I. Munro, "Self Organizing Search Trees," *Journal of the ACM,* 25 (1978), 526–535.

4. R. A. Baeza-Yates, "Expected Behaviour of B^+-trees under Random Insertions," *Acta Informatica,* 26 (1989), 439–471.
5. R. A. Baeza-Yates, "A Trivial Algorithm Whose Analysis Isn't: A Continuation," *BIT,* 29 (1989), 88–113.
6. R. Bayer and E. M. McCreight, "Organization and Maintenance of Large Ordered Indices," *Acta Informatica,* 1 (1972), 173–189.
7. J. L. Bentley, "Multidimensional Binary Search Trees Used for Associative Searching," *Communications of the ACM,* 18 (1975), 509–517.
8. J. L. Bentley and J. H. Friedman, "Data Structures for Range Searching," *Computing Surveys,* 11 (1979), 397–409.
9. J. R. Bitner, "Heuristics that Dynamically Organize Data Structures," *SIAM Journal on Computing,* 8 (1979), 82–110.
10. D. Comer, "The Ubiquitous B-tree," *Computing Surveys,* 11 (1979), 121–137.
11. J. Culberson and J. I. Munro, "Explaining the Behavior of Binary Search Trees under Prolonged Updates: A Model and Simulations," *Computer Journal,* 32 (1989), 68–75.
12. J. Culberson and J. I. Munro, "Analysis of the Standard Deletion Algorithms in Exact Fit Domain Binary Search Trees," *Algorithmica,* 5 (1990) 295–311.
13. K. Culik, T. Ottman, and D. Wood, "Dense Multiway Trees," *ACM Transactions on Database Systems,* 6 (1981), 486–512.
14. B. Eisenbath, N. Ziviana, G. H. Gonnet, K. Melhorn, and D. Wood, "The Theory of Fringe Analysis and its Application to 2–3 Trees and B-trees," *Information and Control,* 55 (1982), 125–174.
15. J. L. Eppinger, "An Empirical Study of Insertion and Deletion in Binary Search Trees," *Communications of the ACM,* 26 (1983), 663–669.
16. P. Flajolet and A. Odlyzko, "The Average Height of Binary Trees and Other Simple Trees," *Journal of Computer and System Sciences,* 25 (1982), 171–213.
17. G. H. Gonnet and R. Baeza-Yates, *Handbook of Algorithms and Data Structures,* 2d ed., Addison-Wesley, Reading, Mass., 1991.
18. E. Gudes and S. Tsur, "Experiments with B-tree Reorganization," *Proceedings of ACM SIGMOD Symposium on Management of Data* (1980), 200–206.
19. L. J. Guibas and R. Sedgewick, "A Dichromatic Framework for Balanced Trees," *Proceedings of the Nineteenth Annual IEEE Symposium on Foundations of Computer Science* (1978), 8–21.
20. T. H. Hibbard, "Some Combinatorial Properties of Certain Trees with Applications to Searching and Sorting," *Journal of the ACM,* 9 (1962), 13–28.
21. A. T. Jonassen and D. E. Knuth, "A Trivial Algorithm Whose Analysis Isn't," *Journal of Computer and System Sciences,* 16 (1978), 301–322.
22. P. L. Karlton, S. H. Fuller, R. E. Scroggs, and E. B. Kaehler, "Performance of Height Balanced Trees," *Communications of the ACM,* 19 (1976), 23–28.
23. D. E. Knuth, *The Art of Computer Programming: Vol. 1: Fundamental Algorithms,* 2d ed., Addison-Wesley, Reading, Mass., 1973.
24. D. E. Knuth, *The Art of Computer Programming: Vol. 3: Sorting and Searching,* Addison-Wesley, Reading, Mass., 1973.
25. K. Melhorn, "A Partial Analysis of Height-Balanced Trees under Random Insertions and Deletions," *SIAM Journal of Computing,* 11 (1982), 748–760.
26. K. Melhorn, *Data Structures and Algorithms 1: Sorting and Searching,* Springer-Verlag, Berlin, 1984.
27. J. Nievergelt and E. M. Reingold, "Binary Search Trees of Bounded Balance," *SIAM Journal on Computing,* 2 (1973), 33–43.

28. A. J. Perlis and C. Thornton, "Symbol Manipulation in Threaded Lists," *Communications of the ACM*, 3 (1960), 195–204.
29. D. D. Sleator and R. E. Tarjan, "Self-adjusting Binary Search Trees," *Journal of ACM*, 32 (1985), 652–686.
30. D. D. Sleator, R. E. Tarjan, and W. P. Thurston, "Rotation Distance, Triangulations, and Hyperbolic Geometry," *Journal of AMS* (1988), 647–682.
31. H. F. Smith, *Data Structures—Form and Function*, Harcourt Brace Jovanovich, Orlando, Fla., 1987.
32. R. E. Tarjan, "Sequential Access in Splay Trees Takes Linear Time," *Combinatorica*, 5 (1985), 367–378.
33. A. C. Yao, "On Random 2–3 Trees," *Acta Informatica*, 9 (1978), 159–170.

CHAPTER 5

Hashing

In Chapter 4, we discussed the search tree ADT, which allowed various operations on a set of elements. In this chapter, we discuss the *hash table* ADT, which supports only a subset of the operations allowed by binary search trees.

The implementation of hash tables is frequently called *hashing*. Hashing is a technique used for performing insertions, deletions, and finds in constant average time. Tree operations that require any ordering information among the elements are not supported efficiently. Thus, operations such as *FindMin, FindMax,* and the printing of the entire table in sorted order in linear time are not supported.

The central data structure in this chapter is the *hash table*. We will

- See several methods of implementing the hash table.
- Compare these methods analytically.
- Show numerous applications of hashing.
- Compare hash tables with binary search trees.

5.1. General Idea

The ideal hash table data structure is merely an array of some fixed size, containing the keys. Typically, a key is a string with an associated value (for instance, salary information). We will refer to the table size as *TableSize,* with the understanding that this is part of a hash data structure and not merely some variable floating around globally. The common convention is to have the table run from 0 to *TableSize* − 1; we will see why shortly.

Each key is mapped into some number in the range 0 to *TableSize* − 1 and placed in the appropriate cell. The mapping is called a *hash function,* which ideally should be simple to compute and should ensure that any two distinct keys get different cells. Since there are a finite number of cells and a virtually inexhaustible supply of keys, this is clearly impossible, and thus we seek a hash function that distributes the keys evenly among the cells. Figure 5.1 is typical of a perfect situation. In this example, *john* hashes to 3, *phil* hashes to 4, *dave* hashes to 6, and *mary* hashes to 7.

0	
1	
2	
3	john 25000
4	phil 31250
5	
6	dave 27500
7	mary 28200
8	
9	

Figure 5.1 An ideal hash table

This is the basic idea of hashing. The only remaining problems deal with choosing a function, deciding what to do when two keys hash to the same value (this is known as a *collision*), and deciding on the table size.

5.2. Hash Function

If the input keys are integers, then simply returning *Key* mod *TableSize* is generally a reasonable strategy, unless *Key* happens to have some undesirable properties. In this case, the choice of hash function needs to be carefully considered. For instance, if the table size is 10 and the keys all end in zero, then the standard hash function is a bad choice. For reasons we shall see later, and to avoid situations like the one above, it is usually a good idea to ensure that the table size is prime. When the input keys are random integers, then this function is not only very simple to compute but also distributes the keys evenly.

Usually, the keys are strings; in this case, the hash function needs to be chosen carefully.

One option is to add up the ASCII values of the characters in the string. In Figure 5.2 we declare the type *Index,* which is returned by the hash function. The routine in Figure 5.3 implements this strategy and uses the typical C method of stepping through a string.

The hash function depicted in Figure 5.3 is simple to implement and computes an answer quickly. However, if the table size is large, the function does not distribute

Figure 5.2 Type returned by hash function

```
        typedef unsigned int Index;
```

```
Index
Hash( const char *Key, int TableSize )
{
    unsigned int HashVal = 0;

    while( *Key != '\0' )
        HashVal += *Key++;

    return HashVal % TableSize;
}
```

Figure 5.3 A simple hash function

the keys well. For instance, suppose that *TableSize* = 10,007 (10,007 is a prime number). Suppose all the keys are eight or fewer characters long. Since a *char* has an integer value that is always at most 127, the hash function can only assume values between 0 and 1,016, which is 127 ∗ 8. This is clearly not an equitable distribution!

Another hash function is shown in Figure 5.4. This hash function assumes that *Key* has at least two characters plus the *NULL* terminator. The value 27 represents the number of letters in the English alphabet, plus the blank, and 729 is 27^2. This function examines only the first three characters, but if these are random and the table size is 10,007, as before, then we would expect a reasonably equitable distribution. Unfortunately, English is not random. Although there are $26^3 = 17{,}576$ possible combinations of three characters (ignoring blanks), a check of a reasonably large on-line dictionary reveals that the number of different combinations is actually only 2,851. Even if none of *these* combinations collide, only 28 percent of the table can actually be hashed to. Thus this function, although easily computable, is also not appropriate if the hash table is reasonably large.

Figure 5.5 shows a third attempt at a hash function. This hash function involves all characters in the key and can generally be expected to distribute well (it computes $\sum_{i=0}^{KeySize-1} Key[KeySize - i - 1] \cdot 32^i$, and brings the result into proper range). The code computes a polynomial function (of 32) by use of Horner's rule. For instance, another way of computing $h_k = k_1 + 27k_2 + 27^2k_3$ is by the formula $h_k = ((k_3) * 27 + k_2) * 27 + k_1$. Horner's rule extends this to an *n*th degree polynomial.

Figure 5.4 Another possible hash function—
not too good

```
Index
Hash( const char *Key, int TableSize )
{
    return ( Key[ 0 ] + 27 * Key[ 1 ] + 729 * Key[ 2 ] )
                    % TableSize;
}
```

```
        Index
        Hash( const char *Key, int TableSize )
        {
            unsigned int HashVal = 0;

/* 1*/      while( *Key != '\0' )
/* 2*/          HashVal = ( HashVal << 5 ) + *Key++;

/* 3*/      return HashVal % TableSize;
        }
```

Figure 5.5 A good hash function

We have used 32 instead of 27, because multiplication by 32 is not really a multiplication, but amounts to bit-shifting by 5. In line 2, the addition could be replaced with a bitwise Exclusive Or, for increased speed.

The hash function described in Figure 5.5 is not necessarily the best with respect to table distribution, but does have the merit of extreme simplicity (and speed if overflows are allowed). If the keys are very long, the hash function will take too long to compute. Furthermore, the early characters will wind up being left-shifted out of the eventual answer. A common practice in this case is not to use all the characters. The length and properties of the keys would then influence the choice. For instance, the keys could be a complete street address. The hash function might include a couple of characters from the street address and perhaps a couple of characters from the city name and ZIP code. Some programmers implement their hash function by using only the characters in the odd spaces, with the idea that the time saved computing the hash function will make up for a slightly less evenly distributed function.

The main programming detail left is collision resolution. If, when an element is inserted, it hashes to the same value as an already inserted element, then we have a *collision* and need to resolve it. There are several methods for dealing with this. We will discuss two of the simplest: separate chaining and open addressing.

5.3. Separate Chaining

The first strategy, commonly known as *separate chaining*, is to keep a list of all elements that hash to the same value. For convenience, our lists have headers. This makes the list implementation the same as in Chapter 3. If space is tight, it might be preferable to avoid their use. We assume for this section that the keys are the first 10 perfect squares and that the hashing function is simply $Hash(X) = X \bmod 10$. (The table size is not prime but is used here for simplicity.) Figure 5.6 should make this clear.

To perform a *Find*, we use the hash function to determine which list to traverse. We then traverse this list in the normal manner, returning the position where the

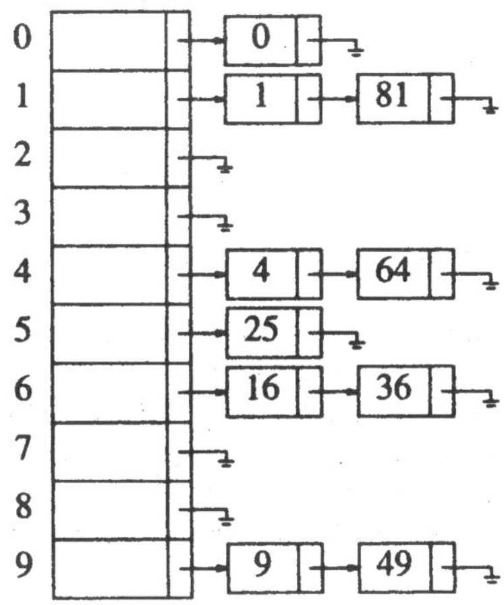

Figure 5.6 A separate chaining hash table

item is found. To perform an *Insert,* we traverse down the appropriate list to check whether the element is already in place (if duplicates are expected, an extra field is usually kept, and this field would be incremented in the event of a match). If the element turns out to be new, it is inserted either at the front of the list or at the end of the list, whichever is easiest. This is an issue most easily addressed while the code is being written. Sometimes new elements are inserted at the front of the list, since it is convenient and also because frequently it happens that recently inserted elements are the most likely to be accessed in the near future.

The type declarations required to implement separate chaining are shown in Figure 5.7. The *ListNode* structure is the same as the linked list declarations of Chapter 3. The hash table structure contains an array of linked lists (and the number of lists in the array), which are dynamically allocated when the table is initialized. The *HashTable* type is just a pointer to this structure.

Notice that the *TheLists* field is actually a pointer to a pointer to a *ListNode* structure. If typedefs and abstraction are not used, this can be quite confusing.

Figure 5.8 shows the initialization function, which uses the same ideas that were seen in the array implementation of stacks. Lines 4 through 6 allocate a hash table structure. If space is available, then *H* will point to a structure containing an integer and a pointer to a list. Line 7 sets the table size to a prime number, and lines 8 through 10 attempt to allocate an array of lists. Since a *List* is defined to be a pointer, the result is an array of pointers.

If our *List* implementation were not using headers, we could stop here. Since our implementation uses headers, we must allocate one header per list and set its *Next* field to *NULL*. This is done in lines 11 through 15. Of course, lines 12 through 15 could be replaced with the statement

```
H->TheLists[ i ] = MakeEmpty( );
```

Although we have not used this option, because in this instance it is preferable to make the code as self-contained as possible, it is certainly worth considering. An

```
#ifndef _HashSep_H

struct ListNode;
typedef struct ListNode *Position;
struct HashTbl;
typedef struct HashTbl *HashTable;

HashTable InitializeTable( int TableSize );
void DestroyTable( HashTable H );
Position Find( ElementType Key, HashTable H );
void Insert( ElementType Key, HashTable H );
ElementType Retrieve( Position P );
/* Routines such as Delete and MakeEmpty are omitted */

#endif  /* _HashSep_H */


/* Place in the implementation file */
struct ListNode
{
    ElementType Element;
    Position    Next;
};

typedef Position List;

/* List *TheList will be an array of lists, allocated later */
/* The lists use headers (for simplicity), */
/* though this wastes space */
struct HashTbl
{
    int TableSize;
    List *TheLists;
};
```

Figure 5.7 Type declaration for separate chaining hash table

inefficiency of our code is that the *malloc* on line 12 is performed *H->TableSize* times. This can be avoided by replacing line 12 with one call to *malloc* before the loop occurs:

```
H->TheLists = malloc (H->TableSize * sizeof (struct ListNode));
```

Line 16 returns *H*.

The call *Find*(*Key*, *H*) will return a pointer to the cell containing *Key*. The code to implement this is shown in Figure 5.9. Notice that lines 2 through 5 are identical to the code to perform a *Find* that is given in Chapter 3. Thus, the list ADT implementation in Chapter 3 could be used here. Remember that if *ElementType*

```
HashTable
InitializeTable( int TableSize )
{
    HashTable H;
    int i;

    if( TableSize < MinTableSize )
    {
        Error( "Table size too small" );
        return NULL;
    }

    /* Allocate table */
    H = malloc( sizeof( struct HashTbl ) );
    if( H == NULL )
        FatalError( "Out of space!!!" );

    H->TableSize = NextPrime( TableSize );

    /* Allocate array of lists */
    H->TheLists = malloc( sizeof( List ) * H->TableSize );
    if( H->TheLists == NULL )
        FatalError( "Out of space!!!" );

    /* Allocate list headers */
    for( i = 0; i < H->TableSize; i++ )
    {
        H->TheLists[ i ] = malloc( sizeof( struct ListNode ) );
        if( H->TheLists[ i ] == NULL )
            FatalError( "Out of space!!!" );
        else
            H->TheLists[ i ]->Next = NULL;
    }

    return H;
}
```

Figure 5.8 Initialization routine for separate chaining hash table

is a string, comparison and assignment must be done with *strcmp* and *strcpy*, respectively.

Next comes the insertion routine. If the item to be inserted is already present, then we do nothing; otherwise we place it at the front of the list (see Fig. 5.10).* The element can be placed anywhere in the list; this is most convenient in our case.

*Since the table in Figure 5.6 was created by inserting at the end of the list, the code in Figure 5.10 will produce a table with the lists in Figure 5.6 reversed.

```
Position
Find( ElementType Key, HashTable H )
{
    Position P;
    List L;

    L = H->TheLists[ Hash( Key, H->TableSize ) ];
    P = L->Next;
    while( P != NULL && P->Element != Key )
                            /* Probably need strcmp!! */
        P = P->Next;
    return P;
}
```

Figure 5.9 *Find* routine for separate chaining hash table

Notice that the code to insert at the front of the list is essentially identical to the code in Chapter 3 that implements a *Push* using linked lists. Again, if the ADTs in Chapter 3 have already been carefully implemented, they can be used here.

The insertion routine coded in Figure 5.10 is somewhat poorly coded, because it computes the hash function twice. Redundant calculations are always bad, so this code should be rewritten if it turns out that the hash routines account for a significant portion of a program's running time.

Figure 5.10 *Insert* routine for separate chaining hash table

```
void
Insert( ElementType Key, HashTable H )
{
    Position Pos, NewCell;
    List L;

    Pos = Find( Key, H );
    if( Pos == NULL )   /* Key is not found */
    {
        NewCell = malloc( sizeof( struct ListNode ) );
        if( NewCell == NULL )
            FatalError( "Out of space!!!" );
        else
        {
            L = H->TheLists[ Hash( Key, H->TableSize ) ];
            NewCell->Next = L->Next;
            NewCell->Element = Key;  /* Probably need strcpy! */
            L->Next = NewCell;
        }
    }
}
```

The deletion routine is a straightforward implementation of deletion in a linked list, so we will not bother with it here. If the repertoire of hash routines does not include deletions, it is probably best to not use headers, since their use would provide no simplification and would waste considerable space. We leave this as an exercise, too.

Any scheme could be used besides linked lists to resolve the collisions; a binary search tree or even another hash table would work, but we expect that if the table is large and the hash function is good, all the lists should be short, so it is not worthwhile to try anything complicated.

We define the load factor, λ, of a hash table to be the ratio of the number of elements in the hash table to the table size. In the example above, $\lambda = 1.0$. The average length of a list is λ. The effort required to perform a search is the constant time required to evaluate the hash function plus the time to traverse the list. In an unsuccessful search, the number of links to traverse is λ (excluding the final *NULL* link) on average. A successful search requires that about $1 + (\lambda/2)$ links be traversed; there is a guarantee that one link must be traversed (since the search is successful), and we also expect to go halfway down a list to find our match. This analysis shows that the table size is not really important, but the load factor is. The general rule for separate chaining hashing is to make the table size about as large as the number of elements expected (in other words, let $\lambda \approx 1$). It is also a good idea, as mentioned before, to keep the table size prime to ensure a good distribution.

5.4. Open Addressing

Separate chaining hashing has the disadvantage of requiring pointers. This tends to slow the algorithm down a bit because of the time required to allocate new cells, and also essentially requires the implementation of a second data structure. *Open addressing hashing* is an alternative to resolving collisions with linked lists. In an open addressing hashing system, if a collision occurs, alternative cells are tried until an empty cell is found. More formally, cells $h_0(X), h_1(X), h_2(X), \ldots$ are tried in succession, where $h_i(X) = (Hash(X) + F(i)) \bmod TableSize$, with $F(0) = 0$. The function, F, is the collision resolution strategy. Because all the data go inside the table, a bigger table is needed for open addressing hashing than for separate chaining hashing. Generally, the load factor should be below $\lambda = 0.5$ for open addressing hashing. We now look at three common collision resolution strategies.

5.4.1. *Linear Probing*

In linear probing, F is a linear function of i, typically $F(i) = i$. This amounts to trying cells sequentially (with wraparound) in search of an empty cell. Figure 5.11 shows the result of inserting keys {89, 18, 49, 58, 69} into a hash table using the same hash function as before and the collision resolution strategy, $F(i) = i$.

The first collision occurs when 49 is inserted; it is put in the next available spot, namely, spot 0, which is open. The key 58 collides with 18, 89, and then 49

	Empty Table	After 89	After 18	After 49	After 58	After 69
0				49	49	49
1					58	58
2						69
3						
4						
5						
6						
7						
8			18	18	18	18
9		89	89	89	89	89

Figure 5.11 Open addressing hash table with linear probing, after each insertion

before an empty cell is found three away. The collision for 69 is handled in a similar manner. As long as the table is big enough, a free cell can always be found, but the time to do so can get quite large. Worse, even if the table is relatively empty, blocks of occupied cells start forming. This effect, known as *primary clustering,* means that any key that hashes into the cluster will require several attempts to resolve the collision, and then it will add to the cluster.

Although we will not perform the calculations here, it can be shown that the expected number of probes using linear probing is roughly $\frac{1}{2}(1 + 1/(1 - \lambda)^2)$ for insertions and unsuccessful searches, and $\frac{1}{2}(1 + 1/(1 - \lambda))$ for successful searches. The calculations are somewhat involved. It is easy to see from the code that insertions and unsuccessful searches require the same number of probes. A moment's thought suggests that, on average, successful searches should take less time than unsuccessful searches.

The corresponding formulas, if clustering is not a problem, are fairly easy to derive. We will assume a very large table and that each probe is independent of the previous probes. These assumptions are satisfied by a *random* collision resolution strategy and are reasonable unless λ is very close to 1. First, we derive the expected number of probes in an unsuccessful search. This is just the expected number of probes until we find an empty cell. Since the fraction of empty cells is $1 - \lambda$, the number of cells we expect to probe is $1/(1 - \lambda)$. The number of probes for a successful search is equal to the number of probes required when the particular element was inserted. When an element is inserted, it is done as a result of an unsuccessful search. Thus, we can use the cost of an unsuccessful search to compute the average cost of a successful search.

The caveat is that λ changes from 0 to its current value, so that earlier insertions are cheaper and should bring the average down. For instance, in the table above, $\lambda = 0.5$, but the cost of accessing 18 is determined when 18 is inserted. At that point, $\lambda = 0.2$. Since 18 was inserted into a relatively empty table, accessing it should be easier than accessing a recently inserted element such as 69. We can estimate the average by using an integral to calculate the mean value of the insertion time, obtaining

$$I(\lambda) = \frac{1}{\lambda}\int_0^\lambda \frac{1}{1-x}dx = \frac{1}{\lambda}\ln\frac{1}{1-\lambda}$$

These formulas are clearly better than the corresponding formulas for linear probing. Clustering is not only a theoretical problem but actually occurs in real implementations. Figure 5.12 compares the performance of linear probing (dashed curves) with what would be expected from more random collision resolution. Successful searches are indicated by an *S*, and unsuccessful searches and insertions are marked with *U* and *I*, respectively.

If $\lambda = 0.75$, then the formula above indicates that 8.5 probes are expected for an insertion in linear probing. If $\lambda = 0.9$, then 50 probes are expected, which is unreasonable. This compares with 4 and 10 probes for the respective load factors if clustering were not a problem. We see from these formulas that linear probing can be a bad idea if the table is expected to be more than half full. If $\lambda = 0.5$, however, only 2.5 probes are required on average for insertion, and only 1.5 probes are required, on average, for a successful search.

Figure 5.12 Number of probes plotted against load factor for linear probing (dashed) and random strategy (*S* is successful search, *U* is unsuccessful search, and *I* is insertion)

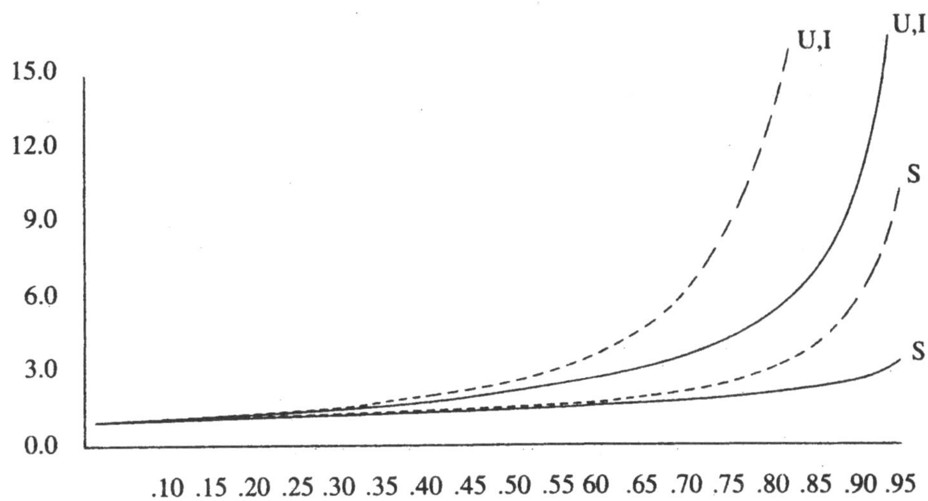

5.4.2. Quadratic Probing

Quadratic probing is a collision resolution method that eliminates the primary clustering problem of linear probing. Quadratic probing is what you would expect—the collision function is quadratic. The popular choice is $F(i) = i^2$. Figure 5.13 shows the resulting open addressing hash table with this collision function on the same input used in the linear probing example.

When 49 collides with 89, the next position attempted is one cell away. This cell is empty, so 49 is placed there. Next 58 collides at position 8. Then the cell one away is tried, but another collision occurs. A vacant cell is found at the next cell tried, which is $2^2 = 4$ away. The key 58 is thus placed in cell 2. The same thing happens for 69.

For linear probing it is a bad idea to let the hash table get nearly full, because performance degrades. For quadratic probing, the situation is even more drastic: There is no guarantee of finding an empty cell once the table gets more than half full, or even before the table gets half full if the table size is not prime. This is because at most half of the table can be used as alternative locations to resolve collisions.

Indeed, we prove now that if the table is half empty and the table size is prime, then we are always guaranteed to be able to insert a new element.

THEOREM 5.1.

If quadratic probing is used, and the table size is prime, then a new element can always be inserted if the table is at least half empty.

Figure 5.13 Open addressing hash table with quadratic probing, after each insertion

	Empty Table	After 89	After 18	After 49	After 58	After 69
0				49	49	49
1						
2					58	58
3						69
4						
5						
6						
7						
8			18	18	18	18
9		89	89	89	89	89

PROOF:

Let the table size, *TableSize,* be an (odd) prime greater than 3. We show that the first $\lfloor TableSize/2 \rfloor$ alternative locations are all distinct. Two of these locations are $h(X) + i^2 (\text{mod } TableSize)$ and $h(X) + j^2 (\text{mod } TableSize)$, where $0 < i, j \leq \lfloor TableSize/2 \rfloor$. Suppose, for the sake of contradiction, that these locations are the same, but $i \neq j$. Then

$$h(X) + i^2 = h(X) + j^2 \qquad (\text{mod } TableSize)$$
$$i^2 = j^2 \qquad (\text{mod } TableSize)$$
$$i^2 - j^2 = 0 \qquad (\text{mod } TableSize)$$
$$(i - j)(i + j) = 0 \qquad (\text{mod } TableSize)$$

Since *TableSize* is prime, it follows that either $(i - j)$ or $(i + j)$ is equal to 0 (mod *TableSize*). Since i and j are distinct, the first option is not possible. Since $0 < i, j < \lfloor TableSize/2 \rfloor$, the second option is also impossible. Thus, the first $\lfloor TableSize/2 \rfloor$ alternative locations are distinct. Since the element to be inserted can also be placed in the cell to which it hashes (if there are no collisions), any element has $\lceil TableSize/2 \rceil$ locations into which it can go. If at most $\lfloor TableSize/2 \rfloor$ positions are taken, then an empty spot can always be found.

If the table is even one more than half full, the insertion could fail (although this is extremely unlikely). Therefore, it is important to keep this in mind. It is also crucial that the table size be prime.* If the table size is not prime, the number of alternative locations can be severely reduced. As an example, if the table size were 16, then the only alternative locations would be at distances 1, 4, or 9 away.

Standard deletion cannot be performed in an open addressing hash table, because the cell might have caused a collision to go past it. For instance, if we remove 89, then virtually all of the remaining *Find*s will fail. Thus, open addressing hash tables require lazy deletion, although in this case there really is no laziness implied.

The type declarations required to implement open addressing hashing are shown in Figure 5.14. Instead of an array of lists, we have an array of hash table entry cells, which, as in separate chaining hashing, are allocated dynamically. Initializing the table (Fig. 5.15) consists of allocating space (lines 1 through 10) and then setting the *Info* field to *Empty* for each cell.

As with separate chaining hashing, *Find*(*Key*, *H*) will return the position of *Key* in the hash table. If *Key* is not present, then *Find* will return the last cell. This cell is where *Key* would be inserted if needed. Further, because it is marked *Empty,* it is easy to tell that the *Find* failed. We assume for convenience that the hash table is at least twice as large as the number of elements in the table, so quadratic resolution will always work. Otherwise, we would need to test i (*CollisionNum*) before line 4.

*If the table size is a prime of the form $4k + 3$, and the quadratic collision resolution strategy $F(i) = \pm i^2$ is used, then the entire table can be probed. The cost is a slightly more complicated routine.

```
#ifndef _HashQuad_H

typedef unsigned int Index;
typedef Index Position;

struct HashTbl;
typedef struct HashTbl *HashTable;

HashTable InitializeTable( int TableSize );
void DestroyTable( HashTable H );
Position Find( ElementType Key, HashTable H );
void Insert( ElementType Key, HashTable H );
ElementType Retrieve( Position P, HashTable H );
HashTable Rehash( HashTable H );
/* Routines such as Delete and MakeEmpty are omitted */

#endif  /* _HashQuad_H */


/* Place in the implementation file */
enum KindOfEntry { Legitimate, Empty, Deleted };

struct HashEntry
{
    ElementType      Element;
    enum KindOfEntry Info;
};

typedef struct HashEntry Cell;

/* Cell *TheCells will be an array of */
/* HashEntry cells, allocated later */
struct HashTbl
{
    int TableSize;
    Cell *TheCells;
};
```

Figure 5.14 Type declaration for open addressing hash tables

In the implementation in Figure 5.16, elements that are marked as deleted count as being in the table. This can cause problems, because the table can get too full prematurely. We shall discuss this item presently.

Lines 4 through 6 represent the fast way of doing quadratic resolution. From the definition of the quadratic resolution function, $F(i) = F(i-1) + 2i - 1$, so the next cell to try can be determined with a multiplication by 2 (really a bit

```
        HashTable
        InitializeTable( int TableSize )
        {
            HashTable H;
            int i;

/* 1*/      if( TableSize < MinTableSize )
            {
/* 2*/          Error( "Table size too small" );
/* 3*/          return NULL;
            }

            /* Allocate table */
/* 4*/      H = malloc( sizeof( struct HashTbl ) );
/* 5*/      if( H == NULL )
/* 6*/          FatalError( "Out of space!!!" );

/* 7*/      H->TableSize = NextPrime( TableSize );

            /* Allocate array of Cells */
/* 8*/      H->TheCells = malloc( sizeof( Cell ) * H->TableSize );
/* 9*/      if( H->TheCells == NULL )
/*10*/          FatalError( "Out of space!!!" );

/*11*/      for( i = 0; i < H->TableSize; i++ )
/*12*/          H->TheCells[ i ].Info = Empty;

/*13*/      return H;
        }
```

Figure 5.15 Routine to initialize open addressing hash table

shift) and a decrement. If the new location is past the array, it can be put back in range by subtracting *TableSize*. This is faster than the obvious method, because it avoids the multiplication and division that seem to be required. An important warning: The order of testing at line 3 is important. Don't switch it!

The final routine is insertion. As with separate chaining hashing, we do nothing if *Key* is already present. It is a simple modification to do something else. Otherwise, we place it at the spot suggested by the *Find* routine. The code is shown in Figure 5.17.

Although quadratic probing eliminates primary clustering, elements that hash to the same position will probe the same alternative cells. This is known as *secondary clustering*. Secondary clustering is a slight theoretical blemish. Simulation results suggest that it generally causes less than an extra half probe per search. The following technique eliminates this, but does so at the cost of extra multiplications and divisions.

```
Position
Find( ElementType Key, HashTable H )
{
    Position CurrentPos;
    int CollisionNum;

    CollisionNum = 0;
    CurrentPos = Hash( Key, H->TableSize );
    while( H->TheCells[ CurrentPos ].Info != Empty &&
           H->TheCells[ CurrentPos ].Element != Key )
                    /* Probably need strcmp!! */
    {
        CurrentPos += 2 * ++CollisionNum - 1;
        if( CurrentPos >= H->TableSize )
            CurrentPos -= H->TableSize;
    }
    return CurrentPos;
}
```

Figure 5.16 *Find* routine for hashing with quadratic probing

5.4.3. *Double Hashing*

The last collision resolution method we will examine is *double hashing*. For double hashing, one popular choice is $F(i) = i \cdot hash_2(X)$. This formula says that we apply a second hash function to X and probe at a distance $hash_2(X), 2hash_2(X), \ldots,$ and so on. A poor choice of $hash_2(X)$ would be disastrous. For instance, the obvious choice $hash_2(X) = X \bmod 9$ would not help if 99 were inserted into the input in the previous examples. Thus, the function must never evaluate to zero. It is also important to make sure all cells can be probed (this is not possible in the example below, because

Figure 5.17 *Insert* routine for hash tables with quadratic probing

```
void
Insert( ElementType Key, HashTable H )
{
    Position Pos;

    Pos = Find( Key, H );
    if( H->TheCells[ Pos ].Info != Legitimate )
    {
                /* OK to insert here */
        H->TheCells[ Pos ].Info = Legitimate;
        H->TheCells[ Pos ].Element = Key;
                /* Probably need strcpy! */
    }
}
```

	Empty Table	After 89	After 18	After 49	After 58	After 69
0						69
1						
2						
3					58	58
4						
5						
6				49	49	49
7						
8			18	18	18	18
9		89	89	89	89	89

Figure 5.18 Open addressing hash table with double hashing, after each insertion

the table size is not prime). A function such as $hash_2(X) = R - (X \bmod R)$, with R a prime smaller than *TableSize*, will work well. If we choose $R = 7$, then Figure 5.18 shows the results of inserting the same keys as before.

The first collision occurs when 49 is inserted. $hash_2(49) = 7 - 0 = 7$, so 49 is inserted in position 6. $hash_2(58) = 7-2 = 5$, so 58 is inserted at location 3. Finally, 69 collides and is inserted at a distance $hash_2(69) = 7 - 6 = 1$ away. If we tried to insert 60 in position 0, we would have a collision. Since $hash_2(60) = 7 - 4 = 3$, we would then try positions 3, 6, 9, and then 2 until an empty spot is found. It is generally possible to find some bad case, but there are not too many here.

As we have said before, the size of our sample hash table is not prime. We have done this for convenience in computing the hash function, but it is worth seeing why it is important to make sure the table size is prime when double hashing is used. If we attempt to insert 23 into the table, it would collide with 58. Since $hash_2(23) = 7 - 2 = 5$, and the table size is 10, we essentially have only one alternative location, and it is already taken. Thus, if the table size is not prime, it is possible to run out of alternative locations prematurely. However, if double hashing is correctly implemented, simulations imply that the expected number of probes is almost the same as for a random collision resolution strategy. This makes double hashing theoretically interesting. Quadratic probing, however, does not require the use of a second hash function and is thus likely to be simpler and faster in practice.

5.5. Rehashing

If the table gets too full, the running time for the operations will start taking too long and *Inserts* might fail for open addressing hashing with quadratic resolution. This can happen if there are too many removals intermixed with insertions. A solution, then, is to build another table that is about twice as big (with an associated new hash function) and scan down the entire original hash table, computing the new hash value for each (nondeleted) element and inserting it in the new table.

As an example, suppose the elements 13, 15, 24, and 6 are inserted into an open addressing hash table of size 7. The hash function is $h(X) = X \bmod 7$. Suppose linear probing is used to resolve collisions. The resulting hash table appears in Figure 5.19.

If 23 is inserted into the table, the resulting table in Fig. 5.20 will be over 70 percent full. Because the table is so full, a new table is created. The size of this table is 17, because this is the first prime that is twice as large as the old table size. The new hash function is then $h(X) = X \bmod 17$. The old table is scanned, and elements 6, 15, 23, 24, and 13 are inserted into the new table. The resulting table appears in Figure 5.21.

This entire operation is called *rehashing*. This is obviously a very expensive operation; the running time is $O(N)$, since there are N elements to rehash and the

Figure 5.19 Open addressing hash table with linear probing with input 13, 15, 6, 24

0	6
1	15
2	
3	24
4	
5	
6	13

Figure 5.20 Open addressing hash table with linear probing after 23 is inserted

0	6
1	15
2	23
3	24
4	
5	
6	13

0	
1	
2	
3	
4	
5	
6	6
7	23
8	24
9	
10	
11	
12	
13	13
14	
15	15
16	

Figure 5.21 Open addressing hash table after rehashing

table size is roughly $2N$, but it is actually not all that bad, because it happens very infrequently. In particular, there must have been $N/2$ *Inserts* prior to the last rehash, so it essentially adds a constant cost to each insertion.* If this data structure is part of the program, the effect is not noticeable. On the other hand, if the hashing is performed as part of an interactive system, then the unfortunate user whose insertion caused a rehash could see a slowdown.

Rehashing can be implemented in several ways with quadratic probing. One alternative is to rehash as soon as the table is half full. The other extreme is to rehash only when an insertion fails. A third, middle-of-the-road strategy is to rehash when the table reaches a certain load factor. Since performance does degrade as the load factor increases, the third strategy, implemented with a good cutoff, could be best.

Rehashing frees the programmer from worrying about the table size and is important because hash tables cannot be made arbitrarily large in complex programs. The exercises ask you to investigate the use of rehashing in conjunction with lazy

*This is why the new table is made twice as large as the old table.

```
        HashTable
        Rehash( HashTable H )
        {
            int i, OldSize;
            Cell *OldCells;

            OldCells = H->TheCells;
            OldSize  = H->TableSize;

            /* Get a new, empty table */
            H = InitializeTable( 2 * OldSize );

            /* Scan through old table, reinserting into new */
            for( i = 0; i < OldSize; i++ )
                if( OldCells[ i ].Info == Legitimate )
                    Insert( OldCells[ i ].Element, H );

            free( OldCells );

            return H;
        }
```

Figure 5.22 Rehashing for open addressing hash tables

deletion. Rehashing can be used in other data structures as well. For instance, if the queue data structure of Chapter 3 became full, we could declare a double-sized array and copy everything over, freeing the original.

Figure 5.22 shows that rehashing is simple to implement.

5.6. Extendible Hashing

Our last topic in this chapter deals with the case where the amount of data is too large to fit in main memory. As we saw in Chapter 4, the main consideration then is the number of disk accesses required to retrieve data.

As before, we assume that at any point we have N records to store; the value of N changes over time. Furthermore, at most M records fit in one disk block. We will use $M = 4$ in this section.

If either open addressing hashing or separate chaining hashing is used, the major problem is that collisions could cause several blocks to be examined during a *Find*, even for a well-distributed hash table. Furthermore, when the table gets too full, an extremely expensive rehashing step must be performed, which requires $O(N)$ disk accesses.

A clever alternative, known as extendible hashing, allows a *Find* to be performed in two disk accesses. Insertions also require few disk accesses.

We recall from Chapter 4 that a B-tree has depth $O(\log_{M/2} N)$. As M increases, the depth of a B-tree decreases. We could in theory choose M to be so large that the depth of the B-tree would be 1. Then any *Find* after the first would take one disk access, since, presumably, the root node could be stored in main memory. The problem with this strategy is that the branching factor is so high that it would take considerable processing to determine which leaf the data was in. If the time to perform this step could be reduced, then we would have a practical scheme. This is exactly the strategy used by extendible hashing.

Let us suppose, for the moment, that our data consists of several six-bit integers. Figure 5.23 shows an extendible hashing scheme for these data. The root of the "tree" contains four pointers determined by the leading two bits of the data. Each leaf has up to $M = 4$ elements. It happens that in each leaf the first two bits are identical; this is indicated by the number in parentheses. To be more formal, D will represent the number of bits used by the root, which is sometimes known as the *directory*. The number of entries in the directory is thus 2^D. d_L is the number of leading bits that all the elements of some leaf L have in common. d_L will depend on the particular leaf, and $d_L \leq D$.

Suppose that we want to insert the key 100100. This would go into the third leaf, but as the third leaf is already full, there is no room. We thus split this leaf into two leaves, which are now determined by the first *three* bits. This requires increasing the directory size to 3. These changes are reflected in Figure 5.24.

Notice that all of the leaves not involved in the split are now pointed to by two adjacent directory entries. Thus, although an entire directory is rewritten, none of the other leaves is actually accessed.

If the key 000000 is now inserted, then the first leaf is split, generating two leaves with $d_L = 3$. Since $D = 3$, the only change required in the directory is the updating of the 000 and 001 pointers. See Figure 5.25.

This very simple strategy provides quick access times for *Insert* and *Find* operations on large databases. There are a few important details we have not considered.

Figure 5.23 Extendible hashing: original data

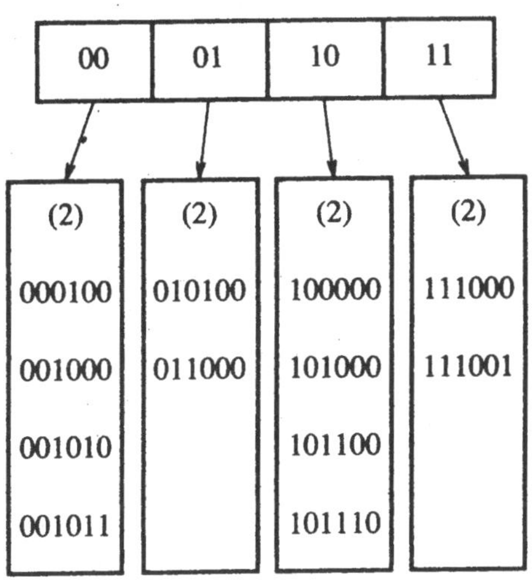

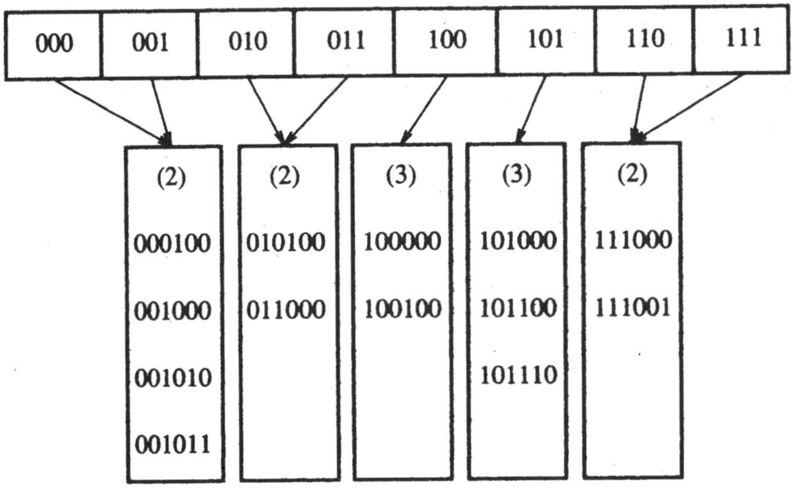

Figure 5.24 Extendible hashing: after insertion of 100100 and directory split

First, it is possible that several directory splits will be required if the elements in a leaf agree in more than $D + 1$ leading bits. For instance, starting at the original example, with $D = 2$, if 111010, 111011, and finally 111100 are inserted, the directory size must be increased to 4 to distinguish between the five keys. This is an easy detail to take care of, but must not be forgotten. Second, there is the possibility of duplicate keys; if there are more than M duplicates, then this algorithm does not work at all. In this case, some other arrangements need to be made.

These possibilities suggest that it is important for the bits to be fairly random. This can be accomplished by hashing the keys into a reasonably long integer—hence the name.

Figure 5.25 Extendible hashing: after insertion of 000000 and leaf split

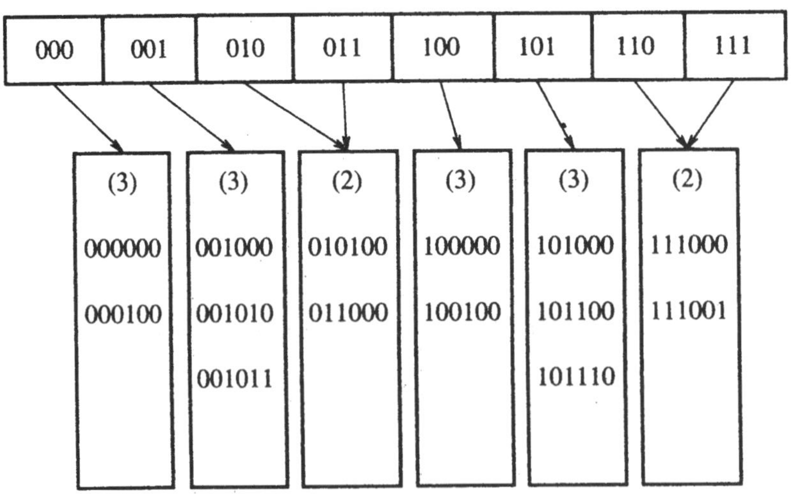

We close by mentioning some of the performance properties of extendible hashing, which are derived after a very difficult analysis. These results are based on the reasonable assumption that the bit patterns are uniformly distributed.

The expected number of leaves is $(N/M)\log_2 e$. Thus the average leaf is $\ln 2 = 0.69$ full. This is the same as for B-trees, which is not entirely surprising, since for both data structures new nodes are created when the $(M + 1)$th entry is added.

The more surprising result is that the expected size of the directory (in other words, 2^D) is $O(N^{1+1/M}/M)$. If M is very small, then the directory can get unduly large. In this case, we can have the leaves contain pointers to the records instead of the actual records, thus increasing the value of M. This adds a second disk access to each *Find* operation in order to maintain a smaller directory. If the directory is too large to fit in main memory, the second disk access would be needed anyway.

Summary

Hash tables can be used to implement the *Insert* and *Find* operations in constant average time. It is especially important to pay attention to details such as load factor when using hash tables, since otherwise the time bounds are not valid. It is also important to choose the hash function carefully when the key is not a short string or integer.

For separate chaining hashing, the load factor should be close to 1, although performance does not significantly degrade unless the load factor becomes very large. For open addressing hashing, the load factor should not exceed 0.5, unless this is completely unavoidable. If linear probing is used, performance degenerates rapidly as the load factor approaches 1. Rehashing can be implemented to allow the table to grow (and shrink), thus maintaining a reasonable load factor. This is important if space is tight and it is not possible just to declare a huge hash table.

Binary search trees can also be used to implement *Insert* and *Find* operations. Although the resulting average time bounds are $O(\log N)$, binary search trees also support routines that require order and are thus more powerful. Using a hash table, it is not possible to find the minimum element. It is not possible to search efficiently for a string unless the exact string is known. A binary search tree could quickly find all items in a certain range; this is not supported by hash tables. Furthermore, the $O(\log N)$ bound is not necessarily that much more than $O(1)$, especially since no multiplications or divisions are required by search trees.

On the other hand, the worst case for hashing generally results from an implementation error, whereas sorted input can make binary trees perform poorly. Balanced search trees are quite expensive to implement, so if no ordering information is required and there is any suspicion that the input might be sorted, then hashing is the data structure of choice.

Hashing applications are abundant. Compilers use hash tables to keep track of declared variables in source code. The data structure is known as a *symbol* table. Hash tables are the ideal application for this problem because only *Insert*s and *Find*s are performed. Identifiers are typically short, so the hash function can be computed quickly.

A hash table is useful for any graph theory problem where the nodes have real names instead of numbers. Here, as the input is read, vertices are assigned integers from 1 onward by order of appearance. Again, the input is likely to have large groups of alphabetized entries. For example, the vertices could be computers. Then if one particular installation lists its computers as *ibm1, ibm2, ibm3,...*, there could be a dramatic effect on efficiency if a search tree is used.

A third common use of hash tables is in programs that play games. As the program searches through different lines of play, it keeps track of positions it has seen by computing a hash function based on the position (and storing its move for that position). If the same position reoccurs, usually by a simple transposition of moves, the program can avoid expensive recomputation. This general feature of all game-playing programs is known as the *transposition table.*

Yet another use of hashing is in on-line spelling checkers. If misspelling detection (as opposed to correction) is important, an entire dictionary can be prehashed and words can be checked in constant time. Hash tables are well suited for this, because it is not important to alphabetize words; printing out misspellings in the order they occurred in the document is certainly acceptable.

We close this chapter by returning to the word puzzle problem of Chapter 1. If the second algorithm described in Chapter 1 is used, and we assume that the maximum word size is some small constant, then the time to read in the dictionary containing W words and put it in a hash table is $O(W)$. This time is likely to be dominated by the disk I/O and not the hashing routines. The rest of the algorithm would test for the presence of a word for each ordered quadruple (*row, column, orientation, number of characters*). As each lookup would be $O(1)$, and there are only a constant number of orientations (8) and characters per word, the running time of this phase would be $O(R \cdot C)$. The total running time would be $O(R \cdot C + W)$, which is a distinct improvement over the original $O(R \cdot C \cdot W)$. We could make further optimizations, which would decrease the running time in practice; these are described in the exercises.

Exercises

5.1 Given input {4371, 1323, 6173, 4199, 4344, 9679, 1989} and a hash function $h(X) = X \pmod{10}$, show the resulting:

a. Separate chaining hash table.

b. Open addressing hash table using linear probing.

c. Open addressing hash table using quadratic probing.

d. Open addressing hash table with second hash function $h_2(X) = 7 - (X \bmod 7)$.

5.2 Show the result of rehashing the hash tables in Exercise 5.1.

5.3 Write a program to compute the number of collisions required in a long random sequence of insertions using linear probing, quadratic probing, and double hashing.

5.4 A large number of deletions in a separate chaining hash table can cause the table to be fairly empty, which wastes space. In this case, we can rehash to a table half as large. Assume that we rehash to a larger table when there are twice as many elements as the table size. How empty should the table be before we rehash to a smaller table?

5.5 An alternative collision resolution strategy is to define a sequence, $F(i) = r_i$, where $r_0 = 0$ and $r_1, r_2, \ldots, r_N$ is a random permutation of the first N integers (each integer appears exactly once).

a. Prove that under this strategy, if the table is not full, then the collision can always be resolved.

b. Would this strategy be expected to eliminate clustering?

c. If the load factor of the table is λ, what is the expected time to perform an insert?

d. If the load factor of the table is λ, what is the expected time for a successful search?

e. Give an efficient algorithm (theoretically as well as practically) to generate the random sequence. Explain why the rules for choosing P are important.

5.6 What are the advantages and disadvantages of the various collision resolution strategies?

5.7 Write a program to implement the following strategy for multiplying two sparse polynomials P_1, P_2 of size M and N, respectively. Each polynomial is represented as a linked list with cells consisting of a coefficient, an exponent, and a *Next* pointer (Exercise 3.7). We multiply each term in P_1 by a term in P_2 for a total of MN operations. One method is to sort these terms and combine like terms, but this requires sorting MN records, which could be expensive, especially in small-memory environments. Alternatively, we could merge terms as they are computed and then sort the result.

a. Write a program to implement the alternative strategy.

b. If the output polynomial has about $O(M + N)$ terms, what is the running time of both methods?

5.8 A spelling checker reads an input file and prints out all words not in some on-line dictionary. Suppose the dictionary contains 30,000 words and the file is large, so that the algorithm can make only one pass through the input file. A simple strategy is to read the dictionary into a hash table and look for each input word as it is read. Assuming that an average word is seven characters and that it is possible to store words of length L in $L + 1$ bytes (so space waste is not much of a consideration), and assuming an open addressing table, how much space does this require?

5.9 If memory is limited and the entire dictionary cannot be stored in a hash table, we can still get an efficient algorithm that almost always works. We declare an array *Table* of bits (initialized to zeros) from 0 to *TableSize* − 1. As we read in a word, we set *Table*[*Hash*(*Word*)] = 1. Which of the following is true?

a. If a word hashes to a location with value 0, the word is not in the dictionary.

b. If a word hashes to a location with value 1, then the word is in the dictionary.

Suppose we choose *TableSize* = 300,007.

c. How much memory does this require?

d. What is the probability of an error in this algorithm?

e. A typical document might have about three actual misspellings per page of 500 words. Is this algorithm usable?

*5.10 Describe a procedure that avoids initializing a hash table (at the expense of memory).

5.11 Suppose we want to find the first occurrence of a string $P_1P_2 \cdots P_k$ in a long input string $A_1A_2 \cdots A_N$. We can solve this problem by hashing the pattern string, obtaining a hash value H_P, and comparing this value with the hash value formed from $A_1A_2 \cdots A_k, A_2A_3 \cdots A_{k+1}, A_3A_4 \cdots A_{k+2}$, and so on until $A_{N-k+1}A_{N-k+2} \cdots A_N$. If we have a match of hash values, we compare the strings character by character to verify the match. We return the position (in A) if the strings actually do match, and we continue in the unlikely event that the match is false.

*a. Show that if the hash value of $A_iA_{i+1} \cdots A_{i+k-1}$ is known, then the hash value of $A_{i+1}A_{i+2} \cdots A_{i+k}$ can be computed in constant time.

b. Show that the running time is $O(k + N)$ plus the time spent refuting false matches.

*c. Show that the expected number of false matches is negligible.

d. Write a program to implement this algorithm.

**e. Describe an algorithm that runs in $O(k + N)$ worst-case time.

**f. Describe an algorithm that runs in $O(N/k)$ average time.

5.12 A BASIC program consists of a series of statements numbered in ascending order. Control is passed by use of a *goto* or *gosub* and a statement number. Write a program that reads in a legal BASIC program and renumbers the statements so that the first starts at number F and each statement has a number D higher than the previous statement. You may assume an upper limit of N statements, but the statement numbers in the input might be as large as a 32-bit integer. Your program must run in linear time.

5.13 a. Implement the word puzzle program using the algorithm described at the end of the chapter.

b. We can get a big speed increase by storing, in addition to each word W, all of W's prefixes. (If one of W's prefixes is another word in the dictionary, it is stored as a real word.) Although this may seem to increase the size of the hash table drastically, it does not, because many words have the same prefixes. When a scan is performed in a particular direction, if the word that is looked up is not even in the hash table as a prefix, then the scan in that direction can be terminated early. Use this idea to write an improved program to solve the word puzzle.

c. If we are willing to sacrifice the sanctity of the hash table ADT, we can speed up the program in part (b) by noting that if, for example, we have just computed the hash function for "excel," we do not need to compute the hash function for "excels" from scratch. Adjust your hash function so that it can take advantage of its previous calculation.

d. In Chapter 2, we suggested using binary search. Incorporate the idea of using prefixes into your binary search algorithm. The modification should be simple. Which algorithm is faster?

5.14 Show the result of inserting the keys 10111101, 00000010, 10011011, 10111110, 01111111, 01010001, 10010110, 00001011, 11001111, 10011110, 11011011, 00101011, 01100001, 11110000, 01101111 into an initially empty extendible hashing data structure with $M = 4$.

5.15 Write a program to implement extendible hashing. If the table is small enough to fit in main memory, how does its performance compare with separate chaining and open addressing hashing?

References

Despite the apparent simplicity of hashing, much of the analysis is quite difficult and there are still many unresolved questions. There are also many interesting theoretical issues, which generally attempt to make it unlikely that the worst-case possibilities of hashing arise.

An early paper on hashing is [17]. A wealth of information on the subject, including an analysis of hashing with linear probing, can be found in [11]. An excellent survey on the subject is [14]; [15] contains suggestions, and pitfalls, for choosing hash functions. Precise analytic and simulation results for all of the methods described in this chapter can be found in [8].

An analysis of double hashing can be found in [9] and [13]. Yet another collision resolution scheme is coalesced hashing, described in [18]. Yao [20] has shown that uniform hashing, in which no clustering exists, is optimal with respect to cost of a successful search.

If the input keys are known in advance, then perfect hash functions, which do not allow collisions, exist [2], [7]. Some more complicated hashing schemes, for which the worst case depends not on the particular input but on random numbers chosen by the algorithm, appear in [3] and [4].

Extendible hashing appears in [5], with analysis in [6] and [19].

One method of implementing Exercise 5.5 is described in [16]. Exercise 5.11 (a–d) is from [10]. Part (e) is from [12], and part (f) is from [1].

1. R. S. Boyer and J. S. Moore, "A Fast String Searching Algorithm," *Communications of the ACM*, 20 (1977), 762–772.
2. J. L. Carter and M. N. Wegman, "Universal Classes of Hash Functions," *Journal of Computer and System Sciences*, 18 (1979), 143–154.

3. M. Dietzfelbinger, A. R. Karlin, K. Melhorn, F. Meyer auf der Heide, H. Rohnert, and R. E. Tarjan, "Dynamic Perfect Hashing: Upper and Lower Bounds," *SIAM Journal on Computing,* 23 (1994), 738–761.
4. R. J. Enbody and H. C. Du, "Dynamic Hashing Schemes," *Computing Surveys,* 20 (1988), 85–113.
5. R. Fagin, J. Nievergelt, N. Pippenger, and H. R. Strong, "Extendible Hashing—A Fast Access Method for Dynamic Files," *ACM Transactions on Database Systems,* 4 (1979), 315–344.
6. P. Flajolet, "On the Performance Evaluation of Extendible Hashing and Trie Searching," *Acta Informatica,* 20 (1983), 345–369.
7. M. L. Fredman, J. Komlos, and E. Szemeredi, "Storing a Sparse Table with O(1) Worst Case Access Time," *Journal of the ACM,* 31 (1984), 538–544.
8. G. H. Gonnet and R. Baeza-Yates, *Handbook of Algorithms and Data Structures,* 2nd ed., Addison-Wesley, Reading, Mass., 1991.
9. L. J. Guibas and E. Szemeredi, "The Analysis of Double Hashing," *Journal of Computer and System Sciences,* 16 (1978), 226–274.
10. R. M. Karp and M. O. Rabin, "Efficient Randomized Pattern-Matching Algorithms," *Aiken Computer Laboratory Report TR-31-81,* Harvard University, Cambridge, Mass., 1981.
11. D. E. Knuth, *The Art of Computer Programming, Vol 3: Sorting and Searching,* Addison-Wesley, Reading, Mass., 1973.
12. D. E. Knuth, J. H. Morris, and V. R. Pratt, "Fast Pattern Matching in Strings," *SIAM Journal on Computing,* 6 (1977), 323–350.
13. G. Lueker and M. Molodowitch, "More Analysis of Double Hashing," *Proceedings of the Twentieth ACM Symposium on Theory of Computing* (1988), 354–359.
14. W. D. Maurer and T. G. Lewis, "Hash Table Methods," *Computing Surveys,* 7 (1975), 5–20.
15. B. J. McKenzie, R. Harries, and T. Bell, "Selecting a Hashing Algorithm," *Software—Practice and Experience,* 20 (1990), 209–224.
16. R. Morris, "Scatter Storage Techniques," *Communications of the ACM,* 11 (1968), 38–44.
17. W. W. Peterson, "Addressing for Random Access Storage," *IBM Journal of Research and Development,* 1 (1957), 130–146.
18. J. S. Vitter, "Implementations for Coalesced Hashing," *Communications of the ACM,* 25 (1982), 911–926.
19. A. C. Yao, "A Note on The Analysis of Extendible Hashing," *Information Processing Letters,* 11 (1980), 84–86.
20. A. C. Yao, "Uniform Hashing Is Optimal," *Journal of the ACM,* 32 (1985), 687–693.

CHAPTER 6

Priority Queues (Heaps)

Although jobs sent to a printer are generally placed on a queue, this might not always be the best thing to do. For instance, one job might be particularly important, so it might be desirable to allow that job to be run as soon as the printer is available. Conversely, if, when the printer becomes available, there are several 1-page jobs and one 100-page job, it might be reasonable to make the long job go last, even if it is not the last job submitted. (Unfortunately, most systems do not do this, which can be particularly annoying at times.)

Similarly, in a multiuser environment, the operating system scheduler must decide which of several processes to run. Generally a process is allowed to run only for a fixed period of time. One algorithm uses a queue. Jobs are initially placed at the end of the queue. The scheduler will repeatedly take the first job on the queue, run it until either it finishes or its time limit is up, and place it at the end of the queue if it does not finish. This strategy is generally not appropriate, because very short jobs will seem to take a long time because of the wait involved to run. Generally, it is important that short jobs finish as fast as possible, so these jobs should have precedence over jobs that have already been running. Furthermore, some jobs that are not short are still very important and should also have precedence.

This particular application seems to require a special kind of queue, known as a *priority queue*. In this chapter, we will discuss

- Efficient implementation of the priority queue ADT.
- Uses of priority queues.
- Advanced implementations of priority queues.

The data structures we will see are among the most elegant in computer science.

6.1. Model

A priority queue is a data structure that allows at least the following two operations: *Insert,* which does the obvious thing; and *DeleteMin*, which finds, returns, and removes the minimum element in the priority queue. The *Insert* operation is the equivalent of *Enqueue*, and *DeleteMin* is the priority queue equivalent of the queue's *Dequeue* operation. The *DeleteMin* function also alters its input. Current

Figure 6.1 Basic model of a priority queue

thinking in the software engineering community suggests that this is no longer a good idea. However, we will continue to use this function for historical reasons; many programmers expect *DeleteMin* to operate this way.

As with most data structures, it is sometimes possible to add other operations, but these are extensions and not part of the basic model depicted in Figure 6.1.

Priority queues have many applications besides operating systems. In Chapter 7, we will see how priority queues are used for external sorting. Priority queues are also important in the implementation of *greedy algorithms*, which operate by repeatedly finding a minimum; we will see specific examples in Chapters 9 and 10. In this chapter we will see a use of priority queues in discrete event simulation.

6.2. Simple Implementations

There are several obvious ways to implement a priority queue. We could use a simple linked list, performing insertions at the front in $O(1)$ and traversing the list, which requires $O(N)$ time, to delete the minimum. Alternatively, we could insist that the list be kept always sorted; this makes insertions expensive ($O(N)$) and *DeleteMins* cheap ($O(1)$). The former is probably the better idea of the two, based on the fact that there are never more *DeleteMins* than insertions.

Another way of implementing priority queues would be to use a binary search tree. This gives an $O(\log N)$ average running time for both operations. This is true in spite of the fact that although the insertions are random, the deletions are not. Recall that the only element we ever delete is the minimum. Repeatedly removing a node that is in the left subtree would seem to hurt the balance of the tree by making the right subtree heavy. However, the right subtree is random. In the worst case, where the *DeleteMins* have depleted the left subtree, the right subtree would have at most twice as many elements as it should. This adds only a small constant to its expected depth. Notice that the bound can be made into a worst-case bound by using a balanced tree; this protects one against bad insertion sequences.

Using a search tree could be overkill because it supports a host of operations that are not required. The basic data structure we will use will not require pointers and will support both operations in $O(\log N)$ worst-case time. Insertion will actually take constant time on average, and our implementation will allow building a priority queue of N items in linear time, if no deletions intervene. We will then discuss how to implement priority queues to support efficient merging. This additional operation seems to complicate matters a bit and apparently requires the use of pointers.

6.3. Binary Heap

The implementation we will use is known as a *binary heap*. Its use is so common for priority queue implementations that when the word *heap* is used without a qualifier, it is generally assumed to be referring to this implementation of the data structure. In this section, we will refer to binary heaps merely as *heaps*. Like binary search trees, heaps have two properties, namely, a structure property and a heap order property. As with AVL trees, an operation on a heap can destroy one of the properties, so a heap operation must not terminate until all heap properties are in order. This turns out to be simple to do.

6.3.1. *Structure Property*

A heap is a binary tree that is completely filled, with the possible exception of the bottom level, which is filled from left to right. Such a tree is known as a *complete binary tree*. Figure 6.2 shows an example.

It is easy to show that a complete binary tree of height h has between 2^h and $2^{h+1} - 1$ nodes. This implies that the height of a complete binary tree is $\lfloor \log N \rfloor$, which is clearly $O(\log N)$.

An important observation is that because a complete binary tree is so regular, it can be represented in an array and no pointers are necessary. The array in Figure 6.3 corresponds to the heap in Figure 6.2.

For any element in array position i, the left child is in position $2i$, the right child is in the cell after the left child $(2i + 1)$, and the parent is in position $\lfloor i/2 \rfloor$. Thus not only are pointers not required, but the operations required to traverse the tree are

Figure 6.2 A complete binary tree

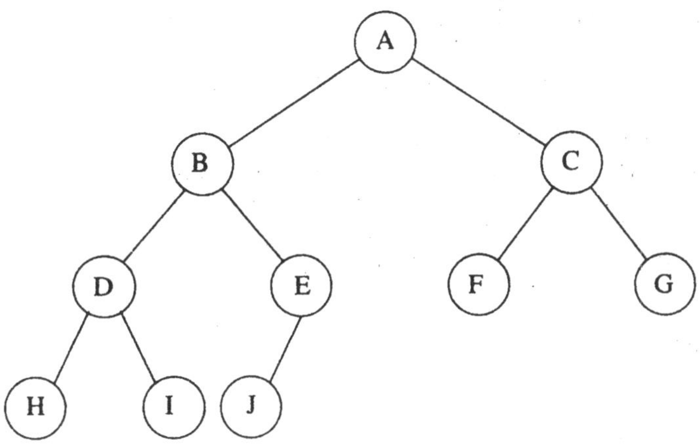

Figure 6.3 Array implementation of complete binary tree

	A	B	C	D	E	F	G	H	I	J			
0	1	2	3	4	5	6	7	8	9	10	11	12	13

extremely simple and likely to be very fast on most computers. The only problem with this implementation is that an estimate of the maximum heap size is required in advance, but typically this is not a problem. In Figure 6.3, the limit on the heap size is 13 elements. The array has a position 0; more on this later.

A heap data structure will, then, consist of an array (of whatever type the key is) and an integer representing the maximum and current heap sizes. Figure 6.4 shows a typical priority queue declaration. Notice the similarity to the stack declaration in Figure 3.47. Figure 6.4a creates an empty heap. Line 11 will be explained later.

Throughout this chapter, we shall draw the heaps as trees, with the implication that an actual implementation will use simple arrays.

6.3.2. *Heap Order Property*

The property that allows operations to be performed quickly is the *heap order* property. Since we want to be able to find the minimum quickly, it makes sense that the smallest element should be at the root. If we consider that any subtree should also be a heap, then any node should be smaller than all of its descendants.

Applying this logic, we arrive at the heap order property. In a heap, for every node X, the key in the parent of X is smaller than (or equal to) the key in X, with

Figure 6.4 Declaration for priority queue

```
#ifndef _BinHeap_H

struct HeapStruct;
typedef struct HeapStruct *PriorityQueue;

PriorityQueue Initialize( int MaxElements );
void Destroy( PriorityQueue H );
void MakeEmpty( PriorityQueue H );
void Insert( ElementType X, PriorityQueue H );
ElementType DeleteMin( PriorityQueue H );
ElementType FindMin( PriorityQueue H );
int IsEmpty( PriorityQueue H );
int IsFull( PriorityQueue H );

#endif


/* Place in implementation file */
struct HeapStruct
{
    int Capacity;
    int Size;
    ElementType *Elements;
};
```

the exception of the root (which has no parent).* In Figure 6.5 the tree on the left is a heap, but the tree on the right is not (the dashed line shows the violation of heap order). As usual, we will assume that the keys are integers, although they could be arbitrarily complex.

Figure 6.4 Declaration for priority queue

```
        PriorityQueue
        Initialize( int MaxElements )
        {
            PriorityQueue H;

            if( MaxElements < MinPQSize )
                Error( "Priority queue size is too small" );

            H = malloc( sizeof( struct HeapStruct ) );
            if( H == NULL )
                FatalError( "Out of space!!!" );

            /* Allocate the array plus one extra for sentinel */
            H->Elements = malloc( ( MaxElements + 1 )
                                    * sizeof( ElementType ) );
            if( H->Elements == NULL )
                FatalError( "Out of space!!!" );

            H->Capacity = MaxElements;
            H->Size = 0;
            H->Elements[ 0 ] = MinData;

            return H;
        }
```

Figure 6.5 Two complete trees (only the left tree is a heap)

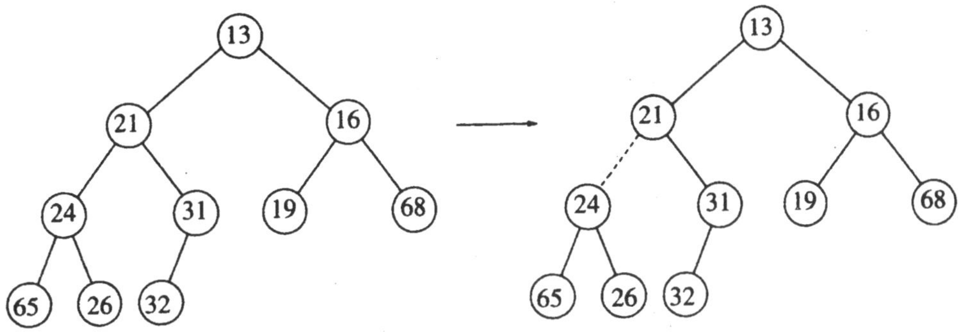

*Analogously, we can declare a (*max*) heap, which enables us to efficiently find and remove the maximum element, by changing the heap order property. Thus, a priority queue can be used to find *either* a minimum or a maximum, but this needs to be decided ahead of time.

By the heap order property, the minimum element can always be found at the root. Thus, we get the extra operation, *FindMin,* in constant time.

6.3.3. *Basic Heap Operations*

It is easy (both conceptually and practically) to perform the two required operations. All the work involves ensuring that the heap order property is maintained.

Insert

To insert an element X into the heap, we create a hole in the next available location, since otherwise the tree will not be complete. If X can be placed in the hole without violating heap order, then we do so and are done. Otherwise we slide the element that is in the hole's parent node into the hole, thus bubbling the hole up toward the root. We continue this process until X can be placed in the hole. Figure 6.6 shows that to insert 14, we create a hole in the next available heap location. Inserting 14 in the hole would violate the heap order property, so 31 is slid down into the hole. This strategy is continued in Figure 6.7 until the correct location for 14 is found.

Figure 6.6 Attempt to insert 14: creating the hole, and bubbling the hole up

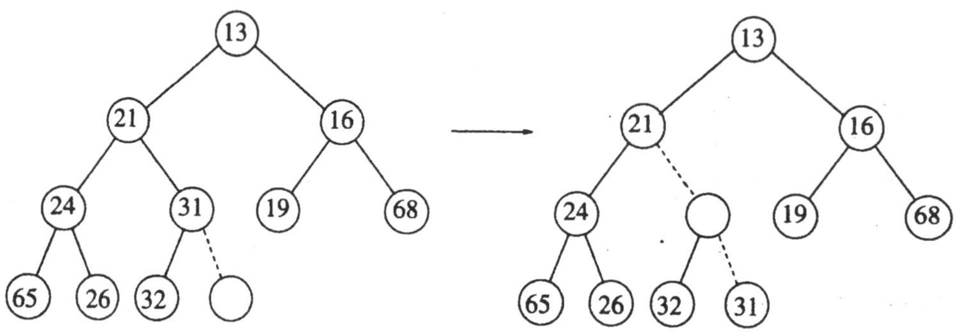

Figure 6.7 The remaining two steps to insert 14 in previous heap

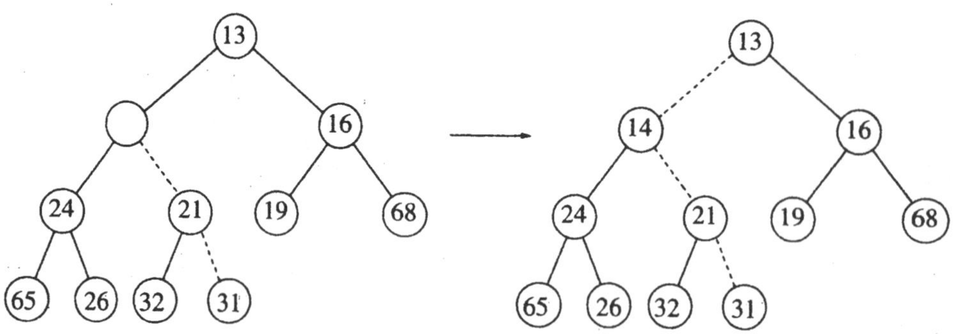

```
/* H->Element[ 0 ] is a sentinel */

void
Insert( ElementType X, PriorityQueue H )
{
    int i;

    if( IsFull( H ) )
    {
        Error( "Priority queue is full" );
        return;
    }

    for( i = ++H->Size; H->Elements[ i / 2 ] > X; i /= 2 )
        H->Elements[ i ] = H->Elements[ i / 2 ];
    H->Elements[ i ] = X;
}
```

Figure 6.8 Procedure to insert into a binary key

This general strategy is known as a *percolate up*; the new element is percolated up the heap until the correct location is found. Insertion is easily implemented with the code shown in Figure 6.8.

We could have implemented the percolation in the *Insert* routine by performing repeated swaps until the correct order was established, but a swap requires three assignment statements. If an element is percolated up d levels, the number of assignments performed by the swaps would be $3d$. Our method uses $d + 1$ assignments.

If the element to be inserted is the new minimum, it will be pushed all the way to the top. At some point, i will be 1 and we will want to break out of the *while* loop. We could do this with an explicit test, but we have chosen to put a very small value in position 0 in order to make the *while* loop terminate. This value must be guaranteed to be smaller than (or equal to) any element in the heap; it is known as a *sentinel*. This idea is similar to the use of header nodes in linked lists. By adding a dummy piece of information, we avoid a test that is executed once per loop iteration, thus saving some time.

The time to do the insertion could be as much as $O(\log N)$, if the element to be inserted is the new minimum and is percolated all the way to the root. On average, the percolation terminates early; it has been shown that 2.607 comparisons are required on average to perform an insert, so the average *Insert* moves an element up 1.607 levels.

DeleteMin

DeleteMins are handled in a similar manner as insertions. Finding the minimum is easy; the hard part is removing it. When the minimum is removed, a hole is created at the root. Since the heap now becomes one smaller, it follows that the last element X in the heap must move somewhere in the heap. If X can be placed in the hole, then we are done. This is unlikely, so we slide the smaller of the hole's children into the hole, thus pushing the hole down one level. We repeat this step until X can be placed in the hole. Thus, our action is to place X in its correct spot along a path from the root containing ***minimum*** children.

In Figure 6.9 the left figure shows a heap prior to the *DeleteMin*. After 13 is removed, we must now try to place 31 in the heap. The value 31 cannot be placed in the hole, because this would violate heap order. Thus, we place the smaller child (14) in the hole, sliding the hole down one level (see Fig. 6.10). We repeat this again, placing 19 into the hole and creating a new hole one level deeper. We then place

Figure 6.9 Creation of the hole at the root

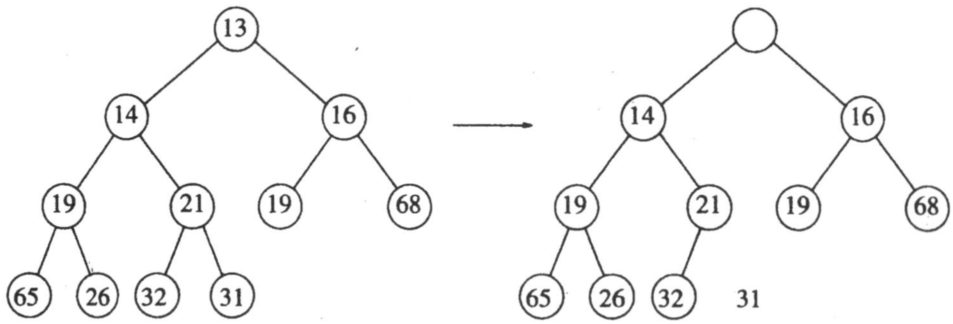

Figure 6.10 Next two steps in *DeleteMin*

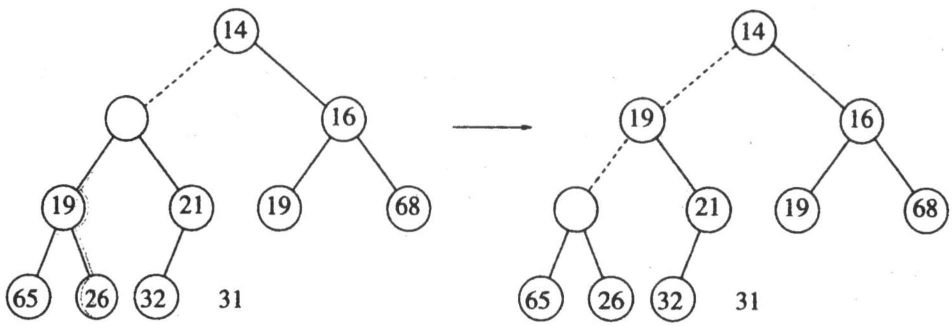

Figure 6.11 Last two steps in *DeleteMin*

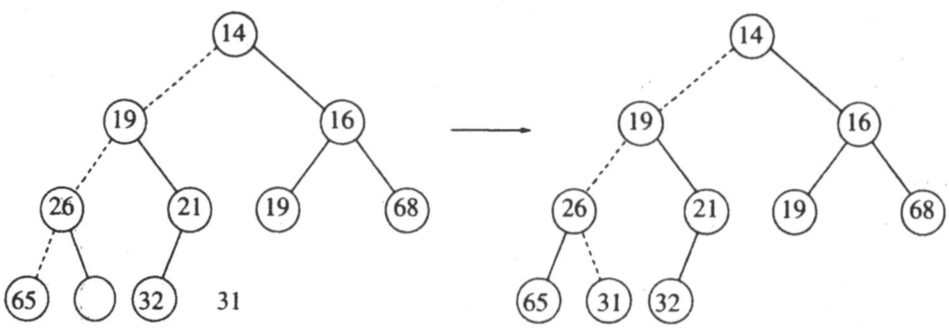

26 in the hole and create a new hole on the bottom level. Finally, we are able to place 31 in the hole (Fig. 6.11). This general strategy is known as a *percolate down*. We use the same technique as in the *Insert* routine to avoid the use of swaps in this routine.

A frequent implementation error in heaps occurs when there are an even number of elements in the heap, and the one node that has only one child is encountered. You must make sure not to assume that there are always two children, so this usually involves an extra test. In the code depicted in Figure 6.12, we've done this test at line 8. One extremely tricky solution is always to ensure that your algorithm *thinks* every node has two children. Do this by placing a sentinel, of value higher than any in the heap, at the spot after the heap ends, at the start of each *percolate down* when the heap size is even. You should think very carefully before attempting this, and you must put in a prominent comment if you do use this technique. Although this

Figure 6.12 Function to perform *DeleteMin* in a binary heap

```
        ElementType
        DeleteMin( PriorityQueue H )
        {
            int i, Child;
            ElementType MinElement, LastElement;

/* 1*/      if( IsEmpty( H ) )
            {
/* 2*/          Error( "Priority queue is empty" );
/* 3*/          return H->Elements[ 0 ];
            }
/* 4*/      MinElement = H->Elements[ 1 ];
/* 5*/      LastElement = H->Elements[ H->Size-- ];

/* 6*/      for( i = 1; i * 2 <= H->Size; i = Child )
            {
                /* Find smaller child */
/* 7*/          Child = i * 2;
/* 8*/          if( Child != H->Size && H->Elements[ Child + 1 ]
/* 9*/                                 < H->Elements[ Child ] )
/*10*/              Child++;

                /* Percolate one level */
/*11*/          if( LastElement > H->Elements[ Child ] )
/*12*/              H->Elements[ i ] = H->Elements[ Child ];
                else
/*13*/              break;
            }
/*14*/      H->Elements[ i ] = LastElement;
/*15*/      return MinElement;
        }
```

eliminates the need to test for the presence of a right child, you cannot eliminate the requirement that you test when you reach the bottom, because this would require a sentinel for every leaf.

The worst-case running time for this operation is $O(\log N)$. On average, the element that is placed at the root is percolated almost to the bottom of the heap (which is the level it came from), so the average running time is $O(\log N)$.

6.3.4. Other Heap Operations

Notice that although finding the minimum can be performed in constant time, a heap designed to find the minimum element (also known as a (*min*)heap) is of no help whatsoever in finding the maximum element. In fact, a heap has very little ordering information, so there is no way to find any particular key without a linear scan through the entire heap. To see this, consider the large heap structure (the elements are not shown) in Figure 6.13, where we see that the only information known about the maximum element is that it is at one of the leaves. Half the elements, though, are contained in leaves, so this is practically useless information. For this reason, if it is important to know where elements are, some other data structure, such as a hash table, must be used in addition to the heap. (Recall that the model does not allow looking inside the heap.)

If we assume that the position of every element is known by some other method, then several other operations become cheap. The three operations below all run in logarithmic worst-case time.

DecreaseKey

The *DecreaseKey*(P, Δ, H) operation lowers the value of the key at position P by a positive amount Δ. Since this might violate the heap order, it must be fixed by a *percolate up*. This operation could be useful to system administrators: They can make their programs run with highest priority.

Figure 6.13 A very large complete binary tree

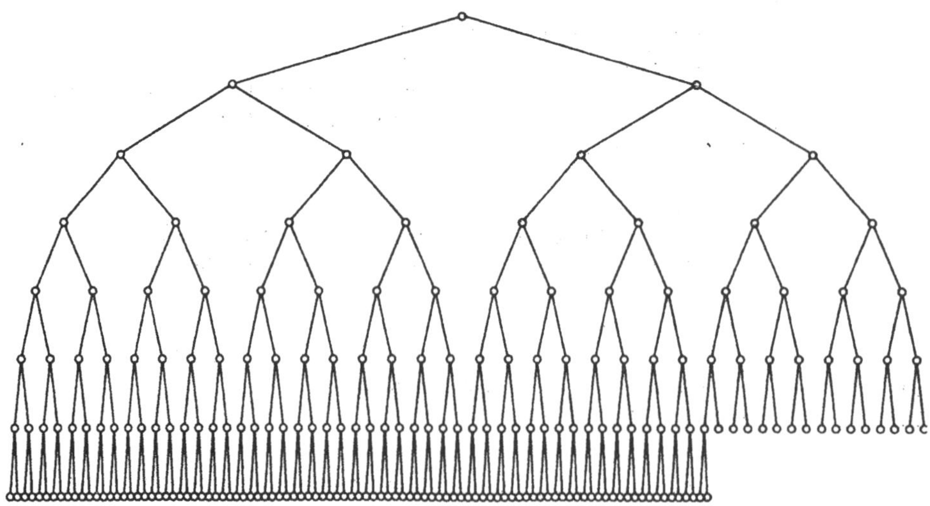

IncreaseKey

The *IncreaseKey*(P, Δ, H) operation increases the value of the key at position P by a positive amount Δ. This is done with a *percolate down*. Many schedulers automatically drop the priority of a process that is consuming excessive CPU time.

Delete

The *Delete*(P, H) operation removes the node at position P from the heap. This is done by first performing *DecreaseKey*(P, ∞, H) and then performing *DeleteMin*(H). When a process is terminated by a user (instead of finishing normally), it must be removed from the priority queue.

BuildHeap

The *BuildHeap*(H) operation takes as input N keys and places them into an empty heap. Obviously, this can be done with N successive *Inserts*. Since each *Insert* will take $O(1)$ average and $O(\log N)$ worst-case time, the total running time of this algorithm would be $O(N)$ average but $O(N \log N)$ worst-case. Since this is a special instruction and there are no other operations intervening, and we already know that the instruction can be performed in linear average time, it is reasonable to expect that with reasonable care a linear time bound can be guaranteed.

The general algorithm is to place the N keys into the tree in any order, maintaining the structure property. Then, if *PercolateDown*(i) percolates down from node i, perform the algorithm in Figure 6.14 to create a heap-ordered tree.

The first tree in Figure 6.15 is the unordered tree. The seven remaining trees in Figures 6.15 through 6.18 show the result of each of the seven *PercolateDowns*. Each dashed line corresponds to two comparisons: one to find the smaller child and one to compare the smaller child with the node. Notice that there are only 10 dashed

Figure 6.14 Sketch of *BuildHeap*

```
for( i = N / 2; i > 0; i-- )
    PercolateDown( i );
```

Figure 6.15 Left: initial heap; right: after *PercolateDown*(7)

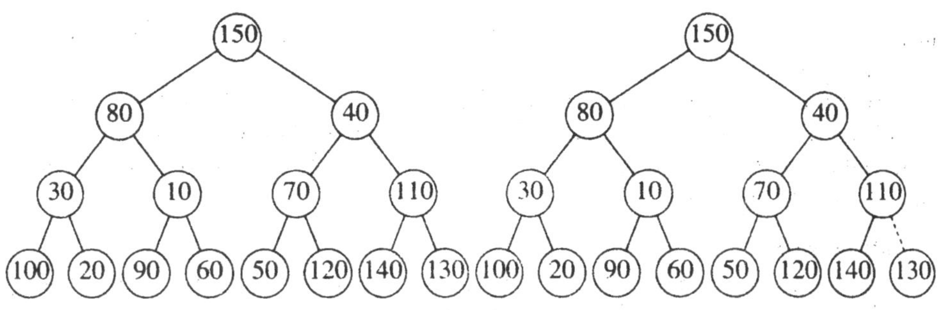

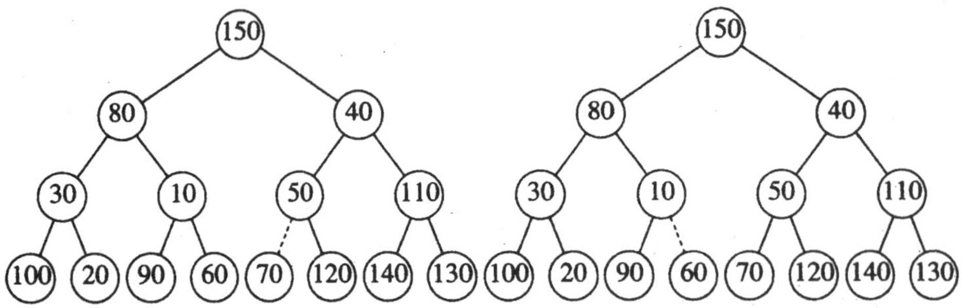

Figure 6.16 Left: after *PercolateDown*(6); right: after *PercolateDown*(5)

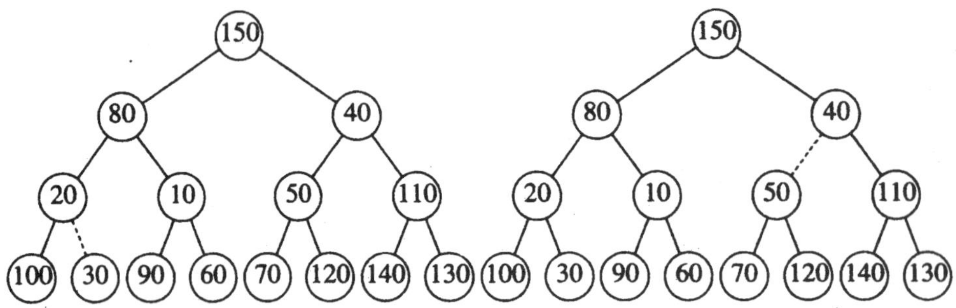

Figure 6.17 Left: after *PercolateDown*(4); right: after *PercolateDown*(3)

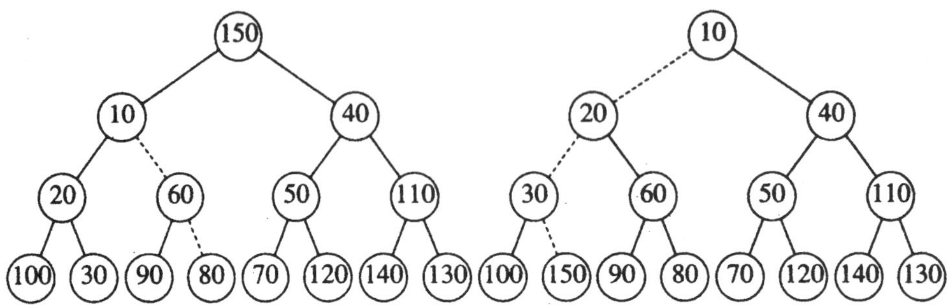

Figure 6.18 Left: after *PercolateDown*(2); right: after *PercolateDown*(1)

lines in the entire algorithm (there could have been an 11th—where?) corresponding to 20 comparisons.

To bound the running time of *BuildHeap*, we must bound the number of dashed lines. This can be done by computing the sum of the heights of all the nodes in the heap, which is the maximum number of dashed lines. What we would like to show is that this sum is $O(N)$.

THEOREM 6.1.

For the perfect binary tree of height h containing $2^{h+1} - 1$ nodes, the sum of the heights of the nodes is $2^{h+1} - 1 - (h + 1)$.

PROOF:

It is easy to see that this tree consists of 1 node at height h, 2 nodes at height $h - 1$, 2^2 nodes at height $h - 2$, and in general 2^i nodes at height $h - i$. The sum of the heights of all the nodes is then

$$S = \sum_{i=0}^{h} 2^i (h - i)$$

$$= h+2(h - 1) + 4(h - 2) + 8(h - 3) + 16(h - 4) + \cdots + 2^{h-1}(1) \quad (6.1)$$

Multiplying by 2 gives the equation

$$2S = 2h + 4(h - 1) + 8(h - 2) + 16(h - 3) + \cdots + 2^h(1) \quad (6.2)$$

We subtract these two equations and obtain Equation (6.3). We find that certain terms almost cancel. For instance, we have $2h - 2(h - 1) = 2$, $4(h - 1) - 4(h - 2) = 4$, and so on. The last term in Equation (6.2), 2^h, does not appear in Equation (6.1); thus, it appears in Equation (6.3). The first term in Equation (6.1), h, does not appear in Equation (6.2); thus, $-h$ appears in Equation (6.3). We obtain

$$S = -h + 2 + 4 + 8 + \cdots + 2^{h-1} + 2^h = (2^{h+1} - 1) - (h + 1) \quad (6.3)$$

which proves the theorem.

A complete tree is not a perfect binary tree, but the result we have obtained is an upper bound on the the sum of the heights of the nodes in a complete tree. Since a complete tree has between 2^h and 2^{h+1} nodes, this theorem implies that this sum is $O(N)$, where N is the number of nodes.

Although the result we have obtained is sufficient to show that *BuildHeap* is linear, the bound on the sum of the heights is not as strong as possible. For a complete tree with $N = 2^h$ nodes, the bound we have obtained is roughly $2N$. The sum of the heights can be shown by induction to be $N - b(N)$, where $b(N)$ is the number of 1s in the binary representation of N.

6.4. Applications of Priority Queues

We have already mentioned how priority queues are used in operating systems design. In Chapter 9, we will see how priority queues are used to implement several graph algorithms efficiently. Here we will show how to use priority queues to obtain solutions to two problems.

6.4.1. *The Selection Problem*

The first problem we will examine is the *selection problem* from Chapter 1. Recall that the input is a list of N elements, which can be totally ordered, and an integer k. The selection problem is to find the kth largest element.

Two algorithms were given in Chapter 1, but neither is very efficient. The first algorithm, which we shall call algorithm 1A, is to read the elements into an array and sort them, returning the appropriate element. Assuming a simple sorting algorithm, the running time is $O(N^2)$. The alternative algorithm, 1B, is to read k elements into an array and sort them. The smallest of these is in the kth position. We process the remaining elements one by one. As an element arrives, it is compared with the kth element in the array. If it is larger, then the kth element is removed, and the new element is placed in the correct place among the remaining $k-1$ elements. When the algorithm ends, the element in the kth position is the answer. The running time is $O(N \cdot k)$ (why?). If $k = \lceil N/2 \rceil$, then both algorithms are $O(N^2)$. Notice that for any k, we can solve the symmetric problem of finding the $(N-k+1)$th smallest element, so $k = \lceil N/2 \rceil$ is really the hardest case for these algorithms. This also happens to be the most interesting case, since this value of k is known as the *median*.

We give two algorithms here, both of which run in $O(N \log N)$ in the extreme case of $k = \lceil N/2 \rceil$, which is a distinct improvement.

Algorithm 6A

For simplicity, we assume that we are interested in finding the kth *smallest* element. The algorithm is simple. We read the N elements into an array. We then apply the *BuildHeap* algorithm to this array. Finally, we perform k *DeleteMin* operations. The last element extracted from the heap is our answer. It should be clear that by changing the heap order property, we could solve the original problem of finding the kth *largest* element.

The correctness of the algorithm should be clear. The worst-case timing is $O(N)$ to construct the heap, if *BuildHeap* is used, and $O(\log N)$ for each *DeleteMin*. Since there are k *DeleteMins*, we obtain a total running time of $O(N + k \log N)$. If $k = O(N/\log N)$, then the running time is dominated by the *BuildHeap* operation and is $O(N)$. For larger values of k, the running time is $O(k \log N)$. If $k = \lceil N/2 \rceil$, then the running time is $\Theta(N \log N)$.

Notice that if we run this program for $k = N$ and record the values as they leave the heap, we will have essentially sorted the input file in $O(N \log N)$ time. In Chapter 7, we will refine this idea to obtain a fast sorting algorithm known as *heapsort*.

Algorithm 6B

For the second algorithm, we return to the original problem and find the kth *largest* element. We use the idea from algorithm 1B. At any point in time we will maintain a set S of the k largest elements. After the first k elements are read, when a new element is read it is compared with the kth largest element, which we denote by S_k. Notice that S_k is the smallest element in S. If the new element is larger, then it replaces S_k in S. S will then have a new smallest element, which may or may not be the newly added element. At the end of the input, we find the smallest element in S and return it as the answer.

This is essentially the same algorithm described in Chapter 1. Here, however, we will use a heap to implement S. The first k elements are placed into the heap in total time $O(k)$ with a call to *BuildHeap*. The time to process each of the remaining elements is $O(1)$, to test if the element goes into S, plus $O(\log k)$, to delete S_k and insert the new element if this is necessary. Thus, the total time is $O(k+(N-k)\log k) = O(N \log k)$. This algorithm also gives a bound of $\Theta(N \log N)$ for finding the median.

In Chapter 7, we will see how to solve this problem in $O(N)$ average time. In Chapter 10, we will see an elegant, albeit impractical, algorithm to solve this problem in $O(N)$ worst-case time.

6.4.2. *Event Simulation*

In Section 3.4.3, we described an important queuing problem. Recall that we have a system, such as a bank, where customers arrive and wait on a line until one of k tellers is available. Customer arrival is governed by a probability distribution function, as is the service time (the amount of time to be served once a teller is available). We are interested in statistics such as how long on average a customer has to wait or how long the line might be.

With certain probability distributions and values of k, these answers can be computed exactly. However, as k gets larger, the analysis becomes considerably more difficult, so it is appealing to use a computer to simulate the operation of the bank. In this way, the bank officers can determine how many tellers are needed to ensure reasonably smooth service.

A simulation consists of processing events. The two events here are (a) a customer arriving and (b) a customer departing, thus freeing up a teller.

We can use the probability functions to generate an input stream consisting of ordered pairs of arrival time and service time for each customer, sorted by arrival time. We do not need to use the exact time of day. Rather, we can use a quantum unit, which we will refer to as a *tick*.

One way to do this simulation is to start a simulation clock at zero ticks. We then advance the clock one tick at a time, checking to see if there is an event. If there is, then we process the event(s) and compile statistics. When there are no customers left in the input stream and all the tellers are free, then the simulation is over.

The problem with this simulation strategy is that its running time does not depend on the number of customers or events (there are two events per customer), but instead depends on the number of ticks, which is not really part of the input. To see why this is important, suppose we changed the clock units to milliticks and multiplied all the times in the input by 1,000. The result would be that the simulation would take 1,000 times longer!

The key to avoiding this problem is to advance the clock to the next event time at each stage. This is conceptually easy to do. At any point, the next event that can occur is either (a) the next customer in the input file arrives or (b) one of the customers at a teller leaves. Since all the times when the events will happen are available, we just need to find the event that happens nearest in the future and process that event.

If the event is a departure, processing includes gathering statistics for the departing customer and checking the line (queue) to see whether there is another customer waiting. If so, we add that customer, process whatever statistics are required, compute the time when that customer will leave, and add that departure to the set of events waiting to happen.

If the event is an arrival, we check for an available teller. If there is none, we place the arrival on the line (queue); otherwise we give the customer a teller, compute the customer's departure time, and add the departure to the set of events waiting to happen.

The waiting line for customers can be implemented as a queue. Since we need to find the event *nearest* in the future, it is appropriate that the set of departures waiting to happen be organized in a priority queue. The next event is thus the next arrival or next departure (whichever is sooner); both are easily available.

It is then straightforward, although possibly time-consuming, to write the simulation routines. If there are C customers (and thus $2C$ events) and k tellers, then the running time *of the simulation* would be $O(C \log(k + 1))$* because computing and processing each event takes $O(\log H)$, where $H = k + 1$ is the size of the heap.

6.5. *d*-Heaps

Binary heaps are so simple that they are almost always used when priority queues are needed. A simple generalization is a *d-heap,* which is exactly like a binary heap except that all nodes have d children (thus, a binary heap is a 2-heap).

Figure 6.19 shows a 3-heap. Notice that a d-heap is much shallower than a binary heap, improving the running time of *Inserts* to $O(\log_d N)$. However, for large d, the *DeleteMin* operation is more expensive, because even though the tree is shallower, the minimum of d children must be found, which takes $d - 1$ comparisons using a standard algorithm. This raises the time for this operation to $O(d \log_d N)$. If d is a constant, both running times are, of course, $O(\log N)$. Although an array can still be used, the multiplications and divisions to find children and parents are now by d, which, unless d is a power of 2, seriously increases the running time, because we can no longer implement division by a bit shift. d-heaps are interesting in theory, because there are many algorithms where the number of insertions is much greater than the number of *DeleteMins* (and thus a theoretical speedup is possible). They are also of interest when the priority queue is too large to fit entirely in main memory. In this case, a d-heap can be advantageous in much the same way as B-trees. Finally, there is evidence suggesting that 4-heaps may outperform binary heaps in practice.

Figure 6.19 A d-heap

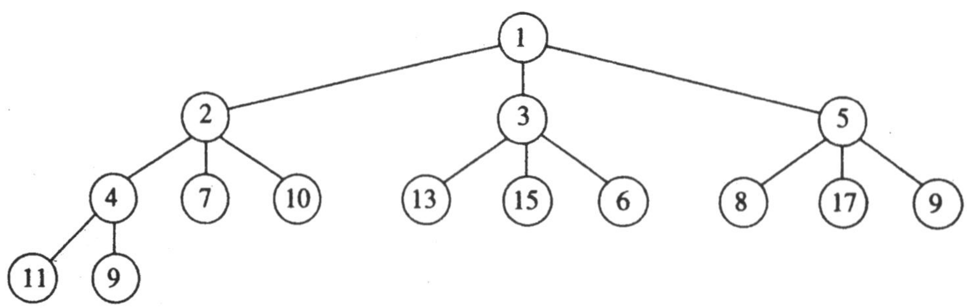

*We use $O(C \log(k + 1))$ instead of $O(C \log k)$ to avoid confusion for the $k = 1$ case.

The most glaring weakness of the heap implementation, aside from the inability to perform *Finds*, is that combining two heaps into one is a hard operation. This extra operation is known as a *Merge*. There are quite a few ways of implementing heaps so that the running time of a *Merge* is $O(\log N)$. We will now discuss three data structures, of various complexity, that support the *Merge* operation efficiently. We will defer any complicated analysis until Chapter 11.

6.6. Leftist Heaps

It seems difficult to design a data structure that efficiently supports merging (that is, processes a *Merge* in $O(N)$ time) and uses only an array, as in a binary heap. The reason for this is that merging would seem to require copying one array into another, which would take $\Theta(N)$ time for equal-sized heaps. For this reason, all the advanced data structures that support efficient merging require the use of pointers. In practice, we can expect that this will make all the other operations slower; pointer manipulation is generally more time-consuming than multiplication and division by 2.

Like a binary heap, a *leftist heap* has both a structural property and an ordering property. Indeed, a leftist heap, like virtually all heaps used, has the same heap order property we have already seen. Furthermore, a leftist heap is also a binary tree. The only difference between a leftist heap and a binary heap is that leftist heaps are not perfectly balanced, but actually attempt to be very unbalanced.

6.6.1. Leftist Heap Property

We define the *null path length, Npl(X),* of any node X to be the length of the shortest path from X to a node without two children. Thus, the *Npl* of a node with zero or one child is 0, while $Npl\ (NULL) = -1$. In the tree in Figure 6.20, the null path lengths are indicated inside the tree nodes.

Notice that the null path length of any node is 1 more than the minimum of the null path lengths of its children. This applies to nodes with less than two children because the null path length of *NULL* is -1.

The leftist heap property is that for every node X in the heap, the null path length of the left child is at least as large as that of the right child. This property is

Figure 6.20 Null path lengths for two trees; only the left tree is leftist

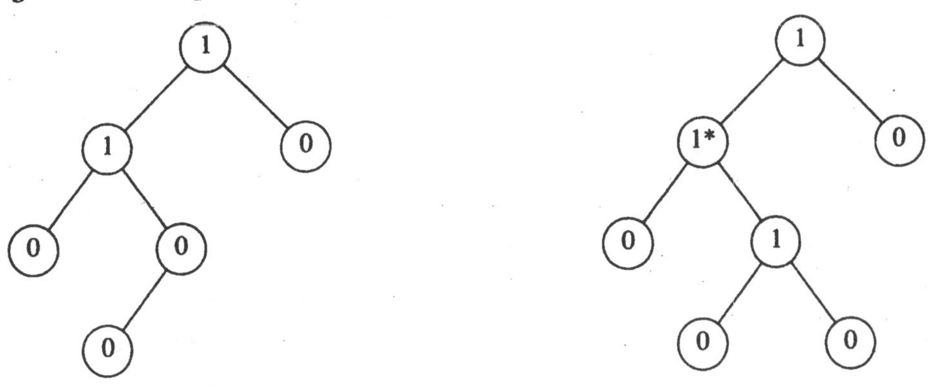

satisfied by only one of the trees in Figure 6.20, namely, the tree on the left. This property actually goes out of its way to ensure that the tree is unbalanced, because it clearly biases the tree to get deep toward the left. Indeed, a tree consisting of a long path of left nodes is possible (and actually preferable to facilitate merging)—hence the name *leftist heap*.

Because leftist heaps tend to have deep left paths, it follows that the right path ought to be short. Indeed, the right path down a leftist heap is as short as any in the heap. Otherwise, there would be a path that goes through some node X and takes the left child. Then X would violate the leftist property.

THEOREM 6.2.

A leftist tree with r nodes on the right path must have at least $2^r - 1$ *nodes.*

PROOF:

The proof is by induction. If $r = 1$, there must be at least one tree node. Otherwise, suppose that the theorem is true for $1, 2, \ldots, r$. Consider a leftist tree with $r + 1$ nodes on the right path. Then the root has a right subtree with r nodes on the right path, and a left subtree with at least r nodes on the right path (otherwise it would not be leftist). Applying the inductive hypothesis to these subtrees yields a minimum of $2^r - 1$ nodes in each subtree. This plus the root gives at least $2^{r+1} - 1$ nodes in the tree, proving the theorem.

From this theorem, it follows immediately that a leftist tree of N nodes has a right path containing at most $\lfloor \log(N + 1) \rfloor$ nodes. The general idea for the leftist heap operations is to perform all the work on the right path, which is guaranteed to be short. The only tricky part is that performing *Inserts* and *Merges* on the right path could destroy the leftist heap property. It turns out to be extremely easy to restore the property.

6.6.2. *Leftist Heap Operations*

The fundamental operation on leftist heaps is merging. Notice that insertion is merely a special case of merging, since we may view an insertion as a *Merge* of a one-node heap with a larger heap. We will first give a simple recursive solution and then show how this might be done nonrecursively. Our input is the two leftist heaps, H_1 and H_2, in Figure 6.21. You should check that these heaps really are leftist. Notice that the smallest elements are at the roots. In addition to space for the data and left and right pointers, each cell will have an entry that indicates the null path length.

If either of the two heaps is empty, then we can return the other heap. Otherwise, to merge the two heaps, we compare their roots. First, we recursively merge the heap with the larger root with the right subheap of the heap with the smaller root. In our example, this means we recursively merge H_2 with the subheap of H_1 rooted at 8, obtaining the heap in Figure 6.22.

Since this tree is formed recursively, and we have not yet finished the description of the algorithm, we cannot at this point show how this heap was obtained. However, it is reasonable to assume that the resulting tree is a leftist heap, because it was obtained via a recursive step. This is much like the inductive hypothesis in a proof by induction. Since we can handle the base case (which occurs when one tree is empty), we can assume that the recursive step works as long as we can finish the merge; this

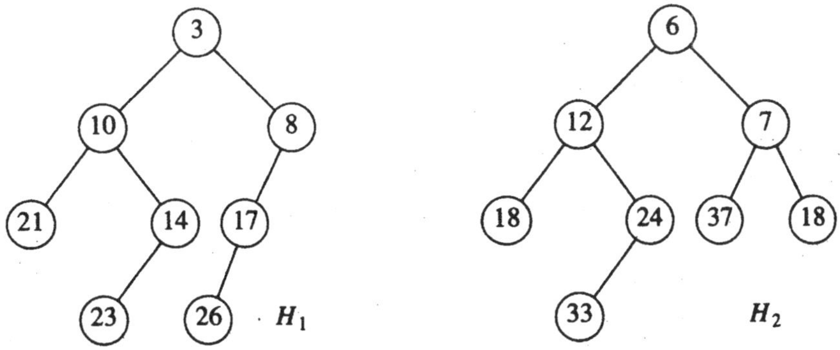

Figure 6.21 Two leftist heaps H_1 and H_2

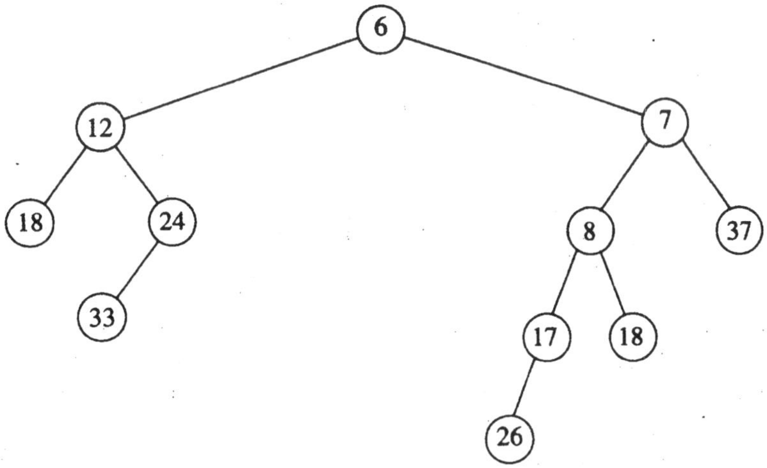

Figure 6.22 Result of merging H_2 with H_1's right subheap

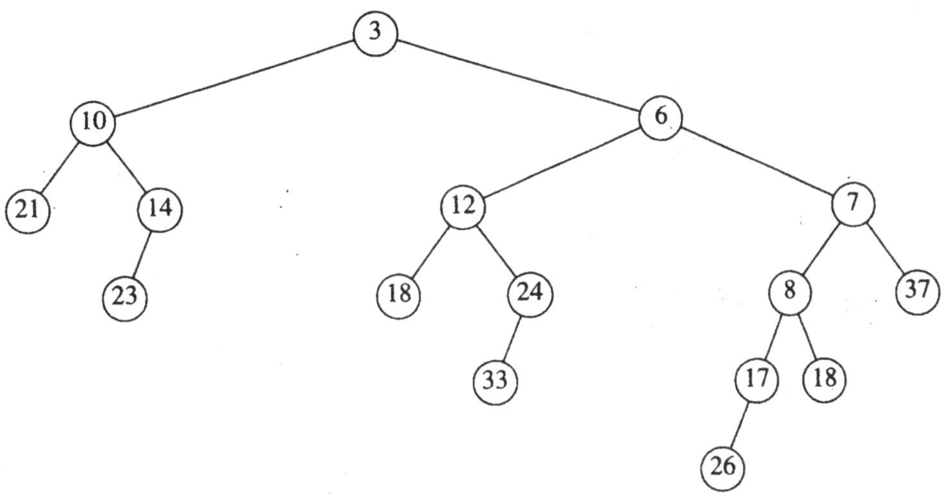

Figure 6.23 Result of attaching leftist heap of previous figure as H_1's right child

is rule 3 of recursion, which we discussed in Chapter 1. We now make this new heap the right child of the root of H_1 (see Figure 6.23).

Although the resulting heap satisfies the heap order property, it is not leftist because the left subtree of the root has a null path length of 1 whereas the right subtree has a null path length of 2. Thus, the leftist property is violated at the root. However, it is easy to see that the remainder of the tree must be leftist. The right subtree of the root is leftist, because of the recursive step. The left subtree of the root has not been changed, so it too must still be leftist. Thus, we need only to fix the root. We can make the entire tree leftist by merely swapping the root's left and right children (Figure 6.24) and updating the null path length—the new null path length is 1 plus the null path length of the new right child—completing the *Merge*. Notice that if the null path length is not updated, then all null path lengths will be 0, and the heap will not be leftist but merely random. In this case, the algorithm will work, but the time bound we will claim will no longer be valid.

The description of the algorithm translates directly into code. The type definition (Fig. 6.25) is the same as the binary tree, except that it is augmented with the *Npl* (null path length) field. We have seen in Chapter 4 that when an element is inserted into an empty binary tree, the pointer to the root will need to change. The easiest way to implement this is to have the insertion routine return a pointer to the new tree. Unfortunately, this will make the leftist heap *Insert* incompatible with the binary heap *Insert* (which does not return anything). The last line in Figure 6.25 represents one way out of this quandary. The leftist heap insertion routine which returns the new tree will be called *Insert1*; the *Insert* macro will make an insertion compatible with binary heaps. Using macros this way may not be the best or safest course, but the alternative, declaring a *PriorityQueue* as a pointer to a *TreeNode*, will flood the code with extra asterisks.*

Because *Insert* is a macro and is textually substituted by the preprocessor, any routine that calls *Insert* must be able to see the macro definition. Figure 6.25 would

Figure 6.24 Result of swapping children of H_1's root

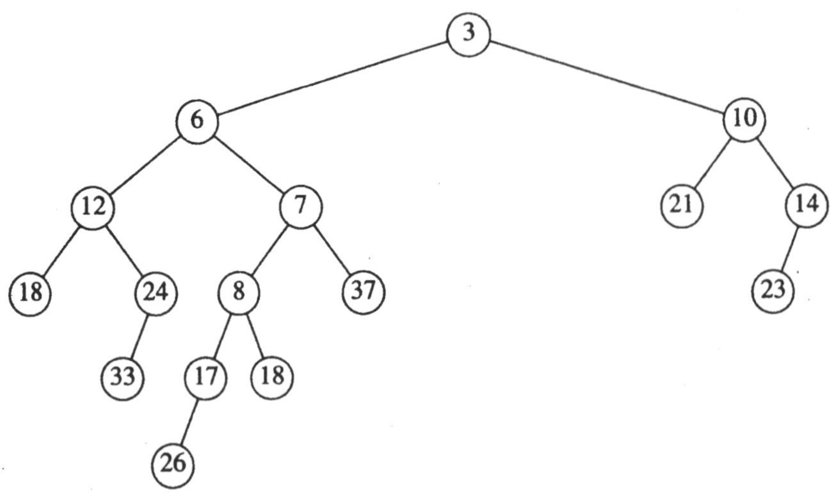

*Another possibility is to accept the incompatible interfaces as a necessary evil.

```
#ifndef _LeftHeap_H

struct TreeNode;
typedef struct TreeNode *PriorityQueue;

/* Minimal set of priority queue operations */
/* Note that nodes will be shared among several */
/* leftist heaps after a merge; the user must */
/* make sure to not use the old leftist heaps */

PriorityQueue Initialize( void );
ElementType FindMin( PriorityQueue H );
int IsEmpty( PriorityQueue H );
PriorityQueue Merge( PriorityQueue H1, PriorityQueue H2 );

#define Insert( X, H ) ( H = Insert1( ( X ), H ) )
/* DeleteMin macro is left as an exercise */

PriorityQueue Insert1( ElementType X, PriorityQueue H );
PriorityQueue DeleteMin1( PriorityQueue H );

#endif


/* Place in implementation file */
struct TreeNode
{
    ElementType   Element;
    PriorityQueue Left;
    PriorityQueue Right;
    int           Npl;
};
```

Figure 6.25 Leftist heap type declarations

typically be a header file, so placing the macro declaration there is the only reasonable course. As we will see later, *DeleteMin* also needs to be written as a macro.

The routine to merge (Fig. 6.26) is a driver designed to remove special cases and ensure that H_1 has the smaller root. The actual merging is performed in *Merge1* (Fig. 6.27). Note that the original leftist heaps should never be used again; changes in them will affect the merged result.

The time to perform the merge is proportional to the sum of the length of the right paths, because constant work is performed at each node visited during the recursive calls. Thus we obtain an $O(\log N)$ time bound to merge two leftist heaps. We can also perform this operation nonrecursively by essentially performing two passes. In the first pass, we create a new tree by merging the right paths of both heaps. To do this, we arrange the nodes on the right paths of H_1 and H_2 in sorted order, keeping their respective left children. In our example, the new right path is

```
PriorityQueue
Merge( PriorityQueue H1, PriorityQueue H2 )
{
    if( H1 == NULL )
        return H2;
    if( H2 == NULL )
        return H1;
    if( H1->Element < H2->Element )
        return Merge1( H1, H2 );
    else
        return Merge1( H2, H1 );
}
```

Figure 6.26 Driving routine for merging leftist heaps

```
static PriorityQueue
Merge1( PriorityQueue H1, PriorityQueue H2 )
{
    if( H1->Left == NULL )  /* Single node */
        H1->Left = H2;      /* H1->Right is already NULL,
                               H1->Npl is already 0 */
    else
    {
        H1->Right = Merge( H1->Right, H2 );
        if( H1->Left->Npl < H1->Right->Npl )
            SwapChildren( H1 );

        H1->Npl = H1->Right->Npl + 1;
    }
    return H1;
}
```

Figure 6.27 Actual routine to merge leftist heaps

3, 6, 7, 8, 18 and the resulting tree is shown in Figure 6.28. A second pass is made up the heap, and child swaps are performed at nodes that violate the leftist heap property. In Figure 6.28, there is a swap at nodes 7 and 3, and the same tree as before is obtained. The nonrecursive version is simpler to visualize but harder to code. We leave it to the reader to show that the recursive and nonrecursive procedures do the same thing.

As mentioned above, we can carry out insertions by making the item to be inserted a one-node heap and performing a *Merge*. To perform a *DeleteMin*, we merely destroy the root, creating two heaps, which can then be merged. Thus, the time to perform a *DeleteMin* is $O(\log N)$. These two routines are coded in Figure 6.29 and Figure 6.30. *DeleteMin* can be written as a macro that calls *DeleteMin1* and *FindMin*. This is left as an exercise to the reader.

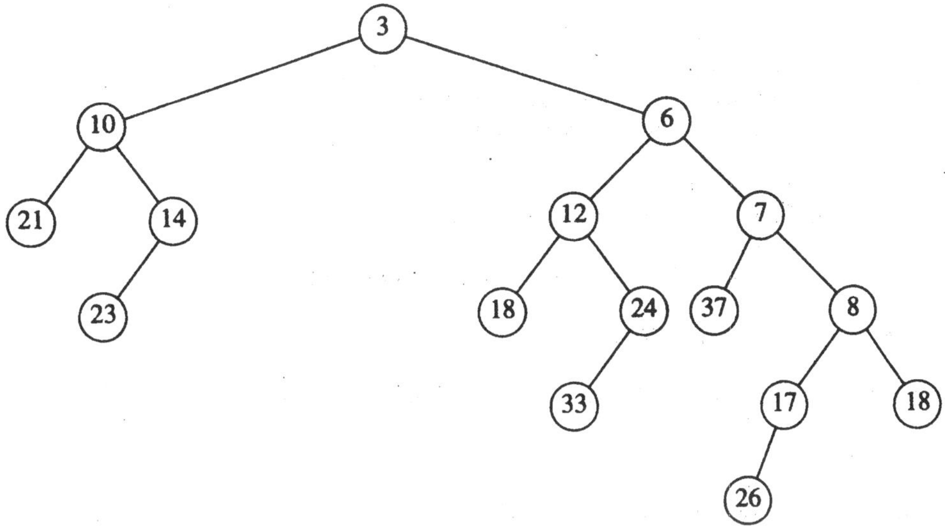

Figure 6.28 Result of merging right paths of H_1 and H_2

```
        PriorityQueue
        Insert1( ElementType X, PriorityQueue H )
        {
            PriorityQueue SingleNode;

/* 1*/      SingleNode = malloc( sizeof( struct TreeNode ) );
/* 2*/      if( SingleNode == NULL )
/* 3*/          FatalError( "Out of space!!!" );
            else
            {
/* 4*/          SingleNode->Element = X; SingleNode->Npl = 0;
/* 5*/          SingleNode->Left = SingleNode->Right = NULL;
/* 6*/          H = Merge( SingleNode, H );
            }
/* 7*/      return H;
        }
```

Figure 6.29 Insertion routine for leftist heaps

Finally, we can build a leftist heap in $O(N)$ time by building a binary heap (obviously using a pointer implementation). Although a binary heap is clearly leftist, this is not necessarily the best solution, because the heap we obtain is the worst possible leftist heap. Furthermore, traversing the tree in reverse-level order is not as easy with pointers. The *BuildHeap* effect can be obtained by recursively building the left and right subtrees and then percolating the root down. The exercises contain an alternative solution.

```
/* DeleteMin1 returns the new tree; */
/* To get the minimum, use FindMin */
/* This is for convenience */

PriorityQueue
DeleteMin1( PriorityQueue H )
{
    PriorityQueue LeftHeap, RightHeap;

    if( IsEmpty( H ) )
    {
        Error( "Priority queue is empty" );
        return H;
    }

    LeftHeap = H->Left;
    RightHeap = H->Right;
    free( H );
    return Merge( LeftHeap, RightHeap );
}
```

Figure 6.30 *DeleteMin* routine for leftist heaps

6.7. Skew Heaps

A *skew heap* is a self-adjusting version of a leftist heap that is incredibly simple to implement. The relationship of skew heaps to leftist heaps is analogous to the relation between splay trees and AVL trees. Skew heaps are binary trees with heap order, but there is no structural constraint on these trees. Unlike leftist heaps, no information is maintained about the null path length of any node. The right path of a skew heap can be arbitrarily long at any time, so the worst-case running time of all operations is $O(N)$. However, as with splay trees, it can be shown (see Chapter 11) that for any M consecutive operations, the total worst-case running time is $O(M \log N)$. Thus, skew heaps have $O(\log N)$ amortized cost per operation.

As with leftist heaps, the fundamental operation on skew heaps is merging. The *Merge* routine is once again recursive, and we perform the exact same operations as before, with one exception. The difference is that for leftist heaps, we check to see whether the left and right children satisfy the leftist heap order property and swap them if they do not. For skew heaps, the swap is unconditional; we *always* do it, with the one exception that the largest of all the nodes on the right paths does not have its children swapped. This one exception is what happens in the natural recursive implementation, so it is not really a special case at all. Furthermore, it is not necessary to prove the bounds, but since this node is guaranteed not to have a right child, it would be silly to perform the swap and give it one. (In our example,

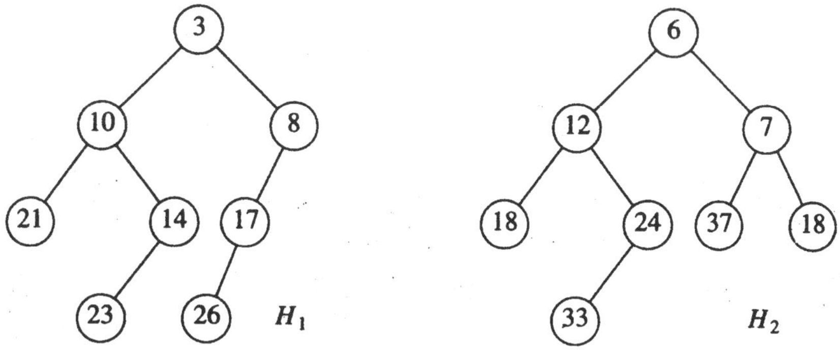

Figure 6.31 Two skew heaps H_1 and H_2

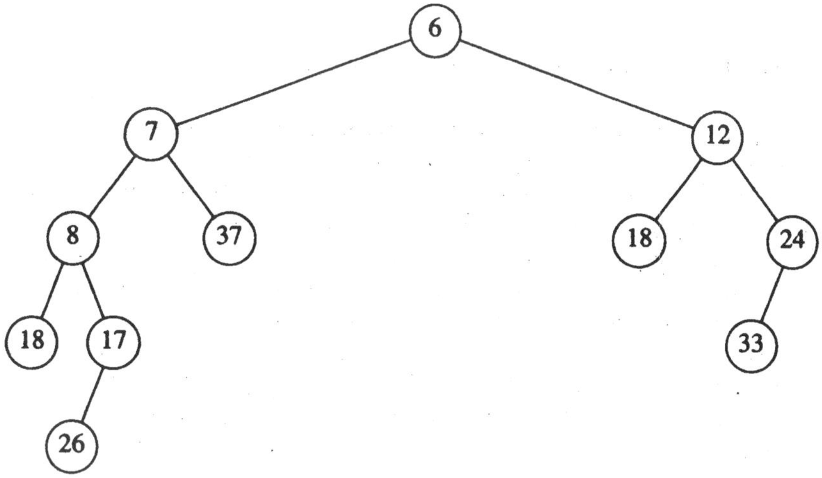

Figure 6.32 Result of merging H_2 with H_1's right subheap

there are no children of this node, so we do not worry about it.) Again, suppose our input is the same two heaps as before, Figure 6.31.

If we recursively merge H_2 with the subheap of H_1 rooted at 8, we will get the heap in Figure 6.32.

Again, this is done recursively, so by the third rule of recursion (Section 1.3) we need not worry about how it was obtained. This heap happens to be leftist, but there is no guarantee that this is always the case. We make this heap the new left child of H_1, and the old left child of H_1 becomes the new right child (see Fig. 6.33).

The entire tree is leftist, but it is easy to see that that is not always true: Inserting 15 into this new heap would destroy the leftist property.

We can perform all operations nonrecursively, as with leftist heaps, by merging the right paths and swapping left and right children for every node on the right path, with the exception of the last. After a few examples, it becomes clear that since all but the last node on the right path have their children swapped, the net effect is that

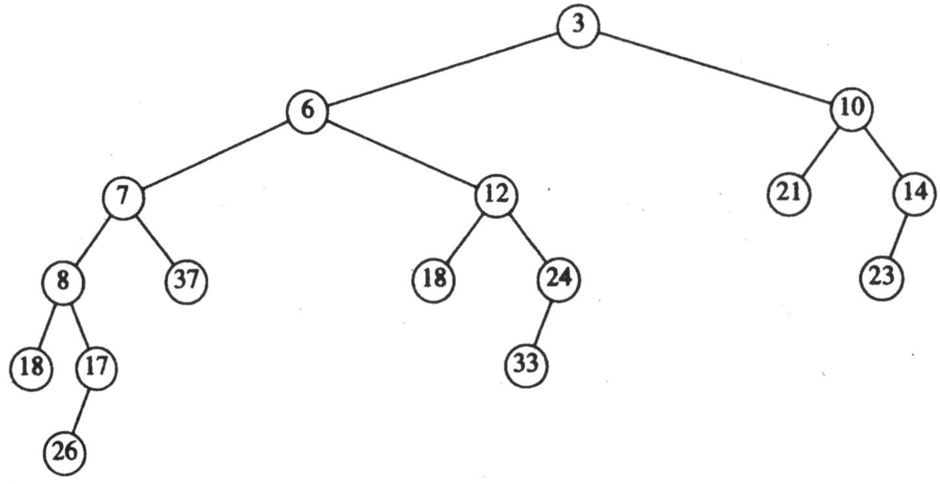

Figure 6.33 Result of merging skew heaps H_1 and H_2

this becomes the new left path (see the preceding example to convince yourself). This makes it very easy to merge two skew heaps visually.*

The implementation of skew heaps is left as a (trivial) exercise. Note that because a right path could be long, a recursive implementation could fail because of lack of stack space, even though performance would otherwise be acceptable. Skew heaps have the advantage that no extra space is required to maintain path lengths and no tests are required to determine when to swap children. It is an open problem to determine precisely the expected right path length of both leftist and skew heaps (the latter is undoubtedly more difficult). Such a comparison would make it easier to determine whether the slight loss of balance information is compensated by the lack of testing.

6.8. Binomial Queues

Although both leftist and skew heaps support merging, insertion, and *DeleteMin* all effectively in $O(\log N)$ time per operation, there is room for improvement because we know that binary heaps support insertion in *constant average* time per operation. Binomial queues support all three operations in $O(\log N)$ worst-case time per operation, but insertions take constant time on average.

6.8.1. *Binomial Queue Structure*

Binomial queues differ from all the priority queue implementations that we have seen in that a binomial queue is not a heap-ordered tree but rather a *collection* of heap-ordered trees, known as a *forest*. Each of the heap-ordered trees is of a

*This is not exactly the same as the recursive implementation (but yields the same time bounds). If we only swap children for nodes on the right path that are above the point where the merging of right paths terminated due to exhaustion of one heap's right path, we get the same result as the recursive version.

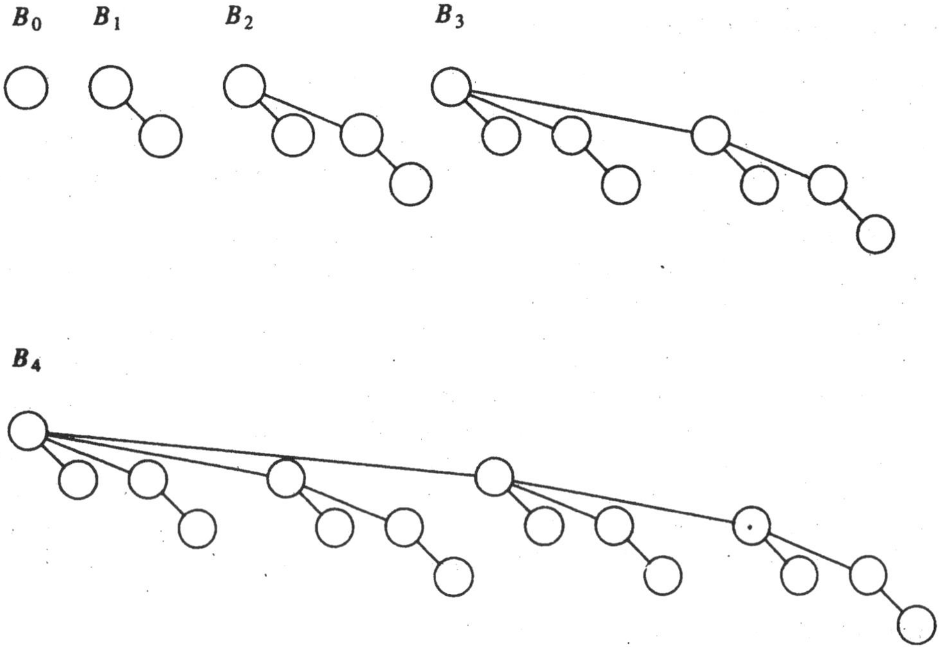

Figure 6.34 Binomial trees B_0, B_1, B_2, B_3, and B_4

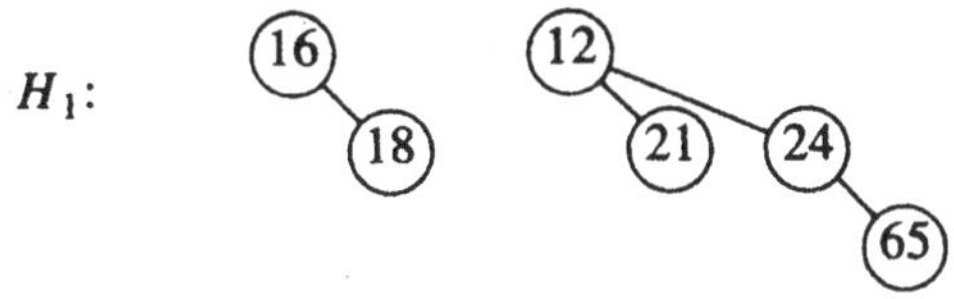

Figure 6.35 Binomial queue H_1 with six elements

constrained form known as a *binomial tree* (the reason for the name will be obvious later). There is at most one binomial tree of every height. A binomial tree of height 0 is a one-node tree; a binomial tree, B_k, of height k is formed by attaching a binomial tree, B_{k-1}, to the root of another binomial tree, B_{k-1}. Figure 6.34 shows binomial trees B_0, B_1, B_2, B_3, and B_4.

From the diagram we see that a binomial tree, B_k, consists of a root with children $B_0, B_1, \ldots, B_{k-1}$. Binomial trees of height k have exactly 2^k nodes, and the number of nodes at depth d is the binomial coefficient $\binom{k}{d}$. If we impose heap order on the binomial trees and allow at most one binomial tree of any height, we can uniquely represent a priority queue of any size by a collection of binomial trees. For instance, a priority queue of size 13 could be represented by the forest B_3, B_2, B_0. We might write this representation as 1101, which not only represents 13 in binary but also represents the fact that B_3, B_2, and B_0 are present in the representation and B_1 is not.

As an example, a priority queue of six elements could be represented as in Figure 6.35.

6.8.2. Binomial Queue Operations

The minimum element can then be found by scanning the roots of all the trees. Since there are at most $\log N$ different trees, the minimum can be found in $O(\log N)$ time. Alternatively, we can maintain knowledge of the minimum and perform the operation in $O(1)$ time, if we remember to update the minimum when it changes during other operations.

Merging two binomial queues is a conceptually easy operation, which we will describe by example. Consider the two binomial queues, H_1 and H_2, with six and seven elements, respectively, pictured in Figure 6.36.

The merge is performed by essentially adding the two queues together. Let H_3 be the new binomial queue. Since H_1 has no binomial tree of height 0 and H_2 does, we can just use the binomial tree of height 0 in H_2 as part of H_3. Next, we add binomial trees of height 1. Since both H_1 and H_2 have binomial trees of height 1, we merge them by making the larger root a subtree of the smaller, creating a binomial tree of height 2, shown in Figure 6.37. Thus, H_3 will not have a binomial tree of height 1. There are now three binomial trees of height 2, namely, the original trees of H_1 and H_2 plus the tree formed by the previous step. We keep one binomial tree

Figure 6.36 Two binomial queues H_1 and H_2

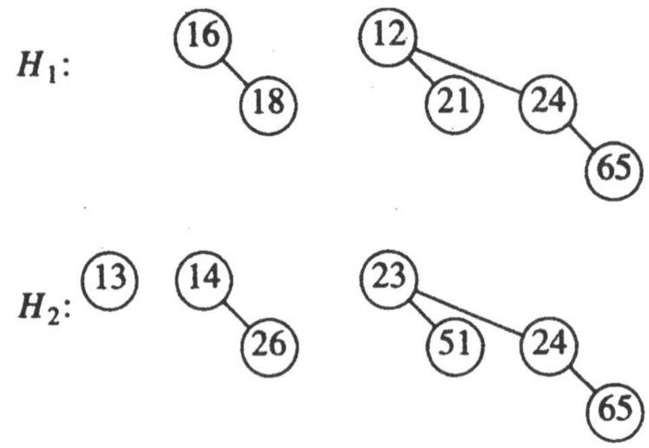

Figure 6.37 Merge of the two B_1 trees in H_1 and H_2

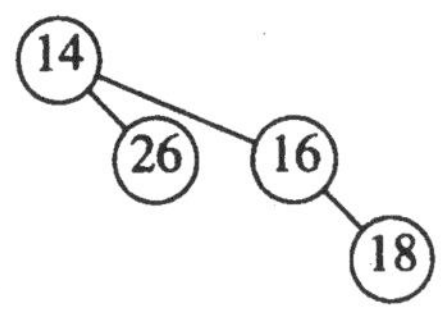

Figure 6.38 Binomial queue H_3: the result of merging H_1 and H_2

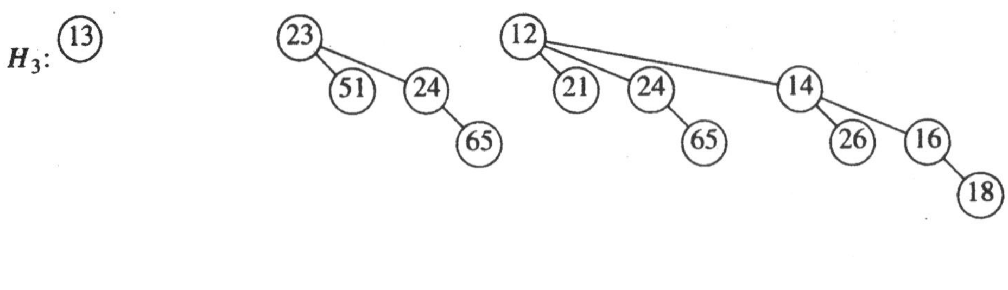

of height 2 in H_3 and merge the other two, creating a binomial tree of height 3. Since H_1 and H_2 have no trees of height 3, this tree becomes part of H_3 and we are finished. The resulting binomial queue is shown in Figure 6.38.

Since merging two binomial trees takes constant time with almost any reasonable implementation, and there are $O(\log N)$ binomial trees, the merge takes $O(\log N)$ time in the worst case. To make this operation efficient, we need to keep the trees in the binomial queue sorted by height, which is certainly a simple thing to do.

Insertion is just a special case of merging, since we merely create a one-node tree and perform a merge. The worst-case time of this operation is likewise $O(\log N)$. More precisely, if the priority queue into which the element is being inserted has the property that the smallest nonexistent binomial tree is B_i, the running time is proportional to $i + 1$. For example, H_3 (Fig. 6.38) is missing a binomial tree of height 1, so the insertion will terminate in two steps. Since each tree in a binomial queue is present with probability $\frac{1}{2}$, it follows that we expect an insertion to terminate in two steps, so the average time is constant. Furthermore, an analysis will show that performing N *Inserts* on an initially empty binomial queue will take $O(N)$ worst-case time. Indeed, it is possible to do this operation using only $N - 1$ comparisons; we leave this as an exercise.

As an example, we show in Figures 6.39 through 6.45 the binomial queues that are formed by inserting 1 through 7 in order. Inserting 4 shows off a bad case. We merge 4 with B_0, obtaining a new tree of height 1. We then merge this tree with B_1, obtaining a tree of height 2, which is the new priority queue. We count this as three steps (two tree merges plus the stopping case). The next insertion after 7 is inserted is another bad case and would require three tree merges.

A *DeleteMin* can be performed by first finding the binomial tree with the smallest root. Let this tree be B_k, and let the original priority queue be H. We remove the binomial tree B_k from the forest of trees in H, forming the new binomial queue H'. We also remove the root of B_k, creating binomial trees $B_0, B_1, \ldots, B_{k-1}$, which collectively form priority queue H''. We finish the operation by merging H' and H''.

As an example, suppose we perform a *DeleteMin* on H_3, which is shown again in Figure 6.46. The minimum root is 12, so we obtain the two priority queues H' and H'' in Figure 6.47 and Figure 6.48. The binomial queue that results from merging H' and H'' is the final answer and is shown in Figure 6.49.

For the analysis, note first that the *DeleteMin* operation breaks the original binomial queue into two. It takes $O(\log N)$ time to find the tree containing the minimum element and to create the queues H' and H''. Merging these two queues takes $O(\log N)$ time, so the entire *DeleteMin* operation takes $O(\log N)$ time.

6.8.3. *Implementation of Binomial Queues*

The *DeleteMin* operation requires the ability to find all the subtrees of the root quickly, so the standard representation of general trees is required: The children of each node are kept in a linked list, and each node has a pointer to its first child (if any). This operation also requires that the children be ordered by the size of their subtrees. We also need to make sure that it is easy to merge two trees. When two trees are merged, one of the trees is added as a child to the other. Since this new tree will be the largest subtree, it makes sense to maintain the subtrees in decreasing

Figure 6.39 After 1 is inserted

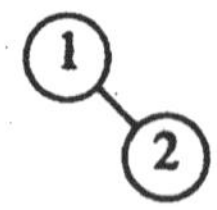

Figure 6.40 After 2 is inserted

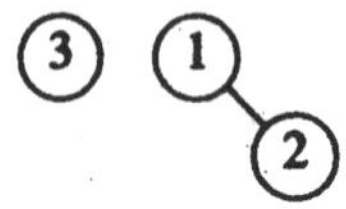

Figure 6.41 After 3 is inserted

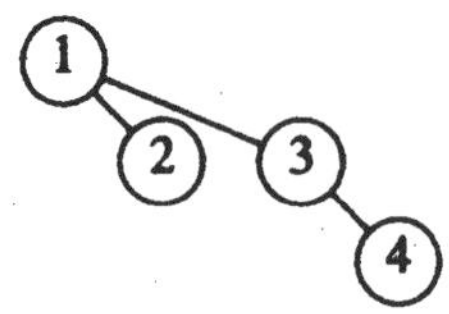

Figure 6.42 After 4 is inserted

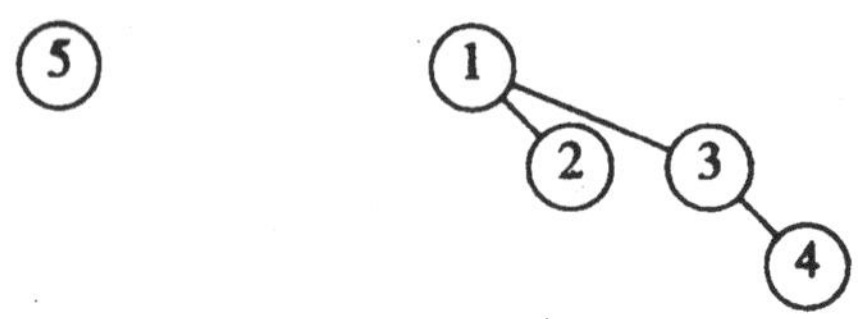

Figure 6.43 After 5 is inserted

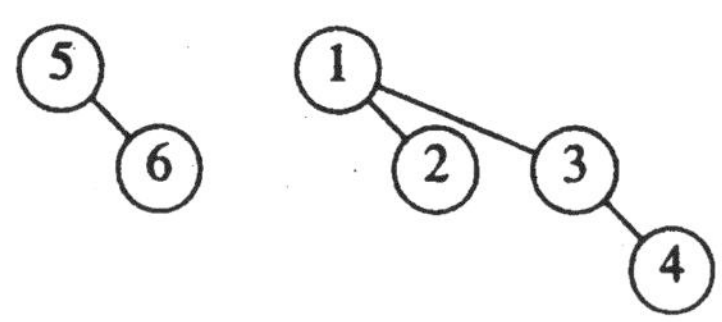

Figure 6.44 After 6 is inserted

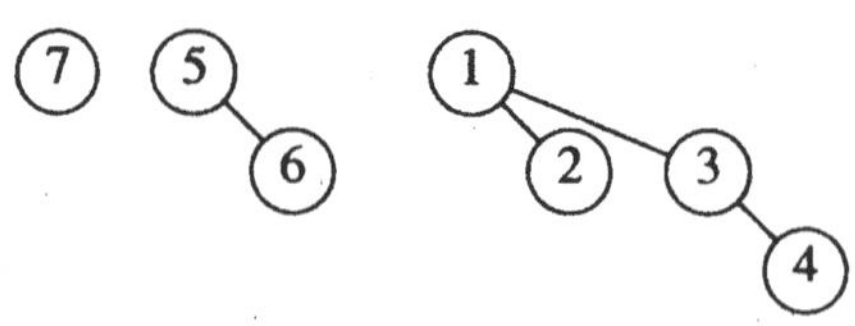

Figure 6.45 After 7 is inserted

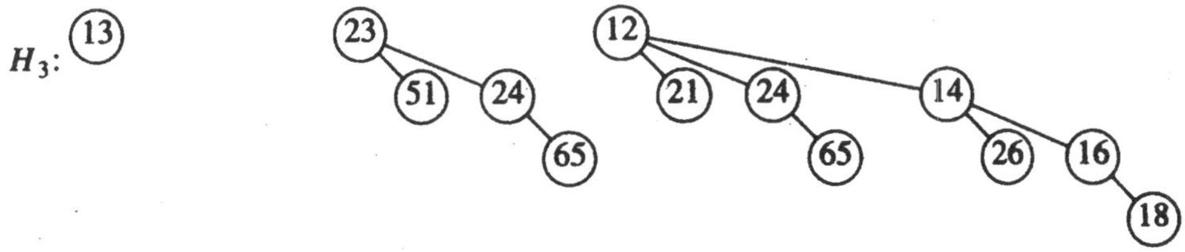

Figure 6.46 Binomial queue H_3

Figure 6.47 Binomial queue H', containing all the binomial trees in H_3 except B_3

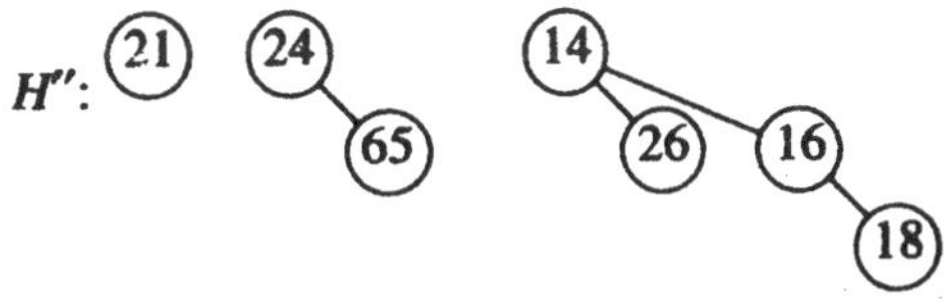

Figure 6.48 Binomial queue H'': B_3 with 12 removed

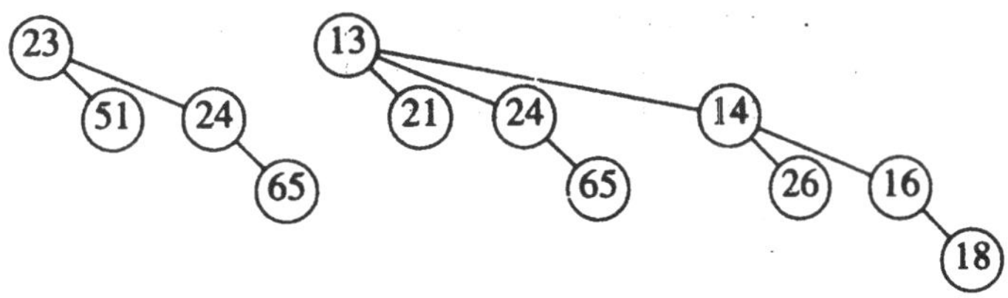

Figure 6.49 Result of *DeleteMin*(H_3)

sizes. Only then will we be able to merge two binomial trees, and thus two binomial queues, efficiently. The binomial queue will be an array of binomial trees.

To summarize, then, each node in a binomial tree will contain the data, first child, and right sibling. The children in a binomial tree are arranged in decreasing rank.

Figure 6.51 shows how the binomial queue in Figure 6.50 is represented. Figure 6.52 shows the type declarations for a node in the binomial tree.

In order to merge two binomial queues, we need a routine to merge two binomial trees of the same size. Figure 6.53 shows how the pointers change when two binomial trees are merged. The code to do this is simple and is shown in Figure 6.54.

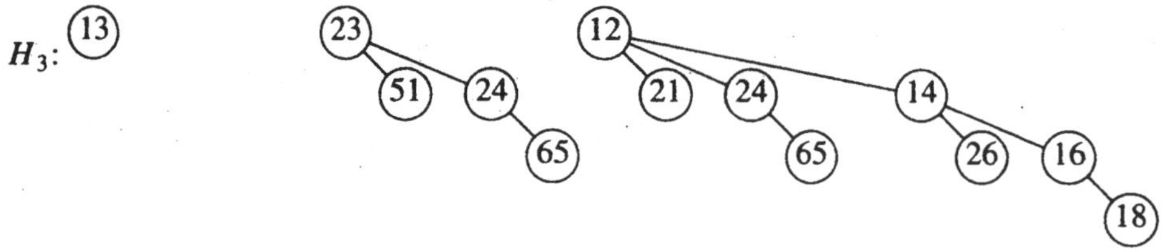

Figure 6.50 Binomial queue H_3 drawn as a forest

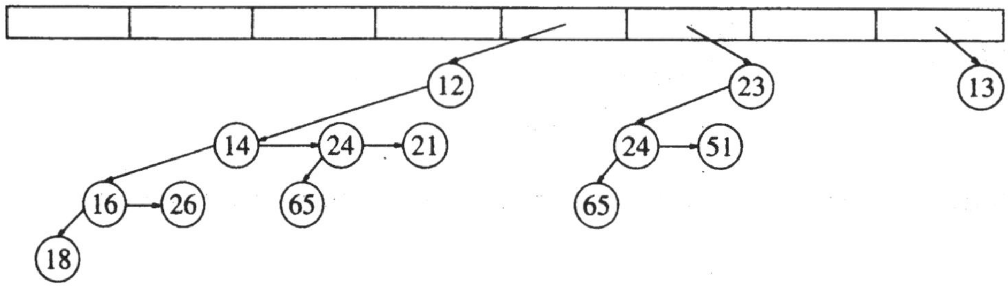

Figure 6.51 Representation of binomial queue H_3

```
typedef struct BinNode *Position;
typedef struct Collection *BinQueue;

struct BinNode
{
    ElementType Element;
    Position    LeftChild;
    Position    NextSibling;
};

struct Collection
{
    int CurrentSize;
    BinTree TheTrees[ MaxTrees ];
};
```

Figure 6.52 Binomial queue type declarations

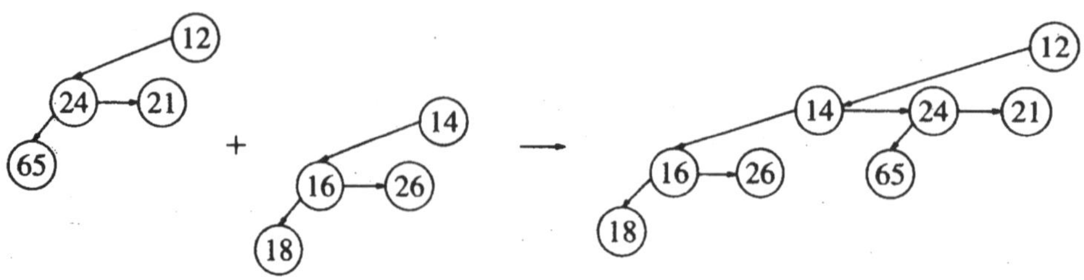

Figure 6.53 Merging two binomial trees

```
/* Return the result of merging equal-sized T1 and T2 */

BinTree
CombineTrees( BinTree T1, BinTree T2 )
{
    if( T1->Element > T2->Element )
        return CombineTrees( T2, T1 );
    T2->NextSibling = T1->LeftChild;
    T1->LeftChild = T2;
    return T1;
}
```

Figure 6.54 Routine to merge two equal-sized binomial trees

We provide a simple implementation of the *Merge* routine. The routine combines H_1 and H_2, placing the result in H_1 and making H_2 empty. At any point we are dealing with trees of rank i. T_1 and T_2 are the trees in H_1 and H_2, respectively, and *Carry* is the tree carried from a previous step (it might be *NULL*). $!!T_1$ is 1 if T_1 exists and is 0 otherwise, and the same is true for the other trees. Depending on each of the eight possible cases, the tree that results for rank i and the *Carry* tree of rank $i + 1$ is formed. This process proceeds from rank 0 to the last rank in the resulting binomial queue. The code is shown in Figure 6.55.

The *DeleteMin* routine for binary queues is given in Figure 6.56.

We can extend binomial queues to support some of the nonstandard operations that binary heaps allow, such as *DecreaseKey* and *Delete*, when the position of the affected element is known. A *DecreaseKey* is a *PercolateUp*, which can be performed in $O(\log N)$ time if we add a field to each node pointing to its parent. An arbitrary *Delete* can be performed by a combination of *DecreaseKey* and *DeleteMin* in $O(\log N)$ time.

Figure 6.55 Routine to merge two priority queues

```
/* Merge two binomial queues */
/* Not optimized for early termination */
/* H1 contains merged result */

BinQueue
Merge( BinQueue H1, BinQueue H2 )
{
    BinTree T1, T2, Carry = NULL;
    int i, j;

    if( H1->CurrentSize + H2->CurrentSize > Capacity )
        Error( "Merge would exceed capacity" );
```

Figure 6.55 *(continued)*

```
        H1->CurrentSize += H2->CurrentSize;
        for( i = 0, j = 1; j <= H1->CurrentSize; i++, j *= 2 )
        {
            T1 = H1->TheTrees[ i ]; T2 = H2->TheTrees[ i ];

            switch( !!T1 + 2 * !!T2 + 4 * !!Carry )
            {
                case 0: /* No trees */
                case 1: /* Only H1 */
                    break;
                case 2: /* Only H2 */
                    H1->TheTrees[ i ] = T2;
                    H2->TheTrees[ i ] = NULL;
                    break;
                case 4: /* Only Carry */
                    H1->TheTrees[ i ] = Carry;
                    Carry = NULL;
                    break;
                case 3: /* H1 and H2 */
                    Carry = CombineTrees( T1, T2 );
                    H1->TheTrees[ i ] = H2->TheTrees[ i ] = NULL;
                    break;
                case 5: /* H1 and Carry */
                    Carry = CombineTrees( T1, Carry );
                    H1->TheTrees[ i ] = NULL;
                    break;
                case 6: /* H2 and Carry */
                    Carry = CombineTrees( T2, Carry );
                    H2->TheTrees[ i ] = NULL;
                    break;
                case 7: /* All three */
                    H1->TheTrees[ i ] = Carry;
                    Carry = CombineTrees( T1, T2 );
                    H2->TheTrees[ i ] = NULL;
                    break;
            }
        }
        return H1;
    }
```

```
ElementType
DeleteMin( BinQueue H )
{
    int i, j;
    int MinTree;    /* The tree with the minimum item */
    BinQueue DeletedQueue;
    Position DeletedTree, OldRoot;
    ElementType MinItem;

    if( IsEmpty( H ) )
    {
        Error( "Empty binomial queue" );
        return -Infinity;
    }

    MinItem = Infinity;
    for( i = 0; i < MaxTrees; i++ )
    {
        if( H->TheTrees[ i ] &&
            H->TheTrees[ i ]->Element < MinItem )
        {
            /* Update minimum */
            MinItem = H->TheTrees[ i ] ->Element;
            MinTree = i;
        }
    }

    DeletedTree = H->TheTrees[ MinTree ];
    OldRoot = DeletedTree;
    DeletedTree = DeletedTree->LeftChild;
    free( OldRoot );

    DeletedQueue = Initialize( );
    DeletedQueue->CurrentSize = ( 1 << MinTree ) - 1;
    for( j = MinTree - 1; j >= 0; j-- )
    {
        DeletedQueue->TheTrees[ j ] = DeletedTree;
        DeletedTree = DeletedTree->NextSibling;
        DeletedQueue->TheTrees[ j ]->NextSibling = NULL;
    }

    H->TheTrees[ MinTree ] = NULL;
    H->CurrentSize -= DeletedQueue->CurrentSize + 1;

    Merge( H, DeletedQueue );
    return MinItem;
}
```

Figure 6.56 *DeleteMin* for binomial queues

Summary

In this chapter we have seen various implementations and uses of the priority queue ADT. The standard binary heap implementation is elegant because of its simplicity and speed. It requires no pointers and only a constant amount of extra space, yet supports the priority queue operations efficiently.

We considered the additional *Merge* operation and developed three implementations, each of which is unique in its own way. The leftist heap is a wonderful example of the power of recursion. The skew heap represents a remarkable data structure because of the lack of balance criteria. Its analysis, which we will perform in Chapter 11, is interesting in its own right. The binomial queue shows how a simple idea can be used to achieve a good time bound.

We have also seen several uses of priority queues, ranging from operating systems scheduling to simulation. We will see their use again in Chapters 7, 9, and 10.

Exercises

6.1 Suppose that we replace the *DeleteMin* function with *FindMin*. Can both *Insert* and *FindMin* be implemented in constant time?

6.2 a. Show the result of inserting 10, 12, 1, 14, 6, 5, 8, 15, 3, 9, 7, 4, 11, 13, and 2, one at a time, into an initially empty binary heap.

b. Show the result of using the linear-time algorithm to build a binary heap using the same input.

6.3 Show the result of performing three *DeleteMin* operations in the heap of the previous exercise.

6.4 Write the routines to do a *percolate up* and a *percolate down* in a binary heap.

6.5 Write and test a program that performs the operations *Insert, DeleteMin, BuildHeap, FindMin, DecreaseKey, Delete,* and *IncreaseKey* in a binary heap.

6.6 How many nodes are in the large heap in Figure 6.13?

6.7 a. Prove that for binary heaps, *BuildHeap* does at most $2N - 2$ comparisons between elements.

b. Show that a heap of eight elements can be constructed in eight comparisons between heap elements.

**c. Give an algorithm to build a binary heap in $\frac{13}{8}N + O(\log N)$ element comparisons.

**6.8 Show that the expected depth of the kth smallest element in a large complete heap (you may assume $N = 2^k - 1$) is bounded by $\log k$.

6.9 *a. Give an algorithm to find all nodes less than some value, X, in a binary heap. Your algorithm should run in $O(K)$, where K is the number of nodes output.

b. Does your algorithm extend to any of the other heap structures discussed in this chapter?

*c. Give an algorithm that finds an arbitrary item X in a binary heap using at most roughly $3N/4$ comparisons.

**6.10 Propose an algorithm to insert M nodes into a binary heap on N elements in $O(M + \log N \log\log N)$ time. Prove your time bound.

6.11 Write a program to take N elements and do the following:

a. Insert them into a heap one by one.

b. Build a heap in linear time.

Compare the running time of both algorithms for sorted, reverse-ordered, and random inputs.

6.12 Each *DeleteMin* operation uses $2 \log N$ comparisons in the worst case.

*a. Propose a scheme so that the *DeleteMin* operation uses only $\log N + \log\log N + O(1)$ comparisons between elements. This need not imply less data movement.

**b. Extend your scheme in part (a) so that only $\log N + \log\log\log N + O(1)$ comparisons are performed.

**c. How far can you take this idea?

d. Do the savings in comparisons compensate for the increased complexity of your algorithm?

6.13 If a *d*-heap is stored as an array, for an entry located in position i, where are the parents and children?

6.14 Suppose we need to perform M *PercolateUps* and N *DeleteMins* on a *d*-heap that initially has N elements.

a. What is the total running time of all operations in terms of M, N, and d?

b. If $d = 2$, what is the running time of all heap operations?

c. If $d = \Theta(N)$, what is the total running time?

*d. What choice of d minimizes the total running time?

6.15 A *min-max heap* is a data structure that supports both *DeleteMin* and *DeleteMax* in $O(\log N)$ per operation. The structure is identical to a binary heap, but the heap order property is that for any node, X, at even depth, the key stored at X is smaller than the parent but larger than the grandparent (where this makes sense), and for any node X at odd depth, the key stored at X is larger than the parent but smaller than the grandparent. See Figure 6.57.

a. How do we find the minimum and maximum elements?

*b. Give an algorithm to insert a new node into the min-max heap.

*c. Give an algorithm to perform *DeleteMin* and *DeleteMax*.

*d. Can you build a min-max heap in linear time?

**e. Suppose we would like to support *DeleteMin*, *DeleteMax*, and *Merge*. Propose a data structure to support all operations in $O(\log N)$ time.

6.16 Merge the two leftist heaps in Figure 6.58.

6.17 Show the result of inserting keys 1 to 15 in order into an initially empty leftist heap.

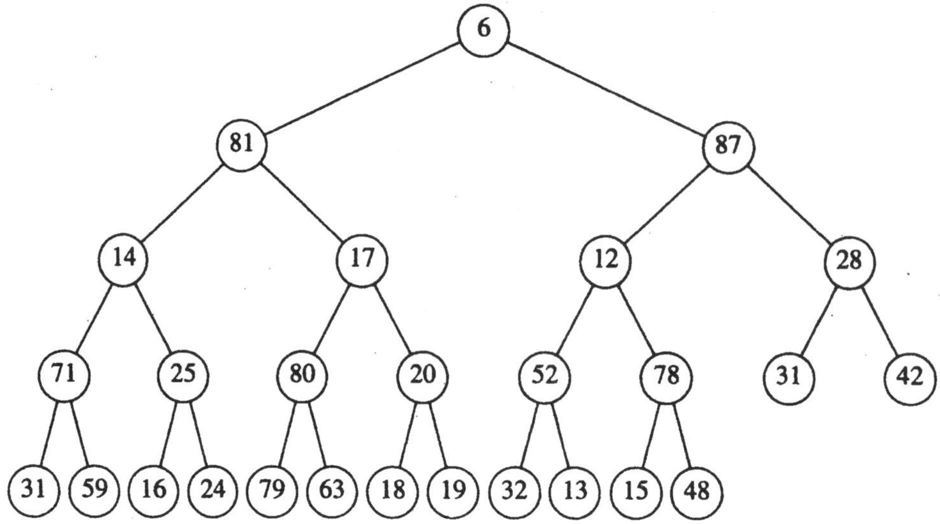

Figure 6.57 Min-max heap

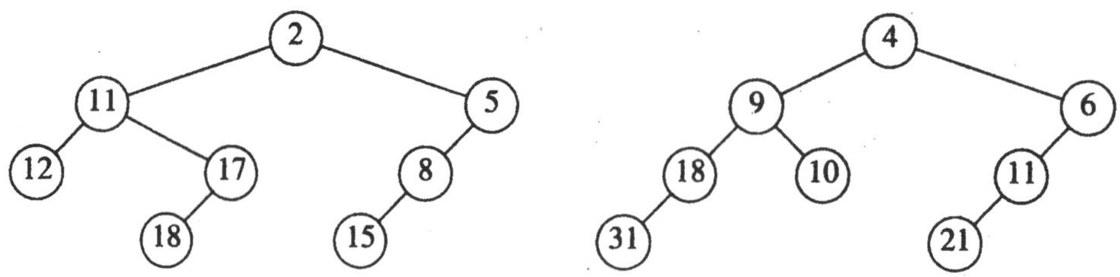

Figure 6.58

6.18 Prove or disprove: A perfectly balanced tree forms if keys 1 to $2^k - 1$ are inserted in order into an initially empty leftist heap.

6.19 Give an example of input that generates the best leftist heap.

6.20 a. Can leftist heaps efficiently support *DecreaseKey?*

b. What changes, if any (if possible), are required to do this?

6.21 One way to delete nodes from a known position in a leftist heap is to use a lazy strategy. To delete a node, merely mark it deleted. When a *FindMin* or *DeleteMin* is performed, there is a potential problem if the root is marked deleted, since then the node has to be actually deleted and the real minimum needs to be found, which may involve deleting other marked nodes. In this strategy, *Delete*s cost one unit, but the cost of a *DeleteMin* or *FindMin* depends on the number of nodes that are marked deleted. Suppose that after a *DeleteMin* or *FindMin* there are k fewer marked nodes than before the operation.

*a. Show how to perform the *DeleteMin* in $O(k \log N)$ time.

**b. Propose an implementation, with an analysis to show that the time to perform the *DeleteMin* is $O(k \log(2N/k))$.

6.22 We can perform *BuildHeap* in linear time for leftist heaps by considering each element as a one-node leftist heap, placing all these heaps on a queue, and performing the following step: Until only one heap is on the queue, dequeue two heaps, merge them, and enqueue the result.

a. Prove that this algorithm is $O(N)$ in the worst case.

b. Why might this algorithm be preferable to the algorithm described in the text?

6.23 Merge the two skew heaps in Figure 6.58.

6.24 Show the result of inserting keys 1 to 15 in order into a skew heap.

6.25 Prove or disprove: A perfectly balanced tree forms if the keys 1 to $2^k - 1$ are inserted in order into an initially empty skew heap.

6.26 A skew heap of N elements can be built using the standard binary heap algorithm. Can we use the same merging strategy described in Exercise 6.22 for skew heaps to get an $O(N)$ running time?

6.27 Prove that a binomial tree B_k has binomial trees $B_0, B_1, \ldots, B_{k-1}$ as children of the root.

6.28 Prove that a binomial tree of height k has $\binom{k}{d}$ nodes at depth d.

6.29 Merge the two binomial queues in Figure 6.59.

6.30 a. Show that N *Inserts* into an initially empty binomial queue takes $O(N)$ time in the worst case.

b. Give an algorithm to build a binomial queue of N elements, using at most $N - 1$ comparisons between elements.

*c. Propose an algorithm to insert M nodes into a binomial queue of N elements in $O(M + \log N)$ worst-case time. Prove your bound.

6.31 Write an efficient routine to perform *Insert* using binomial queues. Do not call *Merge*.

Figure 6.59

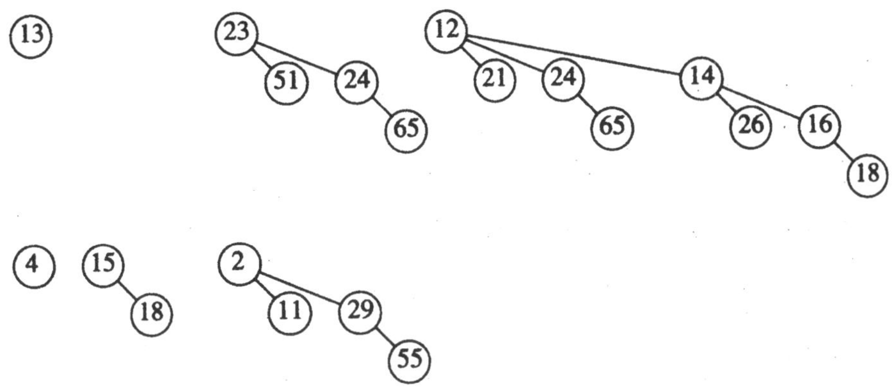

6.32 For the binomial queue:

a. What happens when the call *Merge*(H, H) is made? Modify the code to fix this problem.

b. Modify the *Merge* routine to terminate merging if there are no trees left in H_2 and the *Carry* tree is $NULL$.

c. Modify the *Merge* so that the smaller tree is always merged into the larger.

**6.33 Suppose we extend binomial queues to allow at most two trees of the same height per structure. Can we obtain $O(1)$ worst-case time for insertion while retaining $O(\log N)$ for the other operations?

6.34 Suppose you have a number of boxes, each of which can hold total weight C and items $i_1, i_2, i_3, \ldots, i_N$, which weigh $w_1, w_2, w_3, \ldots, w_N$, respectively. The object is to pack all the items without placing more weight in any box than its capacity and using as few boxes as possible. For instance, if $C = 5$, and the items have weights 2, 2, 3, 3, then we can solve the problem with two boxes.

In general, this problem is very hard, and no efficient solution is known. Write programs to implement efficiently the following approximation strategies:

*a. Place the weight in the first box for which it fits (creating a new box if there is no box with enough room). (This strategy and all that follow would give three boxes, which is suboptimal.)

b. Place the weight in the box with the most room for it.

*c. Place the weight in the most filled box that can accept it without overflowing.

**d. Are any of these strategies enhanced by presorting the items by weight?

6.35 Suppose we want to add the *DecreaseAllKeys*(Δ) operation to the heap repertoire. The result of this operation is that all keys in the heap have their value decreased by an amount Δ. For the heap implementation of your choice, explain the necessary modifications so that all other operations retain their running times and *DecreaseAllKeys* runs in $O(1)$.

6.36 Which of the two selection algorithms has the better time bound?

References

The binary heap was first described in [27]. The linear-time algorithm for its construction is from [14].

The first description of d-heaps was in [19]. Leftist heaps were invented by Crane [11] and described in Knuth [21]. Skew heaps were developed by Sleator and Tarjan [23]. Binomial queues were invented by Vuillemin [26]; Brown provided a detailed analysis and empirical study showing that they perform well in practice [4], if carefully implemented.

Exercise 6.7(b–c) is taken from [17]. Exercise 6.9(c) is from [6]. A method for constructing binary heaps that uses about $1.52N$ comparisons on average is described in [22]. Lazy deletion in leftist heaps (Exercise 6.21) is from [10]. A solution to Exercise 6.33 can be found in [8].

Min-max heaps (Exercise 6.15) were originally described in [1]. A more efficient implementation of the operations is given in [18] and [24]. Alternative representations for double-ended priority queues are the *deap* and *diamond dequeue*. Details can be found in [5], [7], and [9]. Solutions to 6.15(e) are given in [12] and [20].

A theoretically interesting priority queue representation is the *Fibonacci heap* [16], which we will describe in Chapter 11. The Fibonacci heap allows all operations to be performed in $O(1)$ amortized time, except for deletions, which are $O(\log N)$. *Relaxed heaps* [13] achieve identical bounds in the worst case (with the exception of *Merge*). The procedure of [3] achieves optimal worst-case bounds for all operations. Another interesting implementation is the *pairing heap* [15], which is described in Chapter 12. Finally, priority queues that work when the data consist of small integers are described in [2] and [25].

1. M. D. Atkinson, J. R. Sack, N. Santoro, and T. Strothotte, "Min-Max Heaps and Generalized Priority Queues," *Communications of the ACM*, 29 (1986), 996–1000.
2. J. D. Bright, "Range Restricted Mergeable Priority Queues," *Information Processing Letters*, 47 (1993), 159–164.
3. G. S. Brodal, "Worst-Case Efficient Priority Queues," *Proceedings of the Seventh Annual ACM-SIAM Symposium on Discrete Algorithms* (1996), 52–58.
4. M. R. Brown, "Implementation and Analysis of Binomial Queue Algorithms," *SIAM Journal on Computing*, 7 (1978), 298–319.
5. S. Carlsson, "The Deap—A Double-Ended Heap to Implement Double-Ended Priority Queues," *Information Processing Letters*, 26 (1987), 33–36.
6. S. Carlsson and J. Chen, "The Complexity of Heaps," *Proceedings of the Third Symposium on Discrete Algorithms* (1992), 393–402.
7. S. Carlsson, J. Chen, and T. Strothotte, "A Note on the Construction of the Data Structure 'Deap'," *Information Processing Letters*, 31 (1989), 315–317.
8. S. Carlsson, J. I. Munro, and P. V. Poblete, "An Implicit Binomial Queue with Constant Insertion Time," *Proceedings of First Scandinavian Workshop on Algorithm Theory* (1988), 1–13.
9. S. C. Chang and M. W. Due, "Diamond Deque: A Simple Data Structure for Priority Deques," *Information Processing Letters*, 46 (1993), 231–237.
10. D. Cheriton and R. E. Tarjan, "Finding Minimum Spanning Trees," *SIAM Journal on Computing*, 5 (1976), 724–742.
11. C. A. Crane, "Linear Lists and Priority Queues as Balanced Binary Trees," *Technical Report STAN-CS-72-259*, Computer Science Department, Stanford University, Stanford, Calif., 1972.
12. Y. Ding and M. A. Weiss, "The Relaxed Min-Max Heap: A Mergeable Double-Ended Priority Queue," *Acta Informatica*, 30 (1993), 215–231.
13. J. R. Driscoll, H. N. Gabow, R. Shrairman, and R. E. Tarjan, "Relaxed Heaps: An Alternative to Fibonacci Heaps with Applications to Parallel Computation," *Communications of the ACM*, 31 (1988), 1343–1354.
14. R. W. Floyd, "Algorithm 245: Treesort 3," *Communications of the ACM*, 7 (1964), 701.
15. M. L. Fredman, R. Sedgewick, D. D. Sleator, and R. E. Tarjan, "The Pairing Heap: A New Form of Self-adjusting Heap," *Algorithmica*, 1 (1986), 111–129.
16. M. L. Fredman and R. E. Tarjan, "Fibonacci Heaps and Their Uses in Improved Network Optimization Algorithms," *Journal of the ACM*, 34 (1987), 596–615.

17. G. H. Gonnet and J. I. Munro, "Heaps on Heaps," *SIAM Journal on Computing*, 15 (1986), 964–971.
18. A. Hasham and J. R. Sack, "Bounds for Min-max Heaps," *BIT*, 27 (1987), 315–323.
19. D. B. Johnson, "Priority Queues with Update and Finding Minimum Spanning Trees," *Information Processing Letters*, 4 (1975), 53–57.
20. C. M. Khoong and H. W. Leong, "Double-Ended Binomial Queues," *Proceedings of the Fourth Annual International Symposium on Algorithms and Computation* (1993).
21. D. E. Knuth, *The Art of Computer Programming, Vol 3: Sorting and Searching*, Addison-Wesley, Reading, Mass., 1973.
22. C. J. H. McDiarmid and B. A. Reed, "Building Heaps Fast," *Journal of Algorithms*, 10 (1989), 352–365.
23. D. D. Sleator and R. E. Tarjan, "Self-adjusting Heaps," *SIAM Journal on Computing*, 15 (1986), 52–69.
24. T. Strothotte, P. Eriksson, and S. Vallner, "A Note on Constructing Min-max Heaps," *BIT*, 29 (1989), 251–256.
25. P. van Emde Boas, R. Kaas, and E. Zijlstra, "Design and Implementation of an Efficient Priority Queue," *Mathematical Systems Theory*, 10 (1977), 99–127.
26. J. Vuillemin, "A Data Structure for Manipulating Priority Queues," *Communications of the ACM*, 21 (1978), 309–314.
27. J. W. J. Williams, "Algorithm 232: Heapsort," *Communications of the ACM*, 7 (1964), 347–348.

CHAPTER 7

Sorting

In this chapter we discuss the problem of sorting an array of elements. To simplify matters, we will assume in our examples that the array contains only integers, although, obviously, more complicated structures are possible. For most of this chapter, we will also assume that the entire sort can be done in main memory, so that the number of elements is relatively small (less than a million). Sorts that cannot be performed in main memory and must be done on disk or tape are also quite important. This type of sorting, known as external sorting, will be discussed at the end of the chapter.

Our investigation of internal sorting will show that

- There are several easy algorithms to sort in $O(N^2)$, such as insertion sort.
- There is an algorithm, Shellsort, that is very simple to code, runs in $o(N^2)$, and is efficient in practice.
- There are slightly more complicated $O(N \log N)$ sorting algorithms.
- Any general-purpose sorting algorithm requires $\Omega(N \log N)$ comparisons.

The rest of this chapter will describe and analyze the various sorting algorithms. These algorithms contain interesting and important ideas for code optimization as well as algorithm design. Sorting is also an example where the analysis can be precisely performed. Be forewarned that where appropriate, we will do as much analysis as possible.

7.1. Preliminaries

The algorithms we describe will all be interchangeable. Each will be passed an array containing the elements and an integer containing the number of elements.

We will assume that N, the number of elements passed to our sorting routines, has already been checked and is legal. In accordance with C conventions, the data will start at position 0 for all the sorts.

We will also assume the existence of the "<" and ">" operators, which can be used to place a consistent ordering on the input. Besides the assignment operator, these are the only operations allowed on the input data. Sorting under these conditions is known as *comparison-based* sorting.

7.2. Insertion Sort

7.2.1. *The Algorithm*

One of the simplest sorting algorithms is the *insertion sort*. Insertion sort consists of $N - 1$ *passes*. For pass $P = 1$ through $N - 1$, insertion sort ensures that the elements in positions 0 through P are in sorted order. Insertion sort makes use of the fact that elements in positions 0 through $P - 1$ are already known to be in sorted order. Figure 7.1 shows a sample array after each pass of insertion sort.

Figure 7.1 shows the general strategy. In pass P, we move the element in position P left until its correct place is found among the first $P + 1$ elements. The code in Figure 7.2 implements this strategy. Lines 2 through 5 implement that data movement without the explicit use of swaps. The element in position P is saved in *Tmp*, and all larger elements (prior to position P) are moved one spot to the right. Then *Tmp* is placed in the correct spot. This is the same technique that was used in the implementation of binary heaps.

Figure 7.1 Insertion sort after each pass

Original	**34**	**8**	**64**	**51**	**32**	**21**	**Positions Moved**
After $p = 1$	8	34	64	51	32	21	1
After $p = 2$	8	34	64	51	32	21	0
After $p = 3$	8	34	51	64	32	21	1
After $p = 4$	8	32	34	51	64	21	3
After $p = 5$	8	21	32	34	51	64	4

Figure 7.2 Insertion sort routine

```
        void
        InsertionSort( ElementType A[ ], int N )
        {
            int j, P;

            Element Type Tmp;
/* 1*/      for( P = 1; P < N; P++ )
            {
/* 2*/          Tmp = A[ P ];
/* 3*/          for( j = P; j > 0 && A[ j - 1 ] > Tmp; j-- )
/* 4*/              A[ j ] = A[ j - 1 ];
/* 5*/          A[ j ] = Tmp;
            }
        }
```

7.2.2. *Analysis of Insertion Sort*

Because of the nested loops, each of which can take N iterations, insertion sort is $O(N^2)$. Furthermore, this bound is tight, because input in reverse order can achieve this bound. A precise calculation shows that the test at line 4 can be executed at most $P + 1$ times for each value of P. Summing over all P gives a total of

$$\sum_{i=2}^{N} i = 2 + 3 + 4 + \cdots + N = \Theta(N^2)$$

On the other hand, if the input is presorted, the running time is $O(N)$, because the test in the inner *for* loop always fails immediately. Indeed, if the input is almost sorted (this term will be more rigorously defined in the next section), insertion sort will run quickly. Because of this wide variation, it is worth analyzing the average-case behavior of this algorithm. It turns out that the average case is $\Theta(N^2)$ for insertion sort, as well as for a variety of other sorting algorithms, as the next section shows.

7.3. A Lower Bound for Simple Sorting Algorithms

An *inversion* in an array of numbers is any ordered pair (i, j) having the property that $i < j$ but $A[i] > A[j]$. In the example of the last section, the input list 34, 8, 64, 51, 32, 21 had nine inversions, namely (34, 8), (34, 32), (34, 21), (64, 51), (64, 32), (64, 21), (51, 32), (51, 21), and (32, 21). Notice that this is exactly the number of swaps that needed to be (implicitly) performed by insertion sort. This is always the case, because swapping two adjacent elements that are out of place removes exactly one inversion, and a sorted array has no inversions. Since there is $O(N)$ other work involved in the algorithm, the running time of insertion sort is $O(I + N)$, where I is the number of inversions in the original array. Thus, insertion sort runs in linear time if the number of inversions is $O(N)$.

We can compute precise bounds on the average running time of insertion sort by computing the average number of inversions in a permutation. As usual, defining *average* is a difficult proposition. We will assume that there are no duplicate elements (if we allow duplicates, it is not even clear what the average number of duplicates is). Using this assumption, we can assume that the input is some permutation of the first N integers (since only relative ordering is important) and that all are equally likely. Under these assumptions, we have the following theorem:

THEOREM 7.1.

The average number of inversions in an array of N distinct numbers is $N(N-1)/4$.

PROOF:

For any list, L, of numbers, consider L_r, the list in reverse order. The reverse list of the example is 21, 32, 51, 64, 8, 34. Consider any pair of two numbers in the list (x, y), with $y > x$. Clearly, in exactly one of L and L_r this ordered pair represents an inversion. The total number of these pairs in a list L and its reverse L_r is $N(N-1)/2$. Thus, an average list has half this amount, or $N(N-1)/4$ inversions.

This theorem implies that insertion sort is quadratic on average. It also provides a very strong lower bound about any algorithm that only exchanges adjacent elements.

THEOREM 7.2.

Any algorithm that sorts by exchanging adjacent elements requires $\Omega(N^2)$ time on average.

PROOF:

The average number of inversions is initially $N(N-1)/4 = \Omega(N^2)$. Each swap removes only one inversion, so $\Omega(N^2)$ swaps are required.

This is an example of a lower-bound proof. It is valid not only for insertion sort, which performs adjacent exchanges implicitly, but also for other simple algorithms such as bubble sort and selection sort, which we will not describe here. In fact, it is valid over an entire *class* of sorting algorithms, including those undiscovered, that perform only adjacent exchanges. Because of this, this proof cannot be confirmed empirically. Although this lower-bound proof is rather simple, in general proving lower bounds is much more complicated than proving upper bounds and in some cases resembles voodoo.

This lower bound shows us that in order for a sorting algorithm to run in subquadratic, or $o(N^2)$, time, it must do comparisons and, in particular, exchanges between elements that are far apart. A sorting algorithm makes progress by eliminating inversions, and to run efficiently, it must eliminate more than just one inversion per exchange.

7.4. Shellsort

Shellsort, named after its inventor, Donald Shell, was one of the first algorithms to break the quadratic time barrier, although it was not until several years after its initial discovery that a subquadratic time bound was proven. As suggested in the previous section, it works by comparing elements that are distant; the distance between comparisons decreases as the algorithm runs until the last phase, in which adjacent elements are compared. For this reason, Shellsort is sometimes referred to as *diminishing increment* sort.

Shellsort uses a sequence, $h_1, h_2, \ldots, h_t$, called the *increment sequence*. Any increment sequence will do as long as $h_1 = 1$, but some choices are better than others (we will discuss that question later). After a *phase*, using some increment h_k, for every i, we have $A[i] \leq A[i + h_k]$ (where this makes sense); all elements spaced h_k apart are sorted. The file is then said to be *h_k-sorted*. For example, Figure 7.3 shows an array after several phases of Shellsort. An important property of Shellsort (which we state without proof) is that an h_k-sorted file that is then h_{k-1}-sorted

Original	**81**	**94**	**11**	**96**	**12**	**35**	**17**	**95**	**28**	**58**	**41**	**75**	**15**
After 5-sort	35	17	11	28	12	41	75	15	96	58	81	94	95
After 3-sort	28	12	11	35	15	41	58	17	94	75	81	96	95
After 1-sort	11	12	15	17	28	35	41	58	75	81	94	95	96

Figure 7.3 Shellsort after each pass

remains h_k-sorted. If this were not the case, the algorithm would likely be of little value, since work done by early phases would be undone by later phases.

The general strategy to h_k-sort is for each position, i, in $h_k, h_k + 1, \ldots, N - 1$, place the element in the correct spot among $i, i - h_k, i - 2h_k$, etc. Although this does not affect the implementation, a careful examination shows that the action of an h_k-sort is to perform an insertion sort on h_k independent subarrays. This observation will be important when we analyze the running time of Shellsort.

A popular (but poor) choice for increment sequence is to use the sequence suggested by Shell: $h_t = \lfloor N/2 \rfloor$, and $h_k = \lfloor h_{k+1}/2 \rfloor$. Figure 7.4 contains a program that implements Shellsort using this sequence. We shall see later that there are increment sequences that give a significant improvement in the algorithm's running time; even a minor change can drastically affect performance (Exercise 7.10).

The program in Figure 7.4 avoids the explicit use of swaps in the same manner as our implementation of insertion sort.

Figure 7.4 Shellsort routine using Shell's increments (better increments are possible)

```
void
Shellsort( ElementType A[ ], int N )
{
    int i, j, Increment;
    ElementType Tmp;

    for( Increment = N / 2; Increment > 0; Increment /= 2 )
        for( i = Increment; i < N; i++ )
        {
            Tmp = A[ i ];
            for( j = i; j >= Increment; j -= Increment )
                if( Tmp < A[ j - Increment ] )
                    A[ j ] = A[ j - Increment ];
                else
                    break;
            A[ j ] = Tmp;
        }
}
```

7.4.1. *Worst-Case Analysis of Shellsort*

Although Shellsort is simple to code, the analysis of its running time is quite another story. The running time of Shellsort depends on the choice of increment sequence, and the proofs can be rather involved. The average-case analysis of Shellsort is a long-standing open problem, except for the most trivial increment sequences. We will prove tight worst-case bounds for two particular increment sequences.

THEOREM 7.3.

The worst-case running time of Shellsort, using Shell's increments, is $\Theta(N^2)$.

PROOF:

The proof requires showing not only an upper bound on the worst-case running time but also showing that there exists some input that actually takes $\Omega(N^2)$ time to run. We prove the lower bound first, by constructing a bad case. First, we choose N to be a power of 2. This makes all the increments even, except for the last increment, which is 1. Now, we will give as input an array, *InputData*, with the $N/2$ largest numbers in the even positions and the $N/2$ smallest numbers in the odd positions (for this proof, the first position is position 1). As all the increments except the last are even, when we come to the last pass, the $N/2$ largest numbers are still all in even positions and the $N/2$ smallest numbers are still all in odd positions. The ith smallest number ($i \le N/2$) is thus in position $2i - 1$ before the beginning of the last pass. Restoring the ith element to its correct place requires moving it $i - 1$ spaces in the array. Thus, to merely place the $N/2$ smallest elements in the correct place requires at least $\sum_{i=1}^{N/2} i - 1 = \Omega(N^2)$ work. As an example, Figure 7.5 shows a bad (but not the worst) input when $N = 16$. The number of inversions remaining after the 2-sort is exactly $1 + 2 + 3 + 4 + 5 + 6 + 7 = 28$; thus, the last pass will take considerable time.

To finish the proof, we show the upper bound of $O(N^2)$. As we have observed before, a pass with increment h_k consists of h_k insertion sorts of about N/h_k elements. Since insertion sort is quadratic, the total cost of a pass is $O(h_k(N/h_k)^2) = O(N^2/h_k)$. Summing over all passes gives a total bound of $O(\sum_{i=1}^{t} N^2/h_i) = O(N^2 \sum_{i=1}^{t} 1/h_i)$. Because the increments form a geometric series with common ratio 2, and the largest term in the series is $h_1 = 1$, $\sum_{i=1}^{t} 1/h_i < 2$. Thus we obtain a total bound of $O(N^2)$.

The problem with Shell's increments is that pairs of increments are not necessarily relatively prime, and thus the smaller increment can have little effect. Hibbard

Figure 7.5 Bad case for Shellsort with Shell's increments (positions are numbered 1 to 16)

Start	**1**	**9**	**2**	**10**	**3**	**11**	**4**	**12**	**5**	**13**	**6**	**14**	**7**	**15**	**8**	**16**
After 8-sort	1	9	2	10	3	11	4	12	5	13	6	14	7	15	8	16
After 4-sort	1	9	2	10	3	11	4	12	5	13	6	14	7	15	8	16
After 2-sort	1	9	2	10	3	11	4	12	5	13	6	14	7	15	8	16
After 1-sort	1	2	3	4	5	6	7	8	9	10	11	12	13	14	15	16

suggested a slightly different increment sequence, which gives better results in practice (and theoretically). His increments are of the form $1, 3, 7, \ldots, 2^k - 1$. Although these increments are almost identical, the key difference is that consecutive increments have no common factors. We now analyze the worst-case running time of Shellsort for this increment sequence. The proof is rather complicated.

THEOREM 7.4.

The worst-case running time of Shellsort using Hibbard's increments is $\Theta(N^{3/2})$.

PROOF:

We will prove only the upper bound and leave the proof of the lower bound as an exercise. The proof requires some well-known results from additive number theory. References to these results are provided at the end of the chapter.

For the upper bound, as before, we bound the running time of each pass and sum over all passes. For increments $h_k > N^{1/2}$, we will use the bound $O(N^2/h_k)$ from the previous theorem. Although this bound holds for the other increments, it is too large to be useful. Intuitively, we must take advantage of the fact that *this* increment sequence is *special*. What we need to show is that for any element A_P in position P, when it is time to perform an h_k-sort, there are only a few elements to the left of position P that are larger than A_P.

When we come to h_k-sort the input array, we know that it has already been h_{k+1}- and h_{k+2}-sorted. Prior to the h_k-sort, consider elements in positions P and $P - i$, $i \leq P$. If i is a multiple of h_{k+1} or h_{k+2}, then clearly $A[P-i] < A[P]$. We can say more, however. If i is expressible as a linear combination (in nonnegative integers) of h_{k+1} and h_{k+2}, then $A[P-i] < A[P]$. As an example, when we come to 3-sort, the file is already 7- and 15-sorted. 52 is expressible as a linear combination of 7 and 15, because $52 = 1 * 7 + 3 * 15$. Thus, $A[100]$ cannot be larger than $A[152]$ because $A[100] \leq A[107] \leq A[122] \leq A[137] \leq A[152]$.

Now, $h_{k+2} = 2h_{k+1} + 1$, so h_{k+1} and h_{k+2} cannot share a common factor. In this case, it is possible to show that all integers that are at least as large as $(h_{k+1} - 1)(h_{k+2} - 1) = 8h_k^2 + 4h_k$ can be expressed as a linear combination of h_{k+1} and h_{k+2} (see the reference at the end of the chapter).

This tells us that the body of the *for* loop at line 4 can be executed at most $8h_k + 4 = O(h_k)$ times for each of the $N - h_k$ positions. This gives a bound of $O(Nh_k)$ per pass.

Using the fact that about half the increments satisfy $h_k < \sqrt{N}$, and assuming that t is even, the total running time is then

$$O\left(\sum_{k=1}^{t/2} Nh_k + \sum_{k=t/2+1}^{t} N^2/h_k\right) = O\left(N\sum_{k=1}^{t/2} h_k + N^2 \sum_{k=t/2+1}^{t} 1/h_k\right)$$

Because both sums are geometric series, and since $h_{t/2} = \Theta(\sqrt{N})$, this simplifies to

$$= O(Nh_{t/2}) + O\left(\frac{N^2}{h_{t/2}}\right) = O(N^{3/2})$$

The average-case running time of Shellsort, using Hibbard's increments, is thought to be $O(N^{5/4})$, based on simulations, but nobody has been able to prove this. Pratt has shown that the $\Theta(N^{3/2})$ bound applies to a wide range of increment sequences.

Sedgewick has proposed several increment sequences that give an $O(N^{4/3})$ worst-case running time (also achievable). The average running time is conjectured to be $O(N^{7/6})$ for these increment sequences. Empirical studies show that these sequences perform significantly better in practice than Hibbard's. The best of these is the sequence $\{1, 5, 19, 41, 109, \ldots\}$, in which the terms are either of the form $9 \cdot 4^i - 9 \cdot 2^i + 1$ or $4^i - 3 \cdot 2^i + 1$. This is most easily implemented by placing these values in an array. This increment sequence is the best known in practice, although there is a lingering possibility that some increment sequence might exist that could give a significant improvement in the running time of Shellsort.

There are several other results on Shellsort that (generally) require difficult theorems from number theory and combinatorics and are mainly of theoretical interest. Shellsort is a fine example of a very simple algorithm with an extremely complex analysis.

The performance of Shellsort is quite acceptable in practice, even for N in the tens of thousands. The simplicity of the code makes it the algorithm of choice for sorting up to moderately large input.

7.5. Heapsort

As mentioned in Chapter 6, priority queues can be used to sort in $O(N \log N)$ time. The algorithm based on this idea is known as *heapsort* and gives the best Big-Oh running time we have seen so far. In practice however, it is slower than a version of Shellsort that uses Sedgewick's increment sequence.

Recall, from Chapter 6, that the basic strategy is to build a binary heap of N elements. This stage takes $O(N)$ time. We then perform N *DeleteMin* operations. The elements leave the heap smallest first, in sorted order. By recording these elements in a second array and then copying the array back, we sort N elements. Since each *DeleteMin* takes $O(\log N)$ time, the total running time is $O(N \log N)$.

The main problem with this algorithm is that it uses an extra array. Thus, the memory requirement is doubled. This could be a problem in some instances. Notice that the extra time spent copying the second array back to the first is only $O(N)$, so that this is not likely to affect the running *time* significantly. The problem is space.

A clever way to avoid using a second array makes use of the fact that after each *DeleteMin*, the heap shrinks by 1. Thus the cell that was last in the heap can be used to store the element that was just deleted. As an example, suppose we have a heap with six elements. The first *DeleteMin* produces A_1. Now the heap has only five elements, so we can place A_1 in position 6. The next *DeleteMin* produces A_2. Since the heap will now only have four elements, we can place A_2 in position 5.

Using this strategy, after the last *DeleteMin* the array will contain the elements in *decreasing* sorted order. If we want the elements in the more typical *increasing* sorted order, we can change the ordering property so that the parent has a larger key than the child. Thus we have a (*max*)heap.

In our implementation, we will use a (*max*)heap, but avoid the actual ADT for the purposes of speed. As usual, everything is done in an array. The first step builds the heap in linear time. We then perform $N - 1$ *DeleteMaxes* by swapping the last element in the heap with the first, decrementing the heap size, and percolating down. When the algorithm terminates, the array contains the elements in sorted order. For instance, consider the input sequence $31, 41, 59, 26, 53, 58, 97$. The resulting heap is shown in Figure 7.6.

Figure 7.7 shows the heap that results after the first *DeleteMax*. As the figures imply, the last element in the heap is 31; 97 has been placed in a part of the heap array that is technically no longer part of the heap. After 5 more *DeleteMax*

Figure 7.6 (*Max*) heap after *BuildHeap* phase

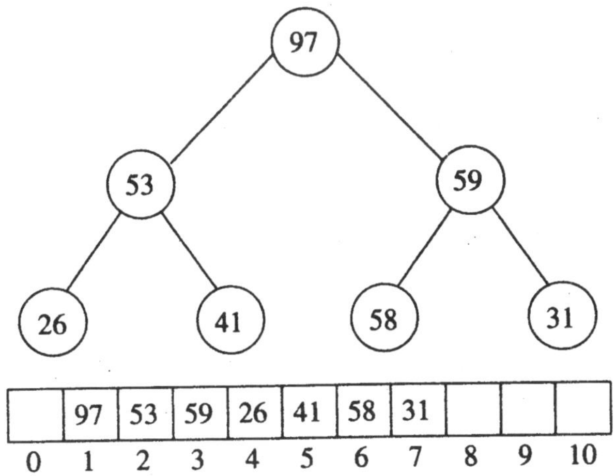

Figure 7.7 Heap after first *DeleteMax*

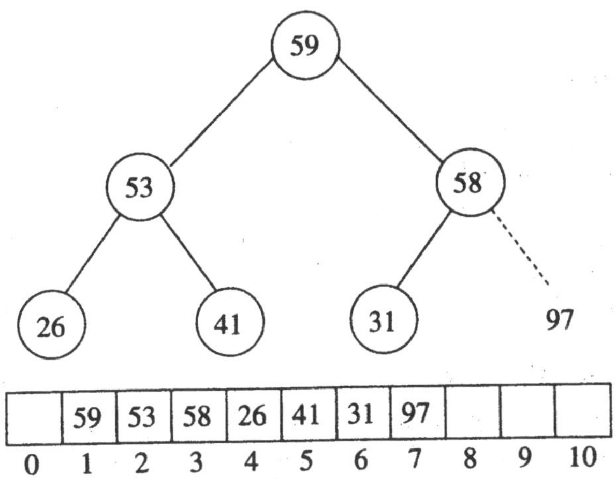

```
#define LeftChild( i ) ( 2 * ( i ) + 1 )

void
PercDown( ElementType A[ ], int i, int N )
{
    int Child;
    ElementType Tmp;

    for( Tmp = A[ i ]; LeftChild( i ) < N; i = Child )
    {
        Child = LeftChild( i );
        if( Child != N - 1 && A[ Child + 1 ] > A[ Child ] )
            Child++;
        if( Tmp < A[ Child ] )
            A[ i ] = A[ Child ];
        else
            break;
    }
    A[ i ] = Tmp;
}

void
Heapsort( ElementType A[ ], int N )
{
    int i;

    for( i = N / 2; i >= 0; i-- ) /* BuildHeap */
        PercDown( A, i, N );
    for( i = N - 1; i > 0; i-- )
    {
        Swap( &A[ 0 ], &A[ i ] ); /* DeleteMax */
        PercDown( A, 0, i );
    }
}
```

Figure 7.8 Heapsort

operations, the heap will actually have only one element, but the elements left in the heap array will be in sorted order.

The code to perform heapsort is given in Figure 7.8. The slight complication is that, unlike the binary heap, where the data begin at array index 1, the array for heapsort contains data in position 0. Thus the code is a little different from the binary heap code. The changes are minor.

7.5.1. *Analysis of Heapsort*

As we saw in Chapter 6, the first phase, which constitutes the building of the heap, uses at most $2N$ comparisons. In the second phase, the ith *DeleteMax* uses at most $2\lfloor \log i \rfloor$ comparisons, for a total of at most $2N \log N - O(N)$ comparisons (assuming

$N \geq 2$). Consequently, in the worst case, at most $2N \log N - O(N)$ comparisons are used by heapsort. Exercise 7.12(b) asks you to show that it is possible for all of the *DeleteMax* operations to achieve their worst case simultaneously.

Experiments have shown that heapsort is an extremely stable algorithm: On average it uses only slightly fewer comparisons than the worst-case bound suggests. Until recently, however, nobody had been able to show nontrivial bounds on heapsort's average running time. The problem, it seems, is that successive *DeleteMax* operations destroy the heap's randomness, making the probability arguments very complex. Recently another approach proved successful.

THEOREM 7.5.

The average number of comparisons used to heapsort a random permutation of N distinct items is $2N \log N - O(N \log\log N)$.

PROOF:

The heap construction phase uses $\Theta(N)$ comparisons on average, and so we only need to prove the bound for the second phase. We assume a permutation of $\{1, 2, \ldots, N\}$.

Suppose the ith *DeleteMax* pushes the root element down d_i levels. Then it uses $2d_i$ comparisons. For heapsort on any input, there is a cost sequence $D : d_1, d_2, \ldots, d_N$ that defines the cost of phase 2. That cost is given by $M_D = \sum_{i=1}^{N} d_i$; the number of comparisons used is thus $2M_D$.

Let $f(N)$ be the number of heaps of N items. One can show (Exercise 7.42) that $f(N) > (N/(4e))^N$ (where $e = 2.71828\ldots$). We will show that only an exponentially small fraction of these heaps (in particular $(N/16)^N$) have a cost smaller than $M = N(\log N - \log\log N - 4)$. When this is shown, it follows that the average value of M_D is at least M minus a term that is $o(1)$, and thus the average number of comparisons is at least $2M$. Consequently, our basic goal is to show that there are very few heaps that have small cost sequences.

Because level d_i has at most 2^{d_i} nodes, there are 2^{d_i} possible places that the root element can go for any d_i. Consequently, for any sequence D, the number of distinct corresponding *DeleteMax* sequences is at most

$$S_D = 2^{d_1} 2^{d_2} \cdots 2^{d_N}$$

A simple algebraic manipulation shows that *for a given sequence D*

$$S_D = 2^{M_D}$$

Because each d_i can assume any value between 1 and $\lfloor \log N \rfloor$, there are at most $(\log N)^N$ possible sequences D. It follows that the number of distinct *DeleteMax* sequences that require cost exactly M is at most the number of cost sequences of total cost M times the number of *DeleteMax* sequences for each of these cost sequences. A bound of $(\log N)^N 2^M$ follows immediately.

The total number of heaps with cost sequence less than M is at most

$$\sum_{i=1}^{M-1} (\log N)^N 2^i < (\log N)^N 2^M$$

If we choose $M = N(\log N - \log\log N - 4)$, then the number of heaps that have cost sequence less than M is at most $(N/16)^N$, and the theorem follows from our earlier comments.

Using a more complex argument, it can be shown that heapsort always uses at least $N \log N - O(N)$ comparisons, and that there are inputs that can achieve this bound. It seems that the average case should also be $2N \log N - O(N)$ comparisons (rather than the more linear second term in Theorem 7.5); whether this is provable (or even true) is open.

7.6. Mergesort

We now turn our attention to *mergesort*. Mergesort runs in $O(N \log N)$ worst-case running time, and the number of comparisons used is nearly optimal. It is a fine example of a recursive algorithm.

The fundamental operation in this algorithm is merging two sorted lists. Because the lists are sorted, this can be done in one pass through the input, if the output is put in a third list. The basic merging algorithm takes two input arrays A and B, an output array C, and three counters, *Aptr*, *Bptr*, and *Cptr*, which are initially set to the beginning of their respective arrays. The smaller of $A[Aptr]$ and $B[Bptr]$ is copied to the next entry in C, and the appropriate counters are advanced. When either input list is exhausted, the remainder of the other list is copied to C. An example of how the merge routine works is provided for the following input.

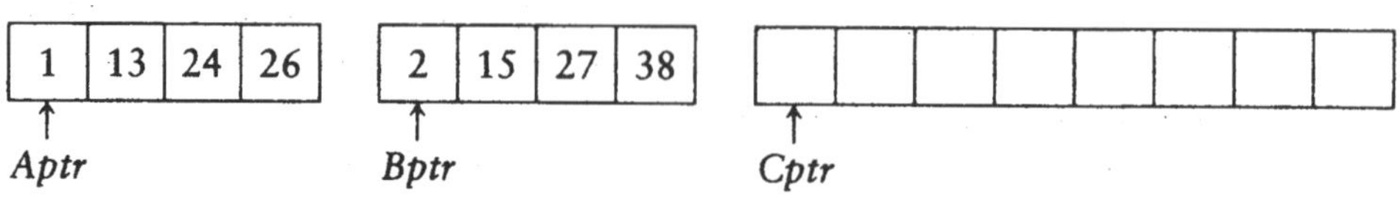

If the array A contains 1, 13, 24, 26, and B contains 2, 15, 27, 38, then the algorithm proceeds as follows: First, a comparison is done between 1 and 2. 1 is added to C, and then 13 and 2 are compared.

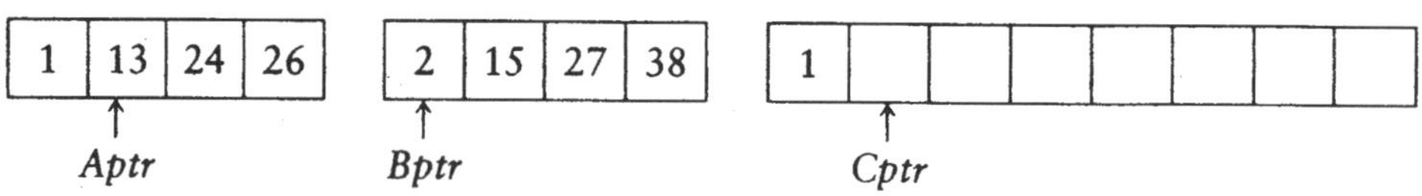

2 is added to C, and then 13 and 15 are compared.

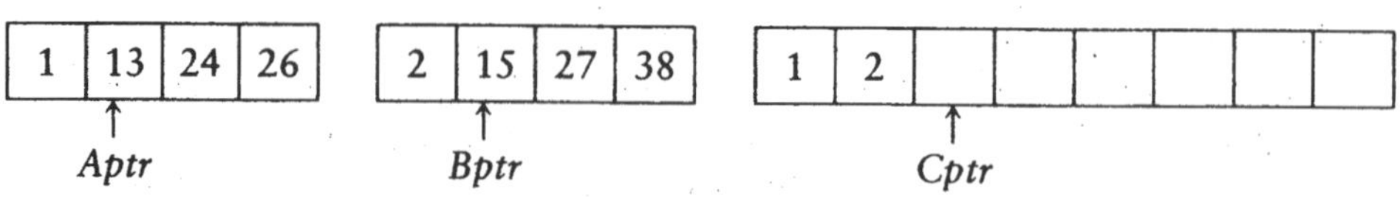

13 is added to C, and then 24 and 15 are compared. This proceeds until 26 and 27 are compared.

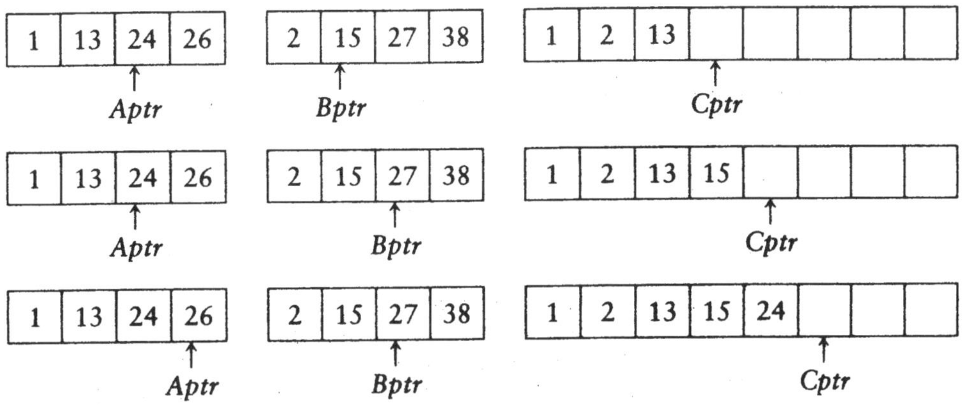

26 is added to *C*, and the *A* array is exhausted.

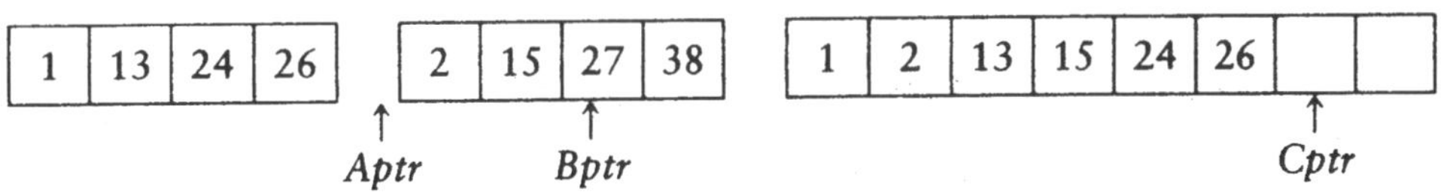

The remainder of the *B* array is then copied to *C*.

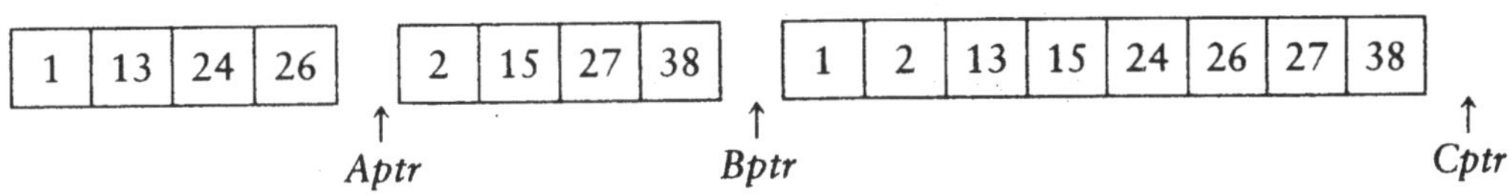

The time to merge two sorted lists is clearly linear, because at most $N - 1$ comparisons are made, where N is the total number of elements. To see this, note that every comparison adds an element to *C*, except the last comparison, which adds at least two.

The mergesort algorithm is therefore easy to describe. If $N = 1$, there is only one element to sort, and the answer is at hand. Otherwise, recursively mergesort the first half and the second half. This gives two sorted halves, which can then be merged together using the merging algorithm described above. For instance, to sort the eight-element array 24, 13, 26, 1, 2, 27, 38, 15, we recursively sort the first four and last four elements, obtaining 1, 13, 24, 26, 2, 15, 27, 38. Then we merge the two halves as above, obtaining the final list 1, 2, 13, 15, 24, 26, 27, 38. This algorithm is a classic divide-and-conquer strategy. The problem is *divided* into smaller problems and solved recursively. The *conquering* phase consists of patching together the answers. Divide-and-conquer is a very powerful use of recursion that we will see many times.

An implementation of mergesort is provided in Figure 7.9. The procedure called *Mergesort* is just a driver for the recursive routine *MSort*.

The *Merge* routine is subtle. If a temporary array is declared locally for each recursive call of *Merge*, then there could be $\log N$ temporary arrays active at any point. This could be fatal on a machine with small memory. On the other hand, if the

```
void
MSort( ElementType A[ ], ElementType TmpArray[ ],
                int Left, int Right )
{
    int Center;

    if( Left < Right )
    {
        Center = ( Left + Right ) / 2;
        MSort( A, TmpArray, Left, Center );
        MSort( A, TmpArray, Center + 1, Right );
        Merge( A, TmpArray, Left, Center + 1, Right );
    }
}

void
Mergesort( ElementType A[ ], int N )
{
    ElementType *TmpArray;

    TmpArray = malloc( N * sizeof( ElementType ) );
    if( TmpArray != NULL )
    {
        MSort( A, TmpArray, 0, N - 1 );
        free( TmpArray );
    }
    else
        FatalError( "No space for tmp array!!!" );
}
```

Figure 7.9 Mergesort routine

merge routine dynamically allocates and frees the minimum amount of temporary memory, considerable time will be used by *malloc*. A close examination shows that since *Merge* is the last line of *MSort*, there only needs to be one temporary array active at any point. Further, we can use any part of the temporary array; we will use the same portion as the input array *A*. This allows the improvement described at the end of this section. Figure 7.10 implements the *Merge* routine.

7.6.1. *Analysis of Mergesort*

Mergesort is a classic example of the techniques used to analyze recursive routines: we have to write a recurrence relation for the running time. We will assume that N is a power of 2, so that we always split into even halves. For $N = 1$, the time to mergesort is constant, which we will denote by 1. Otherwise, the time to mergesort N numbers is equal to the time to do two recursive mergesorts of size $N/2$, plus the

```
/* Lpos = start of left half, Rpos = start of right half */

void
Merge( ElementType A[ ], ElementType TmpArray[ ],
                int Lpos, int Rpos, int RightEnd )
{
    int i, LeftEnd, NumElements, TmpPos;

    LeftEnd = Rpos - 1;
    TmpPos = Lpos;
    NumElements = RightEnd - Lpos + 1;

    /* main loop */
    while( Lpos <= LeftEnd && Rpos <= RightEnd )
        if( A[ Lpos ] <= A[ Rpos ] )
            TmpArray[ TmpPos++ ] = A[ Lpos++ ];
        else
            TmpArray[ TmpPos++ ] = A[ Rpos++ ];

    while( Lpos <= LeftEnd )  /* Copy rest of first half */
        TmpArray[ TmpPos++ ] = A[ Lpos++ ];
    while( Rpos <= RightEnd ) /* Copy rest of second half */
        TmpArray[ TmpPos++ ] = A[ Rpos++ ];

    /* Copy TmpArray back */
    for( i = 0; i < NumElements; i++, RightEnd-- )
        A[ RightEnd ] = TmpArray[ RightEnd ];
}
```

Figure 7.10 *Merge* routine

time to merge, which is linear. The following equations say this exactly:

$$T(1) = 1$$

$$T(N) = 2T(N/2) + N$$

This is a standard recurrence relation, which can be solved several ways. We will show two methods. The first idea is to divide the recurrence relation through by N. The reason for doing this will become apparent soon. This yields

$$\frac{T(N)}{N} = \frac{T(N/2)}{N/2} + 1$$

This equation is valid for any N that is a power of 2, so we may also write

$$\frac{T(N/2)}{N/2} = \frac{T(N/4)}{N/4} + 1$$

and

$$\frac{T(N/4)}{N/4} = \frac{T(N/8)}{N/8} + 1$$

$$\vdots$$

$$\frac{T(2)}{2} = \frac{T(1)}{1} + 1$$

Now add up all the equations. This means that we add all of the terms on the left-hand side and set the result equal to the sum of all of the terms on the right-hand side. Observe that the term $T(N/2)/(N/2)$ appears on both sides and thus cancels. In fact, virtually all the terms appear on both sides and cancel. This is called *telescoping* a sum. After everything is added, the final result is

$$\frac{T(N)}{N} = \frac{T(1)}{1} + \log N$$

because all of the other terms cancel and there are $\log N$ equations, and so all the 1s at the end of these equations add up to $\log N$. Multiplying through by N gives the final answer.

$$T(N) = N \log N + N = O(N \log N)$$

Notice that if we did not divide through by N at the start of the solutions, the sum would not telescope. This is why it was necessary to divide through by N.

An alternative method is to substitute the recurrence relation continually on the right-hand side. We have

$$T(N) = 2T(N/2) + N$$

Since we can substitute $N/2$ into the main equation,

$$2T(N/2) = 2(2(T(N/4)) + N/2) = 4T(N/4) + N$$

we have

$$T(N) = 4T(N/4) + 2N$$

Again, by substituting $N/4$ into the main equation, we see that

$$4T(N/4) = 4(2T(N/8)) + N/4) = 8T(N/8) + N$$

So we have

$$T(N) = 8T(N/8) + 3N$$

Continuing in this manner, we obtain

$$T(N) = 2^k T(N/2^k) + k \cdot N$$

Using $k = \log N$, we obtain

$$T(N) = NT(1) + N \log N = N \log N + N$$

The choice of which method to use is a matter of taste. The first method tends to produce scrap work that fits better on a standard, $8\frac{1}{2} \times 11$ sheet of paper, leading to fewer mathematical errors, but it requires a certain amount of experience to apply. The second method is more of a brute force approach.

Recall that we have assumed $N = 2^k$. The analysis can be refined to handle cases when N is not a power of 2. The answer turns out to be almost identical (this is usually the case).

Although mergesort's running time is $O(N \log N)$, it is hardly ever used for main memory sorts. The main problem is that merging two sorted lists requires linear extra memory, and the additional work spent copying to the temporary array and back, throughout the algorithm, has the effect of slowing down the sort considerably. This copying can be avoided by judiciously switching the roles of A and *TmpArray* at alternate levels of the recursion. A variant of mergesort can also be implemented nonrecursively (Exercise 7.14), but even so, for serious internal sorting applications, the algorithm of choice is quicksort, which is described in the next section. Nevertheless, as we will see later in this chapter, the merging routine is the cornerstone of most external sorting algorithms.

7.7. Quicksort

As its name implies, *quicksort* is the fastest known sorting algorithm in practice. Its average running time is $O(N \log N)$. It is very fast, mainly due to a very tight and highly optimized inner loop. It has $O(N^2)$ worst-case performance, but this can be made exponentially unlikely with a little effort. The quicksort algorithm is simple to understand and prove correct, although for many years it had the reputation of being an algorithm that could in theory be highly optimized but in practice was impossible to code correctly. Like mergesort, quicksort is a divide-and-conquer recursive algorithm. The basic algorithm to sort an array S consists of the following four easy steps:

1. If the number of elements in S is 0 or 1, then return.
2. Pick any element v in S. This is called the *pivot*.
3. *Partition* $S - \{v\}$ (the remaining elements in S) into two disjoint groups: $S_1 = \{x \in S - \{v\} | x \leq v\}$, and $S_2 = \{x \in S - \{v\} | x \geq v\}$.
4. Return {quicksort(S_1) followed by v followed by quicksort(S_2)}.

Since the partition step ambiguously describes what to do with elements equal to the pivot, this becomes a design decision. Part of a good implementation is handling this case as efficiently as possible. Intuitively, we would hope that about half the keys that are equal to the pivot go into S_1 and the other half into S_2, much as we like binary search trees to be balanced.

Figure 7.11 shows the action of quicksort on a set of numbers. The pivot is chosen (by chance) to be 65. The remaining elements in the set are partitioned into two smaller sets. Recursively sorting the set of smaller numbers yields 0, 13, 26, 31, 43, 57 (by rule 3 of recursion). The set of large numbers is similarly sorted. The sorted arrangement of the entire set is then trivially obtained.

It should be clear that this algorithm works, but it is not clear why it is any faster than mergesort. Like mergesort, it recursively solves two subproblems and requires linear additional work (step 3), but, unlike mergesort, the subproblems are not guaranteed to be of equal size, which is potentially bad. The reason that quicksort is faster is that the partitioning step can actually be performed in place

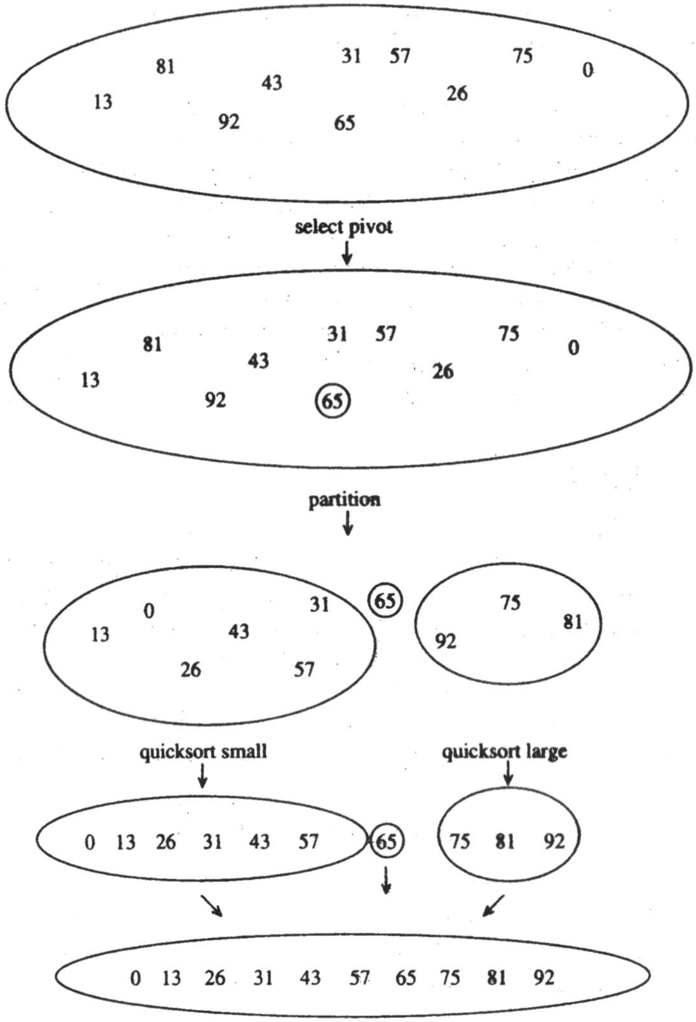

Figure 7.11 The steps of quicksort illustrated by example

and very efficiently. This efficiency more than makes up for the lack of equal-sized recursive calls.

The algorithm as described so far lacks quite a few details, which we now fill in. There are many ways to implement steps 2 and 3; the method presented here is the result of extensive analysis and empirical study and represents a very efficient way to implement quicksort. Even the slightest deviations from this method can cause surprisingly bad results.

7.7.1. *Picking the Pivot*

Although the algorithm as described works no matter which element is chosen as pivot, some choices are obviously better than others.

A Wrong Way

The popular, uninformed choice is to use the first element as the pivot. This is acceptable if the input is random, but if the input is presorted or in reverse order, then the pivot provides a poor partition, because either all the elements go into S_1 or they go into S_2. Worse, this happens consistently throughout the recursive calls. The practical effect is that if the first element is used as the pivot and the input is presorted, then quicksort will take quadratic time to do essentially nothing at all, which is quite embarrassing. Moreover, presorted input (or input with a large presorted section) is quite frequent, so using the first element as pivot *is an absolutely horrible idea* and should be discarded immediately. An alternative is choosing the larger of the first two distinct keys as pivot, but this has the same bad properties as merely choosing the first key. Do not use that pivoting strategy either.

A Safe Maneuver

A safe course is merely to choose the pivot randomly. This strategy is generally perfectly safe, unless the random number generator has a flaw (which is not as uncommon as you might think), since it is very unlikely that a random pivot would consistently provide a poor partition. On the other hand, random number generation is generally an expensive commodity and does not reduce the average running time of the rest of the algorithm at all.

Median-of-Three Partitioning

The median of a group of N numbers is the $\lceil N/2 \rceil$th largest number. The best choice of pivot would be the median of the array. Unfortunately, this is hard to calculate and would slow down quicksort considerably. A good estimate can be obtained by picking three elements randomly and using the median of these three as pivot. The randomness turns out not to help much, so the common course is to use as pivot the median of the left, right, and center elements. For instance, with input $8, 1, 4, 9, 6, 3, 5, 2, 7, 0$ as before, the left element is 8, the right element is 0 and the center (in position $\lfloor (Left + Right)/2 \rfloor$) element is 6. Thus, the pivot would be $v = 6$. Using median-of-three partitioning clearly eliminates the bad case for sorted input (the partitions become equal in this case) and actually reduces the running time of quicksort by about 5 percent.

7.7.2. *Partitioning Strategy*

There are several partitioning strategies used in practice, but the one described here is known to give good results. It is very easy, as we shall see, to do this wrong or inefficiently, but it is safe to use a known method. The first step is to get the pivot element out of the way by swapping it with the last element. i starts at the first element and j starts at the next-to-last element. If the original input was the same as before, the following figure shows the current situation.

8	1	4	9	0	3	5	2	7	6
↑								↑	
i								*j*	

For now we will assume that all the elements are distinct. Later on we will worry about what to do in the presence of duplicates. As a limiting case, our algorithm must do the proper thing if *all* of the elements are identical. It is surprising how easy it is to do the *wrong* thing.

What our partitioning stage wants to do is to move all the small elements to the left part of the array and all the large elements to the right part. "Small" and "large" are, of course, relative to the pivot.

While *i* is to the left of *j*, we move *i* right, skipping over elements that are smaller than the pivot. We move *j* left, skipping over elements that are larger than the pivot. When *i* and *j* have stopped, *i* is pointing at a large element and *j* is pointing at a small element. If *i* is to the left of *j*, those elements are swapped. The effect is to push a large element to the right and a small element to the left. In the example above, *i* would not move and *j* would slide over one place. The situation is as follows.

8	1	4	9	0	3	5	2	7	6
↑							↑		
i							*j*		

We then swap the elements pointed to by *i* and *j* and repeat the process until *i* and *j* cross.

After First Swap

2	1	4	9	0	3	5	8	7	6
↑							↑		
i							*j*		

Before Second Swap

2	1	4	9	0	3	5	8	7	6
			↑			↑			
			i			*j*			

After Second Swap

2	1	4	5	0	3	9	8	7	6
			↑			↑			
			i			*j*			

Before Third Swap									
2	1	4	5	0	3	9	8	7	6
					↑	↑			
					j	i			

At this stage, i and j have crossed, so no swap is performed. The final part of the partitioning is to swap the pivot element with the element pointed to by i.

After Swap with Pivot									
2	1	4	5	0	3	6	8	7	9
						↑			↑
						i			pivot

When the pivot is swapped with i in the last step, we know that every element in a position $P < i$ must be small. This is because either position P contained a small element to start with, or the large element originally in position P was replaced during a swap. A similar argument shows that elements in positions $P > i$ must be large.

One important detail we must consider is how to handle keys that are equal to the pivot. The questions are whether or not i should stop when it sees a key equal to the pivot and whether or not j should stop when it sees a key equal to the pivot. Intuitively, i and j ought to do the same thing, since otherwise the partitioning step is biased. For instance, if i stops and j does not, then all keys that are equal to the pivot will wind up in S_2.

To get an idea of what might be good, we consider the case where all the keys in the array are identical. If both i and j stop, there will be many swaps between identical elements. Although this seems useless, the positive effect is that i and j will cross in the middle, so when the pivot is replaced, the partition creates two nearly equal subarrays. The mergesort analysis tells us that the total running time would then be $O(N \log N)$.

If neither i nor j stops, and code is present to prevent them from running off the end of the array, no swaps will be performed. Although this seems good, a correct implementation would then swap the pivot into the last spot that i touched, which would be the next-to-last position (or last, depending on the exact implementation). This would create very uneven subarrays. If all the keys are identical, the running time is $O(N^2)$. The effect is the same as using the first element as a pivot for presorted input. It takes quadratic time to do nothing!

Thus, we find that it is better to do the unnecessary swaps and create even subarrays than to risk wildly uneven subarrays. Therefore, we will have both i and j stop if they encounter a key equal to the pivot. This turns out to be the only one of the four possibilities that does not take quadratic time for this input.

At first glance it may seem that worrying about an array of identical elements is silly. After all, why would anyone want to sort 5,000 identical elements? However, recall that quicksort is recursive. Suppose there are 100,000 elements, of which 5,000 are identical. Eventually, quicksort will make the recursive call on only these 5,000 elements. Then it really will be important to make sure that 5,000 identical elements can be sorted efficiently.

7.7.3. *Small Arrays*

For very small arrays ($N \leq 20$), quicksort does not perform as well as insertion sort. Furthermore, because quicksort is recursive, these cases will occur frequently. A common solution is not to use quicksort recursively for small arrays, but instead use a sorting algorithm that is efficient for small arrays, such as insertion sort. Using this strategy can actually save about 15 percent in the running time (over doing no cutoff at all). A good cutoff range is $N = 10$, although any cutoff between 5 and 20 is likely to produce similar results. This also saves nasty degenerate cases, such as taking the median of three elements when there are only one or two.

7.7.4. *Actual Quicksort Routines*

The driver for quicksort is shown in Figure 7.12.

The general form of the routines will be to pass the array and the range of the array (*Left* and *Right*) to be sorted. The first routine to deal with is pivot selection. The easiest way to do this is to sort *A*[*Left*], *A*[*Right*], and *A*[*Center*] in place. This has the extra advantage that the smallest of the three winds up in *A*[*Left*], which is where the partitioning step would put it anyway. The largest winds up in *A*[*Right*], which is also the correct place, since it is larger than the pivot. Therefore, we can place the pivot in *A*[*Right* − 1] and initialize i and j to *Left* + 1 and *Right* − 2 in the partition phase. Yet another benefit is that because *A*[*Left*] is smaller than the pivot, it will act as a sentinel for j. Thus, we do not need to worry about j running past the end. Since i will stop on keys equal to the pivot, storing the pivot in *A*[*Right* − 1] provides a sentinel for i. The code in Figure 7.13 does the median-of-three partitioning with all the side effects described. It may seem that it is only slightly inefficient to compute the pivot by a method that does not actually sort *A*[*Left*], *A*[*Center*], and *A*[*Right*], but, surprisingly, this produces bad results (see Exercise 7.38).

The real heart of the quicksort routine is in Figure 7.14. It includes the partitioning and recursive calls. There are several things worth noting in this

Figure 7.12 Driver for quicksort

```
void
Quicksort( ElementType A[ ], int N )
{
    Qsort( A, 0, N - 1 );
}
```

```
/* Return median of Left, Center, and Right */
/* Order these and hide the pivot */

ElementType
Median3( ElementType A[ ], int Left, int Right )
{
    int Center = ( Left + Right ) / 2;

    if( A[ Left ] > A[ Center ] )
        Swap( &A[ Left ], &A[ Center ] );
    if( A[ Left ] > A[ Right ] )
        Swap( &A[ Left ], &A[ Right ] );
    if( A[ Center ] > A[ Right ] )
        Swap( &A[ Center ], &A[ Right ] );

    /* Invariant: A[ Left ] <= A[ Center ] <= A[ Right ] */

    Swap( &A[ Center ], &A[ Right - 1 ] );  /* Hide pivot */
    return A[ Right - 1 ];                 /* Return pivot */
}
```

Figure 7.13 Code to perform median-of-three partitioning

implementation. Line 3 initializes i and j to 1 past their correct values, so that there are no special cases to consider. This initialization depends on the fact that median-of-three partitioning has some side effects; this program will not work if you try to use it without change with a simple pivoting strategy, because i and j start in the wrong place and there is no longer a sentinel for j.

The *Swap* at line 8 is sometimes written explicitly, for speed purposes. For the algorithm to be fast, it is necessary to force the compiler to compile this code in-line. Many compilers will do this automatically, if asked to, but for those that do not the difference can be significant.

Finally, lines 5 and 6 show why quicksort is so fast. The inner loop of the algorithm consists of an increment/decrement (by 1, which is fast), a test, and a jump. There is no extra juggling as there is in mergesort. This code is still surprisingly tricky. It is tempting to replace lines 3 through 9 with the statements in Figure 7.15. This does not work, because there would be an infinite loop if $A[i] = A[j] = Pivot$.

7.7.5. *Analysis of Quicksort*

Like mergesort, quicksort is recursive, and hence, its analysis requires solving a recurrence formula. We will do the analysis for a quicksort, assuming a random pivot (no median-of-three partitioning) and no cutoff for small files. We will take $T(0) = T(1) = 1$, as in mergesort. The running time of quicksort is equal to the running time of the two recursive calls plus the linear time spent in the partition (the

```
        #define Cutoff ( 3 )

        void
        Qsort( ElementType A[ ], int Left, int Right )
        {
            int i, j;
            ElementType Pivot;

/* 1*/      if( Left + Cutoff <= Right )
            {
/* 2*/          Pivot = Median3( A, Left, Right );
/* 3*/          i = Left; j = Right - 1;
/* 4*/          for( ; ; )
                {
/* 5*/              while( A[ ++i ] < Pivot ){ }
/* 6*/              while( A[ --j ] > Pivot ){ }
/* 7*/              if( i < j )
/* 8*/                  Swap( &A[ i ], &A[ j ] );
                    else
/* 9*/                  break;
                }
/*10*/          Swap( &A[ i ], &A[ Right - 1 ] ); /* Restore pivot */

/*11*/          Qsort( A, Left, i - 1 );
/*12*/          Qsort( A, i + 1, Right );
            }
            else /* Do an insertion sort on the subarray */
/*13*/          InsertionSort( A + Left, Right - Left + 1 );
        }
```

Figure 7.14 Main quicksort routine

```
/* 3*/  i = Left + 1; j = Right - 2;
/* 4*/  for( ; ; )
        {
/* 5*/      while( A[ i ] < Pivot ) i++;
/* 6*/      while( A[ j ] > Pivot ) j--;
/* 7*/      if( i < j )
/* 8*/          Swap( &A[ i ], &A[ j ] );
            else
/* 9*/          break;
        }
```

Figure 7.15 A small change to quicksort, which breaks the algorithm

pivot selection takes only constant time). This gives the basic quicksort relation

$$T(N) = T(i) + T(N - i - 1) + cN \tag{7.1}$$

where $i = |S_1|$ is the number of elements in S_1. We will look at three cases.

Worst-Case Analysis

The pivot is the smallest element, all the time. Then $i = 0$ and if we ignore $T(0) = 1$, which is insignificant, the recurrence is

$$T(N) = T(N - 1) + cN, \quad N > 1 \tag{7.2}$$

We telescope, using Equation (7.2) repeatedly. Thus

$$T(N - 1) = T(N - 2) + c(N - 1) \tag{7.3}$$

$$T(N - 2) = T(N - 3) + c(N - 2) \tag{7.4}$$

$$\vdots$$

$$T(2) = T(1) + c(2) \tag{7.5}$$

Adding up all these equations yields

$$T(N) = T(1) + c\sum_{i=2}^{N} i = O(N^2) \tag{7.6}$$

as claimed earlier.

Best-Case Analysis

In the best case, the pivot is in the middle. To simplify the math, we assume that the two subarrays are each exactly half the size of the original, and although this gives a slight overestimate, this is acceptable because we are only interested in a Big-Oh answer.

$$T(N) = 2T(N/2) + cN \tag{7.7}$$

Divide both sides of Equation (7.7) by N.

$$\frac{T(N)}{N} = \frac{T(N/2)}{N/2} + c \tag{7.8}$$

We will telescope using this equation.

$$\frac{T(N/2)}{N/2} = \frac{T(N/4)}{N/4} + c \tag{7.9}$$

$$\frac{T(N/4)}{N/4} = \frac{T(N/8)}{N/8} + c \tag{7.10}$$

$$\vdots$$

$$\frac{T(2)}{2} = \frac{T(1)}{1} + c \tag{7.11}$$

We add all the equations from (7.7) to (7.11) and note that there are $\log N$ of them:

$$\frac{T(N)}{N} = \frac{T(1)}{1} + c\log N \tag{7.12}$$

which yields

$$T(N) = cN\log N + N = O(N\log N) \tag{7.13}$$

Notice that this is the exact same analysis as mergesort, hence we get the same answer.

Average-Case Analysis

This is the most difficult part. For the average case, we assume that each of the file sizes for S_1 is equally likely, and hence has probability $1/N$. This assumption is actually valid for our pivoting and partitioning strategy, but it is not valid for some others. Partitioning strategies that do not preserve the randomness of the subfiles cannot use this analysis. Interestingly, these strategies seem to result in programs that take longer to run in practice.

With this assumption, the average value of $T(i)$, and hence $T(N-i-1)$, is $(1/N)\sum_{j=0}^{N-1} T(j)$. Equation (7.1) then becomes

$$T(N) = \frac{2}{N}\left[\sum_{j=0}^{N-1} T(j)\right] + cN \tag{7.14}$$

If Equation (7.14) is multiplied by N, it becomes

$$NT(N) = 2\left[\sum_{j=0}^{N-1} T(j)\right] + cN^2 \tag{7.15}$$

We need to remove the summation sign to simplify matters. We note that we can telescope with one more equation.

$$(N-1)T(N-1) = 2\left[\sum_{j=0}^{N-2} T(j)\right] + c(N-1)^2 \tag{7.16}$$

If we subtract (7.16) from (7.15), we obtain

$$NT(N) - (N-1)T(N-1) = 2T(N-1) + 2cN - c \tag{7.17}$$

We rearrange terms and drop the insignificant $-c$ on the right, obtaining

$$NT(N) = (N+1)T(N-1) + 2cN \tag{7.18}$$

We now have a formula for $T(N)$ in terms of $T(N-1)$ only. Again the idea is to telescope, but Equation (7.18) is in the wrong form. Divide Equation (7.18) by $N(N+1)$:

$$\frac{T(N)}{N+1} = \frac{T(N-1)}{N} + \frac{2c}{N+1} \tag{7.19}$$

Now we can telescope.

$$\frac{T(N-1)}{N} = \frac{T(N-2)}{N-1} + \frac{2c}{N} \tag{7.20}$$

$$\frac{T(N-2)}{N-1} = \frac{T(N-3)}{N-2} + \frac{2c}{N-1} \tag{7.21}$$

$$\vdots$$

$$\frac{T(2)}{3} = \frac{T(1)}{2} + \frac{2c}{3} \tag{7.22}$$

Adding Equations (7.19) through (7.22) yields

$$\frac{T(N)}{N+1} = \frac{T(1)}{2} + 2c\sum_{i=3}^{N+1}\frac{1}{i} \tag{7.23}$$

The sum is about $\log_e(N+1)+\gamma-\frac{3}{2}$, where $\gamma \approx 0.577$ is known as Euler's constant, so

$$\frac{T(N)}{N+1} = O(\log N) \tag{7.24}$$

And so

$$T(N) = O(N \log N) \tag{7.25}$$

Although this analysis seems complicated, it really is not—the steps are natural once you have seen some recurrence relations. The analysis can actually be taken further. The highly optimized version that was described above has also been analyzed, and this result gets extremely difficult, involving complicated recurrences and advanced mathematics. The effect of equal keys has also been analyzed in detail, and it turns out that the code presented does the right thing.

7.7.6. A Linear-Expected-Time Algorithm for Selection

Quicksort can be modified to solve the *selection problem*, which we have seen in Chapters 1 and 6. Recall that by using a priority queue, we can find the kth largest (or smallest) element in $O(N + k \log N)$. For the special case of finding the median, this gives an $O(N \log N)$ algorithm.

Since we can sort the array in $O(N \log N)$ time, one might expect to obtain a better time bound for selection. The algorithm we present to find the kth smallest element in a set S is almost identical to quicksort. In fact, the first three steps are the same. We will call this algorithm *quickselect*. Let $|S_i|$ denote the number of elements in S_i. The steps of quickselect are

1. If $|S| = 1$, then $k = 1$ and return the element in S as the answer. If a cutoff for small arrays is being used and $|S| \leq CUTOFF$, then sort S and return the kth smallest element.
2. Pick a pivot element, $v \in S$.
3. Partition $S - \{v\}$ into S_1 and S_2, as was done with quicksort.

4. If $k \leq |S_1|$, then the kth smallest element must be in S_1. In this case, return quickselect (S_1, k). If $k = 1 + |S_1|$, then the pivot is the kth smallest element and we can return it as the answer. Otherwise, the kth smallest element lies in S_2, and it is the $(k - |S_1| - 1)$st smallest element in S_2. We make a recursive call and return quickselect $(S_2, k - |S_1| - 1)$.

In contrast to quicksort, quickselect makes only one recursive call instead of two. The worst case of quickselect is identical to that of quicksort and is $O(N^2)$. Intuitively, this is because quicksort's worst case is when one of S_1 and S_2 is empty; thus, quickselect is not really saving a recursive call. The average running time, however, is $O(N)$. The analysis is similar to quicksort's and is left as an exercise.

The implementation of quickselect is even simpler than the abstract description might imply. The code to do this is shown in Figure 7.16. When the algorithm

Figure 7.16 Main quickselect routine

```
/* Places the kth smallest element in the kth position */
/* Because arrays start at 0, this will be index k-1 */
void
Qselect( ElementType A[ ], int k, int Left, int Right )
{
    int i, j;
    ElementType Pivot;

    if( Left + Cutoff <= Right )
    {
        Pivot = Median3( A, Left, Right );
        i = Left; j = Right - 1;
        for( ; ; )
        {
            while( A[ ++i ] < Pivot ){ }
            while( A[ --j ] > Pivot ){ }
            if( i < j )
                Swap( &A[ i ], &A[ j ] );
            else
                break;
        }
        Swap( &A[ i ], &A[ Right - 1 ] ); /* Restore pivot */

        if( k <= i )
            Qselect( A, k, Left, i - 1 );
        else if( k > i + 1 )
            Qselect( A, k, i + 1, Right );
    }
    else  /* Do an insertion sort on the subarray */
        InsertionSort( A + Left, Right - Left + 1 );
}
```

terminates, the kth smallest element is in position k. This destroys the original ordering; if this is not desirable, then a copy must be made.

Using a median-of-three pivoting strategy makes the chance of the worst case occurring almost negligible. By carefully choosing the pivot, however, we can eliminate the quadratic worst case and ensure an $O(N)$ algorithm. The overhead involved in doing this is considerable, so the resulting algorithm is mostly of theoretical interest. In Chapter 10, we will examine the linear-time worst-case algorithm for selection, and we shall also see an interesting technique of choosing the pivot that results in a somewhat faster selection algorithm in practice.

7.8. Sorting Large Structures

Throughout our discussion of sorting, we have assumed that the elements to be sorted are simply integers. Frequently, we need to sort large structures by a certain key. For instance, we might have payroll records, with each record consisting of a name, address, phone number, financial information such as salary, and tax information. We might want to sort this information by one particular field, such as the name. For all of our algorithms, the fundamental operation is the swap, but here swapping two structures can be a very expensive operation, because the structures are potentially large. If this is the case, a practical solution is to have the input array contain pointers to the structures. We sort by comparing the keys the pointers point to, swapping pointers when necessary. This means that all the data movement is essentially the same as if we were sorting integers. This is known as *indirect sorting*; we can use this technique for most of the data structures we have described. This justifies our assumption that complex structures can be handled without tremendous loss of efficiency.

7.9. A General Lower Bound for Sorting

Although we have $O(N \log N)$ algorithms for sorting, it is not clear that this is as good as we can do. In this section, we prove that any algorithm for sorting that uses only comparisons requires $\Omega(N \log N)$ comparisons (and hence time) in the worst case, so that mergesort and heapsort are optimal to within a constant factor. The proof can be extended to show that $\Omega(N \log N)$ comparisons are required, even on average, for any sorting algorithm that uses only comparisons, which means that quicksort is optimal on average to within a constant factor.

Specifically, we will prove the following result: Any sorting algorithm that uses only comparisons requires $\lceil \log(N!) \rceil$ comparisons in the worst case and $\log(N!)$ comparisons on average. We will assume that all N elements are distinct, since any sorting algorithm must work for this case.

7.9.1. Decision Trees

A *decision tree* is an abstraction used to prove lower bounds. In our context, a decision tree is a binary tree. Each node represents a set of possible orderings,

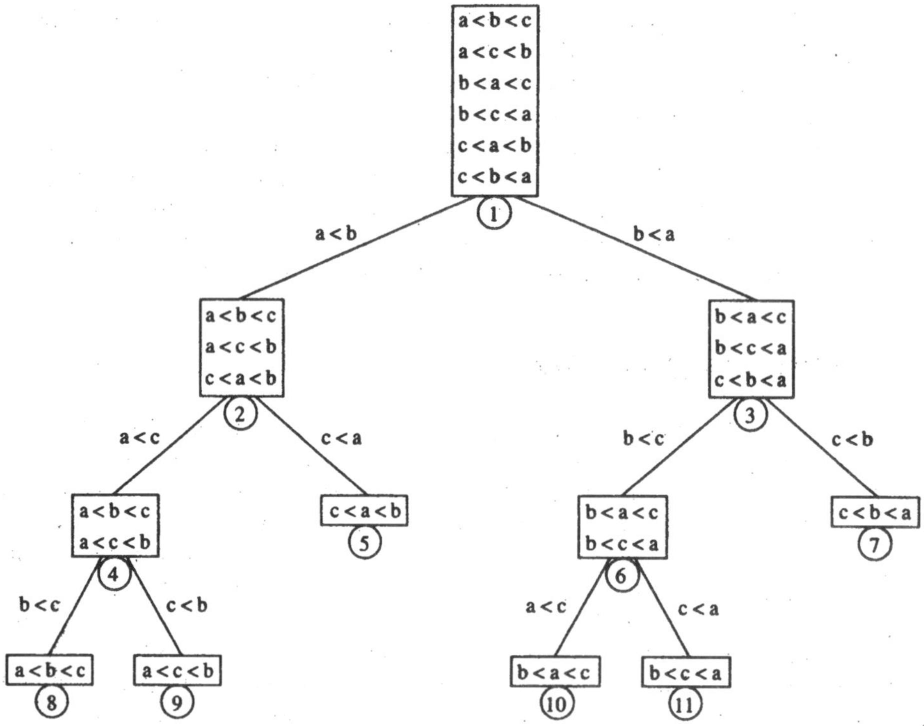

Figure 7.17 A decision tree for three-element insertion sort

consistent with comparisons that have been made, among the elements. The results of the comparisons are the tree edges.

The decision tree in Figure 7.17 represents an algorithm that sorts the three elements a, b, and c. The initial state of the algorithm is at the root. (We will use the terms *state* and *node* interchangeably.) No comparisons have been done, so all orderings are legal. The first comparison that *this particular* algorithm performs compares a and b. The two results lead to two possible states. If $a < b$, then only three possibilities remain. If the algorithm reaches node 2, then it will compare a and c. Other algorithms might do different things; a different algorithm would have a different decision tree. If $a > c$, the algorithm enters state 5. Since there is only one ordering that is consistent, the algorithm can terminate and report that it has completed the sort. If $a < c$, the algorithm cannot do this, because there are two possible orderings and it cannot possibly be sure which is correct. In this case, the algorithm will require one more comparison.

Every algorithm that sorts by using only comparisons can be represented by a decision tree. Of course, it is only feasible to draw the tree for extremely small input sizes. The number of comparisons used by the sorting algorithm is equal to the depth of the deepest leaf. In our case, this algorithm uses three comparisons in the worst case. The average number of comparisons used is equal to the average depth of the leaves. Since a decision tree is large, it follows that there must be some long paths. To prove the lower bounds, all that needs to be shown are some basic tree properties.

LEMMA 7.1.

Let T be a binary tree of depth d. Then T has at most 2^d leaves.

PROOF:

The proof is by induction. If $d = 0$, then there is at most one leaf, so the basis is true. Otherwise, we have a root, which cannot be a leaf, and a left and right subtree, each of depth at most $d - 1$. By the induction hypothesis, they can each have at most 2^{d-1} leaves, giving a total of at most 2^d leaves. This proves the lemma.

LEMMA 7.2.

A binary tree with L leaves must have depth at least $\lceil \log L \rceil$.

PROOF:

Immediate from the preceding lemma.

THEOREM 7.6.

Any sorting algorithm that uses only comparisons between elements requires at least $\lceil \log(N!) \rceil$ *comparisons in the worst case.*

PROOF:

A decision tree to sort N elements must have $N!$ leaves. The theorem follows from the preceding lemma.

THEOREM 7.7.

Any sorting algorithm that uses only comparisons between elements requires $\Omega(N \log N)$ *comparisons.*

PROOF:

From the previous theorem, $\log(N!)$ comparisons are required.

$$\begin{aligned}
\log(N!) &= \log(N(N-1)(N-2)\cdots(2)(1)) \\
&= \log N + \log(N-1) + \log(N-2) + \cdots + \log 2 + \log 1 \\
&\geq \log N + \log(N-1) + \log(N-2) + \cdots + \log N/2 \\
&\geq \frac{N}{2} \log \frac{N}{2} \\
&\geq \frac{N}{2} \log N - \frac{N}{2} \\
&= \Omega(N \log N)
\end{aligned}$$

This type of lower-bound argument, when used to prove a worst-case result, is sometimes known as an *information-theoretic* lower bound. The general theorem says that if there are P different possible cases to distinguish, and the questions are of the form YES/NO, then $\lceil \log P \rceil$ questions are always required in some case by any algorithm to solve the problem. It is possible to prove a similar result for the average-case running time of any comparison-based sorting algorithm. This result is implied by the following lemma, which is left as an exercise: Any binary tree with L leaves has an average depth of at least $\log L$.

7.10. Bucket Sort

Although we proved in the previous section that any general sorting algorithm that uses only comparisons requires $\Omega(N \log N)$ time in the worst case, recall that it is still possible to sort in linear time in some special cases.

A simple example is bucket sort. For bucket sort to work, extra information must be available. The input $A_1, A_2, \ldots, A_N$ must consist of only positive integers smaller than M. (Obviously extensions to this are possible.) If this is the case, then the algorithm is simple: Keep an array called *Count*, of size M, which is initialized to all 0s. Thus, *Count* has M cells, or buckets, which are initially empty. When A_i is read, increment *Count*$[A_i]$ by 1. After all the input is read, scan the *Count* array, printing out a representation of the sorted list. This algorithm takes $O(M + N)$; the proof is left as an exercise. If M is $O(N)$, then the total is $O(N)$.

Although this algorithm seems to violate the lower bound, it turns out that it does not because it uses a more powerful operation than simple comparisons. By incrementing the appropriate bucket, the algorithm essentially performs an M-way comparison in unit time. This is similar to the strategy used in extendible hashing (Section 5.6). This is clearly not in the model for which the lower bound was proven.

This algorithm does, however, question the validity of the model used in proving the lower bound. The model actually is a strong model, because a *general-purpose* sorting algorithm cannot make assumptions about the type of input it can expect to see, but must make decisions based on ordering information only. Naturally, if there is extra information available, we should expect to find a more efficient algorithm, since otherwise the extra information would be wasted.

Although bucket sort seems like much too trivial an algorithm to be useful, it turns out that there are many cases where the input is only small integers, so that using a method like quicksort is really overkill.

7.11. External Sorting

So far, all the algorithms we have examined require that the input fit into main memory. There are, however, applications where the input is much too large to fit into memory. This section will discuss *external sorting* algorithms, which are designed to handle very large inputs.

7.11.1. Why We Need New Algorithms

Most of the internal sorting algorithms take advantage of the fact that memory is directly addressable. Shellsort compares elements $A[i]$ and $A[i - h_k]$ in one time unit. Heapsort compares elements $A[i]$ and $A[i * 2 + 1]$ in one time unit. Quicksort, with median-of-three partitioning, requires comparing $A[Left]$, $A[Center]$, and $A[Right]$ in a constant number of time units. If the input is on a tape, then all these operations lose their efficiency, since elements on a tape can only be accessed sequentially. Even if the data is on a disk, there is still a practical loss of efficiency because of the delay required to spin the disk and move the disk head.

To see how slow external accesses really are, create a random file that is large, but not too big to fit in main memory. Read the file in and sort it using an efficient algorithm. The time it takes to sort the input is certain to be insignificant compared to the time to read the input, even though sorting is an $O(N \log N)$ operation and reading the input is only $O(N)$.

7.11.2. Model for External Sorting

The wide variety of mass storage devices makes external sorting much more device-dependent than internal sorting. The algorithms that we will consider work on tapes, which are probably the most restrictive storage medium. Since access to an element on tape is done by winding the tape to the correct location, tapes can be efficiently accessed only in sequential order (in either direction).

We will assume that we have at least three tape drives to perform the sorting. We need two drives to do an efficient sort; the third drive simplifies matters. If only one tape drive is present, then we are in trouble: any algorithm will require $\Omega(N^2)$ tape accesses.

7.11.3. The Simple Algorithm

The basic external sorting algorithm uses the *Merge* routine from mergesort. Suppose we have four tapes, T_{a1}, T_{a2}, T_{b1}, T_{b2}, which are two input and two output tapes. Depending on the point in the algorithm, the *a* and *b* tapes are either input tapes or output tapes. Suppose the data is initially on T_{a1}. Suppose further that the internal memory can hold (and sort) M records at a time. A natural first step is to read M records at a time from the input tape, sort the records internally, and then write the sorted records alternately to T_{b1} and T_{b2}. We will call each set of sorted records a *run*. When this is done, we rewind all the tapes. Suppose we have the same input as our example for Shellsort.

T_{a1}	81	94	11	96	12	35	17	99	28	58	41	75	15
T_{a2}													
T_{b1}													
T_{b2}													

If $M = 3$, then after the runs are constructed, the tapes will contain the data indicated in the following figure.

T_{a1}							
T_{a2}							
T_{b1}	11	81	94	17	28	99	15
T_{b2}	12	35	96	41	58	75	

Now T_{b1} and T_{b2} contain a group of runs. We take the first run from each tape and merge them, writing the result, which is a run twice as long, onto T_{a1}. Then we take the next run from each tape, merge these, and write the result to T_{a2}. We continue this process, alternating between T_{a1} and T_{a2}, until either T_{b1} or T_{b2} is empty. At this point either both are empty or there is one run left. In the latter case, we copy this run to the appropriate tape. We rewind all four tapes, and repeat the same steps, this time using the *a* tapes as input and the *b* tapes as output. This will give runs of $4M$. We continue the process until we get one run of length N.

This algorithm will require $\lceil \log(N/M) \rceil$ passes, plus the initial run-constructing pass. For instance, if we have 10 million records of 128 bytes each, and four megabytes of internal memory, then the first pass will create 320 runs. We would then need nine more passes to complete the sort. Our example requires $\lceil \log 13/3 \rceil = 3$ more passes, which are shown in the following figure.

T_{a1}	11	12	35	81	94	96	15
T_{a2}	17	28	41	58	75	99	
T_{b1}							
T_{b2}							

T_{a1}												
T_{a2}												
T_{b1}	11	12	17	28	35	51	58	75	81	94	96	99
T_{b2}	15											

T_{a1}	11	12	15	17	28	35	41	58	75	81	94	96	99
T_{a2}													
T_{b1}													
T_{b2}													

7.11.4. *Multiway Merge*

If we have extra tapes, then we can expect to reduce the number of passes required to sort our input. We do this by extending the basic (two-way) merge to a k-way merge.

Merging two runs is done by winding each input tape to the beginning of each run. Then the smaller element is found, placed on an output tape, and the appropriate input tape is advanced. If there are k input tapes, this strategy works the same way, the only difference being that it is slightly more complicated to find the smallest of the k elements. We can find the smallest of these elements by using a priority queue. To obtain the next element to write on the output tape, we perform a *DeleteMin* operation. The appropriate input tape is advanced, and if the run on the input tape is not yet completed, we *Insert* the new element into the priority queue. Using the same example as before, we distribute the input onto the three tapes.

T_{a1}						
T_{a2}						
T_{a3}						
T_{b1}	11	81	94	41	58	75
T_{b2}	12	35	96	15		
T_{b3}	17	28	99			

We then need two more passes of three-way merging to complete the sort.

T_{a1}	11	12	17	28	35	81	94	96	99
T_{a2}	15	41	58	75					
T_{a3}									
T_{b1}									
T_{b2}									
T_{b3}									

T_{a1}													
T_{a2}													
T_{a3}													
T_{b1}	11	12	15	17	28	35	41	58	75	81	94	96	99
T_{b2}													
T_{b3}													

After the initial run construction phase, the number of passes required using k-way merging is $\lceil \log_k(N/M) \rceil$, because the runs get k times as large in each pass. For the example above, the formula is verified, since $\lceil \log_3(13/3) \rceil = 2$. If we have 10 tapes, then $k = 5$, and our large example from the previous section would require $\lceil \log_5 320 \rceil = 4$ passes.

7.11.5. Polyphase Merge

The k-way merging strategy developed in the last section requires the use of $2k$ tapes. This could be prohibitive for some applications. It is possible to get by with only $k + 1$ tapes. As an example, we will show how to perform two-way merging using only three tapes.

Suppose we have three tapes, T_1, T_2, and T_3, and an input file on T_1 that will produce 34 runs. One option is to put 17 runs on each of T_2 and T_3. We could then merge this result onto T_1, obtaining one tape with 17 runs. The problem is that since all the runs are on one tape, we must now put some of these runs on T_2 to perform another merge. The logical way to do this is to copy the first eight runs from T_1 onto T_2 and then perform the merge. This has the effect of adding an extra half pass for every pass we do.

An alternative method is to split the original 34 runs unevenly. Suppose we put 21 runs on T_2 and 13 runs on T_3. We would then merge 13 runs onto T_1 before T_3 was empty. At this point, we could rewind T_1 and T_3, and merge T_1, with 13 runs, and T_2, which has 8 runs, onto T_3. We could then merge 8 runs until T_2 was empty, which would leave 5 runs left on T_1 and 8 runs on T_3. We could then merge T_1 and T_3, and so on. The following table shows the number of runs on each tape after each pass.

	Run Const.	After $T_3 + T_2$	After $T_1 + T_2$	After $T_1 + T_3$	After $T_2 + T_3$	After $T_1 + T_2$	After $T_1 + T_3$	After $T_2 + T_3$
T_1	0	13	5	0	3	1	0	1
T_2	21	8	0	5	2	0	1	0
T_3	13	0	8	3	0	2	1	0

The original distribution of runs makes a great deal of difference. For instance, if 22 runs are placed on T_2, with 12 on T_3, then after the first merge, we obtain 12 runs on T_1 and 10 runs on T_2. After another merge, there are 10 runs on T_1 and 2 runs on T_3. At this point the going gets slow, because we can only merge two sets of runs before T_3 is exhausted. Then T_1 has 8 runs and T_2 has 2 runs. Again, we can only merge two sets of runs, obtaining T_1 with 6 runs and T_3 with 2 runs. After three more passes, T_2 has two runs and the other tapes are empty. We must copy one run to another tape, and then we can finish the merge.

It turns out that the first distribution we gave is optimal. If the number of runs is a Fibonacci number F_N, then the best way to distribute them is to split them into two Fibonacci numbers F_{N-1} and F_{N-2}. Otherwise, it is necessary to pad the tape with dummy runs in order to get the number of runs up to a Fibonacci number. We leave the details of how to place the initial set of runs on the tapes as an exercise.

We can extend this to a k-way merge, in which case we need kth order Fibonacci numbers for the distribution, where the kth order Fibonacci number is defined as $F^{(k)}(N) = F^{(k)}(N-1) + F^{(k)}(N-2) + \cdots + F^{(k)}(N-k)$, with the appropriate initial conditions $F^{(k)}(N) = 0, 0 \le N \le k-2, F^{(k)}(k-1) = 1$.

7.11.6. Replacement Selection

The last item we will consider is construction of the runs. The strategy we have used so far is the simplest possible: We read as many records as possible and sort them, writing the result to some tape. This seems like the best approach possible, until one realizes that as soon as the first record is written to an output tape, the memory it used becomes available for another record. If the next record on the input tape is larger than the record we have just output, then it can be included in the run.

Using this observation, we can give an algorithm for producing runs. This technique is commonly referred to as *replacement selection*. Initially, M records are read into memory and placed in a priority queue. We perform a *DeleteMin*, writing the smallest record to the output tape. We read the next record from the input tape. If it is larger than the record we have just written, we can add it to the priority queue. Otherwise, it cannot go into the current run. Since the priority queue is smaller by one element, we can store this new element in the dead space of the priority queue until the run is completed and use the element for the next run. Storing an element in the dead space is similar to what is done in heapsort. We continue doing this until the size of the priority queue is zero, at which point the run is over. We start a new run by building a new priority queue, using all the elements in the dead space. Figure 7.18 shows the run construction for the small example we have been using, with $M = 3$. Dead elements are indicated by an asterisk.

In this example, replacement selection produces only three runs, compared with the five runs obtained by sorting. Because of this, a three-way merge finishes in one pass instead of two. If the input is randomly distributed, replacement selection can be shown to produce runs of average length $2M$. For our large example, we would

Figure 7.18 Example of run construction

	3 Elements In Heap Array			Output	Next Element Read
	H[0]	H[1]	H[2]		
Run 1	11	94	81	11	96
	81	94	96	81	12*
	94	96	12*	94	35*
	96	35*	12*	96	17*
	17*	35*	12*	End of Run.	Rebuild Heap
Run 2	12	35	17	12	99
	17	35	99	17	28
	28	99	35	28	58
	35	99	58	35	41
	41	99	58	41	15*
	58	99	15*	58	end of tape
	99		15*	99	
			15*	End of Run.	Rebuild Heap
Run 3	15			15	

expect 160 runs instead of 320 runs, so a five-way merge would require four passes. In this case, we have not saved a pass, although we might if we get lucky and have 125 runs or less. Since external sorts take so long, every pass saved can make a significant difference in the running time.

As we have seen, it is possible for replacement selection to do no better than the standard algorithm. However, the input is frequently sorted or nearly sorted to start with, in which case replacement selection produces only a few very long runs. This kind of input is common for external sorts and makes replacement selection extremely valuable.

Summary

For most general internal sorting applications, either insertion sort, Shellsort, or quicksort will be the method of choice, and the decision of which to use will depend mostly on the size of the input. Figure 7.19 shows the running time obtained for each algorithm on various file sizes (on a relatively slow computer).

The data was chosen to be random permutations of N integers, and the times given include only the actual time to sort. The code given in Figure 7.2 was used for insertion sort. Shellsort used the code in Section 7.4 modified to run with Sedgewick's increments. Based on literally millions of sorts, ranging in size from 100 to 25 million, the expected running time of Shellsort with these increments is conjectured to be $O(N^{7/6})$. The heapsort routine is the same as in Section 7.5. Two versions of quicksort are given. The first uses a simple pivoting strategy and does not do a cutoff. Fortunately, the input files were random. The second uses median-of-three partitioning and a cutoff of ten. Further optimizations were possible. We could have coded the median-of-three routine in-line instead of using a function, and we could have written quicksort nonrecursively. There are some other optimizations to the code that are fairly tricky to implement, and of course we could have used an assembly language. We have made an honest attempt to code all routines efficiently, but of course the performance can vary somewhat from machine to machine.

Figure 7.19 Comparison of different sorting algorithms (all times are in seconds)

N	Insertion Sort $O(N^2)$	Shellsort $O(N^{7/6})$(?)	Heapsort $O(N \log N)$	Quicksort $O(N \log N)$	Quicksort (opt.) $O(N \log N)$
10	0.00044	0.00041	0.00057	0.00052	.00046
100	0.00675	0.00171	0.00420	0.00284	.00244
1000	0.59564	0.02927	0.05565	0.03153	.02587
10000	58.864	0.42998	0.71650	0.36765	.31532
100000	NA	5.7298	8.8591	4.2298	3.5882
1000000	NA	71.164	104.68	47.065	41.282

The highly optimized version of quicksort is as fast as Shellsort even for very small input sizes. The improved version of quicksort still has an $O(N^2)$ worst case (one exercise asks you to construct a small example), but the chances of this worst case appearing are so negligible as to not be a factor. If you need to sort large files, quicksort is the method of choice. But never, ever, take the easy way out and use the first element as pivot. It is just not safe to assume that the input will be random. If you do not want to worry about this, use Shellsort. Shellsort will give a small performance penalty but could also be acceptable, especially if simplicity is required. Its worst case is only $O(N^{4/3})$; the chance of that worst case occurring is likewise negligible.

Heapsort, although an $O(N \log N)$ algorithm with an apparently tight inner loop, is slower than Shellsort. A close examination of the algorithm reveals that in order to move data, heapsort does two comparisons. An improvement suggested by Floyd moves data with essentially only one comparison, but implementing this improvement makes the code somewhat longer. We leave it to the reader to decide whether the extra coding effort is worth the increased speed (Exercise 7.40).

Insertion sort is useful only for small or very nearly sorted inputs. We have not included mergesort, because its performance is not as good as quicksort for main memory sorts and it is not any simpler to code. We have seen, however, that merging is the central idea of external sorts.

Exercises

7.1 Sort the sequence 3, 1, 4, 1, 5, 9, 2, 6, 5 using insertion sort.

7.2 What is the running time of insertion sort if all keys are equal?

7.3 Suppose we exchange elements $A[i]$ and $A[i + k]$, which were originally out of order. Prove that at least 1 and at most $2k - 1$ inversions are removed.

7.4 Show the result of running Shellsort on the input 9, 8, 7, 6, 5, 4, 3, 2, 1 using the increments $\{1, 3, 7\}$.

7.5 a. What is the running time of Shellsort using the two-increment sequence $\{1, 2\}$?

b. Show that for any N, there exists a three-increment sequence such that Shellsort runs in $O(N^{5/3})$ time.

c. Show that for any N, there exists a six-increment sequence such that Shellsort runs in $O(N^{3/2})$ time.

7.6 *a. Prove that the running time of Shellsort is $\Omega(N^2)$ using increments of the form $1, c, c^2, \ldots, c^i$ for any integer c.

**b. Prove that for these increments, the average running time is $\Theta(N^{3/2})$.

*7.7 Prove that if a k-sorted file is then h-sorted, it remains k-sorted.

**7.8 Prove that the running time of Shellsort, using the increment sequence suggested by Hibbard, is $\Omega(N^{3/2})$ in the worst case. *Hint:* You can prove the bound by considering the special case of what Shellsort does when all elements are either

0 or 1. Set *InputData*[i] = 1 if i is expressible as a linear combination of h_t, h_{t-1}, ..., $h_{\lfloor t/2 \rfloor+1}$ and 0 otherwise.

7.9 Determine the running time of Shellsort for

a. sorted input

*b. reverse-ordered input

7.10 Do either of the following modifications to the Shellsort routine coded in Fig. 7.4 affect the worst case running time?

a. Before line 2, subtract one from *Increment* if it is even.

b. Before line 2, add one to *Increment* if it is even.

7.11 Show how heapsort processes the input 142, 543, 123, 65, 453, 879, 572, 434, 111, 242, 811, 102.

7.12 a. What is the running time of heapsort for presorted input?

*b. Prove that the worst case bound for heapsort is achievable.

7.13 Sort 3, 1, 4, 1, 5, 9, 2, 6 using mergesort.

7.14 How would you implement mergesort without using recursion?

7.15 Determine the running time of mergesort for

a. sorted input

b. reverse-ordered input

c. random input

7.16 In the analysis of mergesort, constants have been disregarded. Prove that the number of comparisons used in the worst case by mergesort is $N\lceil \log N \rceil - 2^{\lceil \log N \rceil} + 1$.

7.17 Sort 3, 1, 4, 1, 5, 9, 2, 6, 5, 3, 5 using quicksort with median-of-three partitioning and a cutoff of 3.

7.18 Using the quicksort implementation in this chapter, determine the running time of quicksort for

a. sorted input

b. reverse-ordered input

c. random input

7.19 Repeat Exercise 7.18 when the pivot is chosen as

a. the first element

b. the largest of the first two nondistinct keys

c. a random element

*d. the average of all keys in the set

7.20 a. For the quicksort implementation in this chapter, what is the running time when all keys are equal?

b. Suppose we change the partitioning strategy so that neither i nor j stops when an element with the same key as the pivot is found. What fixes need to be made in the code to guarantee that quicksort works, and what is the running time, when all keys are equal?

c. Suppose we change the partitioning strategy so that i stops at an element with the same key as the pivot, but j does not stop in a similar case. What fixes need to be made in the code to guarantee that quicksort works, and when all keys are equal, what is the running time of quicksort?

7.21 Suppose we choose the middle key as pivot. Does this make it unlikely that quicksort will require quadratic time?

7.22 Construct a permutation of 20 elements that is as bad as possible for quicksort using median-of-three partitioning and a cutoff of 3.

7.23 Write a program to implement the selection algorithm.

7.24 Solve the following recurrence: $T(N) = (1/N)[\sum_{i=0}^{N-1} T(i)] + cN$, $T(0) = 0$.

7.25 A sorting algorithm is *stable* if elements with equal keys are left in the same order as they occur in the input. Which of the sorting algorithms in this chapter are stable and which are not? Why?

7.26 Suppose you are given a sorted list of N elements followed by $f(N)$ randomly ordered elements. How would you sort the entire list if

a. $f(N) = O(1)$?

b. $f(N) = O(\log N)$?

c. $f(N) = O(\sqrt{N})$?

*d. How large can $f(N)$ be for the entire list still to be sortable in $O(N)$ time?

7.27 Prove that any algorithm that finds an element X in a sorted list of N elements requires $\Omega(\log N)$ comparisons.

7.28 Using Stirling's formula, $N! \approx (N/e)^N \sqrt{2\pi N}$, give a precise estimate for $\log(N!)$.

7.29 *a. In how many ways can two sorted arrays of N elements be merged?

*b. Give a nontrivial lower bound on the number of comparisons required to merge two sorted lists of N elements.

7.30 Prove that sorting N elements with integer keys in the range $1 \le Key \le M$ takes $O(M + N)$ time using bucket sort.

7.31 Suppose you have an array of N elements containing only two distinct keys, *true* and *false*. Give an $O(N)$ algorithm to rearrange the list so that all *false* elements precede the *true* elements. You may use only constant extra space.

7.32 Suppose you have an array of N elements, containing three distinct keys, *true*, *false*, and *maybe*. Give an $O(N)$ algorithm to rearrange the list so that all *false* elements precede *maybe* elements, which in turn precede *true* elements. You may use only constant extra space.

7.33 a. Prove that any comparison-based algorithm to sort 4 elements requires 5 comparisons.

b. Give an algorithm to sort 4 elements in 5 comparisons.

7.34 a. Prove that 7 comparisons are required to sort 5 elements using any comparison-based algorithm.

*b. Give an algorithm to sort 5 elements with 7 comparisons.

7.35 Write an efficient version of Shellsort and compare performance when the following increment sequences are used:

a. Shell's original sequence

b. Hibbard's increments

c. Knuth's increments: $h_i = \frac{1}{2}(3^i + 1)$

d. Gonnet's increments: $h_t = \lfloor \frac{N}{2.2} \rfloor$, and $h_k = \lfloor \frac{h_{k+1}}{2.2} \rfloor$ (with $h_1 = 1$ if $h_2 = 2$)

e. Sedgewick's increments.

7.36 Implement an optimized version of quicksort and experiment with combinations of the following:

a. Pivot: first element, middle element, random element, median of three, median of five.

b. Cutoff values from 0 to 20.

7.37 Write a routine that reads in two alphabetized files and merges them together, forming a third, alphabetized, file.

7.38 Suppose we implement the median of three routine as follows: Find the median of *A*[*Left*], *A*[*Center*], *A*[*Right*], and swap it with *A*[*Right*]. Proceed with the normal partitioning step starting *i* at *Left* and *j* at *Right* − 1 (instead of *Left* + 1 and *Right* − 2).

a. Suppose the input is 2, 3, 4, ..., $N - 1$, N, 1. For this input, what is the running time of this version of quicksort?

b. Suppose the input is in reverse order. For this input, what is the running time of this version of quicksort?

7.39 Prove that any comparison-based sorting algorithm requires $\Omega(N \log N)$ comparisons on average.

7.40 Consider the following strategy for *PercolateDown*. We have a hole at node *X*. The normal routine is to compare *X*'s children and then move the child up to *X* if it is larger (in the case of a (*max*)heap) than the element we are trying to place, thereby pushing the hole down; we stop when it is safe to place the new element in the hole. The alternate strategy is to move elements up and the hole down as far as possible, without testing whether the new cell can be inserted. This would place the new cell in a leaf and probably violate the heap order; to fix the heap order, percolate the new cell up in the normal manner. Write a routine to include this idea, and compare the running time with a standard implementation of heapsort.

7.41 Propose an algorithm to sort a large file using only two tapes.

7.42 a. Show that a lower bound of $N!/2^{2N}$ on the number of heaps is implied by the fact that build-heap uses at most $2N$ comparisons.

b. Use Stirling's formula to expand this bound.

7.43 ANSI C requires the routine *qsort* to be present in C libraries. *qsort* is typically implemented by quicksort (but this is not required). Experiment with various inputs to see if *qsort* can be driven to quadratic behavior. Try random 0s and 1s.

References

Knuth's book [13] is a comprehensive, though somewhat dated, reference for sorting. Gonnet and Baeza-Yates [5] has some more recent results, as well as a huge bibliography.

The original paper detailing Shellsort is [22]. The paper by Hibbard [6] suggested the use of the increments $2^k - 1$ and tightened the code by avoiding swaps. Theorem 7.4 is from [13]. Pratt's lower bound, which uses a more complex method than that suggested in the text, can be found in [15]. Improved increment sequences and upper bounds appear in [10], [21], and [24]; matching lower bounds have been shown in [25]. A recent result shows that no increment sequence gives an $O(N \log N)$ worst-case running time [14]. The average-case running time for Shellsort is still unresolved. Yao [27] has performed an extremely complex analysis for the three-increment case. The result has yet to be extended to more increments. Experiments with various increment sequences appear in [23].

Heapsort was invented by Williams [26]; Floyd [2] provided the linear-time algorithm for heap construction. Theorem 7.5 is from [16].

An exact average-case analysis of mergesort has been claimed in [4]; the paper detailing the results is forthcoming. An algorithm to perform merging in linear time without extra space is described in [9].

Quicksort is from Hoare [7]. This paper analyzes the basic algorithm, describes most of the improvements, and includes the selection algorithm. A detailed analysis and empirical study was the subject of Sedgewick's dissertation [20]. Many of the important results appear in the three papers [17], [18], and [19]. [1] provides a detailed C implementation with some additional improvements, and points out that most implementations of the *qsort* library routine are easily driven to quadratic behavior.

Decision trees and sorting optimality are discussed in Ford and Johnson [3]. This paper also provides an algorithm that almost meets the lower bound in terms of number of comparisons (but not other operations). This algorithm was eventually shown to be slightly suboptimal by Manacher [12].

External sorting is covered in detail in [11]. Stable sorting, described in Exercise 7.25, has been addressed by Horvath [8].

1. J. L. Bentley and M. D. McElroy, "Engineering a Sort Function," *Software—Practice and Experience*, 23 (1993), 1249–1265.
2. R. W. Floyd, "Algorithm 245: Treesort 3," *Communications of the ACM*, 7 (1964), 701.
3. L. R. Ford and S. M. Johnson, "A Tournament Problem," *American Mathematics Monthly*, 66 (1959), 387–389.
4. M. Golin and R. Sedgewick, "Exact Analysis of Mergesort," *Fourth SIAM Conference on Discrete Mathematics*, 1988.
5. G. H. Gonnet and R. Baeza-Yates, *Handbook of Algorithms and Data Structures*, 2nd ed., Addison-Wesley, Reading, Mass., 1991.
6. T. H. Hibbard, "An Empirical Study of Minimal Storage Sorting," *Communications of the ACM*, 6 (1963), 206–213.
7. C. A. R. Hoare, "Quicksort," *Computer Journal*, 5 (1962), 10–15.

8. E. C. Horvath, "Stable Sorting in Asymptotically Optimal Time and Extra Space," *Journal of the ACM*, 25 (1978), 177–199.
9. B. Huang and M. Langston, "Practical In-place Merging," *Communications of the ACM*, 31 (1988), 348–352.
10. J. Incerpi and R. Sedgewick, "Improved Upper Bounds on Shellsort," *Journal of Computer and System Sciences*, 31 (1985), 210–224.
11. D. E. Knuth, *The Art of Computer Programming. Volume 3: Sorting and Searching*, Addison-Wesley, Reading, Mass., 1973.
12. G. K. Manacher, "The Ford-Johnson Sorting Algorithm Is Not Optimal," *Journal of the ACM*, 26 (1979), 441–456.
13. A. A. Papernov and G. V. Stasevich, "A Method of Information Sorting in Computer Memories," *Problems of Information Transmission*, 1 (1965), 63–75.
14. C. G. Plaxton, B. Pooren, and T. Suel, "Improved Lower Bounds for Shellsort," Proceedings of the Thirty-third Annual Symposium on the Foundations of Computer Science (1992), 226–235.
15. V. R. Pratt, *Shellsort and Sorting Networks*, Garland Publishing, New York, 1979. (Originally presented as the author's Ph.D. thesis, Stanford University, 1971.)
16. R. Schaffer and R. Sedgewick, "The Analysis of Heapsort," *Journal of Algorithms*, 14 (1993), 76–100.
17. R. Sedgewick, "Quicksort with Equal Keys," *SIAM Journal on Computing*, 6 (1977), 240–267.
18. R. Sedgewick, "The Analysis of Quicksort Programs," *Acta Informatica*, 7 (1977), 327–355.
19. R. Sedgewick, "Implementing Quicksort Programs," *Communications of the ACM*, 21 (1978), 847–857.
20. R. Sedgewick, *Quicksort*, Garland Publishing, New York, 1978. (Originally presented as the author's Ph.D. thesis, Stanford University, 1975.)
21. R. Sedgewick, "A New Upper Bound for Shellsort," *Journal of Algorithms*, 7 (1986), 159–173.
22. D. L. Shell, "A High-Speed Sorting Procedure," *Communications of the ACM*, 2 (1959), 30–32.
23. M. A. Weiss, "Empirical Results on the Running Time of Shellsort," *Computer Journal*, 34 (1991), 88–91.
24. M. A. Weiss and R. Sedgewick, "More On Shellsort Increment Sequences," *Information Processing Letters*, 34 (1990), 267–270.
25. M. A. Weiss and R. Sedgewick, "Tight Lower Bounds For Shellsort," *Journal of Algorithms*, 11 (1990), 242–251.
26. J. W. J. Williams, "Algorithm 232: Heapsort," *Communications of the ACM*, 7 (1964), 347–348.
27. A. C. Yao, "An Analysis of $(h, k, 1)$ Shellsort," *Journal of Algorithms*, 1 (1980), 14–50.

CHAPTER 8

The Disjoint Set ADT

In this chapter, we describe an efficient data structure to solve the equivalence problem. The data structure is simple to implement. Each routine requires only a few lines of code, and a simple array can be used. The implementation is also extremely fast, requiring constant average time per operation. This data structure is also very interesting from a theoretical point of view, because its analysis is extremely difficult; the functional form of the worst case is unlike any we have yet seen. For the disjoint set ADT, we will

- Show how it can be implemented with minimal coding effort.
- Greatly increase its speed, using just two simple observations.
- Analyze the running time of a fast implementation.
- See a simple application.

8.1. Equivalence Relations

A *relation* R is defined on a set S if for every pair of elements (a, b), $a, b \in S$, $a \, R \, b$ is either true or false. If $a \, R \, b$ is true, then we say that a is related to b.

An *equivalence relation* is a relation R that satisfies three properties:

1. (*Reflexive*) $a \, R \, a$, for all $a \in S$.
2. (*Symmetric*) $a \, R \, b$ if and only if $b \, R \, a$.
3. (*Transitive*) $a \, R \, b$ and $b \, R \, c$ implies that $a \, R \, c$.

We will consider several examples.

The $\leq$ relationship is not an equivalence relationship. Although it is reflexive, since $a \leq a$, and transitive, since $a \leq b$ and $b \leq c$ implies $a \leq c$, it is not symmetric, since $a \leq b$ does not imply $b \leq a$.

Electrical connectivity, where all connections are by metal wires, is an equivalence relation. The relation is clearly reflexive, as any component is connected to itself. If a is electrically connected to b, then b must be electrically connected to a, so the relation is symmetric. Finally, if a is connected to b and b is connected to c, then a is connected to c. Thus electrical connectivity is an equivalence relation.

Two cities are related if they are in the same country. It is easily verified that this is an equivalence relation. Suppose town a is related to b if it is possible to travel from a to b by taking roads. This relation is an equivalence relation if all the roads are two-way.

8.2. The Dynamic Equivalence Problem

Given an equivalence relation $\sim$, the natural problem is to decide, for any a and b, if $a \sim b$. If the relation is stored as a two-dimensional array of booleans, then, of course, this can be done in constant time. The problem is that the relation is usually not explicitly, but rather implicitly, defined.

As an example, suppose the equivalence relation is defined over the five-element set $\{a_1, a_2, a_3, a_4, a_5\}$. Then there are 25 pairs of elements, each of which is either related or not. However, the information $a_1 \sim a_2, a_3 \sim a_4, a_5 \sim a_1, a_4 \sim a_2$ implies that all pairs are related. We would like to be able to infer this quickly.

The *equivalence class* of an element $a \in S$ is the subset of S that contains all the elements that are related to a. Notice that the equivalence classes form a partition of S: every member of S appears in exactly one equivalence class. To decide if $a \sim b$, we need only to check whether a and b are in the same equivalence class. This provides our strategy to solve the equivalence problem.

The input is initially a collection of N sets, each with one element. This initial representation is that all relations (except reflexive relations) are false. Each set has a different element, so that $S_i \cap S_j = \varnothing$; this makes the sets *disjoint*.

There are two permissible operations. The first is *Find*, which returns the name of the set (that is, the equivalence class) containing a given element. The second operation adds relations. If we want to add the relation $a \sim b$, then we first see if a and b are already related. This is done by performing *Find*s on both a and b and checking whether they are in the same equivalence class. If they are not, then we apply *Union*. This operation merges the two equivalence classes containing a and b into a new equivalence class. From a set point of view, the result of $\cup$ is to create a new set $S_k = S_i \cup S_j$, destroying the originals and preserving the disjointness of all the sets. The algorithm to do this is frequently known as the disjoint set *Union/Find algorithm* for this reason.

This algorithm is *dynamic* because, during the course of the algorithm, the sets can change via the *Union* operation. The algorithm must also operate *on-line*: When a *Find* is performed, it must give an answer before continuing. Another possibility would be an *off-line* algorithm. Such an algorithm would be allowed to see the entire sequence of *Unions* and *Finds*. The answer it provides for each *Find* must still be consistent with all the *Unions* that were performed up until the *Find*, but the algorithm can give all its answers after it has seen *all* the questions. The difference is similar to taking a written exam (which is generally off-line—you only have to give the answers before time expires), and an oral exam (which is on-line, because you must answer the current question before proceeding to the next question).

Notice that we do not perform any operations comparing the relative values of elements, but merely require knowledge of their location. For this reason, we can assume that all the elements have been numbered sequentially from 1 to N and that the numbering can be determined easily by some hashing scheme. Thus, initially we have $S_i = \{i\}$ for $i = 1$ through N.

Our second observation is that the name of the set returned by *Find* is actually fairly arbitrary. All that really matters is that *Find*(a) = *Find*(b) if and only if a and b are in the same set.

These operations are important in many graph theory problems and also in compilers which process equivalence (or type) declarations. We will see an application later.

There are two strategies to solve this problem. One ensures that the *Find* instruction can be executed in constant worst-case time, and the other ensures that the *Union* instruction can be executed in constant worst-case time. It has recently been shown that both cannot be done simultaneously in constant worst-case time.

We will now briefly discuss the first approach. For the *Find* operation to be fast, we could maintain, in an array, the name of the equivalence class for each element. Then *Find* is just a simple $O(1)$ lookup. Suppose we want to perform *Union*(a, b). Suppose that a is in equivalence class i and b is in equivalence class j. Then we scan down the array, changing all i's to j. Unfortunately, this scan takes $\Theta(N)$. Thus, a sequence of $N - 1$ *Unions* (the maximum, since then everything is in one set) would take $\Theta(N^2)$ time. If there are $\Omega(N^2)$ *Find* operations, this performance is fine, since the total running time would then amount to $O(1)$ for each *Union* or *Find* operation over the course of the algorithm. If there are fewer *Finds*, this bound is not acceptable.

One idea is to keep all the elements that are in the same equivalence class in a linked list. This saves time when updating, because we do not have to search through the entire array. This by itself does not reduce the asymptotic running time, because it is still possible to perform $\Theta(N^2)$ equivalence class updates over the course of the algorithm.

If we also keep track of the size of each equivalence class, and when performing *Unions* we change the name of the smaller equivalence class to the larger, then the total time spent for $N - 1$ merges is $O(N \log N)$. The reason for this is that each element can have its equivalence class changed at most $\log N$ times, since every time its class is changed, its new equivalence class is at least twice as large as its old. Using this strategy, any sequence of M *Finds* and up to $N - 1$ *Unions* takes at most $O(M + N \log N)$ time.

In the remainder of this chapter, we will examine a solution to the *Union/Find* problem that makes *Unions* easy but *Finds* hard. Even so, the running time for any sequence of at most M *Finds* and up to $N - 1$ *Unions* will be only a little more than $O(M + N)$.

8.3. Basic Data Structure

Recall that the problem does not require that a *Find* operation return any specific name, just that *Finds* on two elements return the same answer if and only if they are in the same set. One idea might be to use a tree to represent each set, since each element in a tree has the same root. Thus, the root can be used to name the set. We will represent each set by a tree. (Recall that a collection of trees is known as a *forest*.) Initially, each set contains one element. The trees we will use are not necessarily binary trees, but their representation is easy, because the only information we will need is a parent pointer. The name of a set is given by the node at the root. Since only the name of the parent is required, we can assume that this tree is stored implicitly in an array: each entry $P[i]$ in the array represents the parent of element i. If i is a root, then $P[i] = 0$. In the forest in Figure 8.1, $P[i] = 0$ for $1 \le i \le 8$. As with heaps, we will draw the trees explicitly, with the understanding that an array is being used. Figure 8.1 shows the explicit representation. We will draw the root's parent pointer vertically for convenience.

To perform a *Union* of two sets, we merge the two trees by making the root pointer of one node point to the root node of the other tree. It should be clear that this operation takes constant time. Figures 8.2, 8.3, and 8.4 represent the forest after each of *Union*(5,6), *Union*(7,8), *Union*(5,7), where we have adopted the convention that the new root after the *Union* (X, Y) is X. The implicit representation of the last forest is shown in Figure 8.5.

A *Find*(X) on element X is performed by returning the root of the tree containing X. The time to perform this operation is proportional to the depth of the node representing X, assuming, of course, that we can find the node representing X in constant time. Using the strategy above, it is possible to create a tree of depth $N - 1$, so the worst-case running time of a *Find* is $O(N)$. Typically, the running time is computed for a *sequence* of M intermixed instructions. In this case, M consecutive operations could take $O(MN)$ time in the worst case.

The code in Figures 8.6 through 8.9 represents an implementation of the basic algorithm, assuming that error checks have already been performed. In our routine, *Unions* are performed on the roots of the trees. Sometimes the operation is performed by passing any two elements, and having the *Union* perform two *Finds* to determine the roots.

The average-case analysis is quite hard to do. The least of the problems is that the answer depends on how to define *average* (with respect to the *Union* operation). For instance, in the forest in Figure 8.4, we could say that since there are five trees, there are $5 \cdot 4 = 20$ equally likely results of the next *Union* (as any two different trees can be *Unioned*). Of course, the implication of this model is that there is only

Figure 8.1 Eight elements, initially in different sets

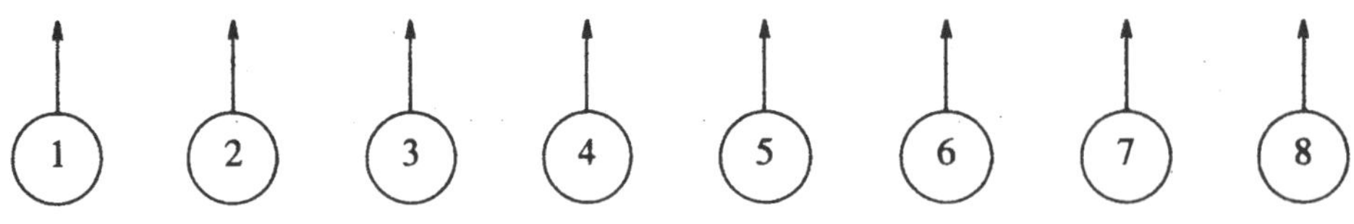

Figure 8.2 After *Union*(5, 6)

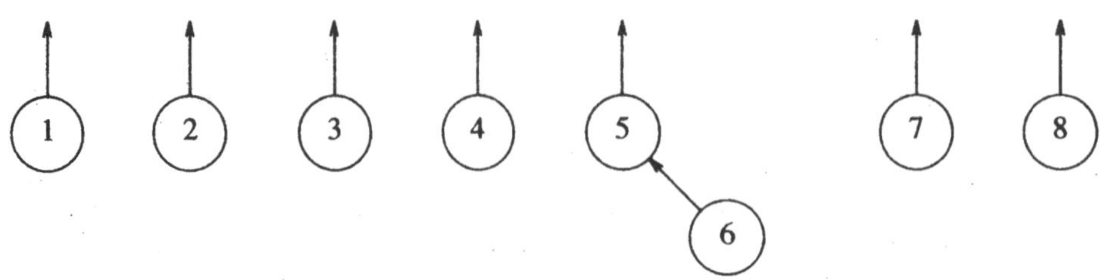

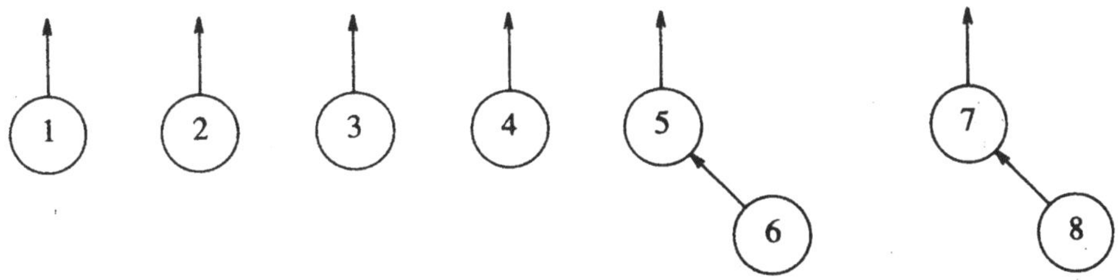

Figure 8.3 After *Union*(7, 8)

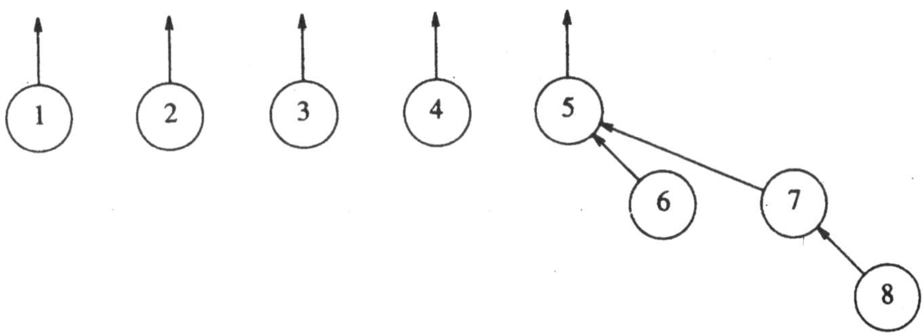

Figure 8.4 After *Union*(5, 7)

0	0	0	0	0	5	5	7
1	2	3	4	5	6	7	8

Figure 8.5 Implicit representation of previous tree

```
#ifndef _DisjSet_H

typedef int DisjSet[ NumSets + 1 ];
typedef int SetType;
typedef int ElementType;

void Initilialize( DisjSet S );
void SetUnion( DisjSet S, SetType Root1, SetType Root2 );
SetType Find( ElementType X, DisjSet S );

#endif /* _DisjSet_H */
```

Figure 8.6 Disjoint set type declaration

```
void
Initialize( DisjSet S)
{
    int i;

    for( i = NumSets; i > 0; i-- )
        S[ i ] = 0;
}
```

Figure 8.7 Disjoint set initialization routine

```
/* Assumes Root1 and Root2 are roots */
/* union is a C keyword, so this routine */
/* is named SetUnion */

void
SetUnion( DisjSet S, SetType Root1, SetType Root2 )
{
    S[ Root2 ] = Root1;
}
```

Figure 8.8 *Union* (not the best way)

```
SetType
Find( ElementType X, DisjSet S )
{
    if( S[ X ] <= 0 )
        return X;
    else
        return Find( S[ X ], S );
}
```

Figure 8.9 A simple disjoint set *Find* algorithm

a $\frac{2}{5}$ chance that the next *Union* will involve the large tree. Another model might say that all *Unions* between any two *elements* in different trees are equally likely, so a larger tree is more likely to be involved in the next *Union* than a smaller tree. In the example above, there is an $\frac{8}{11}$ chance that the large tree is involved in the next union, since (ignoring symmetries) there are 6 ways in which to merge two elements in $\{1, 2, 3, 4\}$, and 16 ways to merge an element in $\{5, 6, 7, 8\}$ with an element in $\{1, 2, 3, 4\}$. There are still more models and no general agreement on which is the best. The average running time depends on the model; $\Theta(M)$, $\Theta(M \log N)$, and $\Theta(MN)$ bounds have actually been shown for three different models, although the latter bound is thought to be more realistic.

Quadratic running time for a sequence of operations is generally unacceptable. Fortunately, there are several ways of easily ensuring that this running time does not occur.

8.4. Smart Union Algorithms

The *Unions* above were performed rather arbitrarily, by making the second tree a subtree of the first. A simple improvement is always to make the smaller tree a subtree of the larger, breaking ties by any method; we call this approach ***union-by-size***. The three *Unions* in the preceding example were all ties, and so we can consider that they were performed by size. If the next operation were *Union*(4, 5), then the forest in Figure 8.10 would form. Had the size heuristic not been used, a deeper tree would have been formed (Fig. 8.11).

Figure 8.10 Result of union-by-size

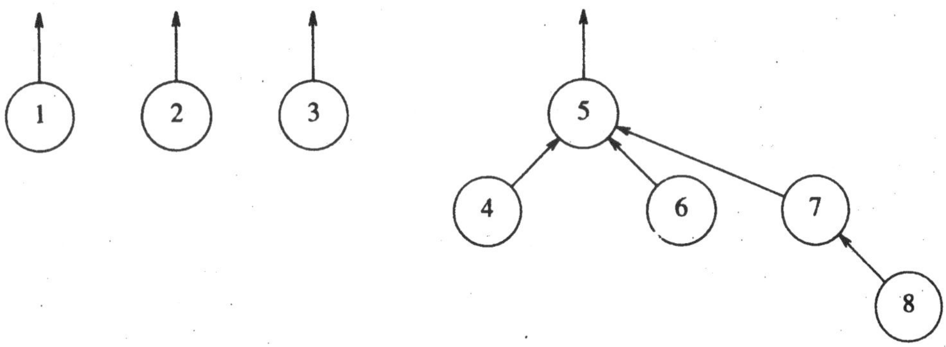

Figure 8.11 Result of an arbitrary *Union*

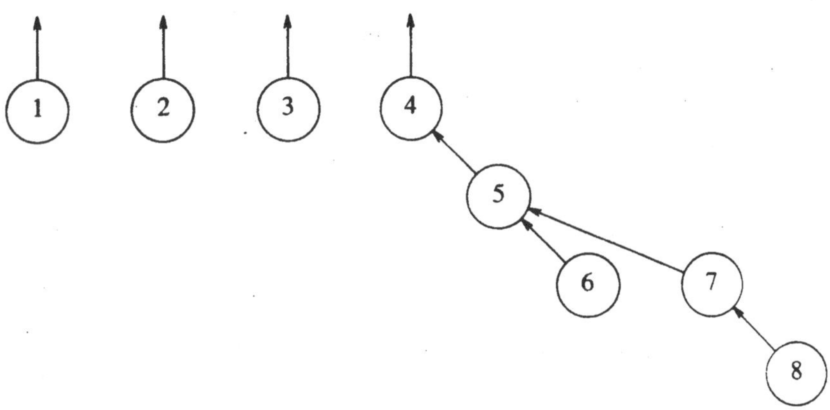

Figure 8.12 Worst-case tree for $N = 16$

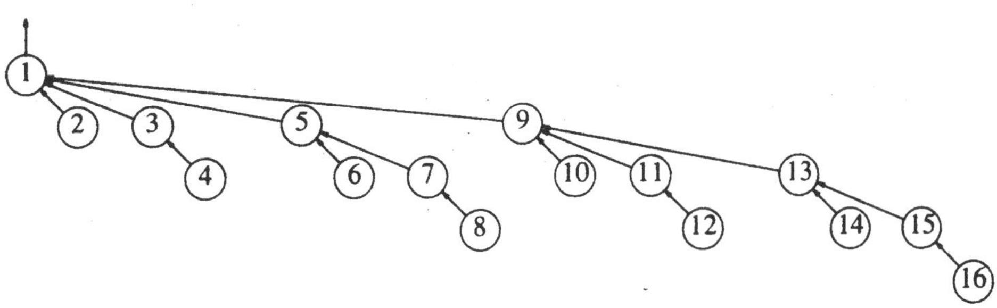

We can prove that if *Unions* are done by size, the depth of any node is never more than $\log N$. To see this, note that a node is initially at depth 0. When its depth increases as a result of a *Union*, it is placed in a tree that is at least twice as large as before. Thus, its depth can be increased at most $\log N$ times. (We used this argument in the quick-find algorithm at the end of Section 8.2.) This implies that the running time for a *Find* operation is $O(\log N)$, and a sequence of M operations takes $O(M \log N)$. The tree in Figure 8.12 shows the worst tree possible after 16 *Unions* and is obtained if all *Unions* are between equal-sized trees (the worst-case trees are binomial trees, discussed in Chapter 6).

To implement this strategy, we need to keep track of the size of each tree. Since we are really just using an array, we can have the array entry of each root contain the *negative* of the size of its tree. Thus, initially the array representation of the tree is all -1s (and Figure 8.7 needs to be changed accordingly). When a *Union* is performed, check the sizes; the new size is the sum of the old. Thus, union-by-size is not at all difficult to implement and requires no extra space. It is also fast, on average. For virtually all reasonable models, it has been shown that a sequence of M operations requires $O(M)$ average time if union-by-size is used. This is because when random *Unions* are performed, generally very small (usually one-element) sets are merged with large sets throughout the algorithm.

An alternative implementation, which also guarantees that all the trees will have depth at most $O(\log N)$, is *union-by-height*. We keep track of the height, instead of the size, of each tree and perform *Unions* by making the shallow tree a subtree of the deeper tree. This is an easy algorithm, since the height of a tree increases only when two equally deep trees are joined (and then the height goes up by one). Thus, union-by-height is a trivial modification of union-by-size.

The following figures show a tree and its implicit representation for both union-by-size and union-by-height. The code in Figure 8.13 implements union-by-height.

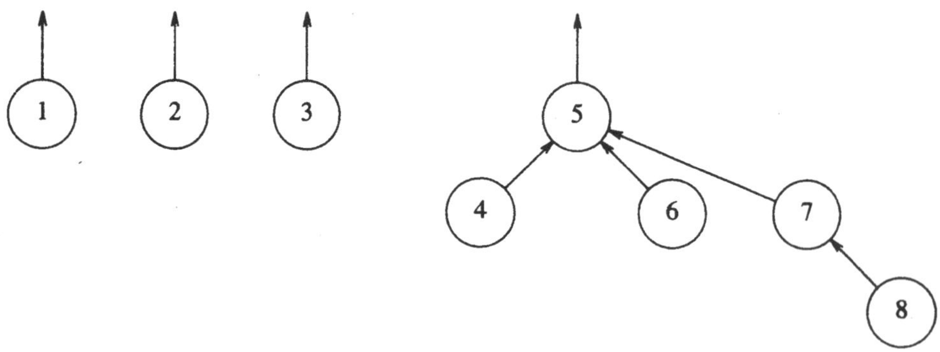

−1	−1	−1	5	−5	5	5	7
1	2	3	4	5	6	7	8

0	0	0	5	−2	5	5	7
1	2	3	4	5	6	7	8

```
/* Assume Root1 and Root2 are roots */
/* union is a C keyword, so this routine */
/* is named SetUnion */

void
SetUnion( DisjSet S, SetType Root1, SetType Root2 )
{
    if( S[ Root2 ] < S[ Root1 ] ) /* Root2 is deeper set */
        S[ Root1 ] = Root2;       /* Make Root2 new root */
    else
    {
        if( S[ Root1 ] == S[ Root2 ] ) /* Same height, */
            S[ Root1 ]--;              /* so update */
        S[ Root2 ] = Root1;
    }
}
```

Figure 8.13 Code for *Union*-by-height (rank)

8.5. Path Compression

The *Union/Find* algorithm, as described so far, is quite acceptable for most cases. It is very simple and linear on average for a sequence of M instructions (under all models). However, the worst case of $O(M \log N)$ can occur fairly easily and naturally. For instance, if we put all the sets on a queue and repeatedly dequeue the first two sets and enqueue the union, the worst case occurs. If there are many more *Finds* than *Unions*, this running time is worse than that of the quick-find algorithm. Moreover, it should be clear that there are probably no more improvements possible for the *Union* algorithm. This is based on the observation that any method to perform the unions will yield the same worst-case trees, since it must break ties arbitrarily. Therefore, the only way to speed the algorithm up, without reworking the data structure entirely, is to do something clever on the *Find* operation.

The clever operation is known as ***path compression.*** Path compression is performed during a *Find* operation and is independent of the strategy used to perform *Unions*. Suppose the operation is *Find*(X). Then the effect of path compression is that *every* node on the path from X to the root has its parent changed to the root. Figure 8.14 shows the effect of path compression after *Find*(15) on the generic worst tree of Figure 8.12.

The effect of path compression is that with an extra two pointer moves, nodes 13 and 14 are now one position closer to the root and nodes 15 and 16 are now two positions closer. Thus, the fast future accesses on these nodes will pay (we hope) for the extra work to do the path compression.

As the code in Figure 8.15 shows, path compression is a trivial change to the basic *Find* algorithm. The only change to the *Find* routine is that $S[X]$ is made equal to the value returned by *Find*; thus after the root of the set

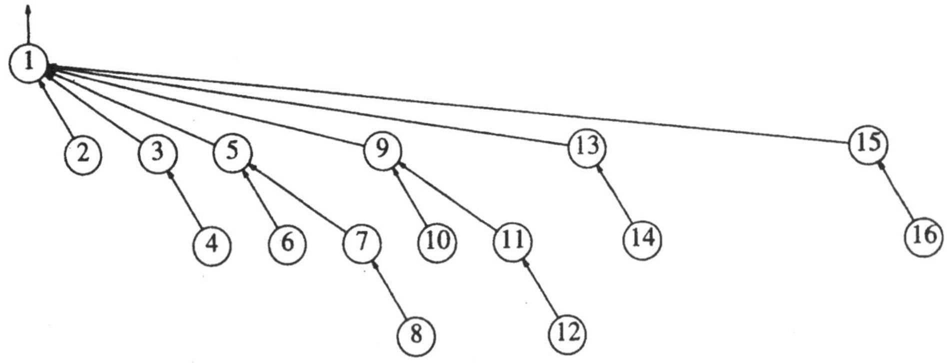

Figure 8.14 An example of path compression

```
SetType
Find( ElementType X, DisjSet S )
{
    if( S[ X ] <= 0 )
        return X;
    else
        return S[ X ] = Find( S[ X ], S );
}
```

Figure 8.15 Code for disjoint set *Find* with path compression

is found recursively, X is made to point directly to it. This occurs recursively to every node on the path to the root, so this implements path compression.

When *Unions* are done arbitrarily, path compression is a good idea, because there is an abundance of deep nodes and these are brought near the root by path compression. It has been proven that when path compression is done in this case, a sequence of M operations requires at most $O(M \log N)$ time. It is still an open problem to determine what the average-case behavior is in this situation.

Path compression is perfectly compatible with union-by-size, and thus both routines can be implemented at the same time. Since doing union-by-size by itself is expected to execute a sequence of M operations in linear time, it is not clear that the extra pass involved in path compression is worthwhile on average. Indeed, this problem is still open. However, as we shall see later, the combination of path compression and a smart union rule guarantees a very efficient algorithm in all cases.

Path compression is not entirely compatible with union-by-height, because path compression can change the heights of the trees. It is not at all clear how to recompute them efficiently. The answer is do not!! Then the heights stored for each tree become estimated heights (sometimes known as *ranks*), but it turns out that union-by-rank (which is what this has now become) is just as efficient in theory as union-by-size. Furthermore, heights are updated less often than sizes. As with union-by-size, it is not clear whether path compression is worthwhile on average. What we will show in the next section is that with either union heuristic, path compression significantly reduces the worst-case running time.

8.6. Worst Case for Union-by-Rank and Path Compression

When both heuristics are used, the algorithm is almost linear in the worst case. Specifically, the time required in the worst case is $\Theta(M\,\alpha(M,N))$ (provided $M \ge N$), where $\alpha(M,N)$ is a functional inverse of Ackermann's function, which is defined below:[†]

$$A(1,j) = 2^j \text{ for } j \ge 1$$

$$A(i,1) = A(i-1,2) \text{ for } i \ge 2$$

$$A(i,j) = A(i-1,A(i,j-1)) \text{ for } i,j \ge 2$$

From this, we define

$$\alpha(M,N) = \min\{i \ge 1 | A(i,\lfloor M/N \rfloor) > \log N\}$$

You may want to compute some values, but for all practical purposes, $\alpha(M,N) \le 4$, which is all that is really important here. The single-variable inverse Ackermann function, sometimes written as $\log^* N$, is the number of times the logarithm of N needs to be applied until $N \le 1$. Thus, $\log^* 65536 = 4$, because $\log\log\log\log 65536 = 1$. $\log^* 2^{65536} = 5$, but keep in mind that 2^{65536} is a 20,000-digit number. $\alpha(M,N)$ actually grows even slower then $\log^* N$. However, $\alpha(M,N)$ is not a constant, so the running time is not linear.

In the remainder of this section, we will prove a slightly weaker result. We will show that any sequence of $M = \Omega(N)$ *Union/Find* operations takes a total of $O(M \log^* N)$ running time. The same bound holds if union-by-rank is replaced with union-by-size. This analysis is probably the most complex in the book and one of the first truly complex worst-case analyses ever performed for an algorithm that is essentially trivial to implement.

8.6.1. Analysis of the Union/Find Algorithm

In this section we establish a fairly tight bound on the running time of a sequence of $M = \Omega(N)$ *Union/Find* operations. The *Union*s and *Find*s may occur in any order, but *Union*s are done by rank and *Find*s are done with path compression.

We begin by establishing some lemmas concerning the number of nodes of *rank* r. Intuitively, because of the union-by-rank rule, there are many more nodes of small rank than large rank. In particular, there can be at most one node of rank $\log N$. What we would like to do is to produce as precise a bound as possible on the number of nodes of any particular rank r. Since ranks only change when *Union*s are

---

[†] Ackermann's function is frequently defined with $A(1,j) = j + 1$ for $j \ge 1$. The form in this text grows faster; thus, the inverse grows more slowly.

performed (and then only when the two trees have the same rank), we can prove this bound by ignoring the path compression.

LEMMA 8.1.

When executing a sequence of Union *instructions, a node of rank r must have at least 2^r descendants (including itself).*

PROOF

By induction. The basis, $r = 0$, is clearly true. Let T be the tree of rank r with the fewest number of descendants and let X be T's root. Suppose the last *Union* X was involved in was between T_1 and T_2. Suppose T_1's root was X. If T_1 had rank r, then T_1 would be a tree of height r with fewer descendants than T, which contradicts the assumption that T is the tree with the smallest number of descendants. Hence the rank of $T_1 \leq r - 1$. The rank of $T_2 \leq$ *rank of* T_1. Since T has rank r and the rank could only increase because of T_2, it follows that the rank of $T_2 = r - 1$. Then the rank of $T_1 = r - 1$. By the induction hypothesis, each tree has at least 2^{r-1} descendants, giving a total of 2^r and establishing the lemma.

Lemma 8.1 tells us that if no path compression is performed, then any node of rank r must have at least 2^r descendants. Path compression can change this, of course, since it can remove descendants from a node. However, when *Unions* are performed, even with path compression, we are using the ranks, which are estimated heights. These ranks behave as though there is no path compression. Thus, when bounding the *number* of nodes of rank r, path compression can be ignored.

Thus, the next lemma is valid with or without path compression.

LEMMA 8.2.

The number of nodes of rank r is at most $N/2^r$.

PROOF

Without path compression, each node of rank r is the root of a subtree of at least 2^r nodes. No node in the subtree can have rank r. Thus all subtrees of nodes of rank r are disjoint. Therefore, there are at most $N/2^r$ disjoint subtrees and hence $N/2^r$ nodes of rank r.

The next lemma seems somewhat obvious but is crucial in the analysis.

LEMMA 8.3.

At any point in the Union/Find *algorithm, the ranks of the nodes on a path from the leaf to a root increase monotonically.*

PROOF

The lemma is obvious if there is no path compression (see the example). If, after path compression, some node v is a descendant of w, then clearly v must have been a descendant of w when only *Unions* were considered. Hence the rank of v is less than the rank of w.

Let us summarize the preliminary results. Lemma 8.2 tells us how many nodes can be assigned rank r. Because ranks are assigned only by *Unions*, which have no idea of path compression, Lemma 8.2 is valid at any stage of the *Union/Find* algorithm—even in the midst of path compression. Figure 8.16 shows that while there are many nodes of ranks 0 and 1, there are fewer nodes of rank r as r gets larger.

Lemma 8.2 is tight, in the sense that it is possible for there to be $N/2^r$ nodes for any rank r. It is slightly loose, because it is not possible for the bound to hold for all ranks r simultaneously. While Lemma 8.2 describes the *number* of nodes in a rank r, Lemma 8.3 tells us their *distribution*. As one would expect, the rank of nodes is strictly increasing along the path from a leaf to the root.

We are now ready to prove the main theorem. Our basic idea is as follows: A *Find* on any node v costs time proportional to the number of nodes on the path from v to the root. Let us, then, charge one unit of cost for every node on the path from v to the root for each *Find*. To help us count the charges, we will deposit an imaginary penny into each node on the path. This is strictly an accounting gimmick, which is not part of the program. When the algorithm is over, we collect all the coins that have been deposited; this is the total cost.

As a further accounting gimmick, we deposit both American and Canadian pennies. We will show that during the execution of the algorithm, we can deposit only a certain number of American pennies during each *Find*. We will also show that we can deposit only a certain number of Canadian pennies to each node. Adding these two totals gives us a bound on the total number of pennies that can be deposited.

We now sketch our accounting scheme in a little more detail. We will divide the nodes by their ranks. We then divide the ranks into rank groups. On each *Find*, we will deposit some American coins into the general kitty and some Canadian coins into specific vertices. To compute the total number of Canadian coins deposited, we will compute the deposits per node. By adding up all the deposits for each node in rank r, we will get the total deposits per rank r. Then we will add up all the deposits for each rank r in group g and thereby obtain the total deposits for each rank group g. Finally, we add up all the deposits for each rank group g to obtain the total number of Canadian coins deposited in the forest. Adding this to the number of American coins in the kitty gives us the answer.

We will partition ranks into groups. Rank r goes into group $G(r)$, and G will be determined later. The largest rank in any rank group g is $F(g)$, where $F = G^{-1}$ is the

Figure 8.16 A large disjoint set tree (numbers below nodes are ranks)

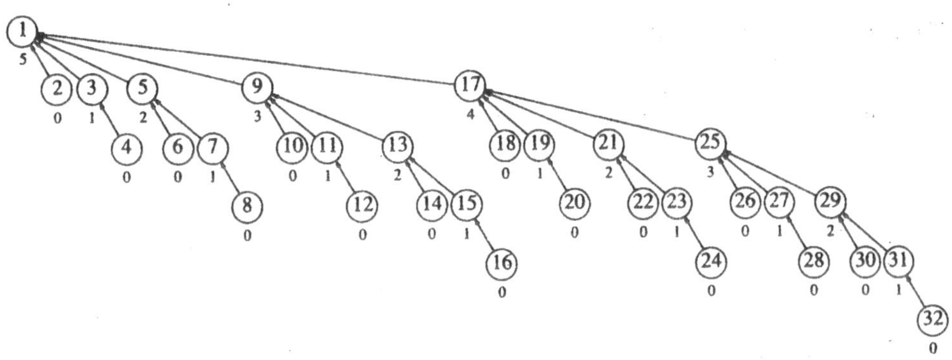

inverse of G. The number of ranks in any rank group, $g > 0$, is thus $F(g) - F(g-1)$. Clearly $G(N)$ is a very loose upper bound on the largest rank group. As an example, suppose that we partitioned the ranks as in Figure 8.17. In this case, $G(r) = \lceil \sqrt{r} \rceil$. The largest rank in group g is $F(g) = g^2$, and observe that group $g > 0$ contains ranks $F(g-1)+1$ through $F(g)$ inclusive. This formula does not apply for rank group 0, so for convenience we will ensure that rank group 0 contains only elements of rank 0. Notice that the groups are made of consecutive ranks.

As mentioned before, each *Union* instruction takes constant time, as long as each root keeps track of how big its subtrees are. Thus, *Unions* are essentially free, as far as this proof goes.

Each *Find(i)* takes time proportional to the number of vertices on the path from the vertex representing i to the root. We will thus deposit one penny for each vertex on the path. If this is all we do, however, we cannot expect much of a bound, because we are not taking advantage of path compression. Thus, we need to take advantage of path compression in our analysis. We will use fancy accounting.

For each vertex, v, on the path from the vertex representing i to the root, we deposit one penny under one of two accounts:

1. If v is the root, or if the parent of v is the root, or if the parent of v is in a different rank group from v, then charge one unit under this rule. This deposits an American penny into the kitty.
2. Otherwise deposit a Canadian penny into the vertex.

LEMMA 8.4.

For any Find(v)*, the total number of pennies deposited, either into the kitty or into a vertex, is exactly equal to the number of nodes on the path from v to the root.*

PROOF

Obvious.

Thus all we need to do is to sum all the American pennies deposited under rule 1 with all the Canadian pennies deposited under rule 2.

We are doing at most M *Finds*. We need to bound the number of pennies that can be deposited into the kitty during a *Find*.

Figure 8.17 Possible partitioning of ranks into groups

Group	Rank
0	0
1	1
2	2,3,4
3	5 through 9
4	10 through 16
i	$(i-1)^2 + 1$ through i^2

LEMMA 8.5.

Over the entire algorithm, the total deposits of American pennies under rule 1 amount to $M(G(N)+2)$.

PROOF

This is easy. For any *Find*, two American pennies are deposited, because of the root and its child. By Lemma 8.3, the vertices going up the path are monotonically increasing in rank, and since there are at most $G(N)$ rank groups, only $G(N)$ other vertices on the path can qualify as a rule 1 deposit for any particular *Find*. Thus, during any one find, at most $G(N)+2$ American pennies can be placed in the kitty. Thus, at most $M(G(N)+2)$ American pennies can be deposited under rule 1 for a sequence of M *Finds*.

To get a good estimate for all the Canadian deposits under rule 2, we will add up the deposits by vertices instead of by *Find* instructions. If a coin is deposited into vertex v under rule 2, v will be moved by path compression and get a new parent of higher rank than its old parent. (This is where we are using the fact that path compression is being done.) Thus, a vertex v in rank group $g>0$ can be moved at most $F(g)-F(g-1)$ times before its parent gets pushed out of rank group g, since that is the size of the rank group.[†] After this happens, all future charges to v will go under rule 1.

LEMMA 8.6.

The number of vertices, $V(g)$*, in rank group* $g>0$ *is at most* $N/2^{F(g-1)}$.

PROOF

By Lemma 8.2, there are at most $N/2^r$ vertices of rank r. Summing over the ranks in group g, we obtain

$$
\begin{aligned}
V(g) &\le \sum_{r=F(g-1)+1}^{F(g)} \frac{N}{2^r} \\
&\le \sum_{r=F(g-1)+1}^{\infty} \frac{N}{2^r} \\
&\le N \sum_{r=F(g-1)+1}^{\infty} \frac{1}{2^r} \\
&\le \frac{N}{2^{F(g-1)+1}} \sum_{s=0}^{\infty} \frac{1}{2^s} \\
&\le \frac{2N}{2^{F(g-1)+1}} \\
&\le \frac{N}{2^{F(g-1)}}
\end{aligned}
$$

[†] This can be reduced by 1. We do not for the sake of clarity; the bound is not improved by being more careful here.

LEMMA 8.7.

The maximum number of Canadian pennies deposited to all vertices in rank group g is at most $NF(g)/2^{F(g-1)}$.

PROOF

Each vertex in the rank group can receive at most $F(g) - F(g-1) \le F(g)$ Canadian pennies while its parent stays in its rank group, and Lemma 8.6 tells how many such vertices there are. The result is obtained by a simple multiplication.

LEMMA 8.8.

The total deposit under rule 2 is at most $N\sum_{g=1}^{G(N)} F(g)/2^{F(g-1)}$ *Canadian pennies.*

PROOF

Because rank group 0 contains only elements of rank 0, it cannot contribute to rule 2 charges (it cannot have a parent in the same rank group). The bound is obtained by summing the other rank groups.

Thus we have the deposits under rules 1 and 2. The total is

$$M(G(N) + 2) + N\sum_{g=1}^{G(N)} F(g)/2^{F(g-1)} \tag{8.1}$$

We still have not specified $G(N)$ or its inverse $F(N)$. Obviously, we are free to choose virtually anything we want, but it makes sense to choose $G(N)$ to minimize the bound above. However, if $G(N)$ is too small, then $F(N)$ will be large, hurting the bound. An apparently good choice is to choose $F(i)$ to be the function recursively defined by $F(0) = 0$ and $F(i) = 2^{F(i-1)}$. This gives $G(N) = 1 + \lfloor \log^* N \rfloor$. Figure 8.18 shows how this partitions the ranks. Notice that group 0 contains only rank 0, which we required in the previous lemma. F is very similar to the single-variable Ackermann function, which differs only in the definition of the base case ($F(0) = 1$).

Figure 8.18 Actual partitioning of ranks into groups used in the proof

Group	Rank
0	0
1	1
2	2
3	3,4
4	5 through 16
5	17 through 2^{16}
6	65537 through 2^{65536}
7	truly huge ranks

THEOREM 8.1.

The running time of M Unions *and* Finds *is* $O(M \log^* N)$.

PROOF

Plug in the definitions of F and G into Equation (8.1). The total number of American pennies is $O(M\,G(N)) = O(M \log^* N)$. The total number of Canadian pennies is $N \sum_{g=1}^{G(N)} F(g)/2^{F(g-1)} = N \sum_{g=1}^{G(N)} 1 = N\,G(N) = O(N \log^* N)$. Since $M = \Omega(N)$, the bound follows.

What the analysis shows is that there are few nodes that could be moved frequently by path compression, and thus the total time spent is relatively small.

8.7. An Application

As an example of how this data structure might be used, consider the following problem. We have a network of computers and a list of bidirectional connections; each of these connections allows a file transfer from one computer to another. Is it possible to send a file from any computer on the network to any other? An extra restriction is that the problem must be solved *on-line*. Thus, the list of connections is presented one at a time, and the algorithm must be prepared to give an answer at any point.

An algorithm to solve this problem can initially put every computer in its own set. Our invariant is that two computers can transfer files if and only if they are in the same set. We can see that the ability to transfer files forms an equivalence relation. We then read connections one at a time. When we read some connection, say (u, v), we test to see whether u and v are in the same set and do nothing if they are. If they are in different sets, we merge their sets. At the end of the algorithm, the graph is connected if and only if there is exactly one set. If there are M connections and N computers, the space requirement is $O(N)$. Using union-by-size and path compression, we obtain a worst-case running time of $O(M\,\alpha(M, N))$, since there are $2M$ *Finds* and at most $N - 1$ *Unions*. This running time is linear for all practical purposes.

We will see a much better application in the next chapter.

Summary

We have seen a very simple data structure to maintain disjoint sets. When the *Union* operation is performed, it does not matter, as far as correctness is concerned, which set retains its name. A valuable lesson that should be learned here is that it can be very important to consider the alternatives when a particular step is not totally specified. The *Union* step is flexible; by taking advantage of this, we are able to get a much more efficient algorithm.

Path compression is one of the earliest forms of *self-adjustment*, which we have seen elsewhere (splay trees, skew heaps). Its use is extremely interesting, especially from a theoretical point of view, because it was one of the first examples of a simple algorithm with a not-so-simple worst-case analysis.

Exercises

8.1 Show the result of the following sequence of instructions: *Union*(1, 2), *Union*(3, 4), *Union*(3, 5), *Union*(1, 7), *Union*(3, 6), *Union*(8, 9), *Union*(1, 8), *Union*(3, 10), *Union*(3, 11), *Union*(3, 12), *Union*(3, 13), *Union*(14, 15), *Union*(16, 17), *Union*(14, 16), *Union*(1, 3), *Union*(1, 14) when the *Unions* are:

a. Performed arbitrarily.

b. Performed by height.

c. Performed by size.

8.2 For each of the trees in the previous exercise, perform a *Find* with path compression on the deepest node.

8.3 Write a program to determine the effects of path compression and the various *Unioning* strategies. Your program should process a long sequence of equivalence operations using all six of the possible strategies.

8.4 Show that if *Unions* are performed by height, then the depth of any tree is $O(\log N)$.

8.5 a. Show that if $M = N^2$, then the running time of M *Union/Find* operations is $O(M)$.

b. Show that if $M = N \log N$, then the running time of M *Union/Find* operations is $O(M)$.

*c. Suppose $M = \Theta(N \log\log N)$. What is the running time of M *Union/Find* operations?

*d. Suppose $M = \Theta(N \log^* N)$. What is the running time of M *Union/Find* operations?

8.6 Show the operation of the program in Section 8.7 on the following graph: (1,2), (3,4), (3,6), (5,7), (4,6), (2,4), (8,9), (5,8). What are the connected components?

8.7 Write a program to implement the algorithm in Section 8.7.

*8.8 Suppose we want to add an extra operation, *Deunion,* which undoes the last *Union* operation that has not been already undone.

a. Show that if we do union-by-height and *Finds* without path compression, then *Deunion* is easy and a sequence of M *Union, Find,* and *Deunion* operations takes $O(M \log N)$ time.

b. Why does path compression make *Deunion* hard?

**c. Show how to implement all three operations so that the sequence of M operations takes $O(M \log N / \log\log N)$ time.

*8.9 Suppose we want to add an extra operation, *Remove(X)*, which removes X from its current set and places it in its own. Show how to modify the *Union/Find* algorithm so that the running time of a sequence of M *Union*, *Find*, and *Remove* operations is $O(M\alpha(M,N))$.

**8.10 Give an algorithm that takes as input an N-vertex tree and a list of N pairs of vertices and determines for each pair (v,w) the closest common ancestor of v and w. Your algorithm should run in $O(N \log^* N)$.

*8.11 Show that if all of the *Unions* precede the *Finds*, then the disjoint set algorithm with path compression requires linear time, even if the *Unions* are done arbitrarily.

**8.12 Prove that if *Unions* are done arbitrarily, but path compression is performed on the *Finds*, then the worst-case running time is $\Theta(M \log N)$.

8.13 Prove that if *Unions* are done by size and path compression is performed, the worst-case running time is $O(M \log^* N)$.

8.14 Suppose we implement partial path compression on *Find(i)* by making every other node on the path from i to the root point to its grandparent (where this makes sense). This is known as *path halving*.

a. Write a procedure to do this.

b. Prove that if path halving is performed on the *Finds* and either union-by-height or union-by-size is used, the worst-case running time is $O(M \log^* N)$.

References

Various solutions to the *Union/Find* problem can be found in [6], [9], and [11]. Hopcroft and Ullman showed the $O(M \log^* N)$ bound of Section 8.6. Tarjan [15] obtained the bound $O(M\alpha(M,N))$. A more precise (but asymptotically identical) bound for $M < N$ appears in [2] and [18]. Various other strategies for path compression and *Unions* also achieve the same bound; see [18] for details.

A lower bound showing that under certain restrictions $\Omega(M\alpha(M,N))$ time is required to process M *Union/Find* operations was given by Tarjan [16]. Identical bounds under less restrictive conditions have been shown in [7] and [14].

Applications of the *Union/Find* data structure appear in [1] and [10]. Certain special cases of the *Union/Find* problem can be solved in $O(M)$ time [8]. This reduces the running time of several algorithms, such as [1], graph dominance, and reducibility (see references in Chapter 9) by a factor of $\alpha(M,N)$. Others, such as [10] and the graph connectivity problem in this chapter, are unaffected. The paper lists 10 examples. Tarjan has used path compression to obtain efficient algorithms for several graph problems [17].

Average case results for the *Union/Find* problem appear in [5], [12], [21], and [3]. Results bounding the running time of any single operation (as opposed to the entire sequence) appear in [4] and [13].

Exercise 8.8 is solved in [20]. A general *Union/Find* structure, supporting more operations, is given in [19].

1. A. V. Aho, J. E. Hopcroft, and J. D. Ullman, "On Finding Lowest Common Ancestors in Trees," *SIAM Journal on Computing*, 5 (1976), 115–132.
2. L. Banachowski, "A Complement to Tarjan's Result about the Lower Bound on the Complexity of the Set Union Problem," *Information Processing Letters*, 11 (1980), 59–65.
3. B. Bollobás and I. Simon, "Probabilistic Analysis of Disjoint Set Union Algorithms," *SIAM Journal on Computing*, 22 (1993), 1053–1086.
4. N. Blum, "On the Single-Operation Worst-Case Time Complexity of the Disjoint Set Union Problem," *SIAM Journal on Computing*, 15 (1986), 1021–1024.
5. J. Doyle and R. L. Rivest, "Linear Expected Time of a Simple Union Find Algorithm," *Information Processing Letters*, 5 (1976), 146–148.
6. M. J. Fischer, "Efficiency of Equivalence Algorithms," in *Complexity of Computer Computation* (eds. R. E. Miller and J. W. Thatcher), Plenum Press, New York 1972, 153–168.
7. M. L. Fredman and M. E. Saks, "The Cell Probe Complexity of Dynamic Data Structures," *Proceedings of the Twenty-first Annual Symposium on Theory of Computing* (1989), 345–354.
8. H. N. Gabow and R. E. Tarjan, "A Linear-Time Algorithm for a Special Case of Disjoint Set Union," *Journal of Computer and System Sciences*, 30 (1985), 209–221.
9. B. A. Galler and M. J. Fischer, "An Improved Equivalence Algorithm," *Communications of the ACM*, 7 (1964), 301–303.
10. J. E. Hopcroft and R. M. Karp, "An Algorithm for Testing the Equivalence of Finite Automata," *Technical Report TR-71-114*, Department of Computer Science, Cornell University, Ithaca, N.Y., 1971.
11. J. E. Hopcroft and J. D. Ullman, "Set Merging Algorithms," *SIAM Journal on Computing*, 2 (1973), 294–303.
12. D. E. Knuth and A. Schonhage, "The Expected Linearity of a Simple Equivalence Algorithm," *Theoretical Computer Science*, 6 (1978), 281–315.
13. J. A. LaPoutre, "New Techniques for the Union-Find Problem," *Proceedings of the First Annual ACM–SIAM Symposium on Discrete Algorithms* (1990), 54–63.
14. J. A. LaPoutre, "Lower Bounds for the Union-Find and the Split-Find Problem on Pointer Machines," *Proceedings of the Twenty-Second Annual ACM Symposium on Theory of Computing* (1990), 34–44.
15. R. E. Tarjan, "Efficiency of a Good but Not Linear Set Union Algorithm," *Journal of the ACM*, 22 (1975), 215–225.
16. R. E. Tarjan, "A Class of Algorithms Which Require Nonlinear Time to Maintain Disjoint Sets," *Journal of Computer and System Sciences*, 18 (1979), 110–127.
17. R. E. Tarjan, "Applications of Path Compression on Balanced Trees," *Journal of the ACM*, 26 (1979), 690–715.
18. R. E. Tarjan and J. van Leeuwen, "Worst Case Analysis of Set Union Algorithms," *Journal of the ACM*, 31 (1984), 245–281.
19. M. J. van Kreveld and M. H. Overmars, "Union-Copy Structures and Dynamic Segment Trees," *Journal of the ACM*, 40 (1993), 635–652.
20. J. Westbrook and R. E. Tarjan, "Amortized Analysis of Algorithms for Set Union with Backtracking," *SIAM Journal on Computing*, 18 (1989), 1–11.
21. A. C. Yao, "On the Average Behavior of Set Merging Algorithms," *Proceedings of Eighth Annual ACM Symposium on the Theory of Computation*, (1976), 192–195.

CHAPTER 9

Graph Algorithms

In this chapter we discuss several common problems in graph theory. Not only are these algorithms useful in practice, they are interesting because in many real-life applications they are too slow unless careful attention is paid to the choice of data structures. We will

- Show several real-life problems, which can be converted to problems on graphs.
- Give algorithms to solve several common graph problems.
- Show how the proper choice of data structures can drastically reduce the running time of these algorithms.
- See an important technique, known as depth-first search, and show how it can be used to solve several seemingly nontrivial problems in linear time.

9.1. Definitions

A *graph* $G = (V, E)$ consists of a set of *vertices*, V, and a set of *edges*, E. Each edge is a pair (v, w), where $v, w \in V$. Edges are sometimes referred to as *arcs*. If the pair is ordered, then the graph is *directed*. Directed graphs are sometimes referred to as *digraphs*. Vertex w is *adjacent* to v if and only if $(v, w) \in E$. In an undirected graph with edge (v, w), and hence (w, v), w is adjacent to v and v is adjacent to w. Sometimes an edge has a third component, known as either a *weight* or a *cost*.

A *path* in a graph is a sequence of vertices $w_1, w_2, w_3, \ldots, w_N$ such that $(w_i, w_{i+1}) \in E$ for $1 \leq i < N$. The *length* of such a path is the number of edges on the path, which is equal to $N - 1$. We allow a path from a vertex to itself; if this path contains no edges, then the path length is 0. This is a convenient way to define an otherwise special case. If the graph contains an edge (v, v) from a vertex to itself, then the path v, v is sometimes referred to as a *loop*. The graphs we will consider will generally be loopless. A *simple* path is a path such that all vertices are distinct, except that the first and last could be the same.

A *cycle* in a directed graph is a path of length at least 1 such that $w_1 = w_N$; this cycle is simple if the path is simple. For undirected graphs, we require that the edges be distinct. The logic of these requirements is that the path u, v, u in an undirected graph should not be considered a cycle, because (u, v) and (v, u) are the same edge. In a directed graph, these are different edges, so it makes sense to call this a cycle. A directed graph is *acyclic* if it has no cycles. A directed acyclic graph is sometimes referred to by its abbreviation, *DAG*.

An undirected graph is *connected* if there is a path from every vertex to every other vertex. A directed graph with this property is called *strongly connected.* If a directed graph is not strongly connected, but the underlying graph (without direction to the arcs) is connected, then the graph is said to be *weakly connected.* A *complete graph* is a graph in which there is an edge between every pair of vertices.

An example of a real-life situation that can be modeled by a graph is the airport system. Each airport is a vertex, and two vertices are connected by an edge if there is a nonstop flight from the airports that are represented by the vertices. The edge could have a weight, representing the time, distance, or cost of the flight. It is reasonable to assume that such a graph is directed, since it might take longer or cost more (depending on local taxes, for example) to fly in different directions. We would probably like to make sure that the airport system is strongly connected, so that it is always possible to fly from any airport to any other airport. We might also like to quickly determine the best flight between any two airports. "Best" could mean the path with the fewest number of edges or could be taken with respect to one, or all, of the weight measures.

Traffic flow can be modeled by a graph. Each street intersection represents a vertex, and each street is an edge. The edge costs could represent, among other things, a speed limit or a capacity (number of lanes). We could then ask for the shortest route or use this information to find the most likely location for bottlenecks.

In the remainder of this chapter, we will see several more applications of graphs. Many of these graphs can be quite large, so it is important that the algorithms we use be efficient.

9.1.1. Representation of Graphs

We will consider directed graphs (undirected graphs are similarly represented).

Suppose, for now, that we can number the vertices, starting at 1. The graph shown in Figure 9.1 represents 7 vertices and 12 edges.

One simple way to represent a graph is to use a two-dimensional array. This is known as an *adjacency matrix* representation. For each edge (u, v), we set $A[u][v] = 1$; otherwise the entry in the array is 0. If the edge has a weight associated with it, then we can set $A[u][v]$ equal to the weight and use either a very large or a very small weight as a sentinel to indicate nonexistent edges. For instance, if we were looking for the cheapest airplane route, we could represent nonexistent flights with a cost of ∞. If we were looking, for some strange reason, for the most expensive airplane route, we could use $-\infty$ (or perhaps 0) to represent nonexistent edges.

Although this has the merit of extreme simplicity, the space requirement is $\Theta(|V|^2)$, which can be prohibitive if the graph does not have very many edges. An adjacency matrix is an appropriate representation if the graph is *dense:* $|E| = \Theta(|V|^2)$. In most of the applications that we shall see, this is not true. For instance, suppose the graph represents a street map. Assume a Manhattan-like orientation, where almost

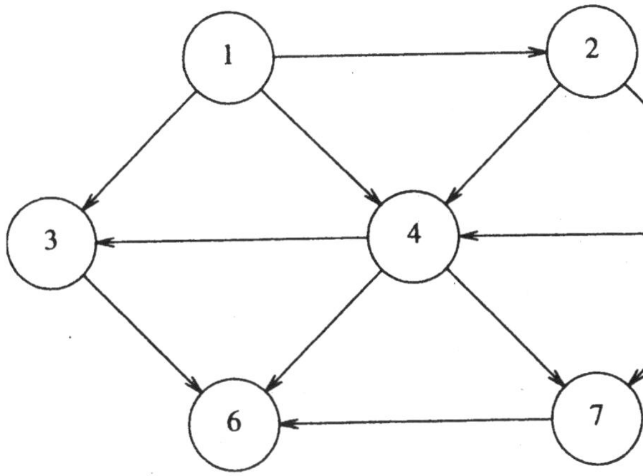

Figure 9.1 A directed graph

all the streets run either north–south or east–west. Therefore, any intersection is attached to roughly four streets, so if the graph is directed and all streets are two-way, then $|E| \approx 4|V|$. If there are 3,000 intersections, then we have a 3,000-vertex graph with 12,000 edge entries, which would require an array of size 9,000,000. Most of these entries would contain zero. This is intuitively bad, because we want our data structures to represent the data that are actually there and not the data that are not present.

If the graph is not dense, in other words, if the graph is *sparse,* a better solution is an *adjacency list* representation. For each vertex, we keep a list of all adjacent vertices. The space requirement is then $O(|E| + |V|)$. The leftmost structure in Figure 9.2 is merely an array of header cells. The representation should be clear from Figure 9.2. If the edges have weights, then this additional information is also stored in the cells.

Adjacency lists are the standard way to represent graphs. Undirected graphs can be similarly represented; each edge (u, v) appears in two lists, so the space usage essentially doubles. A common requirement in graph algorithms is to find all vertices adjacent to some given vertex v, and this can be done, in time proportional to the number of such vertices found, by a simple scan down the appropriate adjacency list.

In most real-life applications, the vertices have names, which are unknown at compile time, instead of numbers. Since we cannot index an array by an unknown name, we must provide a mapping of names to numbers. The easiest way to do this is to use a hash table, in which we store a name and an internal number ranging from 1 to $|V|$ for each vertex. The numbers are assigned as the graph is read. The first number assigned is 1. As each edge is input, we check whether each of the two vertices has been assigned a number, by seeing if it is in the hash table. If so, we use the internal number. Otherwise, we assign to the vertex the next available number and insert the vertex name and number into the hash table.

With this transformation, all the graph algorithms will use only the internal numbers. Since eventually we will need to output the real vertex names and not the internal numbers, we must also record, for each internal number, the corresponding

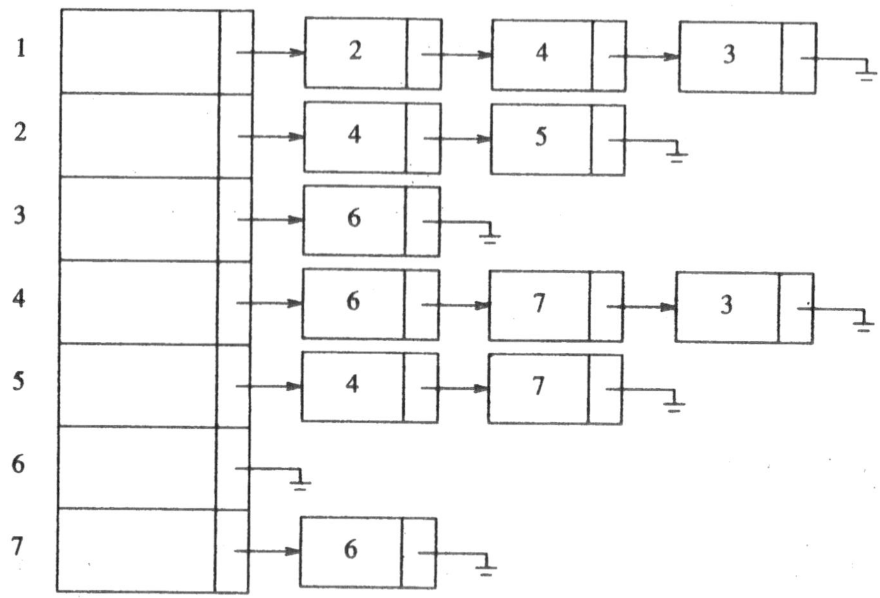

Figure 9.2 An adjacency list representation of a graph

vertex name. One way is to use an array of strings. If the vertex names are long, this can cost considerable space, because the vertex names are stored twice. An alternative is to keep an array of pointers into the hash table. The price of this alternative is a slight loss of the sanctity of the hash table ADT (elements in the hash table are accessible by means other than basic hash table operations).

The code that we present in this chapter will be pseudocode using ADTs as much as possible. We will do this to save space and, of course, to make the algorithmic presentation of the algorithms much clearer.

9.2. Topological Sort

A *topological sort* is an ordering of vertices in a directed acyclic graph, such that if there is a path from v_i to v_j, then v_j appears *after* v_i in the ordering. The graph in Figure 9.3 represents the course prerequisite structure at a state university in Miami. A directed edge (v, w) indicates that course v must be completed before course w may be attempted. A topological ordering of these courses is any course sequence that does not violate the prerequisite requirement.

It is clear that a topological ordering is not possible if the graph has a cycle, since for two vertices v and w on the cycle, v precedes w and w precedes v. Furthermore, the ordering is not necessarily unique; any legal ordering will do. In the graph in Figure 9.4, $v_1, v_2, v_5, v_4, v_3, v_7, v_6$ and $v_1, v_2, v_5, v_4, v_7, v_3, v_6$ are both topological orderings.

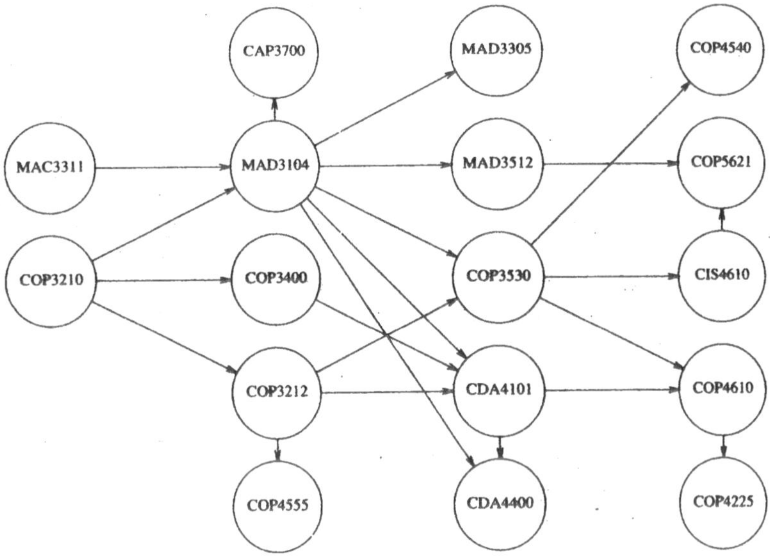

Figure 9.3 An acyclic graph representing course prerequisite structure

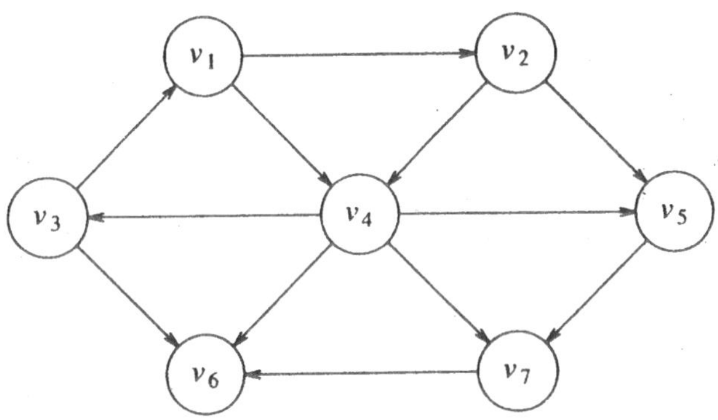

Figure 9.4 An acyclic graph

A simple algorithm to find a topological ordering is first to find any vertex with no incoming edges. We can then print this vertex, and remove it, along with its edges, from the graph. Then we apply this same strategy to the rest of the graph.

To formalize this, we define the *indegree* of a vertex v as the number of edges (u, v). We compute the indegrees of all vertices in the graph. Assuming that the *Indegree* array is initialized and that the graph is read into an adjacency list, we can then apply the algorithm in Figure 9.5 to generate a topological ordering.

The function *FindNewVertexOfIndegreeZero* scans the *Indegree* array looking for a vertex with indegree 0 that has not already been assigned a topological number. It returns *NotAVertex* if no such vertex exists; this indicates that the graph has a cycle.

```
void
Topsort( Graph G )
{
    int Counter;
    Vertex V, W;

    for( Counter = 0; Counter < NumVertex; Counter++ )
    {
        V = FindNewVertexOfDegreeZero( );
        if ( V == NotAVertex )
        {
            Error( "Graph has a cycle" );
            break;
        }
        TopNum[ V ] = Counter;
        for each W adjacent to V
            Indegree[ W ]--;
    }
}
```

Figure 9.5 Simple topological sort pseudocode

Because *FindNewVertexOfIndegreeZero* is a simple sequential scan of the *Indegree* array, each call to it takes $O(|V|)$ time. Since there are $|V|$ such calls, the running time of the algorithm is $O(|V|^2)$.

By paying more careful attention to the data structures, it is possible to do better. The cause of the poor running time is the sequential scan through the *Indegree* array. If the graph is sparse, we would expect that only a few vertices have their indegrees updated during each iteration. However, in the search for a vertex of indegree 0, we look at (potentially) all the vertices, even though only a few have changed.

We can remove this inefficiency by keeping all the (unassigned) vertices of indegree 0 in a special *box*. The *FindNewVertexOfIndegreeZero* function then returns (and removes) any vertex in the box. When we decrement the indegrees of the adjacent vertices, we check each vertex and place it in the box if its indegree falls to 0.

To implement the box, we can use either a stack or a queue. First, the indegree is computed for every vertex. Then all vertices of indegree 0 are placed on an initially empty queue. While the queue is not empty, a vertex v is removed, and all edges adjacent to v have their indegrees decremented. A vertex is put on the queue as soon as its indegree falls to 0. The topological ordering then is the order in which the vertices dequeue. Figure 9.6 shows the status after each phase.

A pseudocode implementation of this algorithm is given in Figure 9.7. As before, we will assume that the graph is already read into an adjacency list and that the indegrees are computed and placed in an array. A convenient way of doing this in practice would be to place the indegree of each vertex in the header cell. We also assume an array *TopNum*, in which to place the topological numbering.

	Indegree Before Dequeue #						
Vertex	**1**	**2**	**3**	**4**	**5**	**6**	**7**
v_1	0	0	0	0	0	0	0
v_2	1	0	0	0	0	0	0
v_3	2	1	1	1	0	0	0
v_4	3	2	1	0	0	0	0
v_5	1	1	0	0	0	0	0
v_6	3	3	3	3	2	1	0
v_7	2	2	2	1	0	0	0
Enqueue	v_1	v_2	v_5	v_4	v_3, v_7		v_6
Dequeue	v_1	v_2	v_5	v_4	v_3	v_7	v_6

Figure 9.6 Result of applying topological sort to the graph in Figure 9.4

```
void
Topsort( Graph G );
{
    Queue Q;
    int Counter = 0;
    Vertex V, W;

    Q = CreateQueue( NumVertex ); MakeEmpty( Q );
    for each vertex V
        if( Indegree[ V ] == 0 )
            Enqueue( V, Q );

    while( !IsEmpty( Q ) )
    {
        V = Dequeue( Q );
        TopNum[ V ] = ++Counter;  /*Assign next number */

        for each W adjacent to V
            if( --Indegree[ W ] == 0 )
                Enqueue( W, Q );
    }

    if ( Counter != NumVertex )
        Error( "Graph has a cycle" );

    DisposeQueue( Q );  /* Free the memory */
}
```

Figure 9.7 Pseudocode to perform topological sort

The time to perform this algorithm is $O(|E| + |V|)$ if adjacency lists are used. This is apparent when one realizes that the body of the *for* loop is executed at most once per edge. The queue operations are done at most once per vertex, and the initialization steps also take time proportional to the size of the graph.

9.3. Shortest-Path Algorithms

In this section we examine various shortest-path problems. The input is a weighted graph: associated with each edge (v_i, v_j) is a cost $c_{i,j}$ to traverse the arc. The cost of a path $v_1 v_2 \ldots v_N$ is $\sum_{i=1}^{N-1} c_{i,i+1}$. This is referred to as the ***weighted path length***. The ***unweighted path length*** is merely the number of edges on the path, namely, $N - 1$.

SINGLE-SOURCE SHORTEST-PATH PROBLEM:

Given as input a weighted graph, $G = (V, E)$, and a distinguished vertex, s, find the shortest weighted path from s to every other vertex in G.

For example, in the graph in Figure 9.8, the shortest weighted path from v_1 to v_6 has a cost of 6 and goes from v_1 to v_4 to v_7 to v_6. The shortest unweighted path between these vertices is 2. Generally, when it is not specified whether we are referring to a weighted or an unweighted path, the path is weighted if the graph is. Notice also that in this graph there is no path from v_6 to v_1.

The graph in the preceding example has no edges of negative cost. The graph in Figure 9.9 shows the problems that negative edges can cause. The path from v_5 to v_4 has cost 1, but a shorter path exists by following the loop v_5, v_4, v_2, v_5, v_4, which has cost -5. This path is still not the shortest, because we could stay in the loop arbitrarily long. Thus, the shortest path between these two points is undefined. Similarly, the shortest path from v_1 to v_6 is undefined, because we can get into the same loop. This loop is known as a *negative-cost cycle;* when one is present in the graph, the shortest paths are not defined. Negative-cost edges are not necessarily bad, as the cycles are, but their presence seems to make the problem harder. For convenience, in the absence of a negative-cost cycle, the shortest path from s to s is zero.

Figure 9.8 A directed graph G

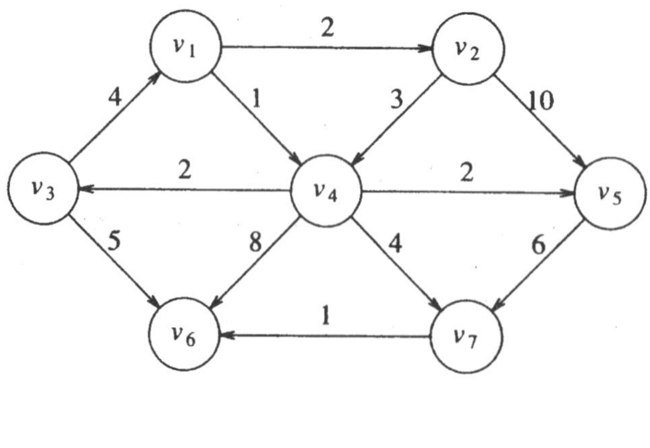

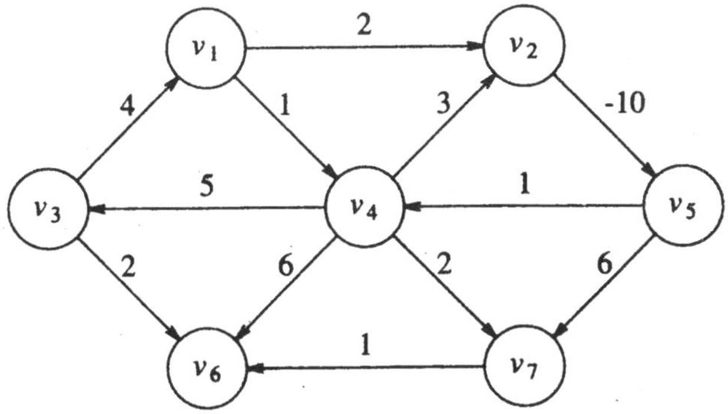

Figure 9.9 A graph with a negative-cost cycle

There are many examples where we might want to solve the shortest-path problem. If the vertices represent computers; the edges represent a link between computers; and the costs represent communication costs (phone bill per 1,000 bytes of data), delay costs (number of seconds required to transmit 1,000 bytes), or a combination of these and other factors, then we can use the shortest-path algorithm to find the cheapest way to send electronic news from one computer to a set of other computers.

We can model airplane or other mass transit routes by graphs and use a shortest-path algorithm to compute the best route between two points. In this and many practical applications, we might want to find the shortest path from one vertex, s, to only one other vertex, t. Currently there are no algorithms in which finding the path from s to one vertex is any faster (by more than a constant factor) than finding the path from s to all vertices.

We will examine algorithms to solve four versions of this problem. First, we will consider the unweighted shortest-path problem and show how to solve it in $O(|E| + |V|)$. Next, we will show how to solve the weighted shortest-path problem if we assume that there are no negative edges. The running time for this algorithm is $O(|E| \log |V|)$ when implemented with reasonable data structures.

If the graph has negative edges, we will provide a simple solution, which unfortunately has a poor time bound of $O(|E| \cdot |V|)$. Finally, we will solve the weighted problem for the special case of acyclic graphs in linear time.

9.3.1. *Unweighted Shortest Paths*

Figure 9.10 shows an unweighted graph, G. Using some vertex, s, which is an input parameter, we would like to find the shortest path from s to all other vertices. We are only interested in the number of edges contained on the path, so there are no weights on the edges. This is clearly a special case of the weighted shortest-path problem, since we could assign all edges a weight of 1.

For now, suppose we are interested only in the length of the shortest paths, not in the actual paths themselves. Keeping track of the actual paths will turn out to be a matter of simple bookkeeping.

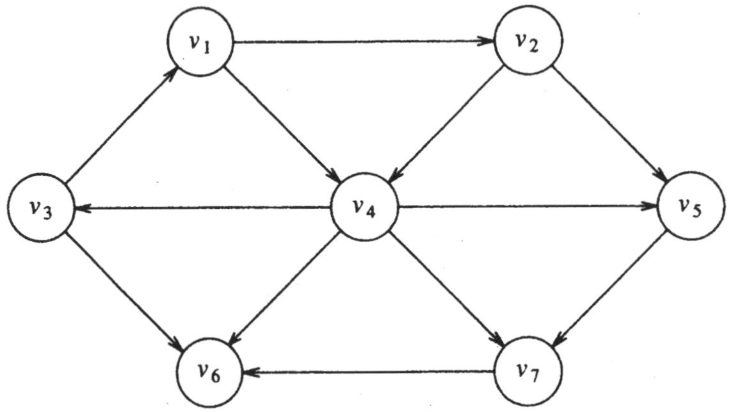

Figure 9.10 An unweighted directed graph G

Suppose we choose s to be v_3. Immediately, we can tell that the shortest path from s to v_3 is then a path of length 0. We can mark this information, obtaining the graph in Figure 9.11.

Now we can start looking for all vertices that are a distance 1 away from s. These can be found by looking at the vertices that are adjacent to s. If we do this, we see that v_1 and v_6 are one edge from s. This is shown in Figure 9.12.

Figure 9.11 Graph after marking the start node as reachable in zero edges

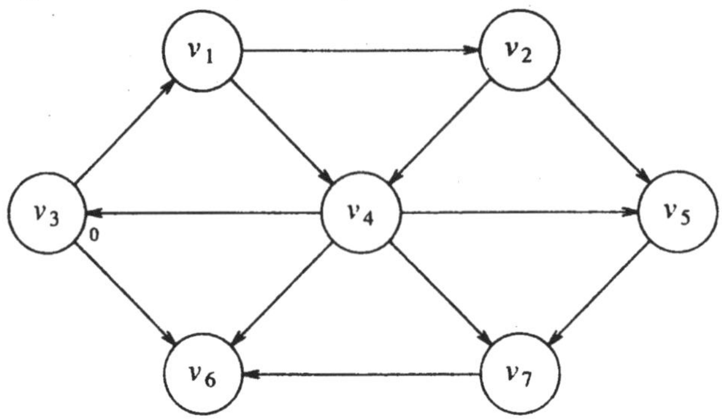

Figure 9.12 Graph after finding all vertices whose path length from s is 1

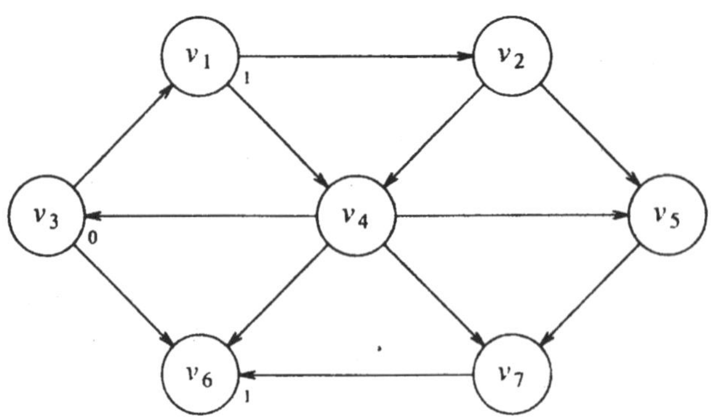

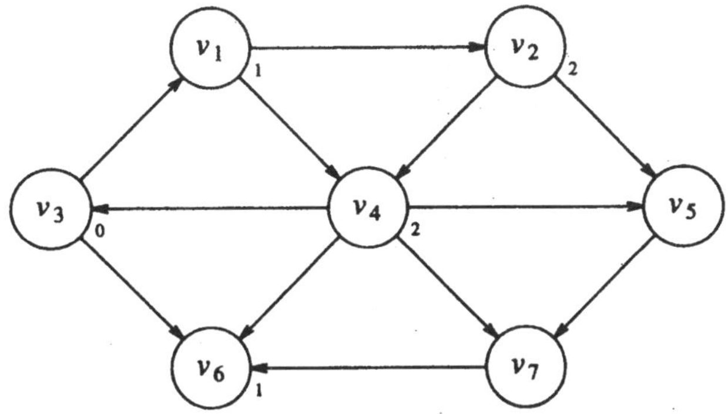

Figure 9.13 Graph after finding all vertices whose shortest path is 2

We can now find vertices whose shortest path from s is exactly 2, by finding all the vertices adjacent to v_1 and v_6 (the vertices at distance 1), whose shortest paths are not already known. This search tells us that the shortest path to v_2 and v_4 is 2. Figure 9.13 shows the progress that has been made so far.

Finally we can find, by examining vertices adjacent to the recently evaluated v_2 and v_4, that v_5 and v_7 have a shortest path of three edges. All vertices have now been calculated, and so Figure 9.14 shows the final result of the algorithm.

This strategy for searching a graph is known as *breadth-first search*. It operates by processing vertices in layers: the vertices closest to the start are evaluated first, and the most distant vertices are evaluated last. This is much the same as a level-order traversal for trees.

Given this strategy, we must translate it into code. Figure 9.15 shows the initial configuration of the table that our algorithm will use to keep track of its progress.

For each vertex, we will keep track of three pieces of information. First, we will keep its distance from s in the entry d_v. Initially all vertices are unreachable except for s, whose path length is 0. The entry in p_v is the bookkeeping variable, which will allow us to print the actual paths. The entry *Known* is set to 1 after a vertex is processed. Initially, all entries are not *Known*, including the start vertex. When

Figure 9.14 Final shortest paths

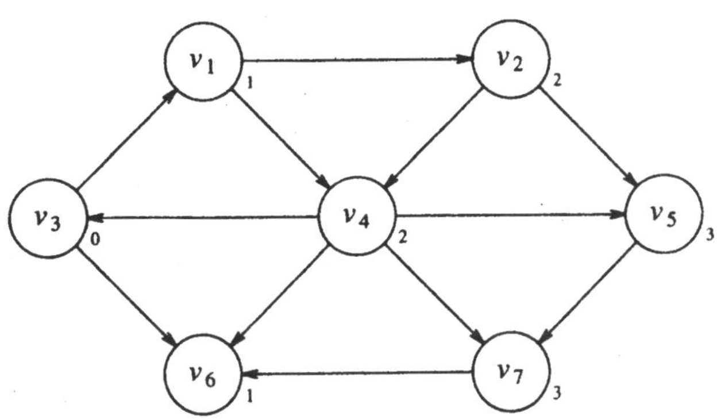

v	Known	d_v	p_v
v_1	0	∞	0
v_2	0	∞	0
v_3	0	0	0
v_4	0	∞	0
v_5	0	∞	0
v_6	0	∞	0
v_7	0	∞	0

Figure 9.15 Initial configuration of table used in unweighted shortest-path computation

a vertex is marked known, we have a guarantee that no cheaper path will ever be found, and so processing for that vertex is essentially complete.

The basic algorithm can be described in Figure 9.16. The algorithm in Figure 9.16 mimics the diagrams by declaring as *Known* the vertices at distance $d = 0$, then $d = 1$, then $d = 2$, and so on, and setting all the adjacent vertices w that still have $d_w = \infty$ to a distance $d_w = d + 1$.

By tracing back through the p_v variable, the actual path can be printed. We will see how when we discuss the weighted case.

The running time of the algorithm is $O(|V|^2)$, because of the doubly nested *for* loops. An obvious inefficiency is that the outside loop continues until

Figure 9.16 Pseudocode for unweighted shortest-path algorithm

```
void
Unweighted( Table T )  /* Assume T is initialized */
{
    int CurrDist;
    Vertex V, W;

    for( CurrDist = 0; CurrDist < NumVertex; CurrDist++ )
        for each vertex V
            if ( !T[ V ].Known && T[ V ].Dist == CurrDist )
            {
                T[ V ].Known = True;
                for each W adjacent to V
                    if( T[ W ].Dist == Infinity )
                    {
                        T[ W ].Dist = CurrDist + 1;
                        T[ W ].Path = V;
                    }
            }
}
```

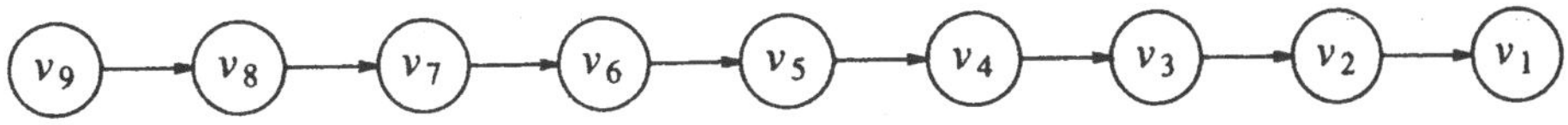

Figure 9.17 A bad case for unweighted shortest-path algorithm using Figure 9.16 (pseudocode)

NumVertex -1, even if all the vertices become known much earlier. Although an extra test could be made to avoid this, it does not affect the worst-case running time, as can be seen by generalizing what happens when the input is the graph in Figure 9.17 with start vertex v_9.

We can remove the inefficiency in much the same way as was done for topological sort. At any point in time, there are only two types of unknown vertices that have $d_v \neq \infty$. Some have $d_v = CurrDist$, and the rest have $d_v = CurrDist + 1$. Because of this extra structure, it is very wasteful to search through the entire table to find a proper vertex at lines 2 and 3.

A very simple but abstract solution is to keep two boxes. Box #1 will have the unknown vertices with $d_v = CurrDist$, and box #2 will have $d_v = CurrDist + 1$. The test at lines 2 and 3 can be replaced by finding any vertex in box #1. After line 8 (inside the *if* block), we can add w to box #2. After the outside *for* loop terminates, box #1 is empty, and box #2 can be transferred to box #1 for the next pass of the *for* loop.

We can refine this idea even further by using just one queue. At the start of the pass, the queue contains only vertices of distance *CurrDist*. When we add adjacent vertices of distance *CurrDist* + 1, since they enqueue at the rear, we are guaranteed that they will not be processed until after all the vertices of distance *CurrDist* have been processed. After the last vertex at distance *CurrDist* dequeues and is processed, the queue only contains vertices of distance *CurrDist* + 1, so this process perpetuates. We merely need to begin the process by placing the start node on the queue by itself.

The refined algorithm is shown in Figure 9.18. In the pseudocode, we have assumed that the start vertex, s, is known somehow and *T[s].Dist* is 0. A C routine might pass s as an argument. Also, it is possible that the queue might empty prematurely, if some vertices are unreachable from the start node. In this case, a distance of *Infinity* will be reported for these nodes, which is perfectly reasonable. Finally, the *Known* field is not used; once a vertex is processed it can never enter the queue again, so the fact that it need not be reprocessed is implicitly marked. Thus, the *Known* field can be discarded. Figure 9.19 shows how the values on the graph we have been using are changed during the algorithm. We keep the *Known* field to make the table easier to follow and for consistency with the rest of this section.

Using the same analysis as was performed for topological sort, we see that the running time is $O(|E| + |V|)$, as long as adjacency lists are used.

9.3.2. Dijkstra's Algorithm

If the graph is weighted, the problem (apparently) becomes harder, but we can still use the ideas from the unweighted case.

```
void
Unweighted( Table T ) /* Assume T is initialized (Fig 9.30) */
{
    Queue Q;
    Vertex V, W;

    Q = CreateQueue( NumVertex ); MakeEmpty( Q );

    /* Enqueue the start vertex S, determined elsewhere */
    Enqueue( S, Q );

    while( !IsEmpty( Q ) )
    {
        V = Dequeue( Q );
        T[ V ].Known = True; /* Not really needed anymore */

        for each W adjacent to V
            if( T[ W ].Dist == Infinity )
            {
                T[ W ].Dist = T[ V ].Dist + 1;
                T[ W ].Path = V;
                Enqueue( W, Q );
            }
    }
    DisposeQueue( Q ); /* Free the memory */
}
```

Figure 9.18 Pseudocode for unweighted shortest-path algorithm

We keep all of the same information as before. Thus, each vertex is marked as either known or unknown. A tentative distance d_v is kept for each vertex, as before. This distance turns out to be the shortest path length from s to v using only known vertices as intermediates. As before, we record p_v, which is the last vertex to cause a change to d_v.

The general method to solve the single-source shortest-path problem is known as *Dijkstra's algorithm*. This thirty-year-old solution is a prime example of a *greedy algorithm*. Greedy algorithms generally solve a problem in stages by doing what appears to be the best thing at each stage. For example, to make change in U.S. currency, most people count out the quarters first, then the dimes, nickels, and pennies. This greedy algorithm gives change using the minimum number of coins. The main problem with greedy algorithms is that they do not always work. The addition of a 12-cent piece breaks the coin-changing algorithm for returning 15 cents, because the answer it gives (one 12-cent piece and three pennies) is not optimal (one dime and one nickel).

Dijkstra's algorithm proceeds in stages, just like the unweighted shortest-path algorithm. At each stage, Dijkstra's algorithm selects a vertex v, which has the smallest

v	Initial State			v_3 Dequeued			v_1 Dequeued			v_6 Dequeued		
	Known	d_v	p_v	Known	d_v	p_v	Known	d_v	p_v	Known	d_v	p_v
v_1	0	∞	0	0	1	v_3	1	1	v_3	1	1	v_3
v_2	0	∞	0	0	∞	0	0	2	v_1	0	2	v_1
v_3	0	0	0	1	0	0	1	0	0	1	0	0
v_4	0	∞	0	0	∞	0	0	2	v_1	0	2	v_1
v_5	0	∞	0	0	∞	0	0	∞	0	0	∞	0
v_6	0	∞	0	0	1	v_3	0	1	v_3	1	1	v_3
v_7	0	∞	0	0	∞	0	0	∞	0	0	∞	0
Q:	v_3			v_1, v_6			v_6, v_2, v_4			v_2, v_4		

v	v_2 Dequeued			v_4 Dequeued			v_5 Dequeued			v_7 Dequeued		
	Known	d_v	p_v	Known	d_v	p_v	Known	d_v	p_v	Known	d_v	p_v
v_1	1	1	v_3	1	1	v_3	1	1	v_3	1	1	v_3
v_2	1	2	v_1	1	2	v_1	1	2	v_1	1	2	v_1
v_3	1	0	0	1	0	0	1	0	0	1	0	0
v_4	0	2	v_1	1	2	v_1	1	2	v_1	1	2	v_1
v_5	0	3	v_2	0	3	v_2	1	3	v_2	1	3	v_2
v_6	1	1	v_3	1	1	v_3	1	1	v_3	1	1	v_3
v_7	0	∞	0	0	3	v_4	0	3	v_4	1	3	v_4
Q:	v_4, v_5			v_5, v_7			v_7			empty		

Figure 9.19 How the data changes during the unweighted shortest-path algorithm

d_v among all the unknown vertices, and declares that the shortest path from s to v is known. The remainder of a stage consists of updating the values of d_w.

In the unweighted case, we set $d_w = d_v + 1$ if $d_w = \infty$. Thus, we essentially lowered the value of d_w if vertex v offered a shorter path. If we apply the same logic to the weighted case, then we should set $d_w = d_v + c_{v,w}$ if this new value for d_w would be an improvement. Put simply, the algorithm decides whether or not it is a good idea to use v on the path to w. The original cost, d_w, is the cost without using v; the cost calculated above is the cheapest path using v (and only known vertices).

The graph in Figure 9.20 is our example. Figure 9.21 represents the initial configuration, assuming that the start node, s, is v_1. The first vertex selected is v_1, with path length 0. This vertex is marked known. Now that v_1 is known, some entries need to be adjusted. The vertices adjacent to v_1 are v_2 and v_4. Both these vertices get their entries adjusted, as indicated in Figure 9.22.

Next, v_4 is selected and marked known. Vertices v_3, v_5, v_6, and v_7 are adjacent, and it turns out that all require adjusting, as shown in Figure 9.23.

Next, v_2 is selected. v_4 is adjacent but already known, so no work is performed on it. v_5 is adjacent but not adjusted, because the cost of going through v_2 is

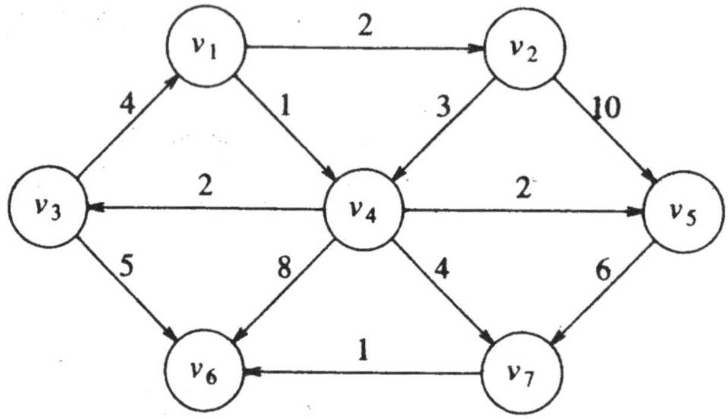

Figure 9.20 The directed graph G (again)

v	Known	d_v	p_v
v_1	0	0	0
v_2	0	∞	0
v_3	0	∞	0
v_4	0	∞	0
v_5	0	∞	0
v_6	0	∞	0
v_7	0	∞	0

Figure 9.21 Initial configuration of table used in Dijkstra's algorithm

v	Known	d_v	p_v
v_1	1	0	0
v_2	0	2	v_1
v_3	0	∞	0
v_4	0	1	v_1
v_5	0	∞	0
v_6	0	∞	0
v_7	0	∞	0

Figure 9.22 After v_1 is declared known

v	Known	d_v	p_v
v_1	1	0	0
v_2	0	2	v_1
v_3	0	3	v_4
v_4	1	1	v_1
v_5	0	3	v_4
v_6	0	9	v_4
v_7	0	5	v_4

Figure 9.23 After v_4 is declared known

v	Known	d_v	p_v
v_1	1	0	0
v_2	1	2	v_1
v_3	0	3	v_4
v_4	1	1	v_1
v_5	0	3	v_4
v_6	0	9	v_4
v_7	0	5	v_4

Figure 9.24 After v_2 is declared known

v	Known	d_v	p_v
v_1	1	0	0
v_2	1	2	v_1
v_3	1	3	v_4
v_4	1	1	v_1
v_5	1	3	v_4
v_6	0	8	v_3
v_7	0	5	v_4

Figure 9.25 After v_5 and then v_3 are declared known

v	Known	d_v	p_v
v_1	1	0	0
v_2	1	2	v_1
v_3	1	3	v_4
v_4	1	1	v_1
v_5	1	3	v_4
v_6	0	6	v_7
v_7	1	5	v_4

Figure 9.26 After v_7 is declared known

$2 + 10 = 12$ and a path of length 3 is already known. Figure 9.24 shows the table after these vertices are selected.

The next vertex selected is v_5 at cost 3. v_7 is the only adjacent vertex, but it is not adjusted, because $3 + 6 > 5$. Then v_3 is selected, and the distance for v_6 is adjusted down to $3 + 5 = 8$. The resulting table is depicted in Figure 9.25.

Next v_7 is selected; v_6 gets updated down to $5 + 1 = 6$. The resulting table is Figure 9.26.

Finally, v_6 is selected. The final table is shown in Figure 9.27. Figure 9.28 graphically shows how edges are marked known and vertices updated during Dijkstra's algorithm.

v	Known	d_v	p_v
v_1	1	0	0
v_2	1	2	v_1
v_3	1	3	v_4
v_4	1	1	v_1
v_5	1	3	v_4
v_6	1	6	v_7
v_7	1	5	v_4

Figure 9.27 After v_6 is declared known and algorithm terminates

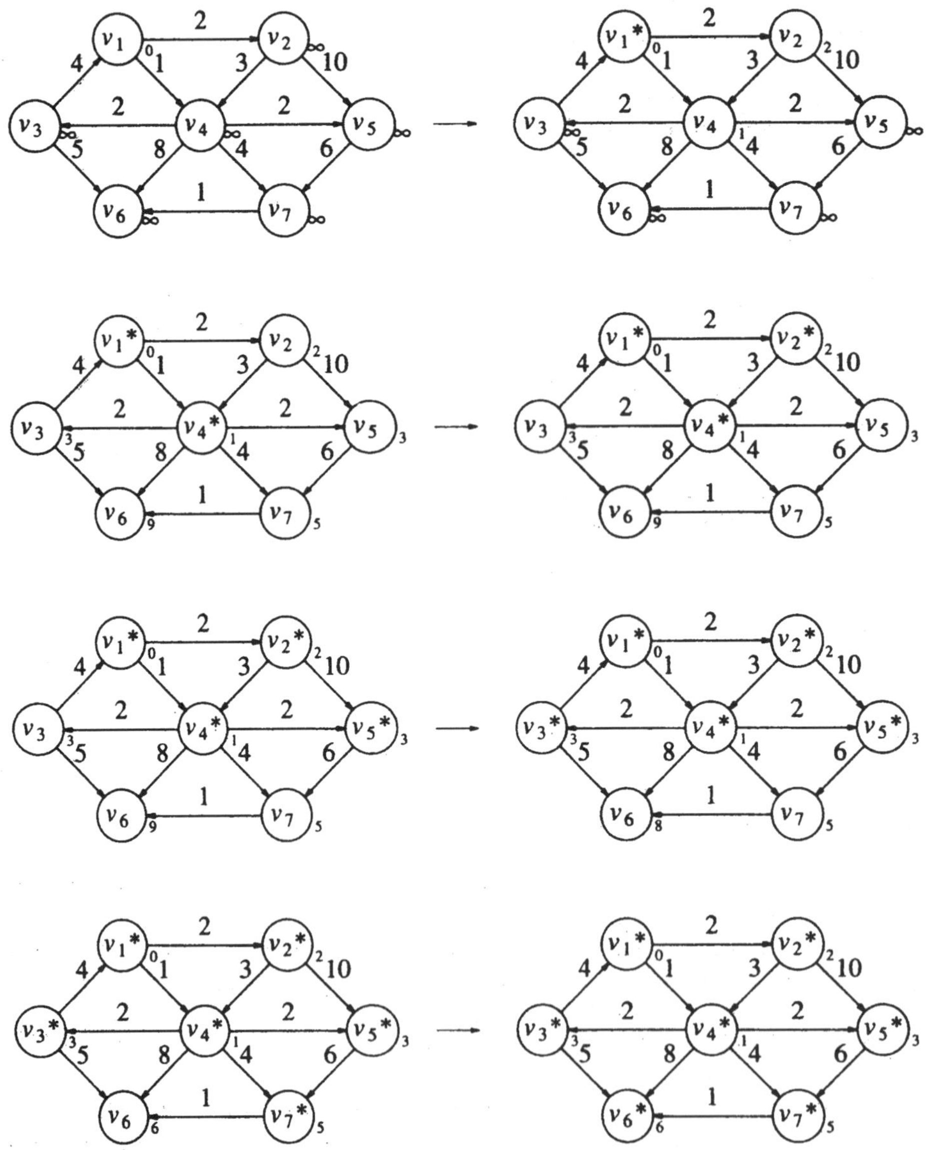

Figure 9.28 Stages of Dijkstra's algorithm

To print out the actual path from a start vertex to some vertex v, we can write a recursive routine to follow the trail left in the p array.

We now give pseudocode to implement Dijkstra's algorithm. We will assume that the vertices are numbered from 0 to $NumVertex - 1$ for convenience (see Fig. 9.29) and that the graph can be read into an adjacency list by the routine *ReadGraph*.

In the routine in Figure 9.30, the start vertex is passed to the initialization routine. This is the only place in the code where the start vertex needs to be known.

The path can be printed out using the recursive routine in Figure 9.31. The routine recursively prints the path all the way up to the vertex before v on the path, and then just prints v. This works because the path is simple.

Figure 9.29 Declarations for Dijkstra's algorithm

```
typedef int Vertex;

struct TableEntry
{
    List     Header; /* Adjacency list */
    int      Known;
    DistType Dist;
    Vertex   Path;
};

/* Vertices are numbered from 0 */
#define NotAVertex (-1)
typedef struct TableEntry Table[ NumVertex ];
```

Figure 9.30 Table initialization routine

```
void
InitTable( Vertex Start, Graph G, Table T )
{
    int i;

    ReadGraph( G, T );  /* Read graph somehow */
    for( i = 0; i < NumVertex; i++ )
    {
        T[ i ].Known = False;
        T[ i ].Dist  = Infinity;
        T[ i ].Path  = NotAVertex;
    }
    T[ Start ].dist = 0;
}
```

```
/* Print shortest path to V after Dijkstra has run */
/* Assume that the path exists */

void
PrintPath( Vertex V, Table T )
{
    if( T[ V ].Path != NotAVertex )
    {
        PrintPath( T[ V ].Path, T );
        printf( " to" );
    }
    printf( "%v", V );  /* %v is pseudocode */
}
```

Figure 9.31 Routine to print the actual shortest path

Figure 9.32 shows the main algorithm, which is just a *for* loop to fill up the table using the greedy selection rule.

A proof by contradiction will show that this algorithm always works as long as no edge has a negative cost. If any edge has negative cost, the algorithm could produce the wrong answer (see Exercise 9.7a). The running time depends on how the table is manipulated, which we have yet to consider. If we use the obvious algorithm of scanning down the table to find the minimum d_v, each phase will take $O(|V|)$ time to find the minimum, and thus $O(|V|^2)$ time will be spent finding the minimum over the course of the algorithm. The time for updating d_w is constant per update, and there is at most one update per edge for a total of $O(|E|)$. Thus, the total running time is $O(|E| + |V|^2) = O(|V|^2)$. If the graph is dense, with $|E| = \Theta(|V|^2)$, this algorithm is not only simple but essentially optimal, since it runs in time linear in the number of edges.

If the graph is sparse, with $|E| = \Theta(|V|)$, this algorithm is too slow. In this case, the distances would need to be kept in a priority queue. There are actually two ways to do this; both are similar.

Lines 2 and 5 combine to form a *DeleteMin* operation, since once the unknown minimum vertex is found, it is no longer unknown and must be removed from future consideration. The update at line 9 can be implemented two ways.

One way treats the update as a *DecreaseKey* operation. The time to find the minimum is then $O(\log|V|)$, as is the time to perform updates, which amount to *DecreaseKey* operations. This gives a running time of $O(|E|\log|V| + |V|\log|V|) = O(|E|\log|V|)$, an improvement over the previous bound for sparse graphs. Since priority queues do not efficiently support the *Find* operation, the location in the priority queue of each value of d_i will need to be maintained and updated whenever d_i changes in the priority queue. If the priority queue is implemented by a binary heap, this will be messy. If a pairing heap (Chapter 12) is used, the code is not too bad.

An alternate method is to insert w and the new value d_w into the priority queue every time line 9 is executed. Thus, there may be more than one representative

```
        void
        Dijkstra( Table T )
        {
            Vertex V, W;

/* 1*/      for( ; ; )
            {
/* 2*/          V = smallest unknown distance vertex;
/* 3*/          if( V == NotAVertex )
/* 4*/              break;

/* 5*/          T[ V ].Known = True;
/* 6*/          for each W adjacent to V
/* 7*/              if( !T[ W ].Known )
/* 8*/                  if( T[ V ].Dist + Cvw < T[ W ].Dist )
                        {   /* Update W */
/* 9*/                      Decrease( T[ W ].Dist to
                                         T[ V ].Dist + Cvw );
/*10*/                      T[ W ].Path = V;
                        }
            }
        }
```

Figure 9.32 Pseudocode for Dijkstra's algoritm

for each vertex in the priority queue. When the *DeleteMin* operation removes the smallest vertex from the priority queue, it must be checked to make sure that it is not already known. Thus, line 2 becomes a loop performing *DeleteMins* until an unknown vertex emerges. Although this method is superior from a software point of view, and is certainly much easier to code, the size of the priority queue could get to be as large as $|E|$. This does not affect the asymptotic time bounds, since $|E| \le |V|^2$ implies that $\log|E| \le 2\log|V|$. Thus, we still get an $O(|E|\log|V|)$ algorithm. However, the space requirement does increase, and this could be important in some applications. Moreover, because this method requires $|E|$ *DeleteMins* instead of only $|V|$, it is likely to be slower in practice.

Notice that for the typical problems, such as computer mail and mass transit commutes, the graphs are typically very sparse because most vertices have only a couple of edges, so it is important in many applications to use a priority queue to solve this problem.

There are better time bounds possible using Dijkstra's algorithm if different data structures are used. In Chapter 11, we will see another priority queue data structure called the Fibonacci heap. When this is used, the running time is $O(|E| + |V|\log|V|)$. Fibonacci heaps have good theoretical time bounds but a fair amount of overhead, so it is not clear whether using Fibonacci heaps is actually better in practice than Dijkstra's algorithm with binary heaps. Needless to say, there are no average-case results for this problem, since it is not even obvious how to model a random graph.

```
void    /* Assume T is initialized as in Fig 9.18 */
WeightedNegative( Table T )
{
    Queue Q;
    Vertex V, W;

    Q = CreateQueue( NumVertex ); MakeEmpty( Q );
    Enqueue( S, Q ); /* Enqueue the start vertex S */

    while( !IsEmpty( Q ) )
    {
        V = Dequeue( Q );
        for each W adjacent to V
            if( T[ V ].Dist + Cvw < T[ W ].Dist )
            {
                /* Update W */
                T[ W ].Dist = T[ V ].Dist + Cvw;
                T[ W ].Path = V;
                if( W is not already in Q )
                    Enqueue( W, Q );
            }
    }
    DisposeQueue( Q );
}
```

Figure 9.33 Pseudocode for weighted shortest-path algorithm with negative edge costs

9.3.3. Graphs with Negative Edge Costs

If the graph has negative edge costs, then Dijkstra's algorithm does not work. The problem is that once a vertex u is declared known, it is possible that from some other, unknown vertex v there is a path back to u that is very negative. In such a case, taking a path from s to v back to u is better than going from s to u without using v. Exercise 9.7(a) asks you to construct an explicit example.

A tempting solution is to add a constant Δ to each edge cost, thus removing negative edges, calculate a shortest path on the new graph, and then use that result on the original. The naive implementation of this strategy does not work because paths with many edges become more weighty than paths with few edges.

A combination of the weighted and unweighted algorithms will solve the problem, but at the cost of a drastic increase in running time. We forget about the concept of known vertices, since our algorithm needs to be able to change its mind. We begin by placing s on a queue. Then, at each stage, we dequeue a vertex v. We find all vertices w adjacent to v such that $d_w > d_v + c_{v,w}$. We update d_w and p_w, and place w on a queue if it is not already there. A bit can be set for each vertex to indicate presence in the queue. We repeat the process until the queue is empty. Figure 9.33 (almost) implements this algorithm.

Although the algorithm works if there are no negative-cost cycles, it is no longer true that the code in lines 6 through 10 is executed once per edge. Each vertex can dequeue at most $|V|$ times, so the running time is $O(|E| \cdot |V|)$ if adjacency lists are used (Exercise 9.7(b)). This is quite an increase from Dijkstra's algorithm, so it is fortunate that, in practice, edge costs are nonnegative. If negative-cost cycles are present, then the algorithm as written will loop indefinitely. By stopping the algorithm after any vertex has dequeued $|V|+1$ times, we can guarantee termination.

9.3.4. *Acyclic Graphs*

If the graph is known to be acyclic, we can improve Dijkstra's algorithm by changing the order in which vertices are declared known, otherwise known as the vertex selection rule. The new rule is to select vertices in topological order. The algorithm can be done in one pass, since the selections and updates can take place as the topological sort is being performed.

This selection rule works because when a vertex v is selected, its distance, d_v, can no longer be lowered, since by the topological ordering rule it has no incoming edges emanating from unknown nodes.

There is no need for a priority queue with this selection rule; the running time is $O(|E| + |V|)$, since the selection takes constant time.

An acyclic graph could model some downhill skiing problem—we want to get from point *a* to *b*, but can only go downhill, so clearly there are no cycles. Another possible application might be the modeling of (nonreversible) chemical reactions. We could have each vertex represent a particular state of an experiment. Edges would represent a transition from one state to another, and the edge weights might represent the energy released. If only transitions from a higher energy state to a lower are allowed, the graph is acyclic.

A more important use of acyclic graphs is *critical path analysis*. The graph in Figure 9.34 will serve as our example. Each node represents an activity that must be performed, along with the time it takes to complete the activity. This graph is thus known as an *activity-node* graph. The edges represent precedence relationships: An edge (v, w) means that activity v must be completed before activity w may begin. Of

Figure 9.34 Activity-node graph

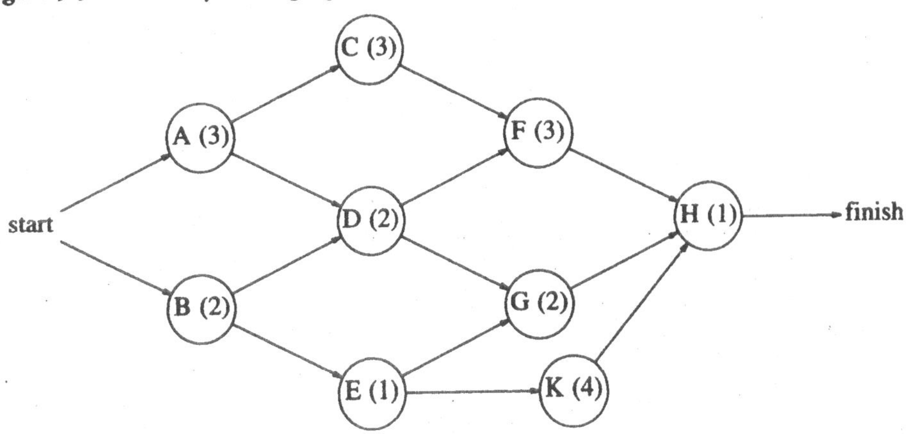

course, this implies that the graph must be acyclic. We assume that any activities that do not depend (either directly or indirectly) on each other can be performed in parallel by different servers.

This type of a graph could be (and frequently is) used to model construction projects. In this case, there are several important questions which would be of interest to answer. First, what is the earliest completion time for the project? We can see from the graph that 10 time units are required along the path A, C, F, H. Another important question is to determine which activities can be delayed, and by how long, without affecting the minimum completion time. For instance, delaying any of A, C, F, or H would push the completion time past 10 units. On the other hand, activity B is less critical and can be delayed up to two time units without affecting the final completion time.

To perform these calculations, we convert the activity-node graph to an *event-node* graph. Each event corresponds to the completion of an activity and all its dependent activities. Events reachable from a node v in the event-node graph may not commence until after the event v is completed. This graph can be constructed automatically or by hand. Dummy edges and nodes may need to be inserted in the case where an activity depends on several others. This is necessary in order to avoid introducing false dependencies (or false lack of dependencies). The event-node graph corresponding to the graph in Figure 9.34 is shown in Figure 9.35.

To find the earliest completion time of the project, we merely need to find the length of the *longest* path from the first event to the last event. For general graphs, the longest-path problem generally does not make sense, because of the possibility of *positive-cost cycles*. These are the equivalent of negative-cost cycles in shortest-path problems. If positive-cost cycles are present, we could ask for the longest *simple* path, but no satisfactory solution is known for this problem. Since the event-node graph is acyclic, we need not worry about cycles. In this case, it is easy to adapt the shortest-path algorithm to compute the earliest completion time for all nodes in the graph. If EC_i is the earliest completion time for node i, then the applicable rules are

$$EC_1 = 0$$

$$EC_w = \max_{(v,w)\in E} (EC_v + c_{v,w})$$

Figure 9.36 shows the earliest completion time for each event in our example event-node graph.

We can also compute the latest time, LC_i, that each event can finish without affecting the final completion time. The formulas to do this are

$$LC_n = EC_n$$

$$LC_v = \min_{(v,w)\in E} (LC_w - c_{v,w})$$

These values can be computed in linear time by maintaining, for each vertex, a list of all adjacent and preceding vertices. The earliest completion times are computed for vertices by their topological order, and the latest completion times are computed by reverse topological order. The latest completion times are shown in Figure 9.37.

The *slack* time for each edge in the event-node graph represents the amount of time that the completion of the corresponding activity can be delayed without

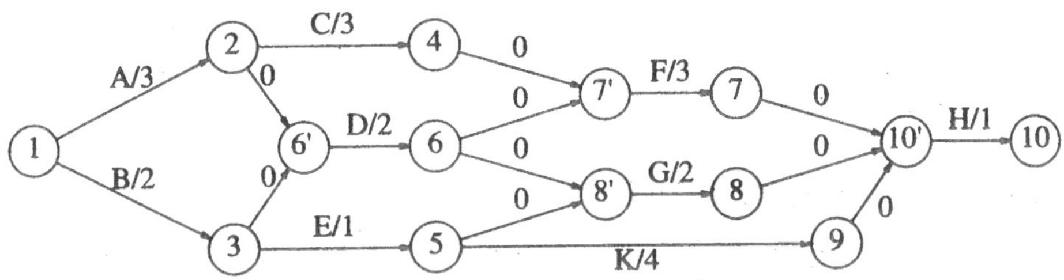

Figure 9.35 Event-node graph

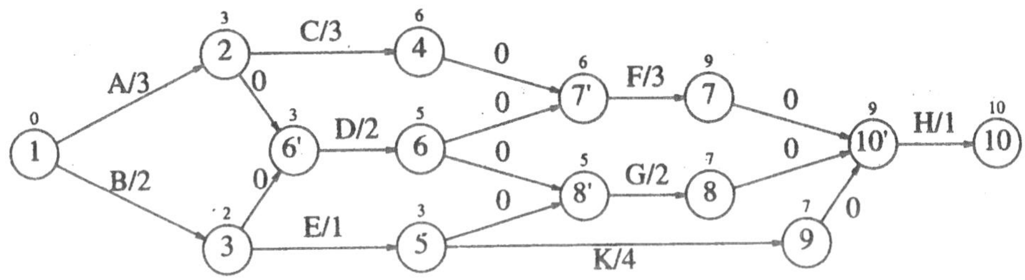

Figure 9.36 Earliest completion times

delaying the overall completion. It is easy to see that

$$Slack_{(v,w)} = LC_w - EC_v - c_{v,w}$$

Figure 9.38 shows the slack (as the third entry) for each activity in the event-node graph. For each node, the top number is the earliest completion time and the bottom entry is the latest completion time.

Figure 9.37 Latest completion times

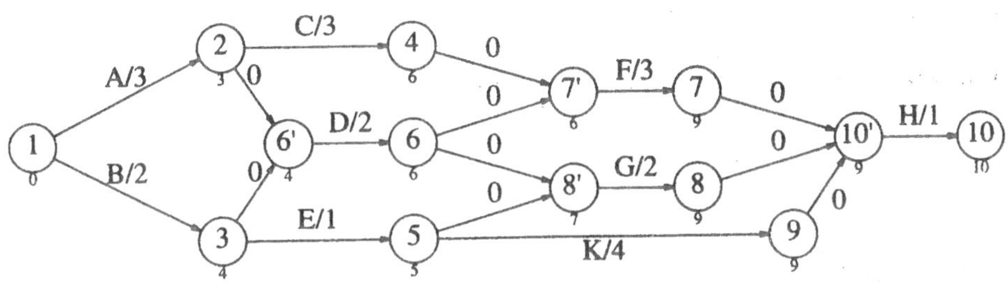

Figure 9.38 Earliest completion time, latest completion time, and slack

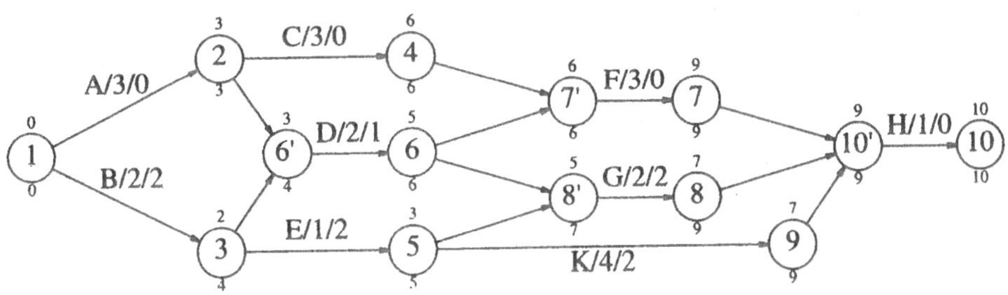

Some activities have zero slack. These are critical activities, which must finish on schedule. There is at least one path consisting entirely of zero-slack edges; such a path is a *critical path*.

9.3.5. All-Pairs Shortest Path

Sometimes it is important to find the shortest paths between all pairs of vertices in the graph. Although we could just run the appropriate single-source algorithm $|V|$ times, we might expect a somewhat faster solution, especially on a dense graph, if we compute all the information at once.

In Chapter 10, we will see an $O(|V|^3)$ algorithm to solve this problem for weighted graphs. Although, for dense graphs, this is the same bound as running a simple (non–priority queue) Dijkstra's algorithm $|V|$ times, the loops are so tight that the specialized all-pairs algorithm is likely to be faster in practice. On sparse graphs, of course, it is faster to run $|V|$ Dijkstra's algorithms coded with priority queues.

9.4. Network Flow Problems

Suppose we are given a directed graph $G = (V, E)$ with edge capacities $c_{v,w}$. These capacities could represent the amount of water that could flow through a pipe or the amount of traffic that could flow on a street between two intersections. We have two vertices: s, which we call the *source*, and t, which is the *sink*. Through any edge, (v, w), at most $c_{v,w}$ units of "flow" may pass. At any vertex, v, that is not either s or t, the total flow coming in must equal the total flow going out. The maximum flow problem is to determine the maximum amount of flow that can pass from s to t. As an example, for the graph in Figure 9.39 on the left the maximum flow is 5, as indicated by the graph on the right.

Figure 9.39 A graph (left) and its maximum flow

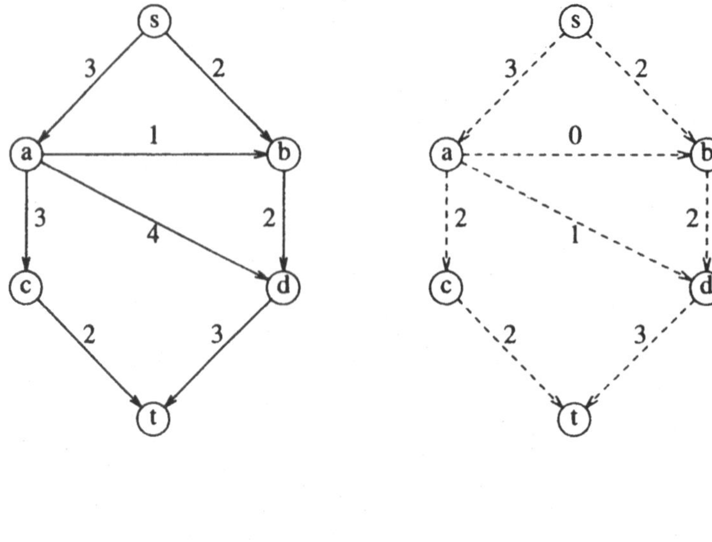

As required by the problem statement, no edge carries more flow than its capacity. Vertex a has three units of flow coming in, which it distributes to c and d. Vertex d takes three units of flow from a and b and combines this, sending the result to t. A vertex can combine and distribute flow in any manner that it likes, as long as edge capacities are not violated and as long as flow conservation is maintained (what goes in must come out).

9.4.1. A Simple Maximum-Flow Algorithm

A first attempt to solve the problem proceeds in stages. We start with our graph, G, and construct a flow graph G_f. G_f tells the flow that has been attained at any stage in the algorithm. Initially all edges in G_f have no flow, and we hope that when the algorithm terminates, G_f contains a maximum flow. We also construct a graph, G_r, called the *residual* graph. G_r tells, for each edge, how much more flow can be added. We can calculate this by subtracting the current flow from the capacity for each edge. An edge in G_r is known as a *residual* edge.

At each stage, we find a path in G_r from s to t. This path is known as an *augmenting path*. The minimum edge on this path is the amount of flow that can be added to every edge on the path. We do this by adjusting G_f and recomputing G_r. When we find no path from s to t in G_r, we terminate. This algorithm is nondeterministic, in that we are free to choose *any* path from s to t; obviously some choices are better than others, and we will address this issue later. We will run this algorithm on our example. The graphs below are G, G_f, G_r, respectively. Keep in mind that there is a slight flaw in this algorithm. The initial configuration is in Figure 9.40.

There are many paths from s to t in the residual graph. Suppose we select s, b, d, t. Then we can send two units of flow through every edge on this path. We will adopt the convention that once we have filled (*saturated*) an edge, it is removed from the residual graph. We then obtain Figure 9.41.

Next, we might select the path s, a, c, t, which also allows two units of flow. Making the required adjustments gives the graphs in Figure 9.42.

Figure 9.40 Initial stages of the graph, flow graph, and residual graph

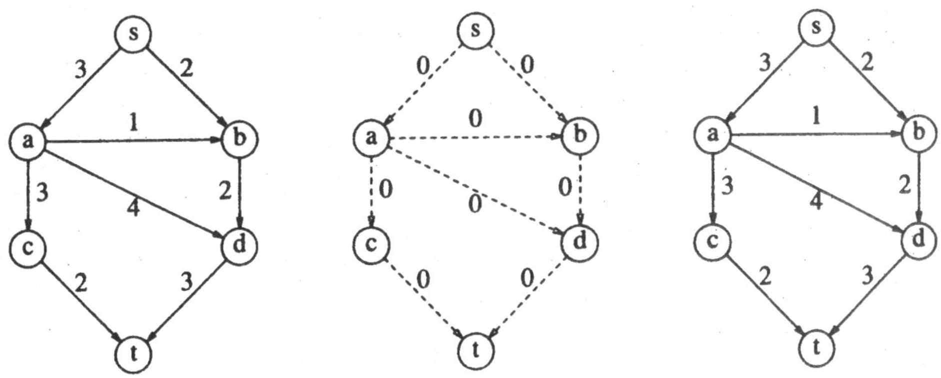

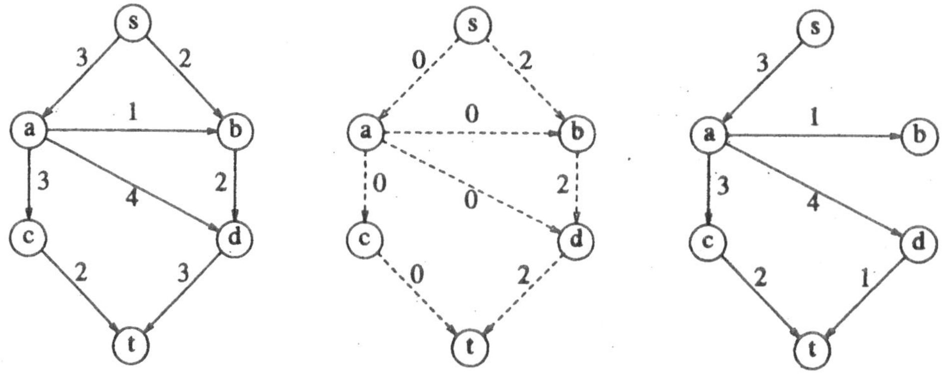

Figure 9.41 G, G_f, G_r after two units of flow added along s, b, d, t

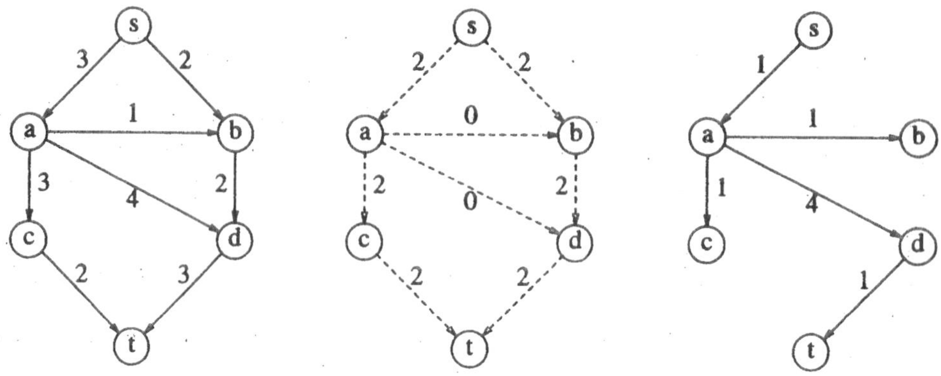

Figure 9.42 G, G_f, G_r after two units of flow added along s, a, c, t

The only path left to select is s, a, d, t, which allows one unit of flow. The resulting graphs are shown in Figure 9.43.

The algorithm terminates at this point, because t is unreachable from s. The resulting flow of 5 happens to be the maximum. To see what the problem is, suppose that with our initial graph, we chose the path s, a, d, t. This path allows three units of flow and thus seems to be a good choice. The result of this choice, however, is that there is now no longer any path from s to t in the residual graph, and thus, our algorithm has failed to find an optimal solution. This is an example of a greedy algorithm that does not work. Figure 9.44 shows why the algorithm fails.

In order to make this algorithm work, we need to allow the algorithm to change its mind. To do this, for every edge (v, w) with flow $f_{v,w}$ in the flow graph, we will add an edge in the residual graph (w, v) of capacity $f_{v,w}$. In effect, we are allowing the algorithm to undo its decisions by sending flow back in the opposite direction. This is best seen by example. Starting from our original graph and selecting the augmenting path s, a, d, t, we obtain the graphs in Figure 9.45.

Notice that in the residual graph, there are edges in both directions between a and d. Either one more unit of flow can be pushed from a to d, or up to three units

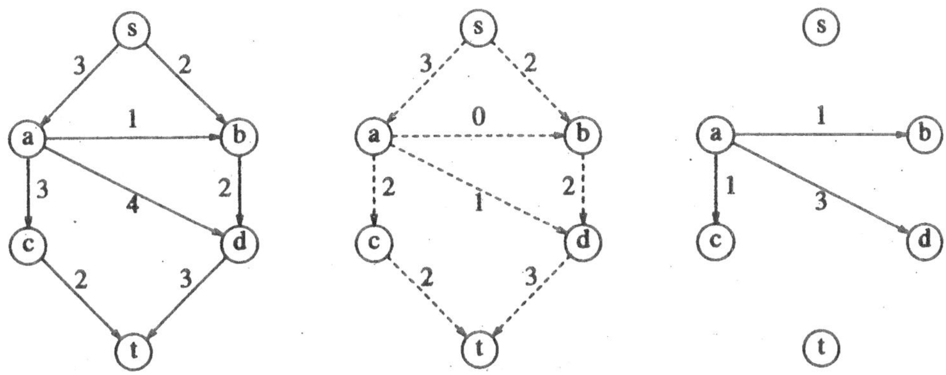

Figure 9.43 G, G_f, G_r after one unit of flow added along s, a, d, t—algorithm terminates

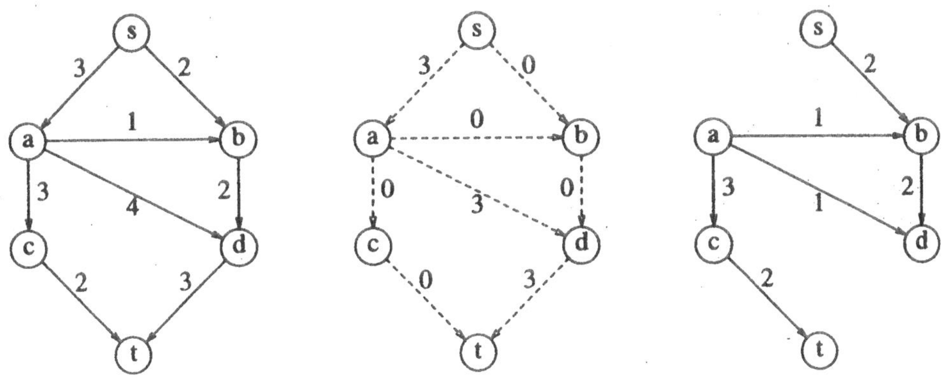

Figure 9.44 G, G_f, G_r if initial action is to add three units of flow along s, a, d, t—algorithm terminates with suboptimal solution

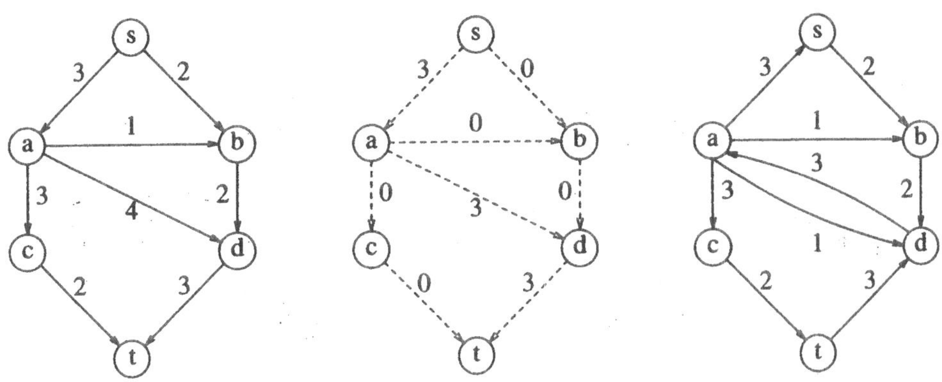

Figure 9.45 Graphs after three units of flow added along s, a, d, t using correct algorithm

can be pushed back—we can undo flow. Now the algorithm finds the augmenting path s, b, d, a, c, t, of flow 2. By pushing two units of flow from d to a, the algorithm takes two units of flow away from the edge (a, d) and is essentially changing its mind. Figure 9.46 shows the new graphs.

There is no augmenting path in this graph, so the algorithm terminates. Surprisingly, it can be shown that if the edge capacities are rational numbers, this algorithm always terminates with a maximum flow. This proof is somewhat difficult and is beyond the scope of this text. Although the example happened to be acyclic, this is not a requirement for the algorithm to work. We have used acyclic graphs merely to keep things simple.

If the capacities are all integers and the maximum flow is f, then, since each augmenting path increases the flow value by at least 1, f stages suffice, and the total running time is $O(f \cdot |E|)$, since an augmenting path can be found in $O(|E|)$ time by an unweighted shortest-path algorithm. The classic example of why this is a bad running time is shown by the graph in Figure 9.47.

Figure 9.46 Graphs after two units of flow added along s, b, d, a, c, t using correct algorithm

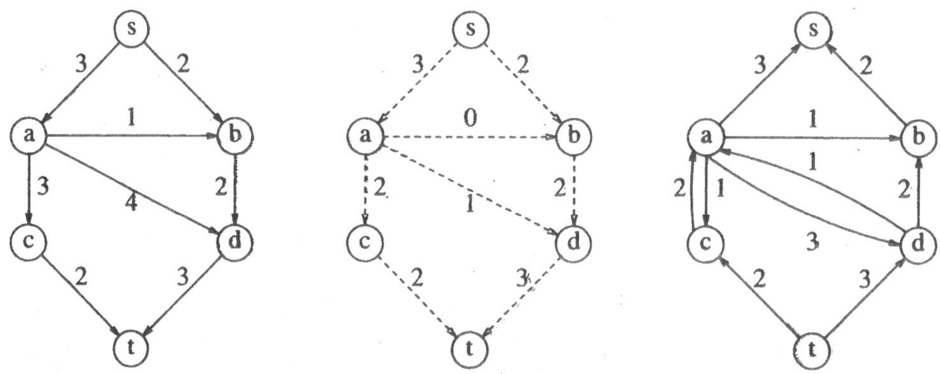

Figure 9.47 The classic bad case for augmenting

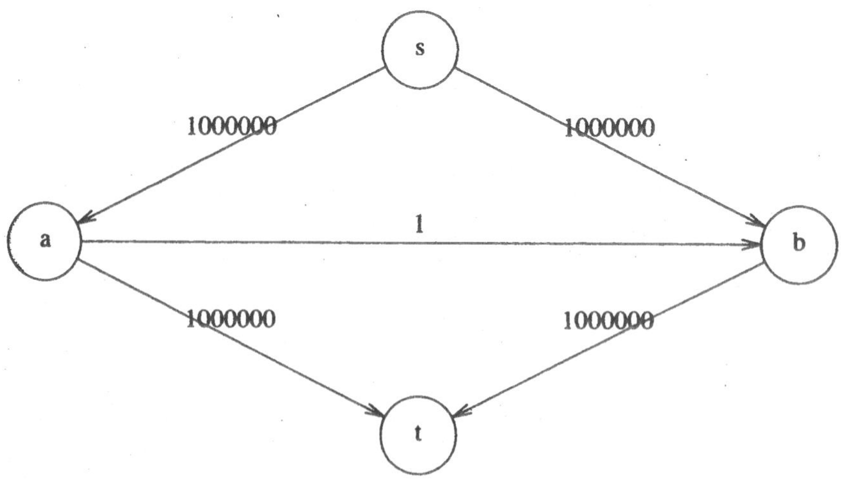

The maximum flow is seen by inspection to be 2,000,000 by sending 1,000,000 down each side. Random augmentations could continually augment along a path that includes the edge connected by a and b. If this were to occur repeatedly, 2,000,000 augmentations would be required, when we could get by with only 2.

A simple method to get around this problem is always to choose the augmenting path that allows the largest increase in flow. Finding such a path is similar to solving a weighted shortest-path problem and a single-line modification to Dijkstra's algorithm will do the trick. If cap_{max} is the maximum edge capacity, then one can show that $O(|E| \log cap_{max})$ augmentations will suffice to find the maximum flow. In this case, since $O(|E| \log |V|)$ time is used for each calculation of an augmenting path, a total bound of $O(|E|^2 \log |V| \log cap_{max})$ is obtained. If the capacities are all small integers, this reduces to $O(|E|^2 \log |V|)$.

Another way to choose augmenting paths is always to take the path with the least number of edges, with the plausible expectation that by choosing a path in this manner, it is less likely that a small, flow-restricting edge will turn up on the path. Using this rule, it can be shown that $O(|E| \cdot |V|)$ augmenting steps are required. Each step takes $O(|E|)$, again using an unweighted shortest-path algorithm, yielding a $O(|E|^2|V|)$ bound on the running time.

Further data structure improvements are possible to this algorithm, and there are several, more complicated, algorithms. A long history of improved bounds has lowered the current best-known bound for this problem. Although no $O(|E||V|)$ algorithm has been reported yet, algorithms with $O(|E||V| \log(|V|^2/|E|))$ and $O(|E||V| + |V|^{2+\epsilon})$ bounds have been discovered (see the references). There are also a host of very good bounds for special cases. For instance, $O(|E||V|^{1/2})$ time finds a maximum flow in a graph, having the property that all vertices except the source and sink have either a single incoming edge of capacity 1 or a single outgoing edge of capacity 1. These graphs occur in many applications.

The analyses required to produce these bounds are rather intricate, and it is not clear how the worst-case results relate to the running times encountered in practice. A related, even more difficult problem is the *min-cost flow* problem. Each edge has not only a capacity but a cost per unit of flow. The problem is to find, among all maximum flows, the one flow of minimum cost. Both of these problems are being actively researched.

9.5. Minimum Spanning Tree

The next problem we will consider is that of finding a *minimum spanning tree* in an undirected graph. The problem makes sense for directed graphs but appears to be more difficult. Informally, a minimum spanning tree of an undirected graph G is a tree formed from graph edges that connects all the vertices of G at lowest total cost. A minimum spanning tree exists if and only if G is connected. Although a robust algorithm should report the case that G is unconnected, we will assume that G is connected and leave the issue of robustness as an exercise to the reader.

In Figure 9.48 the second graph is a minimum spanning tree of the first (it happens to be unique, but this is unusual). Notice that the number of edges in the

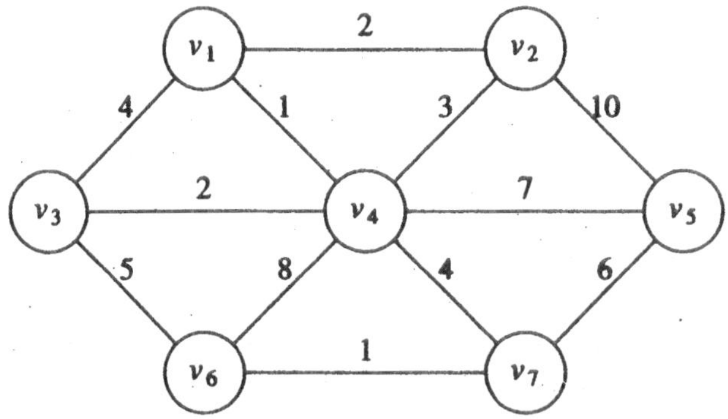

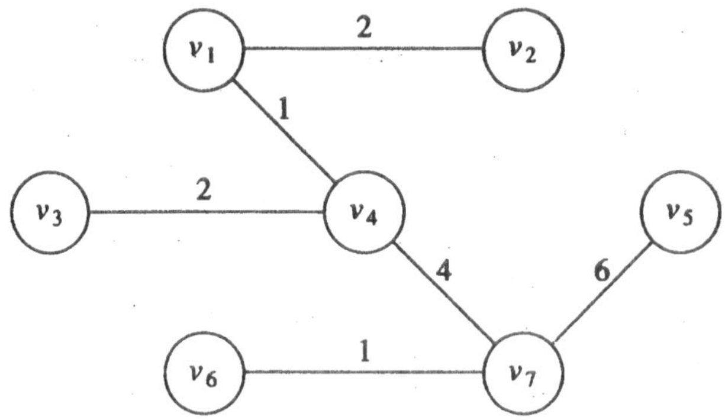

Figure 9.48 A graph G and its minimum spanning tree

minimum spanning tree is $|V| - 1$. The minimum spanning tree is a *tree* because it is acyclic, it is *spanning* because it covers every vertex, and it is *minimum* for the obvious reason. If we need to wire a house with a minimum of cable, then a minimum spanning tree problem needs to be solved.

For any spanning tree T, if an edge e that is not in T is added, a cycle is created. The removal of any edge on the cycle reinstates the spanning tree property. The cost of the spanning tree is lowered if e has lower cost than the edge that was removed. If, as a spanning tree is created, the edge that is added is the one of minimum cost that avoids creation of a cycle, then the cost of the resulting spanning tree cannot be improved, because any replacement edge would have cost at least as much as an edge already in the spanning tree. This shows that greed works for the minimum spanning tree problem. The two algorithms we present differ in how a minimum edge is selected.

9.5.1. *Prim's Algorithm*

One way to compute a minimum spanning tree is to grow the tree in successive stages. In each stage, one node is picked as the root, and we add an edge, and thus an associated vertex, to the tree.

At any point in the algorithm, we can see that we have a set of vertices that have already been included in the tree; the rest of the vertices have not. The algorithm then finds, at each stage, a new vertex to add to the tree by choosing the edge (u, v)

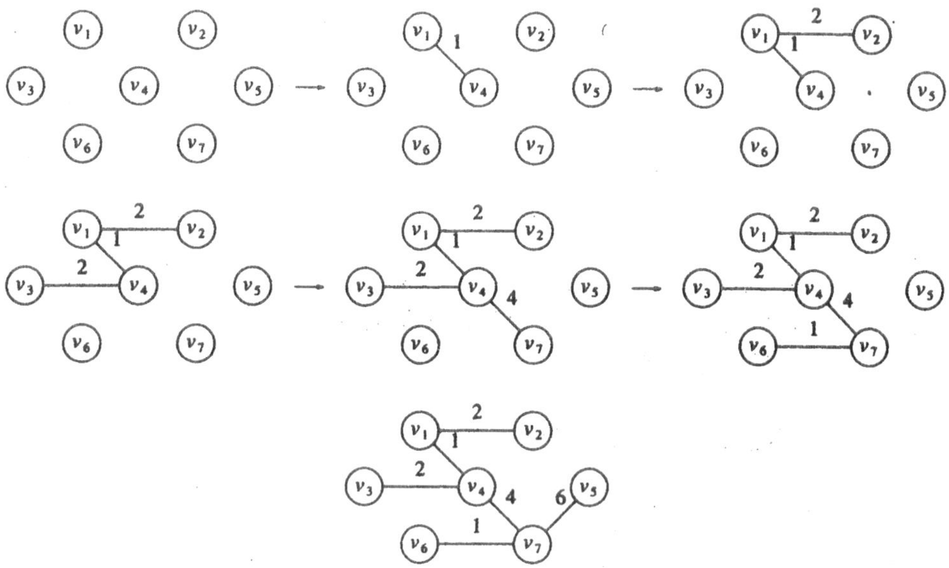

Figure 9.49 Prim's algorithm after each stage

such that the cost of (u, v) is the smallest among all edges where u is in the tree and v is not. Figure 9.49 shows how this algorithm would build the minimum spanning tree, starting from v_1. Initially, v_1 is in the tree as a root with no edges. Each step adds one edge and one vertex to the tree.

We can see that Prim's algorithm is essentially identical to Dijkstra's algorithm for shortest paths. As before, for each vertex we keep values d_v and p_v and an indication of whether it is known or unknown. d_v is the weight of the shortest arc connecting v to a known vertex, and p_v, as before, is the last vertex to cause a change in d_v. The rest of the algorithm is exactly the same, with the exception that since the definition of d_v is different, so is the update rule. For this problem, the update rule is even simpler than before: After a vertex v is selected, for each unknown w adjacent to v, $d_w = \min(d_w, c_{w,v})$.

The initial configuration of the table is shown in Figure 9.50. v_1 is selected, and v_2, v_3, and v_4 are updated. The table resulting from this is shown in Figure 9.51. The next vertex selected is v_4. Every vertex is adjacent to v_4. v_1 is not examined, because it is known. v_2 is unchanged, because it has $d_v = 2$ and the edge cost from v_4 to v_2 is 3; all the rest are updated. Figure 9.52 shows the resulting table. The next vertex chosen is v_2 (arbitrarily breaking a tie). This does not affect any distances. Then v_3 is chosen, which affects the distance in v_6, producing Figure 9.53. Figure 9.54 results from the selection of v_7, which forces v_6 and v_5 to be adjusted. v_6 and then v_5 are selected, completing the algorithm.

The final table is shown in Figure 9.55. The edges in the spanning tree can be read from the table: (v_2, v_1), (v_3, v_4), (v_4, v_1), (v_5, v_7), (v_6, v_7), (v_7, v_4). The total cost is 16.

The entire implementation of this algorithm is virtually identical to that of Dijkstra's algorithm, and everything that was said about the analysis of Dijkstra's algorithm applies here. Be aware that Prim's algorithm runs on undirected graphs,

v	Known	d_v	p_v
v_1	0	0	0
v_2	0	∞	0
v_3	0	∞	0
v_4	0	∞	0
v_5	0	∞	0
v_6	0	∞	0
v_7	0	∞	0

Figure 9.50 Initial configuration of table used in Prim's algorithm

v	Known	d_v	p_v
v_1	1	0	0
v_2	0	2	v_1
v_3	0	4	v_1
v_4	0	1	v_1
v_5	0	∞	0
v_6	0	∞	0
v_7	0	∞	0

Figure 9.51 The table after v_1 is declared known

v	Known	d_v	p_v
v_1	1	0	0
v_2	0	2	v_1
v_3	0	2	v_4
v_4	1	1	v_1
v_5	0	7	v_4
v_6	0	8	v_4
v_7	0	4	v_4

Figure 9.52 The table after v_4 is declared known

so when coding it, remember to put every edge in two adjacency lists. The running time is $O(|V|^2)$ without heaps, which is optimal for dense graphs, and $O(|E| \log |V|)$ using binary heaps, which is good for sparse graphs.

9.5.2. *Kruskal's Algorithm*

A second greedy strategy is continually to select the edges in order of smallest weight and accept an edge if it does not cause a cycle. The action of the algorithm on the graph in the preceding example is shown in Figure 9.56.

v	Known	d_v	p_v
v_1	1	0	0
v_2	1	2	v_1
v_3	1	2	v_4
v_4	1	1	v_1
v_5	0	7	v_4
v_6	0	5	v_3
v_7	0	4	v_4

Figure 9.53 The table after v_2 and then v_3 are declared known

v	Known	d_v	p_v
v_1	1	0	0
v_2	1	2	v_1
v_3	1	2	v_4
v_4	1	1	v_1
v_5	0	6	v_7
v_6	0	1	v_7
v_7	1	4	v_4

Figure 9.54 The table after v_7 is declared known

v	Known	d_v	p_v
v_1	1	0	0
v_2	1	2	v_1
v_3	1	2	v_4
v_4	1	1	v_1
v_5	1	6	v_7
v_6	1	1	v_7
v_7	1	4	v_4

Figure 9.55 The table after v_6 and v_5 are selected (Prim's algorithm terminates)

Formally, Kruskal's algorithm maintains a forest—a collection of trees. Initially, there are $|V|$ single-node trees. Adding an edge merges two trees into one. When the algorithm terminates, there is only one tree, and this is the minimum spanning tree. Figure 9.57 shows the order in which edges are added to the forest.

The algorithm terminates when enough edges are accepted. It turns out to be simple to decide whether edge (u, v) should be accepted or rejected. The appropriate data structure is the *Union/Find* algorithm of the previous chapter.

The invariant we will use is that at any point in the process, two vertices belong to the same set if and only if they are connected in the current spanning forest.

Edge	Weight	Action
(v_1, v_4)	1	Accepted
(v_6, v_7)	1	Accepted
(v_1, v_2)	2	Accepted
(v_3, v_4)	2	Accepted
(v_2, v_4)	3	Rejected
(v_1, v_3)	4	Rejected
(v_4, v_7)	4	Accepted
(v_3, v_6)	5	Rejected
(v_5, v_7)	6	Accepted

Figure 9.56 Action of Kruskal's algorithm on G

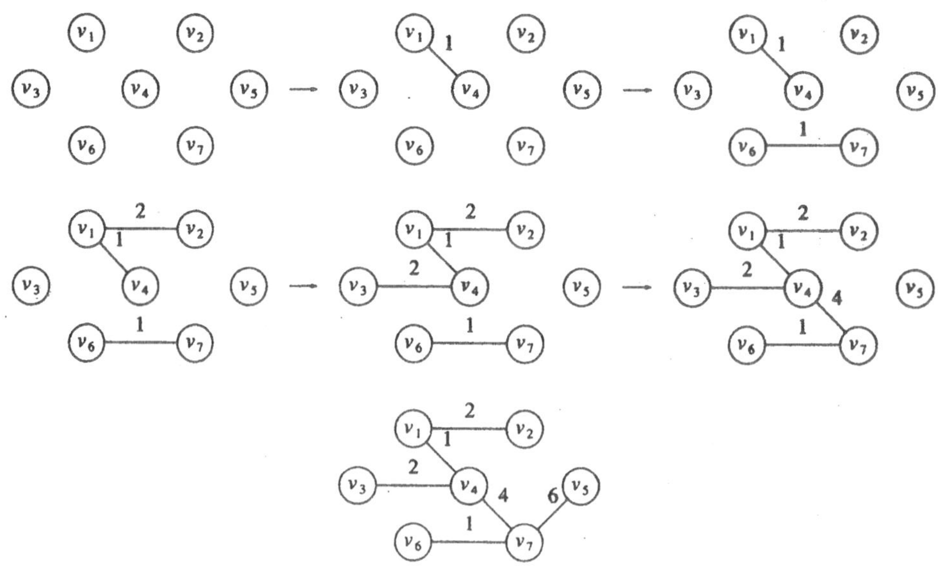

Figure 9.57 Kruskal's algorithm after each stage

Thus, each vertex is initially in its own set. If u and v are in the same set, the edge is rejected, because since they are already connected, adding (u, v) would form a cycle. Otherwise, the edge is accepted, and a *Union* is performed on the two sets containing u and v. It is easy to see that this maintains the set invariant, because once the edge (u, v) is added to the spanning forest, if w was connected to u and x was connected to v, then x and w must now be connected, and thus belong in the same set.

The edges could be sorted to facilitate the selection, but building a heap in linear time is a much better idea. Then *DeleteMins* give the edges to be tested in order. Typically, only a small fraction of the edges need to be tested before the algorithm can terminate, although it is always possible that all the edges must be tried. For instance, if there was an extra vertex v_8 and edge (v_5, v_8) of cost 100, all the edges would have to be examined. Procedure *Kruskal* in Figure 9.58 finds a

```
        void
        Kruskal( Graph G )
        {
            int EdgesAccepted;
            DisjSet S;
            PriorityQueue H;
            Vertex U, V;
            SetType Uset, Vset;
            Edge E;

/* 1*/      Initialize( S );
/* 2*/      ReadGraphIntoHeapArray( G, H );
/* 3*/      BuildHeap( H );

/* 4*/      EdgesAccepted = 0;
/* 5*/      while( EdgesAccepted < NumVertex - 1 )
            {
/* 6*/          E = DeleteMin( H );  /* E = (U,V) */
/* 7*/          Uset = Find( U, S );
/* 8*/          Vset = Find( V, S );
/* 9*/          if( Uset != Vset )
                {
                    /* Accept the edge */
/*10*/              EdgesAccepted++;
/*11*/              SetUnion( S, USet, VSet );
                }
            }
        }
```

Figure 9.58 Pseudocode for Kruskal's algorithm

minimum spanning tree. Because an edge consists of three pieces of data, on some machines it is more efficient to implement the priority queue as an array of pointers to edges, rather than as an array of edges. The effect of this implementation is that, to rearrange the heap, only pointers, not large records, need to be moved.

The worst-case running time of this algorithm is $O(|E| \log |E|)$, which is dominated by the heap operations. Notice that since $|E| = O(|V|^2)$, this running time is actually $O(|E| \log |V|)$. In practice, the algorithm is much faster than this time bound would indicate.

9.6. Applications of Depth-First Search

Depth-first search is a generalization of preorder traversal. Starting at some vertex, v, we process v and then recursively traverse all vertices adjacent to v. If this process is performed on a tree, then all tree vertices are systematically visited in a total of $O(|E|)$ time, since $|E| = \Theta(|V|)$. If we perform this process on an arbitrary graph,

```
void
Dfs( Vertex V )
{
    Visited[ V ] = True;
    for each W adjacent to V
        if( !Visited[ W ] )
            Dfs( W );
}
```

Figure 9.59 Template for depth-first search

we need to be careful to avoid cycles. To do this, when we visit a vertex v, we *mark* it visited, since now we have been there, and recursively call depth-first search on all adjacent vertices that are not already marked. We implicitly assume that for undirected graphs every edge (v, w) appears twice in the adjacency lists: once as (v, w) and once as (w, v). The procedure in Figure 9.59 performs a depth-first search (and does absolutely nothing else) and is a template for the general style.

The (global) boolean array *Visited*[] is initialized to *false*. By recursively calling the procedures only on nodes that have not been visited, we guarantee that we do not loop indefinitely. If the graph is undirected and not connected, or directed and not strongly connected, this strategy might fail to visit some nodes. We then search for an unmarked node, apply a depth-first traversal there, and continue this process until there are no unmarked nodes.* Because this strategy guarantees that each edge is encountered only once, the total time to perform the traversal is $O(|E| + |V|)$, as long as adjacency lists are used.

9.6.1. *Undirected Graphs*

An undirected graph is connected if and only if a depth-first search starting from any node visits every node. Because this test is so easy to apply, we will assume that the graphs we deal with are connected. If they are not, then we can find all the connected components and apply our algorithm on each of these in turn.

As an example of depth-first search, suppose in the graph of Figure 9.60 we start at vertex A. Then we mark A as visited and call $Dfs(B)$ recursively. $Dfs(B)$ marks B as visited and calls $Dfs(C)$ recursively. $Dfs(C)$ marks C as visited and calls $Dfs(D)$ recursively. $Dfs(D)$ sees both A and B, but both of these are marked, so no recursive calls are made. $Dfs(D)$ also sees that C is adjacent but marked, so no recursive call is made there, and $Dfs(D)$ returns back to $Dfs(C)$. $Dfs(C)$ sees B adjacent, ignores it, finds a previously unseen vertex E adjacent, and thus calls $Dfs(E)$. $Dfs(E)$ marks E, ignores A and C, and returns to $Dfs(C)$. $Dfs(C)$ returns to $Dfs(B)$. $Dfs(B)$ ignores both A and D and returns. $Dfs(A)$ ignores both D and E and returns. (We have

*An efficient way of implementing this is to begin the depth-first search at v_1. If we need to restart the depth-first search, we examine the sequence $v_k, v_{k+1}, \ldots$ for an unmarked vertex, where v_{k-1} is the vertex where the last depth-first search was started. This guarantees that throughout the algorithm, only $O(|V|)$ is spent looking for vertices where new depth-first search trees can be started.

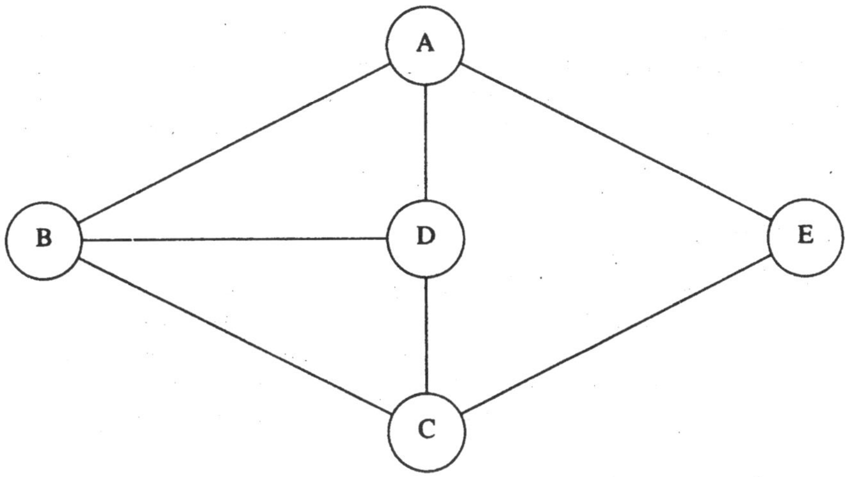

Figure 9.60 An undirected graph

actually touched every edge twice, once as (v, w) and again as (w, v), but this is really once per adjacency list entry.)

We graphically illustrate these steps with a *depth-first spanning tree.* The root of the tree is A, the first vertex visited. Each edge (v, w) in the graph is present in the tree. If, when we process (v, w), we find that w is unmarked, or if, when we process (w, v), we find that v is unmarked, we indicate this with a tree edge. If, when we process (v, w), we find that w is already marked, and when processing (w, v), we find that v is already marked, we draw a dashed line, which we will call a *back edge,* to indicate that this "edge" is not really part of the tree. The depth-first search of the graph in Figure 9.60 is shown in Figure 9.61.

Figure 9.61 Depth-first search of previous graph

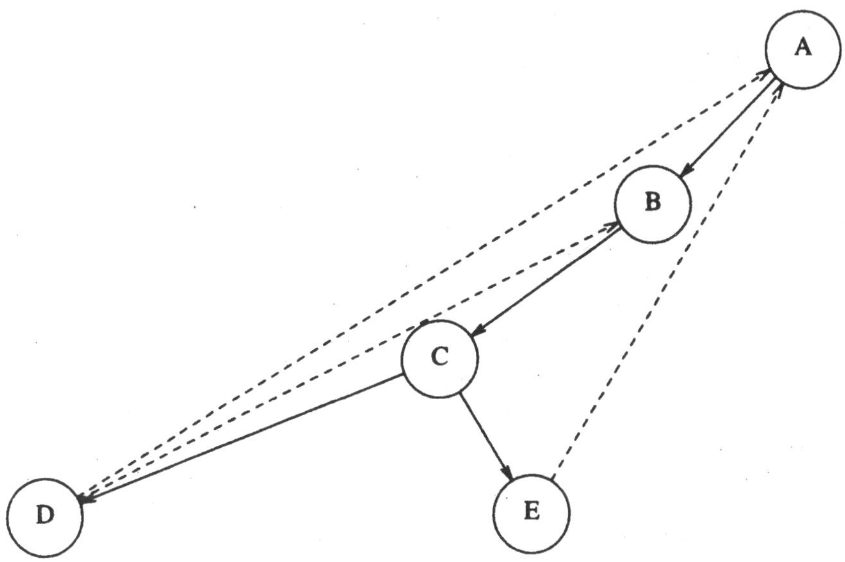

The tree will simulate the traversal we performed. A preorder numbering of the tree, using only tree edges, tells us the order in which the vertices were marked. If the graph is not connected, then processing all nodes (and edges) requires several calls to *Dfs*, and each generates a tree. This entire collection is a *depth-first spanning forest*.

9.6.2. Biconnectivity

A connected undirected graph is *biconnected* if there are no vertices whose removal disconnects the rest of the graph. The graph in the example above is biconnected. If the nodes are computers and the edges are links, then if any computer goes down, network mail is unaffected, except, of course, at the down computer. Similarly, if a mass transit system is biconnected, users always have an alternate route should some terminal be disrupted.

If a graph is not biconnected, the vertices whose removal would disconnect the graph are known as *articulation points*. These nodes are critical in many applications. The graph in Figure 9.62 is not biconnected: C and D are articulation points. The removal of C would disconnect G, and the removal of D would disconnect E and F, from the rest of the graph.

Depth-first search provides a linear-time algorithm to find all articulation points in a connected graph. First, starting at any vertex, we perform a depth-first search and number the nodes as they are visited. For each vertex v, we call this preorder number $Num(v)$. Then, for every vertex v in the depth-first search spanning tree, we compute the lowest-numbered vertex, which we call $Low(v)$, that is reachable from v by taking zero or more tree edges and then possibly one back edge (in that order). The depth-first search tree in Figure 9.63 shows the preorder number first, and then the lowest-numbered vertex reachable under the rule described above.

The lowest-numbered vertex reachable by A, B, and C is vertex 1 (A), because they can all take tree edges to D and then one back edge back to A. We can efficiently compute Low by performing a postorder traversal of the depth-first spanning tree. By the definition of Low, $Low(v)$ is the minimum of

1. $Num(v)$
2. the lowest $Num(w)$ among all back edges (v, w)
3. the lowest $Low(w)$ among all tree edges (v, w)

The first condition is the option of taking no edges, the second way is to choose no tree edges and a back edge, and the third way is to choose some tree edges and possibly a back edge. This third method is succinctly described with a recursive call. Since we need to evaluate Low for all the children of v before we can evaluate $Low(v)$, this is a postorder traversal. For any edge (v, w), we can tell whether it is a tree edge or a back edge merely by checking $Num(v)$ and $Num(w)$. Thus, it is easy to compute $Low(v)$: we merely scan down v's adjacency list, apply the proper rule, and keep track of the minimum. Doing all the computation takes $O(|E| + |V|)$ time.

All that is left to do is to use this information to find articulation points. The root is an articulation point if and only if it has more than one child, because if it has two children, removing the root disconnects nodes in different subtrees, and if it has only one child, removing the root merely disconnects the root. Any other vertex v is

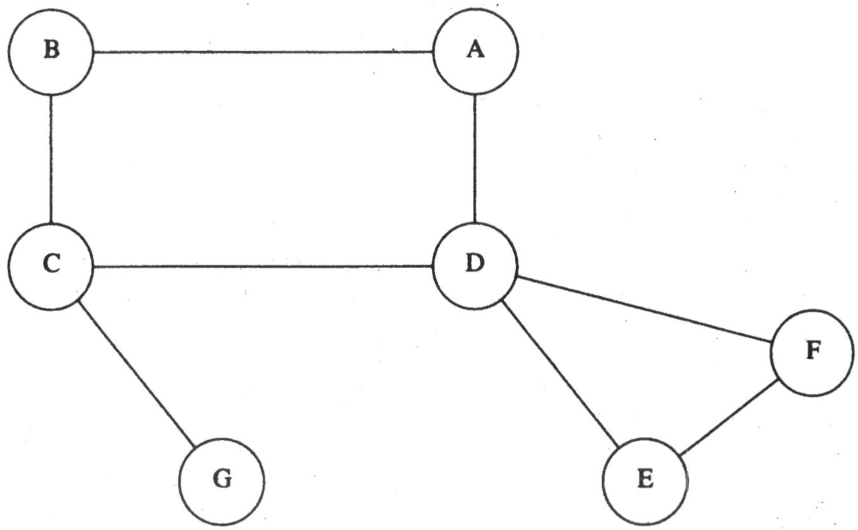

Figure 9.62 A graph with articulation points *C* and *D*

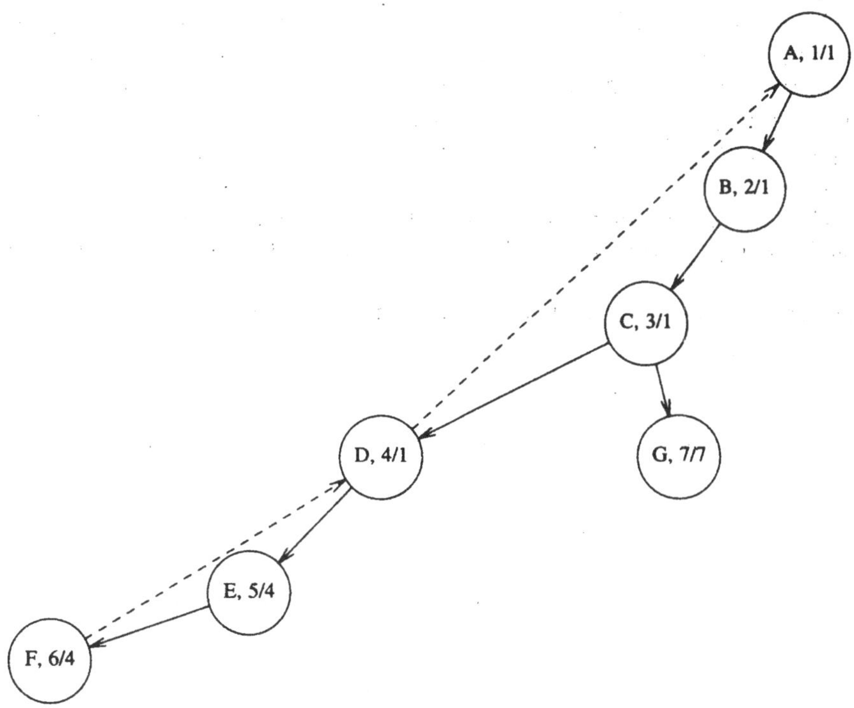

Figure 9.63 Depth-first tree for previous graph, with *Num* and *Low*

an articulation point if and only if v has some child w such that $Low(w) \geq Num(v)$. Notice that this condition is always satisfied at the root; hence the need for a special test.

The *if* part of the proof is clear when we examine the articulation points that the algorithm determines, namely, C and D. D has a child E, and $Low(E) \geq Num(D)$, since both are 4. Thus, there is only one way for E to get to any node above D, and that is by going through D. Similarly, C is an articulation point, because $Low(G) \geq Num(C)$. To prove that this algorithm is correct, one must show that the *only if* part of the assertion is true (that is, this finds *all* articulation points). We leave this as an exercise. As a second example, we show (Fig. 9.64) the result of applying this algorithm on the same graph, starting the depth-first search at C.

We close by giving pseudocode to implement this algorithm. We will assume that the arrays *Visited*[] (initialized to *false*), *Num*[], *Low*[], and *Parent*[] are global to keep the code simple. We will also keep a global variable called *Counter*, which is initialized to 1 to assign the preorder traversal numbers, *Num*[]. This is not normally good programming practice, but including all the declarations and passing the extra parameters would cloud the logic. We also leave out the easily implemented test for the root.

As we have already stated, this algorithm can be implemented by performing a preorder traversal to compute *Num* and then a postorder traversal to compute *Low*. A third traversal can be used to check which vertices satisfy the articulation point criteria. Performing three traversals, however, would be a waste. The first pass is shown in Figure 9.65.

The second and third passes, which are postorder traversals, can be implemented by the code in Figure 9.66. Line 8 handles a special case. If w is adjacent to v, then the recursive call to w will find v adjacent to w. This is not a back edge, only an edge that has already been considered and needs to be ignored. Otherwise, the procedure computes the minimum of the various *Low*[] and *Num*[] entries, as specified by the algorithm.

There is no rule that a traversal must be either preorder or postorder. It is possible to do processing both before and after the recursive calls. The procedure in

Figure 9.64 Depth-first tree that results if depth-first search starts at C

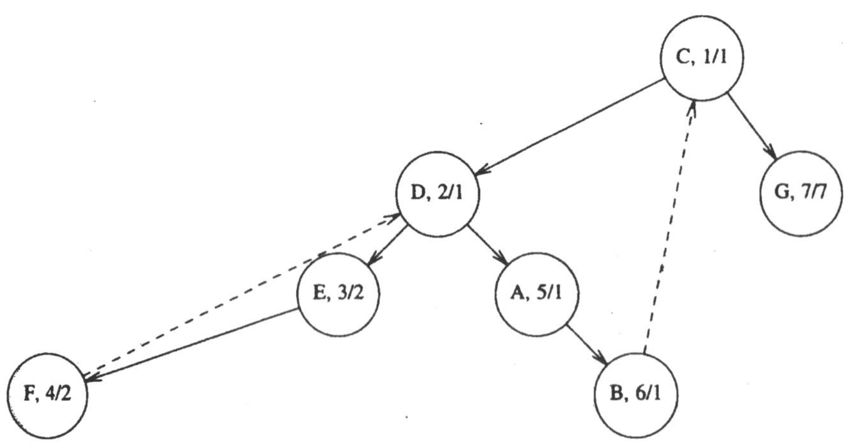

```
/* Assign Num and compute Parents */

void
AssignNum( Vertex V )
{
    Vertex W;

    Num[ V ] = Counter++;
    Visited[ V ] = True;
    for each W adjacent to V
        if( !Visited[ W ] )
        {
            Parent[ W ] = V;
            AssignNum( W );
        }
}
```

Figure 9.65 Routine to assign *Num* to vertices (pseudocode)

```
/* Assign Low; also check for articulation points */

void
AssignLow( Vertex V )
{
    Vertex W;

    Low[ V ] = Num[ V ];  /* Rule 1 */
    for each W adjacent to V
    {
        if( Num[ W ] > Num[ V ] )  /* Forward edge */
        {
            AssignLow( W );
            if( Low[ W ] >= Num[ V ] )
                printf( "%v is an articulation point\n", v );
            Low[ V ] = Min( Low[ V ], Low[ W ] ); /* Rule 3 */
        }
        else
        if( Parent[ V ] != W )  /* Back edge */
            Low[ V ] = Min( Low[ V ], Num[ W ] ); /* Rule 2 */
    }
}
```

Figure 9.66 Pseudocode to compute *Low* and to test for articulation points (test for the root is omitted)

Figure 9.67 combines the two routines *AssignNum* and *AssignLow* in a straightforward manner to produce the procedure *FindArt*.

9.6.3. Euler Circuits

Consider the three figures in Figure 9.68. A popular puzzle is to reconstruct these figures using a pen, drawing each line exactly once. The pen may not be lifted from the paper while the drawing is being performed. As an extra challenge, make the pen finish at the same point at which it started. This puzzle has a surprisingly simple solution. Stop reading if you would like to try to solve it.

The first figure can be drawn only if the starting point is the lower left- or right-hand corner, and it is not possible to finish at the starting point. The second figure is easily drawn with the finishing point the same as the starting point, but the third figure cannot be drawn at all within the parameters of the puzzle.

We can convert this problem to a graph theory problem by assigning a vertex to each intersection. Then the edges can be assigned in the natural manner, as in Figure 9.69.

After this conversion is performed, we must find a path in the graph that visits every edge exactly once. If we are to solve the "extra challenge," then we must find a cycle that visits every edge exactly once. This graph problem was solved in 1736 by Euler and marked the beginning of graph theory. The problem is thus commonly referred to as an *Euler path* (sometimes *Euler tour*) or *Euler circuit problem,* depending on the specific problem statement. The Euler tour and Euler circuit problems, though slightly different, have the same basic solution. Thus, we will consider the Euler circuit problem in this section.

The first observation that can be made is that an Euler circuit, which must end on its starting vertex, is possible only if the graph is connected and each vertex has an even degree (number of edges). This is because, on the Euler circuit, a vertex is entered and then left. If any vertex v has odd degree, then eventually we will reach the point where only one edge into v is unvisited, and taking it will strand us at v. If exactly two vertices have odd degree, an Euler tour, which must visit every edge but need not return to its starting vertex, is still possible if we start at one of the odd-degree vertices and finish at the other. If more than two vertices have odd degree, then an Euler tour is not possible.

The observations of the preceding paragraph provide us with a necessary condition for the existence of an Euler circuit. It does not, however, tell us that all connected graphs that satisfy this property must have an Euler circuit, nor does it give us guidance on how to find one. It turns out that the necessary condition is also sufficient. That is, any connected graph, all of whose vertices have even degree, must have an Euler circuit. Furthermore, a circuit can be found in linear time.

We can assume that we know that an Euler circuit exists, since we can test the necessary and sufficient condition in linear time. Then the basic algorithm is to perform a depth-first search. There are a surprisingly large number of "obvious" solutions that do not work. Some of these are presented in the exercises.

The main problem is that we might visit a portion of the graph and return to the starting point prematurely. If all the edges coming out of the start vertex have

```
void
FindArt( Vertex V )
{
    Vertex W;

    Visited[ V ] = True;
    Low[ V ] = Num[ V ] = Counter++;  /* Rule 1 */
    for each W adjacent to V
    {
        if( !Visited[ W ] )  /* Forward edge */
        {
            Parent[ W ] = V;
            FindArt( W );
            if( Low[ W ] >= Num[ V ] )
                printf( "%v is an articulation point\n", v );
            Low[ V ] = Min( Low[ V ], Low[ W ] ); /* Rule 3 */
        }
        else
        if( Parent[ V ] != W )  /* Back edge */
            Low[ V ] = Min( Low[ V ], Num[ W ] ); /* Rule 2 */
    }
}
```

Figure 9.67 Testing for articulation points in one depth-first search (test for the root is omitted) (pseudocode)

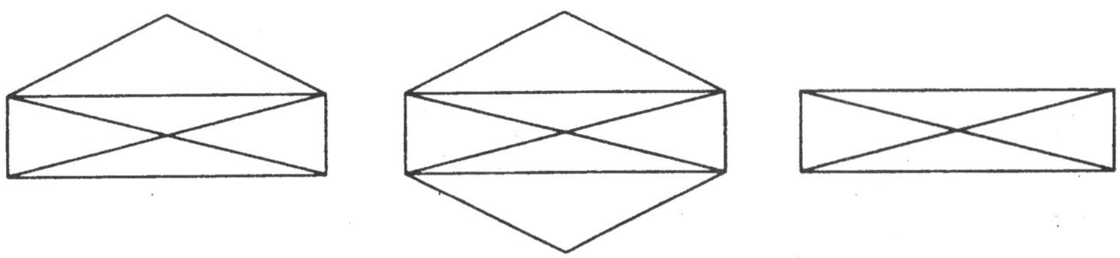

Figure 9.68 Three drawings

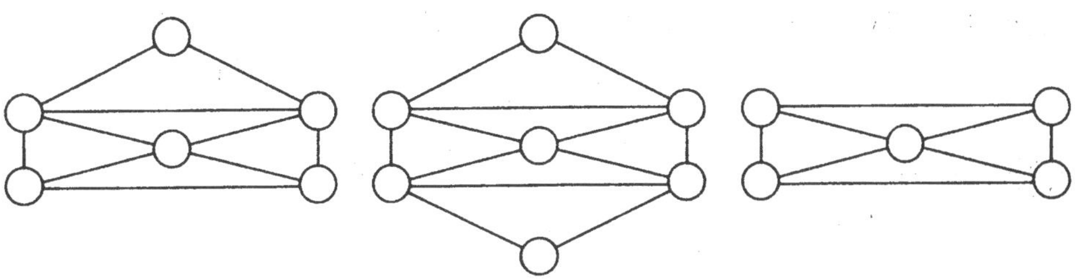

Figure 9.69 Conversion of puzzle to graph

been used up, then part of the graph is untraversed. The easiest way to fix this is to find the first vertex on this path that has an untraversed edge, and perform another depth-first search. This will give another circuit, which can be spliced into the original. This is continued until all edges have been traversed.

As an example, consider the graph in Figure 9.70. It is easily seen that this graph has an Euler circuit. Suppose we start at vertex 5, and traverse the circuit 5, 4, 10, 5. Then we are stuck, and most of the graph is still untraversed. The situation is shown in Figure 9.71.

We then continue from vertex 4, which still has unexplored edges. A depth-first search might come up with the path 4, 1, 3, 7, 4, 11, 10, 7, 9, 3, 4. If we splice this path into the previous path of 5, 4, 10, 5, then we get a new path of 5, 4, 1, 3, 7, 4, 11, 10, 7, 9, 3, 4, 10, 5.

The graph that remains after this is shown in Figure 9.72. Notice that in this graph all the vertices must have even degree, so we are guaranteed to find a cycle to

Figure 9.70 Graph for Euler circuit problem

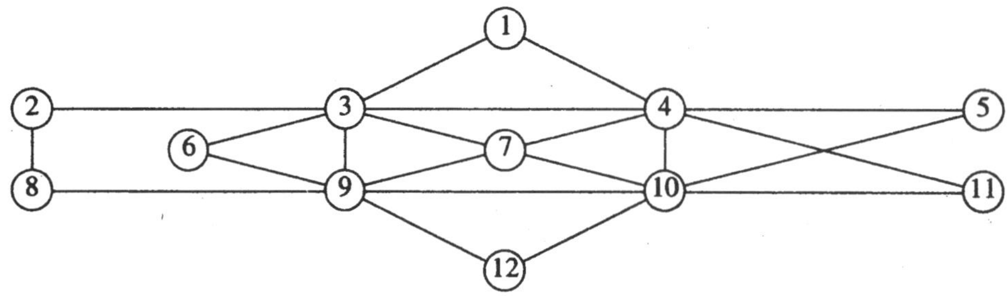

Figure 9.71 Graph remaining after 5, 4, 10, 5

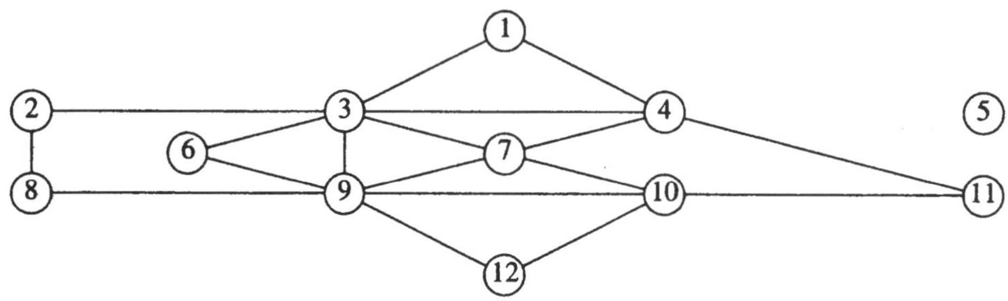

Figure 9.72 Graph after the path 5, 4, 1, 3, 7, 4, 11, 10, 7, 9, 3, 4, 10, 5

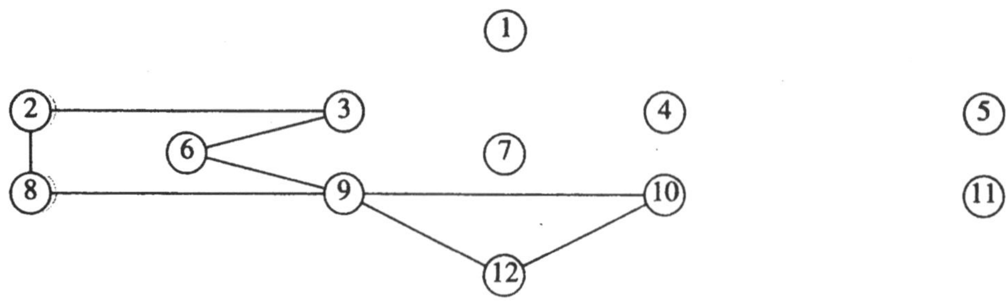

add. The remaining graph might not be connected, but this is not important. The next vertex on the path that has untraversed edges is vertex 3. A possible circuit would then be 3, 2, 8, 9, 6, 3. When spliced in, this gives the path 5, 4, 1, 3, 2, 8, 9, 6, 3, 7, 4, 11, 10, 7, 9, 3, 4, 10, 5.

The graph that remains is in Figure 9.73. On this path, the next vertex with an untraversed edge is 9, and the algorithm finds the circuit 9, 12, 10, 9. When this is added to the current path, a circuit of 5, 4, 1, 3, 2, 8, 9, 12, 10, 9, 6, 3, 7, 4, 11, 10, 7, 9, 3, 4, 10, 5 is obtained. As all the edges are traversed, the algorithm terminates with an Euler circuit.

To make this algorithm efficient, we must use appropriate data structures. We will sketch some of the ideas, leaving the implementation as an exercise. To make splicing simple, the path should be maintained as a linked list. To avoid repetitious scanning of adjacency lists, we must maintain, for each adjacency list, a pointer to the last edge scanned. When a path is spliced in, the search for a new vertex from which to perform the next depth-first search must begin at the start of the splice point. This guarantees that the total work performed on the vertex search phase is $O(|E|)$ during the entire life of the algorithm. With the appropriate data structures, the running time of the algorithm is $O(|E| + |V|)$.

A very similar problem is to find a simple cycle, in an undirected graph, that visits every vertex. This is known as the *Hamiltonian cycle problem*. Although it seems almost identical to the Euler circuit problem, no efficient algorithm for it is known. We shall see this problem again in Section 9.7.

9.6.4. Directed Graphs

Using the same strategy as with undirected graphs, directed graphs can be traversed in linear time, using depth-first search. If the graph is not strongly connected, a depth-first search starting at some node might not visit all nodes. In this case we repeatedly perform depth-first searches, starting at some unmarked node, until all vertices have been visited. As an example, consider the directed graph in Figure 9.74.

We arbitrarily start the depth-first search at vertex B. This visits vertices B, C, A, D, E, and F. We then restart at some unvisited vertex. Arbitrarily, we start at H, which visits I and J. Finally, we start at G, which is the last vertex that needs to be visited. The corresponding depth-first search tree is shown in Figure 9.75.

Figure 9.73 Graph remaining after the path 5, 4, 1, 3, 2, 8, 9, 6, 3, 7, 4, 11, 10, 7, 9, 3, 4, 10, 5

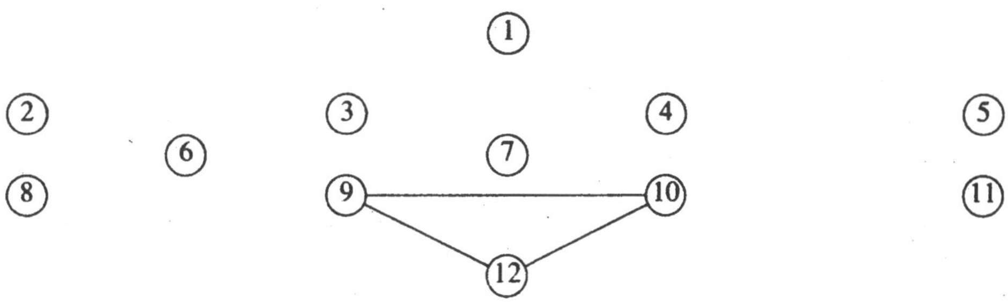

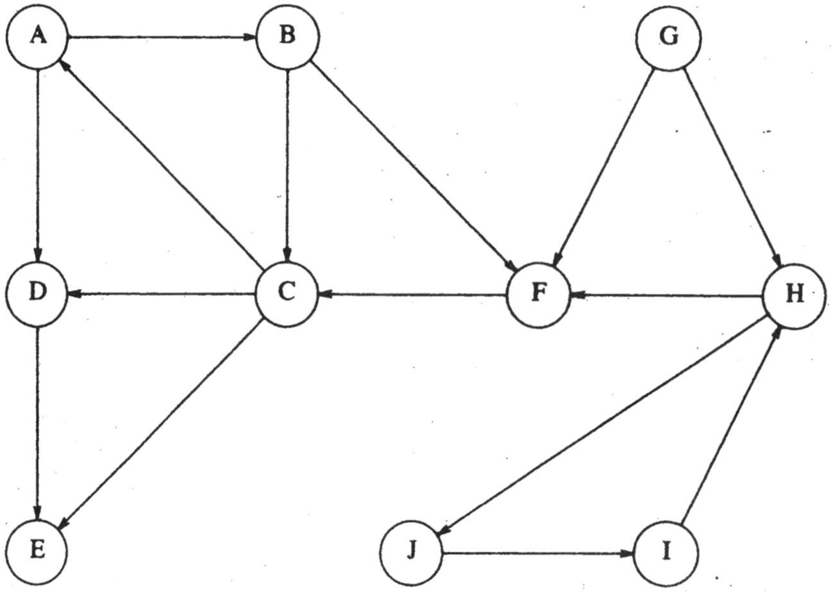

Figure 9.74 A directed graph

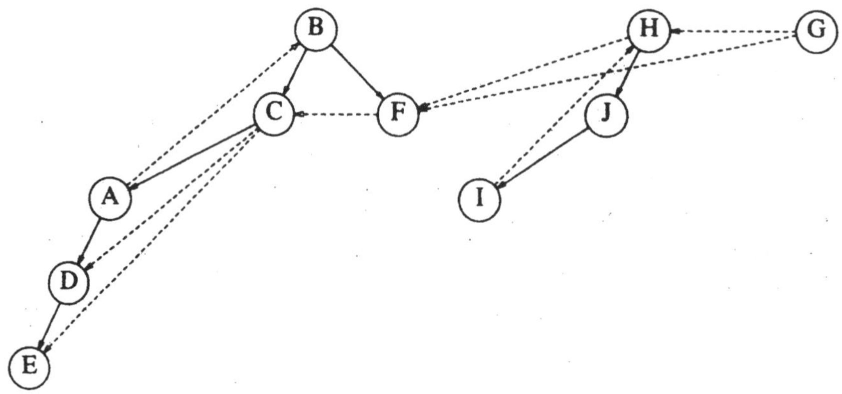

Figure 9.75 Depth-first search of previous graph

The dashed arrows in the depth-first spanning forest are edges (v, w) for which w was already marked at the time of consideration. In undirected graphs, these are always back edges, but, as we can see, there are three types of edges that do not lead to new vertices. First, there are *back edges,* such as (A, B) and (I, H). There are also *forward edges,* such as (C, D) and (C, E), that lead from a tree node to a descendant. Finally, there are *cross edges,* such as (F, C) and (G, F), which connect two tree nodes that are not directly related. Depth-first search forests are generally drawn with children and new trees added to the forest from left to right. In a depth-first search of a directed graph drawn in this manner, cross edges always go from right to left.

Some algorithms that use depth-first search need to distinguish between the three types of nontree edges. This is easy to check as the depth-first search is being performed, and it is left as an exercise.

One use of depth-first search is to test whether or not a directed graph is acyclic. The rule is that a directed graph is acyclic if and only if it has no back edges. (The graph above has back edges, and thus is not acyclic.) The reader may remember that a topological sort can also be used to determine whether a graph is acyclic. Another way to perform topological sorting is to assign the vertices topological numbers $N, N-1, \ldots, 1$ by postorder traversal of the depth-first spanning forest. As long as the graph is acyclic, this ordering will be consistent.

9.6.5. *Finding Strong Components*

By performing two depth-first searches, we can test whether a directed graph is strongly connected, and if it is not, we can actually produce the subsets of vertices that are strongly connected to themselves. This can also be done in only one depth-first search, but the method used here is much simpler to understand.

First, a depth-first search is performed on the input graph G. The vertices of G are numbered by a postorder traversal of the depth-first spanning forest, and then all edges in G are reversed, forming G_r. The graph in Figure 9.76 represents G_r for the graph G shown in Figure 9.74; the vertices are shown with their numbers.

The algorithm is completed by performing a depth-first search on G_r, always starting a new depth-first search at the highest-numbered vertex. Thus, we begin the depth-first search of G_r at vertex G, which is numbered 10. This leads nowhere, so the next search is started at H. This call visits I and J. The next call starts at B and visits A, C, and F. The next calls after this are $Dfs(D)$ and finally $Dfs(E)$. The resulting depth-first spanning forest is shown in Figure 9.77.

Each of the trees (this is easier to see if you completely ignore all nontree edges) in this depth-first spanning forest forms a strongly connected component. Thus, for our example, the strongly connected components are $\{G\}$, $\{H, I, J\}$, $\{B, A, C, F\}$, $\{D\}$, and $\{E\}$.

To see why this algorithm works, first note that if two vertices v and w are in the same strongly connected component, then there are paths from v to w and from w to v in the original graph G, and hence also in G_r. Now, if two vertices v and w are not in the same depth-first spanning tree of G_r, clearly they cannot be in the same strongly connected component.

To prove that this algorithm works, we must show that if two vertices v and w are in the same depth-first spanning tree of G_r, there must be paths from v to w and from w to v. Equivalently, we can show that if x is the root of the depth-first spanning tree of G_r containing v, then there is a path from x to v and from v to x. Applying the same logic to w would then give a path from x to w and from w to x. These paths would imply paths from v to w and w to v (going through x).

Since v is a descendant of x in G_r's depth-first spanning tree, there is a path from x to v in G_r and thus a path from v to x in G. Furthermore, since x is the root, x has the higher postorder number from the first depth-first search. Therefore, during the first depth-first search, all the work processing v was completed before

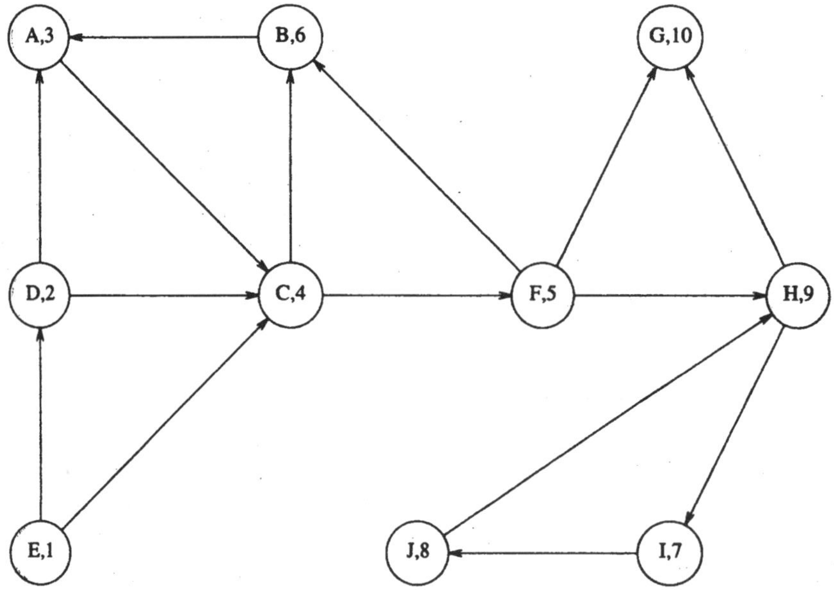

Figure 9.76 G_r numbered by postorder traversal of G (from Fig. 9.74)

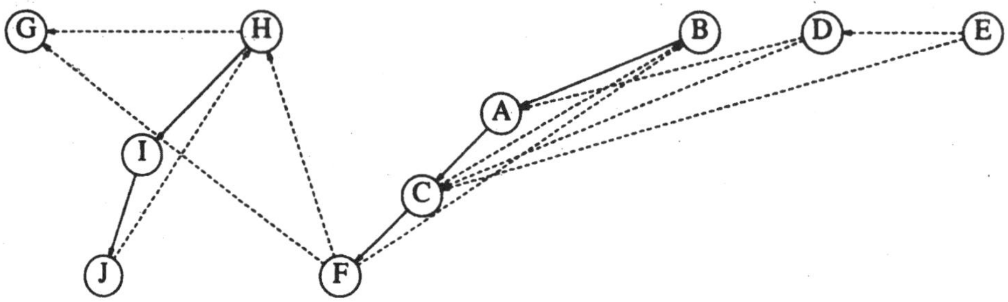

Figure 9.77 Depth-first search of G_r—strong components are $\{G\}$, $\{H, I, J\}$, $\{B, A, C, F\}$, $\{D\}$, $\{E\}$

the work at x was completed. Since there is a path from v to x, it follows that v must be a descendant of x in the spanning tree for G—otherwise v would finish *after* x. This implies a path from x to v in G and completes the proof.

9.7. Introduction to NP-Completeness

In this chapter, we have seen solutions to a wide variety of graph theory problems. All these problems have polynomial running times, and with the exception of the network flow problem, the running time is either linear or only slightly more than linear ($O(|E| \log |E|)$). We have also mentioned, in passing, that for some problems certain variations seem harder than the original.

Recall that the Euler circuit problem, which finds a path that touches every edge exactly once, is solvable in linear time. The Hamiltonian cycle problem asks for a simple cycle that contains every vertex. No linear algorithm is known for this problem.

The single-source unweighted shortest-path problem for directed graphs is also solvable in linear time. No linear-time algorithm is known for the corresponding longest-simple-path problem.

The situation for these problem variations is actually much worse than we have described. Not only are no linear algorithms known for these variations, but there are no known algorithms that are guaranteed to run in polynomial time. The best known algorithms for these problems could take exponential time on some inputs.

In this section we will take a brief look at this problem. This topic is rather complex, so we will only take a quick and informal look at it. Because of this, the discussion may be (necessarily) somewhat imprecise in places.

We will see that there are a host of important problems that are roughly equivalent in complexity. These problems form a class called the *NP-complete* problems. The exact complexity of these *NP*-complete problems has yet to be determined and remains the foremost open problem in theoretical computer science. Either all these problems have polynomial-time solutions or none of them do.

9.7.1. Easy vs. Hard

When classifying problems, the first step is to examine the boundaries. We have already seen that many problems can be solved in linear time. We have also seen some $O(\log N)$ running times, but these either assume some preprocessing (such as input already being read or a data structure already being built) or occur on arithmetic examples. For instance, the *Gcd* algorithm, when applied on two numbers M and N, takes $O(\log N)$ time. Since the numbers consist of $\log M$ and $\log N$ bits respectively, the *Gcd* algorithm is really taking time that is linear in the *amount* or *size* of input. Thus, when we measure running time, we will be concerned with the running time as a function of the amount of input. Generally, we cannot expect better than linear running time.

At the other end of the spectrum lie some truly hard problems. These problems are so hard that they are *impossible*. This does not mean the typical exasperated moan, which means that it would take a genius to solve the problem. Just as real numbers are not sufficient to express a solution to $x^2 < 0$, one can prove that computers cannot solve every problem that happens to come along. These "impossible" problems are called *undecidable* problems.

One particular undecidable problem is the *halting problem*. Is it possible to have your C compiler have an extra feature that not only detects syntax errors but also all infinite loops? This seems like a hard problem, but one might expect that if some very clever programmers spent enough time on it, they could produce this enhancement.

The intuitive reason that this problem is undecidable is that such a program might have a hard time checking itself. For this reason, these problems are sometimes called *recursively undecidable*.

If an infinite loop–checking program could be written, surely it could be used to check itself. We could then produce a program called *LOOP*. *LOOP* takes as input a program *P* and runs *P* on itself. It prints out the phrase *YES* if *P* loops when run on itself. If *P* terminates when run on itself, a natural thing to do would be to print out *NO*. Instead of doing that, we will have *LOOP* go into an infinite loop.

What happens when *LOOP* is given itself as input? Either *LOOP* halts, or it does not halt. The problem is that both these possibilities lead to contradictions, in much the same way as does the phrase "This sentence is a lie."

By our definition, *LOOP(P)* goes into an infinite loop if *P*(*P*) terminates. Suppose that when *P* = *LOOP*, *P*(*P*) terminates. Then, according to the *LOOP* program, *LOOP(P)* is obligated to go into an infinite loop. Thus, we must have *LOOP(LOOP)* terminating ***and*** entering an infinite loop, which is clearly not possible. On the other hand, suppose that when *P* = *LOOP*, *P*(*P*) enters an infinite loop. Then *LOOP(P)* must terminate, and we arrive at the same set of contradictions. Thus, we see that the program *LOOP* cannot possibly exist.

9.7.2. The Class NP

A few steps down from the horrors of undecidable problems lies the class *NP*. *NP* stands for *nondeterministic polynomial-time.* A deterministic machine, at each point in time, is executing an instruction. Depending on the instruction, it then goes to some next instruction, which is unique. A nondeterministic machine has a choice of next steps. It is free to choose any that it wishes, and if one of these steps leads to a solution, it will always choose the correct one. A nondeterministic machine thus has the power of extremely good (optimal) guessing. This probably seems like a ridiculous model, since nobody could possibly build a nondeterministic computer, and because it would seem to be an incredible upgrade to your standard computer (every problem might now seem trivial). We will see that nondeterminism is a very useful theoretical construct. Furthermore, nondeterminism is not as powerful as one might think. For instance, undecidable problems are still undecidable, even if nondeterminism is allowed.

A simple way to check if a problem is in *NP* is to phrase the problem as a yes/no question. The problem is in *NP* if, in polynomial time, we can prove that any "yes" instance is correct. We do not have to worry about "no" instances, since the program always makes the right choice. Thus, for the Hamiltonian cycle problem, a "yes" instance would be any simple circuit in the graph that includes all the vertices. This is in *NP*, since, given the path, it is a simple matter to check that it is really a Hamiltonian cycle. Appropriately phrased questions, such as "Is there a simple path of length > *K*?" can also easily be checked and are in *NP*. Any path that satisfies this property can be checked trivially.

The class *NP* includes all problems that have polynomial-time solutions, since obviously the solution provides a check. One would expect that since it is so much easier to check an answer than to come up with one from scratch, there would be problems in *NP* that do not have polynomial-time solutions. To date no such problem has been found, so it is entirely possible, though not considered likely by experts, that nondeterminism is not such an important improvement. The problem is that proving exponential lower bounds is an extremely difficult task. The information theory bound technique, which we used to show that sorting requires $\Omega(N \log N)$ comparisons, does not seem to be adequate for the task, because the decision trees are not nearly large enough.

Notice also that not all decidable problems are in *NP*. Consider the problem of determining whether a graph *does not* have a Hamiltonian cycle. To prove that a graph has a Hamiltonian cycle is a relatively simple matter—we just need to exhibit one. Nobody knows how to show, in polynomial time, that a graph does not have a Hamiltonian cycle. It seems that one must enumerate all the cycles and check them one by one. Thus the Non–Hamiltonian cycle problem is not known to be in *NP*.

9.7.3. NP-Complete Problems

Among all the problems known to be in *NP*, there is a subset, known as the *NP-complete* problems, which contains the hardest. An *NP*-complete problem has the property that any problem in *NP* can be *polynomially reduced* to it.

A problem P_1 can be reduced to P_2 as follows: Provide a mapping so that any instance of P_1 can be transformed to an instance of P_2. Solve P_2, and then map the answer back to the original. As an example, numbers are entered into a pocket calculator in decimal. The decimal numbers are converted to binary, and all calculations are performed in binary. Then the final answer is converted back to decimal for display. For P_1 to be polynomially reducible to P_2, all the work associated with the transformations must be performed in polynomial time.

The reason that *NP*-complete problems are the hardest *NP* problems is that a problem that is *NP*-complete can essentially be used as a subroutine for *any* problem in *NP*, with only a polynomial amount of overhead. Thus, if any *NP*-complete problem has a polynomial-time solution, then *every* problem in *NP* must have a polynomial-time solution. This makes the *NP*-complete problems the hardest of all *NP* problems.

Suppose we have an *NP*-complete problem P_1. Suppose P_2 is known to be in *NP*. Suppose further that P_1 polynomially reduces to P_2, so that we can solve P_1 by using P_2 with only a polynomial time penalty. Since P_1 is *NP*-complete, every problem in *NP* polynomially reduces to P_1. By applying the closure property of polynomials, we see that every problem in *NP* is polynomially reducible to P_2: We reduce the problem to P_1 and then reduce P_1 to P_2. Thus, P_2 is *NP*-complete.

As an example, suppose that we already know that the Hamiltonian cycle problem is *NP*-complete. The *traveling salesman problem* is as follows.

TRAVELING SALESMAN PROBLEM:

Given a complete graph $G = (V, E)$, with edge costs, and an integer K, is there a simple cycle that visits all vertices and has total cost $\leq K$?

The problem is different from the Hamiltonian cycle problem, because all $|V|(|V| - 1)/2$ edges are present and the graph is weighted. This problem has many important applications. For instance, printed circuit boards need to have holes punched so that chips, resistors, and other electronic components can be placed.

This is done mechanically. Punching the hole is a quick operation; the time-consuming step is positioning the hole puncher. The time required for positioning depends on the distance traveled from hole to hole. Since we would like to punch every hole (and then return to the start for the next board), and minimize the total amount of time spent traveling, what we have is a traveling salesman problem.

The traveling salesman problem is *NP*-complete. It is easy to see that a solution can be checked in polynomial time, so it is certainly in *NP*. To show that it is *NP*-complete, we polynomially reduce the Hamiltonian cycle problem to it. To do this we construct a new graph G'. G' has the same vertices as G. For G', each edge (v, w) has a weight of 1 if $(v, w) \in G$, and 2 otherwise. We choose $K = |V|$. See Figure 9.78.

It is easy to verify that G has a Hamiltonian cycle if and only if G' has a Traveling Salesman tour of total weight $|V|$.

There is now a long list of problems known to be *NP*-complete. To prove that some new problem is *NP*-complete, it must be shown to be in *NP*, and then an appropriate *NP*-complete problem must be transformed into it. Although the transformation to a traveling salesman problem was rather straightforward, most transformations are actually quite involved and require some tricky constructions. Generally, several different *NP*-complete problems are considered before the problem that actually provides the reduction. As we are only interested in the general ideas, we will not show any more transformations; the interested reader can consult the references.

The alert reader may be wondering how the first *NP*-complete problem was actually proven to be *NP*-complete. Since proving that a problem is *NP*-complete requires transforming it from another *NP*-complete problem, there must be some *NP*-complete problem for which this strategy will not work. The first problem that was proven to be *NP*-complete was the *satisfiability* problem. The satisfiability problem takes as input a boolean expression and asks whether the expression has an assignment to the variables that gives a value of 1.

Figure 9.78 Hamiltonian cycle problem transformed to traveling salesman problem

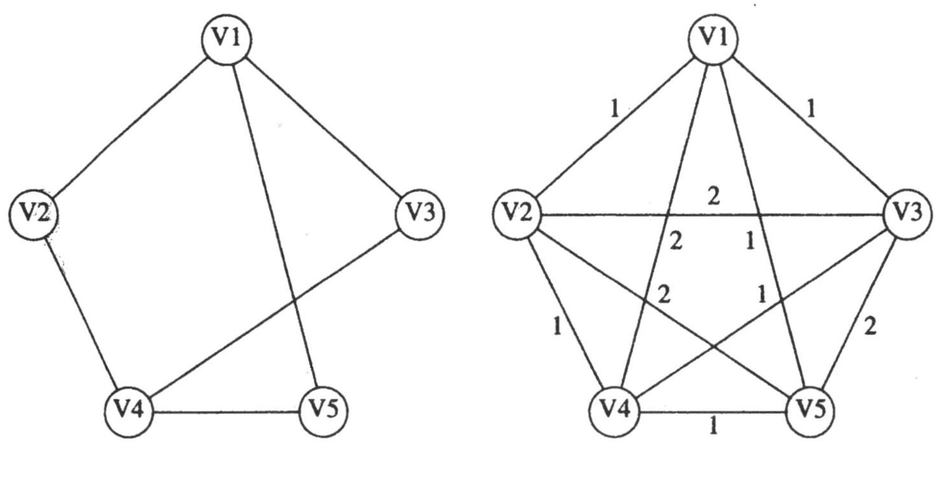

Satisfiability is certainly in *NP*, since it is easy to evaluate a boolean expression and check whether the result is true. In 1971, Cook showed that satisfiability was *NP*-complete by directly proving that all problems that are in *NP* could be transformed to satisfiability. To do this, he used the one known fact about every problem in *NP*: Every problem in *NP* can be solved in polynomial time by a nondeterministic computer. The formal model for a computer is known as a *Turing machine*. Cook showed how the actions of this machine could be simulated by an extremely complicated and long, but still polynomial, boolean formula. This boolean formula would be true if and only if the program which was being run by the Turing machine produced a "yes" answer for its input.

Once satisfiability was shown to be *NP*-complete, a host of new *NP*-complete problems, including some of the most classic problems, were also shown to be *NP*-complete.

In addition to the satisfiability, Hamiltonian circuit, traveling salesman, and longest-path problems, which we have already examined, some of the more well-known *NP*-complete problems which we have not discussed are *bin packing*, *knapsack*, *graph coloring*, and *clique*. The list is quite extensive and includes problems from operating systems (scheduling and security), database systems, operations research, logic, and especially graph theory.

Summary

In this chapter we have seen how graphs can be used to model many real-life problems. Many of the graphs that occur are typically very sparse, so it is important to pay attention to the data structures that are used to implement them.

We have also seen a class of problems that do not seem to have efficient solutions. In Chapter 10, some techniques for dealing with these problems will be discussed.

Exercises

9.1 Find a topological ordering for the graph in Figure 9.79.

9.2 If a stack is used instead of a queue for the topological sort algorithm in Section 9.1, does a different ordering result? Why might one data structure give a "better" answer?

9.3 Write a program to perform a topological sort on a graph.

9.4 An adjacency matrix requires $O(|V|^2)$ merely to initialize using a standard double loop. Propose a method that stores a graph in an adjacency matrix (so that testing for the existence of an edge is $O(1)$) but avoids the quadratic running time.

9.5 a. Find the shortest path from A to all other vertices for the graph in Figure 9.80.

b. Find the shortest unweighed path from B to all other vertices for the graph in Figure 9.80.

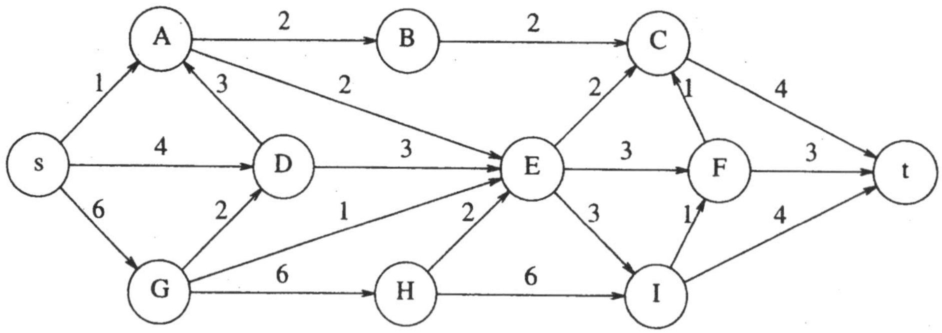

Figure 9.79

9.6 What is the worst-case running time of Dijkstra's algorithm when implemented with d-heaps (Section 6.5)?

9.7 a. Give an example where Dijkstra's algorithm gives the wrong answer in the presence of a negative edge but no negative-cost cycle.

**b. Show that the weighted shortest-path algorithm suggested in Section 9.3.3 works if there are negative-weight edges, but no negative-cost cycles, and that the running time of this algorithm is $O(|E| \cdot |V|)$.

*9.8 Suppose all the edge weights in a graph are integers between 1 and $|E|$. How fast can Dijkstra's algorithm be implemented?

9.9 Write a program to solve the single-source shortest-path problem.

9.10 a. Explain how to modify Dijkstra's algorithm to produce a count of the number of different minimum paths from v to w.

b. Explain how to modify Dijkstra's algorithm so that if there is more than one minimum path from v to w, a path with the fewest number of edges is chosen.

Figure 9.80

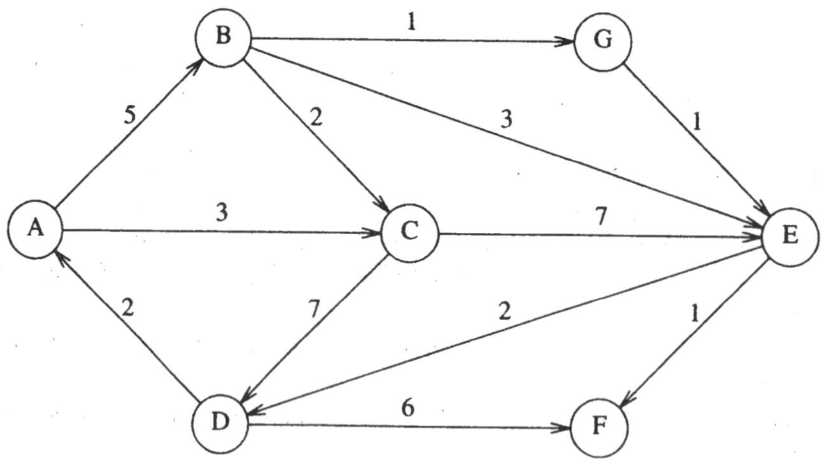

9.11 Find the maximum flow in the network of Figure 9.79.

9.12 Suppose that $G = (V, E)$ is a tree, s is the root, and we add a vertex t and edges of infinite capacity from all leaves in G to t. Give a linear-time algorithm to find a maximum flow from s to t.

9.13 A bipartite graph, $G = (V, E)$, is a graph such that V can be partitioned into two subsets V_1 and V_2 and no edge has both its vertices in the same subset.

a. Give a linear algorithm to determine whether a graph is bipartite.

b. The bipartite matching problem is to find the largest subset E' of E such that no vertex is included in more than one edge. A matching of four edges (indicated by dashed edges) is shown in Figure 9.81. There is a matching of five edges, which is maximum.

Show how the bipartite matching problem can be used to solve the following problem: We have a set of instructors, a set of courses, and a list of courses that each instructor is qualified to teach. If no instructor is required to teach more than one course, and only one instructor may teach a given course, what is the maximum number of courses that can be offered?

c. Show that the network flow problem can be used to solve the bipartite matching problem.

d. What is the time complexity of your solution to part (b)?

9.14 Give an algorithm to find an augmenting path that permits the maximum flow.

9.15 a. Find a minimum spanning tree for the graph in Figure 9.82 using both Prim's and Kruskal's algorithms.

b. Is this minimum spanning tree unique? Why?

9.16 Does either Prim's or Kruskal's algorithm work if there are negative edge weights?

9.17 Show that a graph of V vertices can have V^{V-2} minimum spanning trees.

9.18 Write a program to implement Kruskal's algorithm.

9.19 If all of the edges in a graph have weights between 1 and $|E|$, how fast can the minimum spanning tree be computed?

Figure 9.81 A bipartite graph

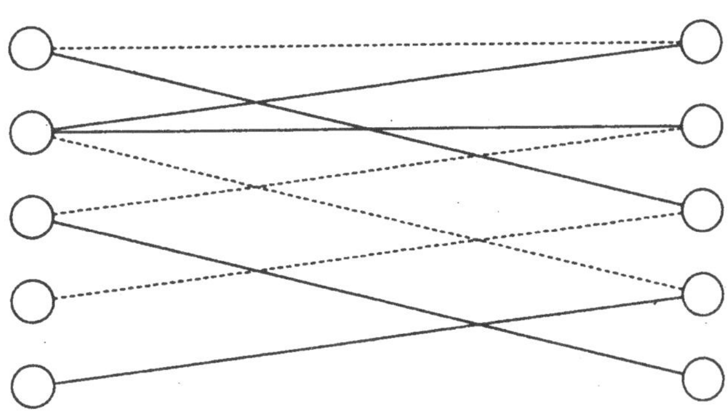

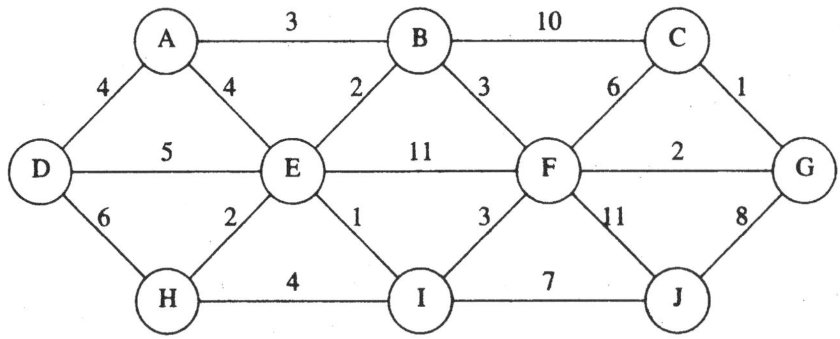

Figure 9.82

9.20 Give an algorithm to find a *maximum* spanning tree. Is this harder than finding a minimum spanning tree?

9.21 Find all the articulation points in the graph in Figure 9.83. Show the depth-first spanning tree and the values of *Num* and *Low* for each vertex.

9.22 Prove that the algorithm to find articulation points works.

9.23 a. Give an algorithm to find the minimum number of edges that need to be removed from an undirected graph so that the resulting graph is acyclic.

*b. Show that this problem is *NP*-complete for directed graphs.

9.24 Prove that in a depth-first spanning forest of a directed graph, all cross edges go from right to left.

9.25 Give an algorithm to decide whether an edge (v, w) in a depth-first spanning forest of a directed graph is a tree, back, cross, or forward edge.

9.26 Find the strongly connected components in the graph of Figure 9.84.

9.27 Write a program to find the strongly connected components in a digraph.

Figure 9.83

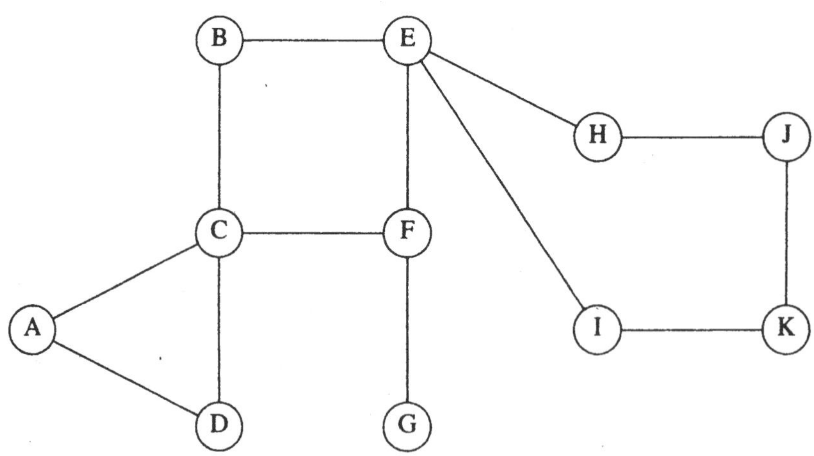

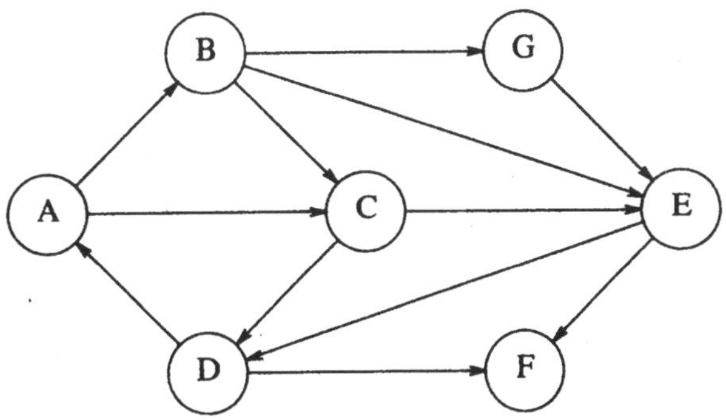

Figure 9.84

*9.28 Give an algorithm that finds the strongly connected components in only one depth-first search. Use an algorithm similar to the biconnectivity algorithm.

9.29 The *biconnected components* of a graph G is a partition of the edges into sets such that the graph formed by each set of edges is biconnected. Modify the algorithm in Figure 9.67 to find the biconnected components instead of the articulation points.

9.30 Suppose we perform a breadth-first search of an undirected graph and build a breadth-first spanning tree. Show that all edges in the tree are either tree edges or cross edges.

9.31 Give an algorithm to find in an undirected (connected) graph a path that goes through every edge exactly once in each direction.

9.32 a. Write a program to find an Euler circuit in a graph if one exists.

b. Write a program to find an Euler tour in a graph if one exists.

9.33 An Euler circuit in a directed graph is a cycle in which every edge is visited exactly once.

*a. Prove that a directed graph has an Euler circuit if and only if it is strongly connected and every vertex has equal indegree and outdegree.

*b. Give a linear-time algorithm to find an Euler circuit in a directed graph where one exists.

9.34 a. Consider the following solution to the Euler circuit problem: Assume that the graph is biconnected. Perform a depth-first search, taking back edges only as a last resort. If the graph is not biconnected, apply the algorithm recursively on the biconnected components. Does this algorithm work?

b. Suppose that when taking back edges, we take the back edge to the nearest ancestor. Does the algorithm work?

9.35 A planar graph is a graph that can be drawn in a plane without any two edges intersecting.

*a. Show that neither of the graphs in Figure 9.85 is planar.

b. Show that in a planar graph, there must exist some vertex which is connected to no more than five nodes.

**c. Show that in a planar graph, $|E| \leq 3|V| - 6$.

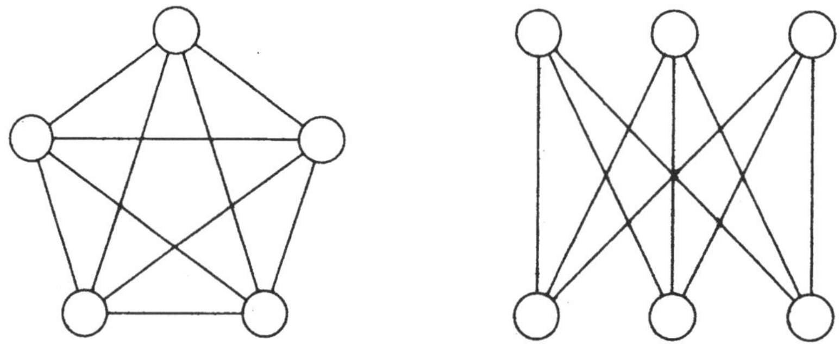

Figure 9.85

9.36 A *multigraph* is a graph in which multiple edges are allowed between pairs of vertices. Which of the algorithms in this chapter work without modification for multigraphs? What modifications need to be done for the others?

*9.37 Let $G = (V, E)$ be an undirected graph. Use depth-first search to design a linear algorithm to convert each edge in G to a directed edge such that the resulting graph is strongly connected, or determine that this is not possible.

9.38 You are given a set of N sticks, which are lying on top of each other in some configuration. Each stick is specified by its two endpoints; each endpoint is an ordered triple giving its x, y, and z coordinates; no stick is vertical. A stick may be picked up only if there is no stick on top of it.

a. Explain how to write a routine that takes two sticks a and b and reports whether a is above, below, or unrelated to b. (This has nothing to do with graph theory.)

b. Give an algorithm that determines whether it is possible to pick up all the sticks, and if so, provides a sequence of stick pickups that accomplishes this.

9.39 The *clique* problem can be stated as follows: Given an undirected graph $G = (V, E)$ and an integer K, does G contain a complete subgraph of at least K vertices?

The *vertex cover* problem can be stated as follows: Given an undirected graph $G = (V, E)$ and an integer K, does G contain a subset $V' \subset V$ such that $|V'| \leq K$ and every edge in G has a vertex in V'? Show that the clique problem is polynomially reducible to vertex cover.

9.40 Assume that the Hamiltonian cycle problem is NP-complete for undirected graphs.

a. Prove that the Hamiltonian cycle problem is NP-complete for directed graphs.

b. Prove that the unweighted simple longest-path problem is NP-complete for directed graphs.

9.41 The *baseball card collector* problem is as follows: Given packets $P_1, P_2, \ldots, P_M$, each of which contains a subset of the year's baseball cards, and an integer K, is it possible to collect all the baseball cards by choosing $\le K$ packets? Show that the baseball card collector problem is *NP*-complete.

References

Good graph theory textbooks include [8], [13], [22], and [37]. More advanced topics, including the more careful attention to running times, are covered in [39], [41], and [48].

Use of adjacency lists was advocated in [24]. The topological sort algorithm is from [29], as described in [34]. Dijkstra's algorithm appeared in [9]. The improvements using *d*-heaps and Fibonacci heaps are described in [28] and [15], respectively. The shortest-path algorithm with negative edge weights is due to Bellman [3]; Tarjan [48] describes a more efficient way to guarantee termination.

Ford and Fulkerson's seminal work on network flow is [14]. The idea of augmenting along shortest paths or on paths admitting the largest flow increase is from [12]. Other approaches to the problem can be found in [10], [32], [21], [6], and [33]. An algorithm for the min-cost flow problem can be found in [19].

An early minimum spanning tree algorithm can be found in [4]. Prim's algorithm is from [42]; Kruskal's algorithm appears in [35]. Two $O(|E| \log\log|V|)$ algorithms are [5] and [49]. The theoretically best-known algorithms appear in [15], [17], and [30]. An empirical study of these algorithms suggests that Prim's algorithm, implemented with *DecreaseKey*, is best in practice on most graphs [40].

The algorithm for biconnectivity is from [44]. The first linear-time strong components algorithm (Exercise 9.28) appears in the same paper. The algorithm presented in the text is due to Kosaraju (unpublished) and Sharir [43]. Other applications of depth-first search appear in [25], [26], [45], and [46] (as mentioned in Chapter 8, the results in [45] and [46] have been improved, but the basic algorithm is unchanged).

The classic reference work for the theory of *NP*-complete problems is [20]. Additional material can be found in [1]. The *NP*-completeness of satisfiability is shown in [7]. The other seminal paper is [31], which showed the *NP*-completeness of 21 problems. An excellent survey of complexity theory is [47]. An approximation algorithm for the traveling salesman problem, which generally gives nearly optimal results, can be found in [38].

A solution to Exercise 9.8 can be found in [2]. Solutions to the bipartite matching problem in Exercise 9.13 can be found in [23] and [36]. The problem can be generalized by adding weights to the edges and removing the restriction that the graph is bipartite. Efficient solutions for the unweighted matching problem for general graphs are quite complex. Details can be found in [11], [16], and [18].

Exercise 9.35 deals with planar graphs, which commonly arise in practice. Planar graphs are very sparse, and many difficult problems are easier on planar graphs. An example is the graph isomorphism problem, which is solvable in linear time for planar graphs [27]. No polynomial time algorithm is known for general graphs.

1. A. V. Aho, J. E. Hopcroft, and J. D. Ullman, *The Design and Analysis of Computer Algorithms,* Addison-Wesley, Reading, Mass., 1974.
2. R. K. Ahuja, K. Melhorn, J. B. Orlin, and R. E. Tarjan, "Faster Algorithms for the Shortest Path Problem," *Journal of the ACM,* 37 (1990), 213–223.
3. R. E. Bellman, "On a Routing Problem," *Quarterly of Applied Mathematics,* 16 (1958), 87–90.
4. O. Borůvka, "Ojistém problému minimálním (On a Minimal Problem)," *Práca Moravské Přírodovědecké Společnosti,* 3 (1926), 37–58.
5. D. Cheriton and R. E. Tarjan, "Finding Minimum Spanning Trees," *SIAM Journal on Computing,* 5 (1976), 724–742.
6. J. Cheriyan and T. Hagerup, "A Randomized Maximum-Flow Algorithm," *SIAM Journal on Computing,* 24 (1995), 203–226.
7. S. Cook, "The Complexity of Theorem Proving Procedures," *Proceedings of the Third Annual ACM Symposium on Theory of Computing* (1971), 151–158.
8. N. Deo, *Graph Theory with Applications to Engineering and Computer Science,* Prentice Hall, Englewood Cliffs, N.J., 1974.
9. E. W. Dijkstra, "A Note on Two Problems in Connexion with Graphs," *Numerische Mathematik,* 1 (1959), 269–271.
10. E. A. Dinic, "Algorithm for Solution of a Problem of Maximum Flow in Networks with Power Estimation," *Soviet Mathematics Doklady,* 11 (1970), 1277–1280.
11. J. Edmonds, "Paths, Trees, and Flowers," *Canadian Journal of Mathematics,* 17 (1965) 449–467.
12. J. Edmonds and R. M. Karp, "Theoretical Improvements in Algorithmic Efficiency for Network Flow Problems," *Journal of the ACM,* 19 (1972), 248–264.
13. S. Even, *Graph Algorithms,* Computer Science Press, Potomac, Md., 1979.
14. L. R. Ford, Jr., and D. R. Fulkerson, *Flows in Networks,* Princeton University Press, Princeton, N.J., 1962.
15. M. L. Fredman and R. E. Tarjan, "Fibonacci Heaps and Their Uses in Improved Network Optimization Algorithms," *Journal of the ACM,* 34 (1987), 596–615.
16. H. N. Gabow, "Data Structures for Weighted Matching and Nearest Common Ancestors with Linking," *Proceedings of First Annual ACM-SIAM Symposium on Discrete Algorithms* (1990), 434–443.
17. H. N. Gabow, Z. Galil, T. H. Spencer, and R. E. Tarjan, "Efficient Algorithms for Finding Minimum Spanning Trees on Directed and Undirected Graphs," *Combinatorica,* 6 (1986), 109–122.
18. Z. Galil, "Efficient Algorithms for Finding Maximum Matchings in Graphs," *ACM Computing Surveys,* 18 (1986), 23–38.
19. Z. Galil and E. Tardos, "An $O(n^2(m + n \log n) \log n)$ Min-Cost Flow Algorithm," *Journal of the ACM,* 35 (1988), 374–386.
20. M. R. Garey and D. S. Johnson, *Computers and Intractability: A Guide to the Theory of NP-Completeness,* Freeman, San Francisco, 1979.
21. A. V. Goldberg and R. E. Tarjan, "A New Approach to the Maximum-Flow Problem," *Journal of the ACM,* 35 (1988), 921–940.
22. F. Harary, *Graph Theory,* Addison-Wesley, Reading, Mass., 1969.
23. J. E. Hopcroft and R. M. Karp, "An $n^{5/2}$ Algorithm for Maximum Matchings in Bipartite Graphs," *SIAM Journal on Computing,* 2 (1973), 225–231.
24. J. E. Hopcroft and R. E. Tarjan, "Algorithm 447: Efficient Algorithms for Graph Manipulation," *Communications of the ACM,* 16 (1973), 372–378.
25. J. E. Hopcroft and R. E. Tarjan, "Dividing a Graph into Triconnected Components," *SIAM Journal on Computing,* 2 (1973), 135–158.

26. J. E. Hopcroft and R. E. Tarjan, "Efficient Planarity Testing," *Journal of the ACM,* 21 (1974), 549–568.
27. J. E. Hopcroft and J. K. Wong, "Linear Time Algorithm for Isomorphism of Planar Graphs," *Proceedings of the Sixth Annual ACM Symposium on Theory of Computing* (1974), 172–184.
28. D. B. Johnson, "Efficient Algorithms for Shortest Paths in Sparse Networks," *Journal of the ACM,* 24 (1977), 1–13.
29. A. B. Kahn, "Topological Sorting of Large Networks," *Communications of the ACM,* 5 (1962), 558–562.
30. D. R. Karger, P. N. Klein, and R. E. Tarjan, "A Randomized Linear-Time Algorithm to Find Minimum Spanning Trees," *Journal of the ACM,* 42 (1995), 321–328.
31. R. M. Karp, "Reducibility among Combinatorial Problems," *Complexity of Computer Computations* (eds. R. E. Miller and J. W. Thatcher), Plenum Press, New York, 1972, 85–103.
32. A. V. Karzanov, "Determining the Maximal Flow in a Network by the Method of Preflows," *Soviet Mathematics Doklady,* 15 (1974), 434–437.
33. V. King, S. Rao, and R. E. Tarjan, "A Faster Deterministic Maximum Flow Algorithm," *Journal of Algorithms,* 17 (1994), 447–474.
34. D. E. Knuth, *The Art of Computer Programming, Vol. 1: Fundamental Algorithms,* second edition, Addison-Wesley, Reading, Mass., 1973.
35. J. B. Kruskal, Jr., "On the Shortest Spanning Subtree of a Graph and the Traveling Salesman Problem," *Proceedings of the American Mathematical Society,* 7 (1956), 48–50.
36. H. W. Kuhn, "The Hungarian Method for the Assignment Problem," *Naval Research Logistics Quarterly,* 2 (1955), 83–97.
37. E. L. Lawler, *Combinatorial Optimization: Networks and Matroids,* Holt, Reinhart, and Winston, New York, 1976.
38. S. Lin and B. W. Kernighan, "An Effective Heuristic Algorithm for the Traveling Salesman Problem," *Operations Research,* 21 (1973), 498–516.
39. K. Melhorn, *Data Structures and Algorithms 2: Graph Algorithms and NP-completeness,* Springer-Verlag, Berlin, 1984.
40. B. M. E. Moret and H. D. Shapiro, "An Empirical Analysis of Algorithms for Constructing a Minimum Spanning Tree," *Proceedings of the Second Workshop on Algorithms and Data Structures* (1991), 400–411.
41. C. H. Papadimitriou and K. Steiglitz, *Combinatorial Optimization: Algorithms and Complexity,* Prentice Hall, Englewood Cliffs, N.J., 1982.
42. R. C. Prim, "Shortest Connection Networks and Some Generalizations," *Bell System Technical Journal,* 36 (1957), 1389–1401.
43. M. Sharir, "A Strong-Connectivity Algorithm and Its Application in Data Flow Analysis," *Computers and Mathematics with Applications,* 7 (1981), 67–72.
44. R. E. Tarjan, "Depth First Search and Linear Graph Algorithms," *SIAM Journal on Computing,* 1 (1972), 146–160.
45. R. E. Tarjan, "Testing Flow Graph Reducibility," *Journal of Computer and System Sciences,* 9 (1974), 355–365.
46. R. E. Tarjan, "Finding Dominators in Directed Graphs," *SIAM Journal on Computing,* 3 (1974), 62–89.
47. R. E. Tarjan, "Complexity of Combinatorial Algorithms," *SIAM Review,* 20 (1978), 457–491.
48. R. E. Tarjan, *Data Structures and Network Algorithms,* Society for Industrial and Applied Mathematics, Philadelphia, 1983.
49. A. C. Yao, "An $O(|E| \log \log |V|)$ Algorithm for Finding Minimum Spanning Trees," *Information Processing Letters,* 4 (1975), 21–23.

CHAPTER 10

Algorithm Design Techniques

So far, we have been concerned with the efficient implementation of algorithms. We have seen that when an algorithm is given, the actual data structures need not be specified. It is up to the programmer to choose the appropriate data structure in order to make the running time as small as possible.

In this chapter, we switch our attention from the *implementation* of algorithms to the *design* of algorithms. Most of the algorithms that we have seen so far are straightforward and simple. Chapter 9 contains some algorithms that are much more subtle, and some require an argument (in some cases lengthy) to show that they are indeed correct. In this chapter, we will focus on five of the common types of algorithms used to solve problems. For many problems, it is quite likely that at least one of these methods will work. Specifically, for each type of algorithm we will

- See the general approach.
- Look at several examples (the exercises at the end of the chapter provide many more examples).
- Discuss, in general terms, the time and space complexity, where appropriate.

10.1. Greedy Algorithms

The first type of algorithm we will examine is the *greedy algorithm*. We have already seen three greedy algorithms in Chapter 9: Dijkstra's, Prim's, and Kruskal's algorithms. Greedy algorithms work in phases. In each phase, a decision is made that appears to be good, without regard for future consequences. Generally, this means that some *local optimum* is chosen. This "take what you can get now" strategy is the source of the name for this class of algorithms. When the algorithm terminates, we hope that the local optimum is equal to the *global optimum*. If this is the case, then the algorithm is correct; otherwise, the algorithm has produced a suboptimal solution. If the absolute best answer is not required, then simple greedy algorithms are sometimes used to generate approximate answers, rather than using the more complicated algorithms generally required to generate an exact answer.

There are several real-life examples of greedy algorithms. The most obvious is the coin-changing problem. To make change in U.S. currency, we repeatedly dispense the largest denomination. Thus, to give out seventeen dollars and sixty-one cents in change, we give out a ten-dollar bill, a five-dollar bill, two one-dollar bills, two quarters, one dime, and one penny. By doing this, we are guaranteed to minimize the number of bills and coins. This algorithm does not work in all monetary systems, but fortunately, we can prove that it does work in the American monetary system. Indeed, it works even if two-dollar bills and fifty-cent pieces are allowed.

Traffic problems provide an example where making locally optimal choices does not always work. For example, during certain rush hour times in Miami, it is best to stay off the prime streets even if they look empty, because traffic will come to a standstill a mile down the road, and you will be stuck. Even more shocking, it is better in some cases to make a temporary detour in the direction opposite your destination in order to avoid all traffic bottlenecks.

In the remainder of this section, we will look at several applications that use greedy algorithms. The first application is a simple scheduling problem. Virtually all scheduling problems are either *NP*-complete (or of similar difficult complexity) or are solvable by a greedy algorithm. The second application deals with file compression and is one of the earliest results in computer science. Finally, we will look at an example of a greedy approximation algorithm.

10.1.1. A Simple Scheduling Problem

We are given jobs $j_1, j_2, \ldots, j_N$, all with known running times $t_1, t_2, \ldots, t_N$, respectively. We have a single processor. What is the best way to schedule these jobs in order to minimize the average completion time? In this entire section, we will assume *nonpreemptive scheduling:* Once a job is started, it must run to completion.

As an example, suppose we have the four jobs and associated running times shown in Figure 10.1. One possible schedule is shown in Figure 10.2. Because j_1 finishes in 15 (time units), j_2 in 23, j_3 in 26, and j_4 in 36, the average completion time is 25. A better schedule, which yields a mean completion time of 17.75, is shown in Figure 10.3.

The schedule given in Figure 10.3 is arranged by shortest job first. We can show that this will always yield an optimal schedule. Let the jobs in the schedule be j_{i_1}, $j_{i_2}, \ldots, j_{i_N}$. The first job finishes in time t_{i_1}. The second job finishes after $t_{i_1} + t_{i_2}$, and the third job finishes after $t_{i_1} + t_{i_2} + t_{i_3}$. From this, we see that the total cost, C, of the schedule is

$$C = \sum_{k=1}^{N}(N - k + 1)t_{i_k} \tag{10.1}$$

$$C = (N + 1)\sum_{k=1}^{N} t_{i_k} - \sum_{k=1}^{N} k \cdot t_{i_k} \tag{10.2}$$

Job	Time
j_1	15
j_2	8
j_3	3
j_4	10

Figure 10.1 Jobs and times

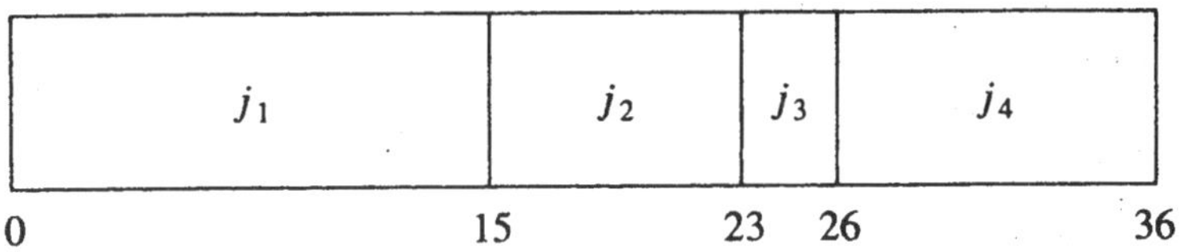

Figure 10.2 Schedule #1

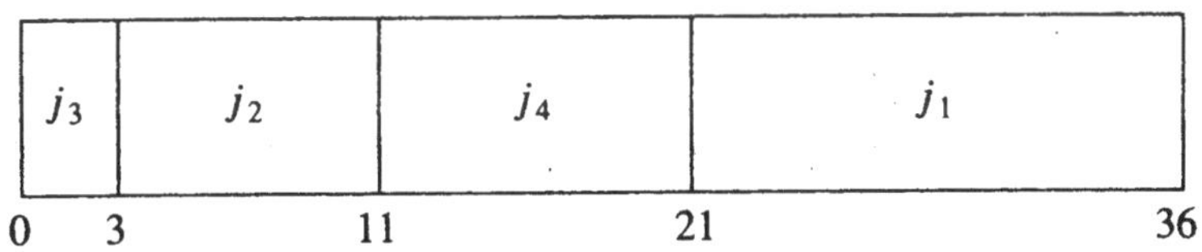

Figure 10.3 Schedule #2 (optimal)

Notice that in Equation (10.2), the first sum is independent of the job ordering, so only the second sum affects the total cost. Suppose that in an ordering there exists some $x > y$ such that $t_{i_x} < t_{i_y}$. Then a calculation shows that by swapping j_{i_x} and j_{i_y}, the second sum increases, decreasing the total cost. Thus, any schedule of jobs in which the times are not monotonically nondecreasing must be suboptimal. The only schedules left are those in which the jobs are arranged by smallest running time first, breaking ties arbitrarily.

This result indicates the reason the operating system scheduler generally gives precedence to shorter jobs.

The Multiprocessor Case

We can extend this problem to the case of several processors. Again we have jobs $j_1, j_2, \ldots, j_N$, with associated running times $t_1, t_2, \ldots, t_N$, and a number P of processors. We will assume without loss of generality that the jobs are ordered, shortest running time first. As an example, suppose $P = 3$, and the jobs are as shown in Figure 10.4.

Figure 10.5 shows an optimal arrangement to minimize mean completion time. Jobs j_1, j_4, and j_7 are run on Processor 1. Processor 2 handles j_2, j_5, and j_8, and Processor 3 runs the remaining jobs. The total time to completion is 165, for an average of $\frac{165}{9} = 18.33$.

The algorithm to solve the multiprocessor case is to start jobs in order, cycling through processors. It is not hard to show that no other ordering can do better,

Job	Time
j_1	3
j_2	5
j_3	6
j_4	10
j_5	11
j_6	14
j_7	15
j_8	18
j_9	20

Figure 10.4 Jobs and times

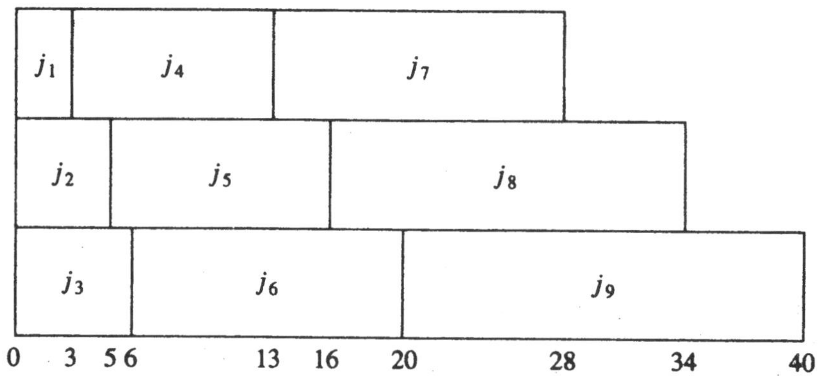

Figure 10.5 An optimal solution for the multiprocessor case

although if the number of processors P evenly divides the number of jobs N, there are many optimal orderings. This is obtained by, for each $0 \le i < N/P$, placing each of the jobs j_{iP+1} through $j_{(i+1)P}$ on a different processor. In our case, Figure 10.6 shows a second optimal solution.

Even if P does not divide N exactly, there can still be many optimal solutions, even if all the job times are distinct. We leave further investigation of this as an exercise.

Minimizing the Final Completion Time

We close this section by considering a very similar problem. Suppose we are only concerned with when the last job finishes. In our two examples above, these completion times are 40 and 38, respectively. Figure 10.7 shows that the minimum final completion time is 34, and this clearly cannot be improved, because every processor is always busy.

Although this schedule does not have minimum mean completion time, it has merit in that the completion time of the entire sequence is earlier. If the same user

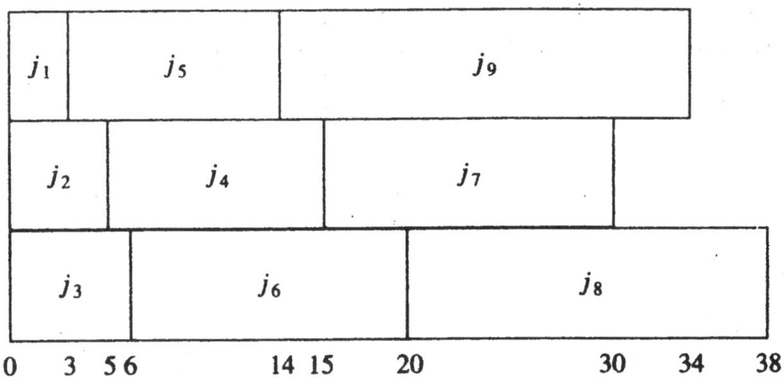

Figure 10.6 A second optimal solution for the multiprocessor case

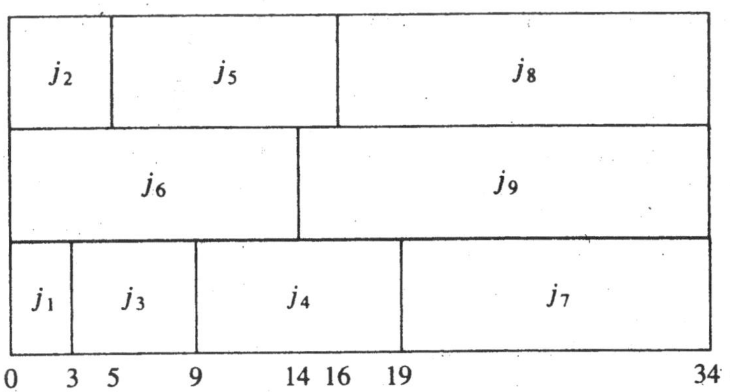

Figure 10.7 Minimizing the final completion time

owns all these jobs, then this is the preferable method of scheduling. Although these problems are very similar, this new problem turns out to be *NP*-complete; it is just another way of phrasing the knapsack or bin-packing problems, which we will encounter later in this section. Thus, minimizing the final completion time is apparently much harder than minimizing the mean completion time.

10.1.2. Huffman Codes

In this section, we consider a second application of greedy algorithms, known as *file compression*.

The normal ASCII character set consists of roughly 100 "printable" characters. In order to distinguish these characters, $\lceil \log 100 \rceil = 7$ bits are required. Seven bits allow the representation of 128 characters, so the ASCII character set adds some other "nonprintable" characters. An eighth bit is added as a parity check. The important point, however, is that if the size of the character set is C, then $\lceil \log C \rceil$ bits are needed in a standard encoding.

Suppose we have a file that contains only the characters *a, e, i, s, t,* plus blank spaces and *newlines*. Suppose further, that the file has ten *a*'s, fifteen *e*'s, twelve

Character	Code	Frequency	Total Bits
a	000	10	30
e	001	15	45
i	010	12	36
s	011	3	9
t	100	4	12
space	101	13	39
newline	110	1	3
Total			174

Figure 10.8 Using a standard coding scheme

i's, three *s*'s, four *t*'s, thirteen blanks, and one *newline*. As the table in Figure 10.8 shows, this file requires 174 bits to represent, since there are 58 characters and each character requires three bits.

In real life, files can be quite large. Many of the very large files are output of some program and there is usually a big disparity between the most frequent and least frequent characters. For instance, many large data files have an inordinately large amount of digits, blanks, and *newline*s, but few *q*'s and *x*'s. We might be interested in reducing the file size in the case where we are transmitting it over a slow phone line. Also, since on virtually every machine disk space is precious, one might wonder if it would be possible to provide a better code and reduce the total number of bits required.

The answer is that this is possible, and a simple strategy achieves 25 percent savings on typical large files and as much as 50 to 60 percent savings on many large data files. The general strategy is to allow the code length to vary from character to character and to ensure that the frequently occurring characters have short codes. Notice that if all the characters occur with the same frequency, then there are not likely to be any savings.

The binary code that represents the alphabet can be represented by the binary tree shown in Figure 10.9.

The tree in Figure 10.9 has data only at the leaves. The representation of each character can be found by starting at the root and recording the path, using a 0 to indicate the left branch and a 1 to indicate the right branch. For instance, *s* is reached by going left, then right, and finally right. This is encoded as 011. This data

Figure 10.9 Representation of the original code in a tree

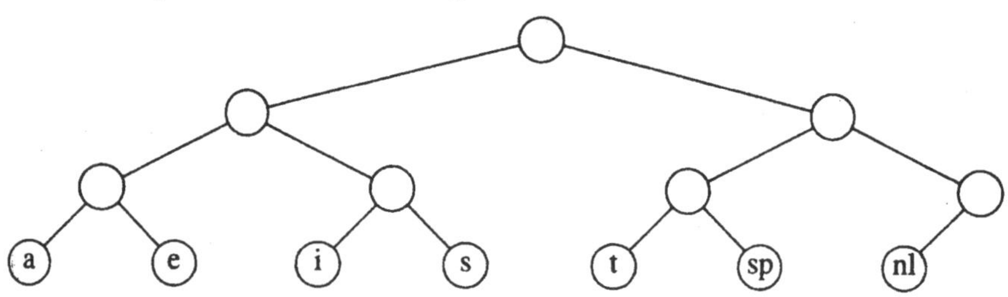

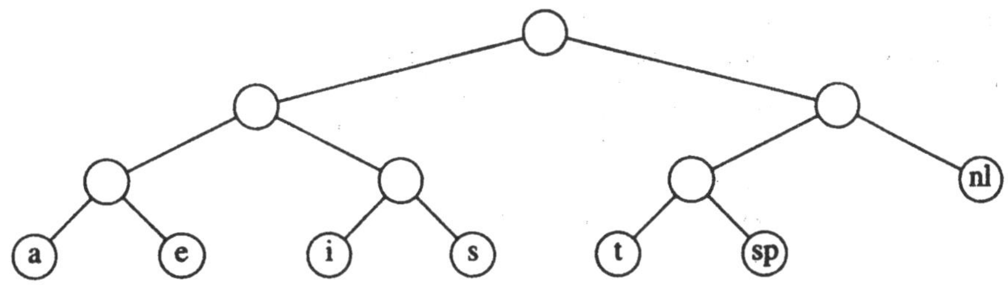

Figure 10.10 A slightly better tree

structure is sometimes referred to as a *trie*. If character c_i is at depth d_i and occurs f_i times, then the *cost* of the code is equal to $\sum d_i f_i$.

A better code than the one given in Figure 10.9 can be obtained by noticing that the *newline* is an only child. By placing the *newline* symbol one level higher at its parent, we obtain the new tree in Figure 10.10. This new tree has cost of 173, but is still far from optimal.

Notice that the tree in Figure 10.10 is a *full tree:* All nodes either are leaves or have two children. An optimal code will always have this property, since otherwise, as we have already seen, nodes with only one child could move up a level.

If the characters are placed only at the leaves, any sequence of bits can always be decoded unambiguously. For instance, suppose the encoded string is 0100111100010110001000111. 0 is not a character code, 01 is not a character code, but 010 represents *i*, so the first character is *i*. Then 011 follows, giving an *s*. Then 11 follows, which is a *newline*. The remainder of the code is *a, space, t, i, e,* and *newline*. Thus, it does not matter if the character codes are different lengths, as long as no character code is a prefix of another character code. Such an encoding is known as a *prefix code*. Conversely, if a character is contained in a nonleaf node, it is no longer possible to guarantee that the decoding will be unambiguous.

Putting these facts together, we see that our basic problem is to find the full binary tree of minimum total cost (as defined above), where all characters are contained in the leaves. The tree in Figure 10.11 shows the optimal tree for our sample alphabet. As can be seen in Figure 10.12, this code uses only 146 bits.

Figure 10.11 Optimal prefix code

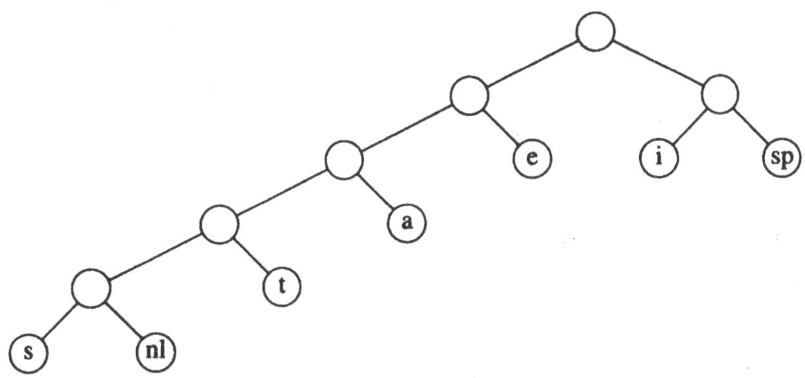

Character	Code	Frequency	Total Bits
a	001	10	30
e	01	15	30
i	10	12	24
s	00000	3	15
t	0001	4	16
space	11	13	26
newline	00001	1	5
Total			146

Figure 10.12 Optimal prefix code

Notice that there are many optimal codes. These can be obtained by swapping children in the encoding tree. The main unresolved question, then, is how the coding tree is constructed. The algorithm to do this was given by Huffman in 1952. Thus, this coding system is commonly referred to as a Huffman code.

Huffman's Algorithm

Throughout this section we will assume that the number of characters is C. Huffman's algorithm can be described as follows: We maintain a forest of trees. The *weight* of a tree is equal to the sum of the frequencies of its leaves. $C - 1$ times, select the two trees, T_1 and T_2, of smallest weight, breaking ties arbitrarily, and form a new tree with subtrees T_1 and T_2. At the beginning of the algorithm, there are C single-node trees—one for each character. At the end of the algorithm there is one tree, and this is the optimal Huffman coding tree.

A worked example will make the operation of the algorithm clear. Figure 10.13 shows the initial forest; the weight of each tree is shown in small type at the root. The two trees of lowest weight are merged together, creating the forest shown in Figure 10.14. We will name the new root $T1$, so that future merges can be stated unambiguously. We have made s the left child arbitrarily; any tiebreaking procedure can be used. The total weight of the new tree is just the sum of the weights of the

Figure 10.13 Initial stage of Huffman's algorithm

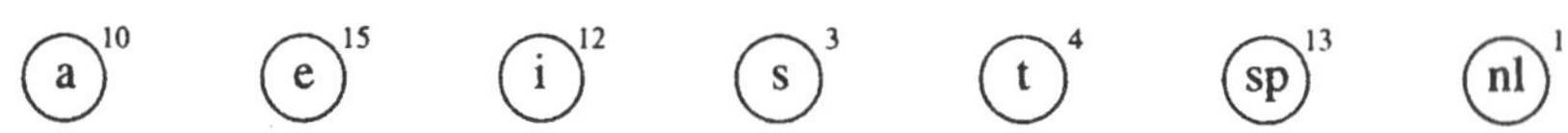

Figure 10.14 Huffman's algorithm after the first merge

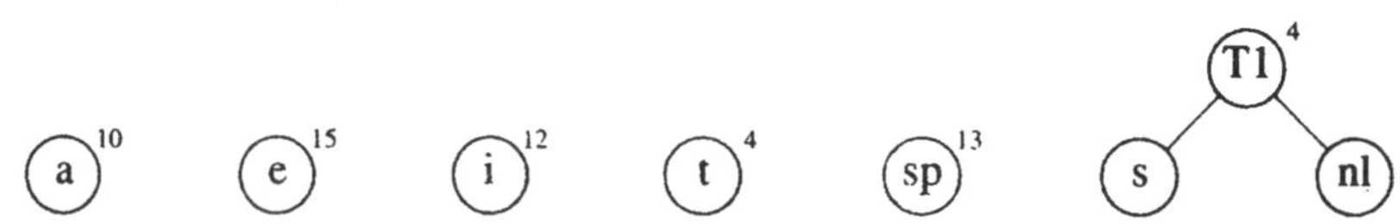

Figure 10.15 Huffman's algorithm after the second merge

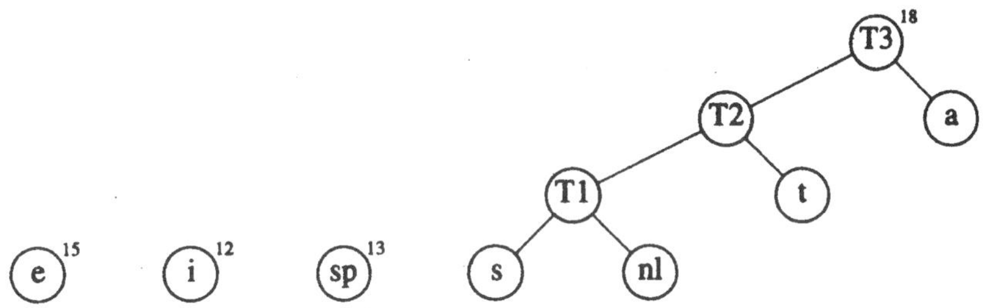

Figure 10.16 Huffman's algorithm after the third merge

old trees, and can thus be easily computed. It is also a simple matter to create the new tree, since we merely need to get a new node, set the left and right pointers, and record the weight.

Now there are six trees, and we again select the two trees of smallest weight. These happen to be $T1$ and t, which are then merged into a new tree with root $T2$ and weight 8. This is shown in Figure 10.15. The third step merges $T2$ and a, creating $T3$, with weight $10 + 8 = 18$. Figure 10.16 shows the result of this operation.

After the third *merge* is completed, the two trees of lowest weight are the single-node trees representing i and the blank space. Figure 10.17 shows how these trees are merged into the new tree with root $T4$. The fifth step is to merge the trees with roots e and $T3$, since these trees have the two smallest weights. The result of this step is shown in Figure 10.18.

Finally, the optimal tree, which was shown in Figure 10.11, is obtained by merging the two remaining trees. Figure 10.19 shows this optimal tree, with root $T6$.

Figure 10.17 Huffman's algorithm after the fourth merge

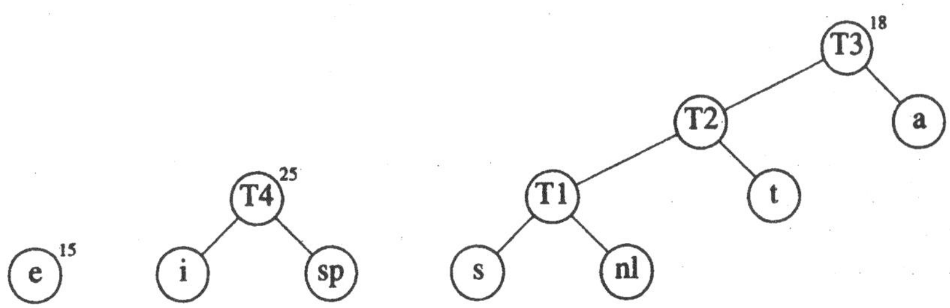

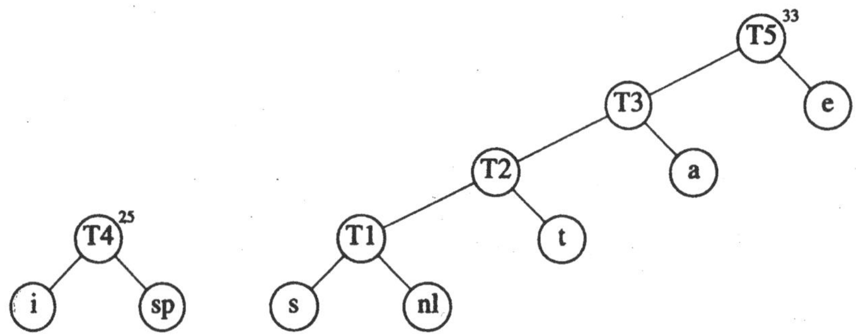

Figure 10.18 Huffman's algorithm after the fifth merge

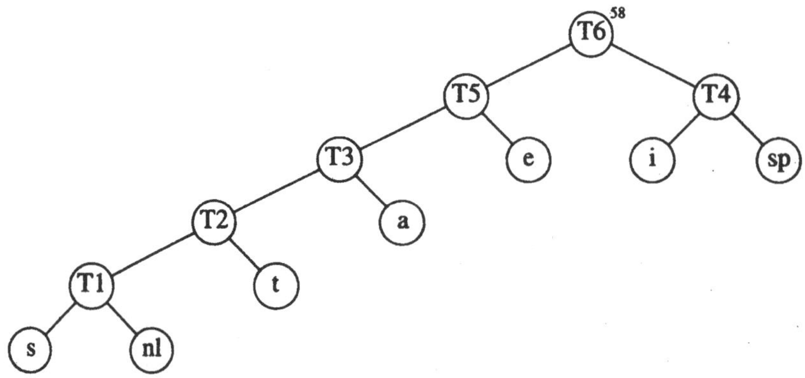

Figure 10.19 Huffman's algorithm after the final merge

We will sketch the ideas involved in proving that Huffman's algorithm yields an optimal code; we will leave the details as an exercise. First, it is not hard to show by contradiction that the tree must be full, since we have already seen how a tree that is not full is improved.

Next, we must show that the two least frequent characters α and β must be the two deepest nodes (although other nodes may be as deep). Again, this is easy to show by contradiction, since if either α or β is not a deepest node, then there must be some γ that is (recall that the tree is full). If α is less frequent than γ, then we can improve the cost by swapping them in the tree.

We can then argue that the characters in any two nodes at the same depth can be swapped without affecting optimality. This shows that an optimal tree can always be found that contains the two least frequent symbols as siblings; thus the first step is not a mistake.

The proof can be completed by using an induction argument. As trees are merged, we consider the new character set to be the characters in the roots. Thus, in our example, after four merges, we can view the character set as consisting of e and the metacharacters $T3$ and $T4$. This is probably the trickiest part of the proof; you are urged to fill in all of the details.

The reason that this is a greedy algorithm is that at each stage we perform a merge without regard to global considerations. We merely select the two smallest trees.

If we maintain the trees in a priority queue, ordered by weight, then the running time is $O(C \log C)$, since there will be one *BuildHeap*, $2C - 2$ *DeleteMins*, and $C - 2$ *Inserts*, on a priority queue that never has more than C elements. A simple implementation of the priority queue, using a linked list, would give an $O(C^2)$ algorithm. The choice of priority queue implementation depends on how large C is. In the typical case of an ASCII character set, C is small enough that the quadratic running time is acceptable. In such an application, virtually all the running time will be spent on the disk I/O required to read the input file and write out the compressed version.

There are two details that must be considered. First, the encoding information must be transmitted at the start of the compressed file, since otherwise it will be impossible to decode. There are several ways of doing this; see Exercise 10.4. For small files, the cost of transmitting this table will override any possible savings in compression, and the result will probably be file expansion. Of course, this can be detected and the original left intact. For large files, the size of the table is not significant.

The second problem is that as described, this is a two-pass algorithm. The first pass collects the frequency data and the second pass does the encoding. This is obviously not a desirable property for a program dealing with large files. Some alternatives are described in the references.

10.1.3. Approximate Bin Packing

In this section, we will consider some algorithms to solve the *bin packing* problem. These algorithms will run quickly but will not necessarily produce optimal solutions. We will prove, however, that the solutions that are produced are not too far from optimal.

We are given N items of sizes $s_1, s_2, \ldots, s_N$. All sizes satisfy $0 < s_i \leq 1$. The problem is to pack these items in the fewest number of bins, given that each bin has unit capacity. As an example, Figure 10.20 shows an optimal packing for an item list with sizes 0.2, 0.5, 0.4, 0.7, 0.1, 0.3, 0.8.

Figure 10.20 Optimal packing for 0.2, 0.5, 0.4, 0.7, 0.1, 0.3, 0.8

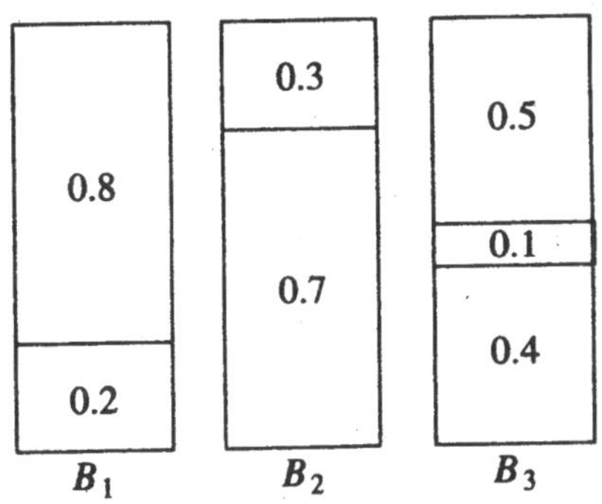

There are two versions of the bin packing problem. The first version is *on-line* bin packing. In this version, each item must be placed in a bin before the next item can be processed. The second version is the *off-line* bin packing problem. In an off-line algorithm, we do not need to do anything until all the input has been read. The distinction between on-line and off-line algorithms was discussed in Section 8.2.

On-line Algorithms

The first issue to consider is whether or not an on-line algorithm can actually always give an optimal answer, even if it is allowed unlimited computation. Remember that even though unlimited computation is allowed, an on-line algorithm must place an item before processing the next item and cannot change its decision.

To show that an on-line algorithm cannot always give an optimal solution, we will give it particularly difficult data to work on. Consider an input sequence I_1 of M small items of weight $\frac{1}{2} - \epsilon$ followed by M large items of weight $\frac{1}{2} + \epsilon$, $0 < \epsilon < 0.01$. It is clear that these items can be packed in M bins if we place one small item and one large item in each bin. Suppose there were an optimal on-line algorithm A that could perform this packing. Consider the operation of algorithm A on the sequence I_2, consisting of only M small items of weight $\frac{1}{2} - \epsilon$. I_2 can be packed in $\lceil M/2 \rceil$ bins. However, A will place each item in a separate bin, since A must yield the same results on I_2 as it does for the first half of I_1, and the first half of I_1 is exactly the same input as I_2. This means that A will use twice as many bins as is optimal for I_2. What we have proven is that there is no optimal algorithm for on-line bin packing.

What the argument above shows is that an on-line algorithm never knows when the input might end, so any performance guarantees it provides must hold at every instant throughout the algorithm. If we follow the foregoing strategy, we can prove the following.

THEOREM 10.1.

There are inputs that force any on-line bin-packing algorithm to use at least $\frac{4}{3}$ the optimal number of bins.

PROOF:

Suppose otherwise, and suppose for simplicity that M is even. Consider any on-line algorithm A running on the input sequence I_1, above. Recall that this sequence consists of M small items followed by M large items. Let us consider what the algorithm A has done after processing the Mth item. Suppose A has already used b bins. At this point in the algorithm, the optimal number of bins is $M/2$, because we can place two elements in each bin. Thus we know that $2b/M < \frac{4}{3}$, by our assumption of a better-than-$\frac{4}{3}$ performance guarantee.

Now consider the performance of algorithm A after all items have been packed. All bins created after the bth bin must contain exactly one item, since all small items are placed in the first b bins, and two large items will not fit in a bin. Since the first b bins can have at most two items each, and the remaining bins have one item each, we see that packing $2M$ items will require at least

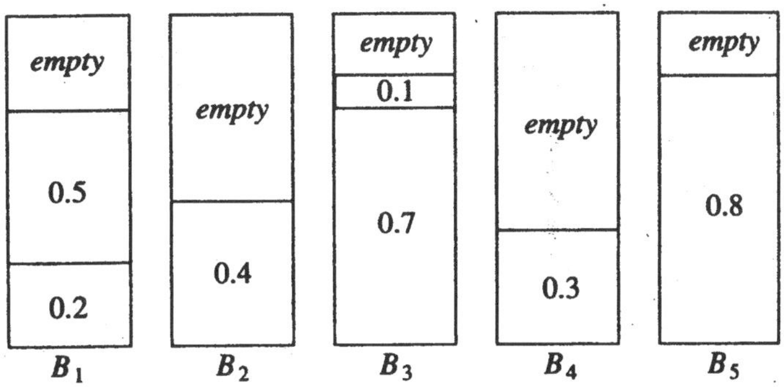

Figure 10.21 Next fit for 0.2, 0.5, 0.4, 0.7, 0.1, 0.3, 0.8

$2M - b$ bins. Since the $2M$ items can be optimally packed using M bins, our performance guarantee assures us that $(2M - b)/M < \frac{4}{3}$.

The first inequality implies that $b/M < \frac{2}{3}$, and the second inequality implies that $b/M > \frac{2}{3}$, which is a contradiction. Thus, no on-line algorithm can guarantee that it will produce a packing with less than $\frac{4}{3}$ the optimal number of bins.

There are three simple algorithms that guarantee that the number of bins used is no more than twice optimal. There are also quite a few more complicated algorithms with better guarantees.

Next Fit

Probably the simplest algorithm is *next fit.* When processing any item, we check to see whether it fits in the same bin as the last item. If it does, it is placed there; otherwise, a new bin is created. This algorithm is incredibly simple to implement and runs in linear time. Figure 10.21 shows the packing produced for the same input as Figure 10.20.

Not only is next fit simple to program, its worst-case behavior is also easy to analyze.

THEOREM 10.2.

Let M be the optimal number of bins required to pack a list I of items. Then next fit never uses more than $2M$ bins. There exist sequences such that next fit uses $2M - 2$ bins.

PROOF:

Consider any adjacent bins B_j and B_{j+1}. The sum of the sizes of all items in B_j and B_{j+1} must be larger than 1, since otherwise all of these items would have been placed in B_j. If we apply this result to all pairs of adjacent bins, we see that at most half of the space is wasted. Thus next fit uses at most twice the number of bins.

To see that this bound is tight, suppose that the N items have size $s_i = 0.5$ if i is odd and $s_i = 2/N$ if i is even. Assume N is divisible by 4. The optimal packing, shown in Figure 10.22, consists of $N/4$ bins, each containing 2 elements

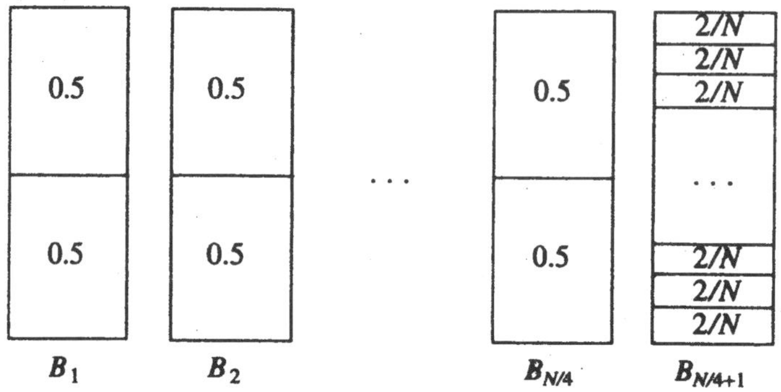

Figure 10.22 Optimal packing for 0.5, $2/N$, 0.5, $2/N$, 0.5, $2/N$, . . .

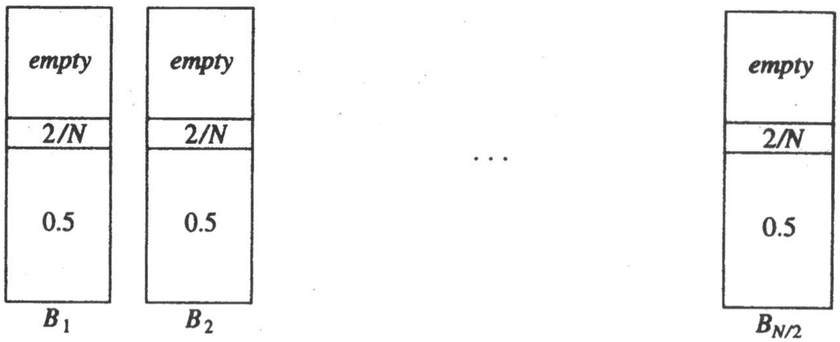

Figure 10.23 Next fit packing for 0.5, $2/N$, 0.5, $2/N$, 0.5, $2/N$, . . .

of size 0.5, and one bin containing the $N/2$ elements of size $2/N$, for a total of $(N/4) + 1$. Figure 10.23 shows that next fit uses $N/2$ bins. Thus, next fit can be forced to use almost twice as many bins as optimal.

First Fit

Although next fit has a reasonable performance guarantee, it performs poorly in practice, because it creates new bins when it does not need to. In the sample run, it could have placed the item of size 0.3 in either B_1 or B_2, rather than create a new bin.

The *first fit* strategy is to scan the bins in order and place the new item in the first bin that is large enough to hold it. Thus, a new bin is created only when the results of previous placements have left no other alternative. Figure 10.24 shows the packing that results from first fit on our standard input.

A simple method of implementing first fit would process each item by scanning down the list of bins sequentially. This would take $O(N^2)$. It is possible to implement first fit to run in $O(N \log N)$; we leave this as an exercise.

A moment's thought will convince you that at any point, at most one bin can be more than half empty, since if a second bin were also half empty, its contents would fit into the first bin. Thus, we can immediately conclude that first fit guarantees a solution with at most twice the optimal number of bins.

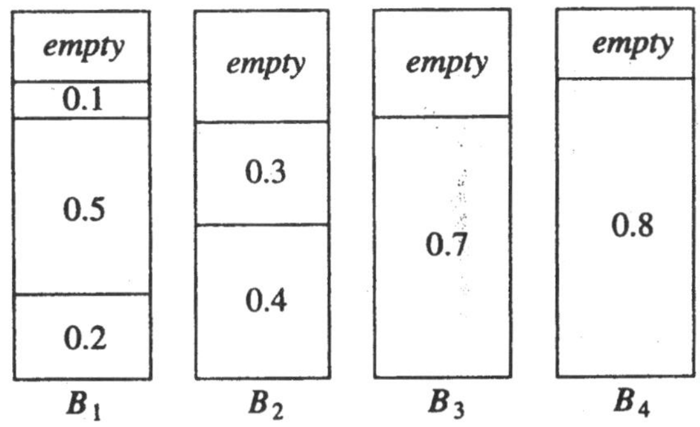

Figure 10.24 First fit for 0.2, 0.5, 0.4, 0.7, 0.1, 0.3, 0.8

On the other hand, the bad case that we used in the proof of next fit's performance bound does not apply for first fit. Thus, one might wonder if a better bound can be proven. The answer is yes, but the proof is complicated.

THEOREM 10.3.

Let M be the optimal number of bins required to pack a list I of items. Then first fit never uses more than $\lceil \frac{17}{10}M \rceil$ bins. There exist sequences such that first fit uses $\frac{17}{10}(M - 1)$ bins.

PROOF:

See the references at the end of the chapter.

An example where first fit does almost as poorly as the previous theorem would indicate is shown in Figure 10.25. The input consists of $6M$ items of size $\frac{1}{7} + \epsilon$, followed by $6M$ items of size $\frac{1}{3} + \epsilon$, followed by $6M$ items of size $\frac{1}{2} + \epsilon$. One simple packing places one item of each size in a bin and requires $6M$ bins. First fit requires $10M$ bins.

When first fit is run on a large number of items with sizes uniformly distributed between 0 and 1, empirical results show that first fit uses roughly 2 percent more bins than optimal. In many cases, this is quite acceptable.

Figure 10.25 A case where first fit uses $10M$ bins instead of $6M$

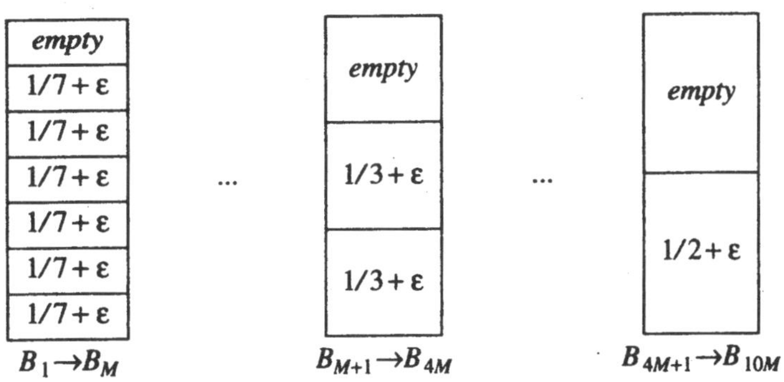

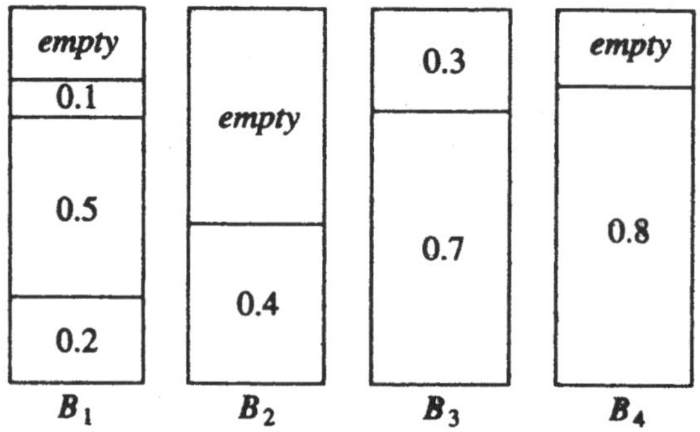

Figure 10.26 Best fit for 0.2, 0.5, 0.4, 0.7, 0.1, 0.3, 0.8

Best Fit

The third on-line strategy we will examine is *best fit*. Instead of placing a new item in the first spot that is found, it is placed in the tightest spot among all bins. A typical packing is shown in Figure 10.26.

Notice that the item of size 0.3 is placed in B_3, where it fits perfectly, instead of B_2. One might expect that since we are now making a more educated choice of bins, the performance guarantee would improve. This is not the case, because the generic bad cases are the same. Best fit is never more than roughly 1.7 times as bad as optimal, and there are inputs for which it (nearly) achieves this bound. Nevertheless, best fit is also simple to code, especially if an $O(N \log N)$ algorithm is required, and it does perform better for random inputs.

Off-line Algorithms

If we are allowed to view the entire item list before producing an answer, then we should expect to do better. Indeed, since we can eventually find the optimal packing by exhaustive search, we already have a theoretical improvement over the on-line case.

The major problem with all the on-line algorithms is that it is hard to pack the large items, especially when they occur late in the input. The natural way around this is to sort the items, placing the largest items first. We can then apply first fit or best fit, yielding the algorithms *first fit decreasing* and *best fit decreasing*, respectively. Figure 10.27 shows that in our case this yields an optimal solution (although, of course, this is not true in general).

In this section, we will deal with first fit decreasing. The results for best fit decreasing are almost identical. Since it is possible that the item sizes are not distinct, some authors prefer to call the algorithm *first fit nonincreasing*. We will stay with the original name. We will also assume, without loss of generality, that input sizes are already sorted.

The first remark we can make is that the bad case, which showed first fit using $10M$ bins instead of $6M$ bins, does not apply when the items are sorted. We will show that if an optimal packing uses M bins, then first fit decreasing never uses more than $(4M + 1)/3$ bins.

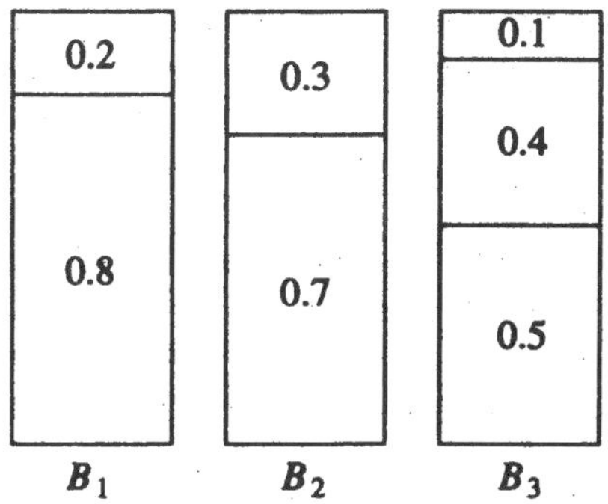

Figure 10.27 First fit for 0.8, 0.7, 0.5, 0.4, 0.3, 0.2, 0.1

The result depends on two observations. First, all the items with weight larger than $\frac{1}{3}$ will be placed in the first M bins. This implies that all the items in the extra bins have weight at most $\frac{1}{3}$. The second observation is that the number of items in the extra bins can be at most $M - 1$. Combining these two results, we find that at most $\lceil (M - 1)/3 \rceil$ extra bins can be required. We now prove these two observations.

LEMMA 10.1.

Let the N items have (sorted in decreasing order) input sizes $s_1, s_2, \ldots, s_N$, respectively, and suppose that the optimal packing is M bins. Then all items that first fit decreasing places in extra bins have size at most $\frac{1}{3}$.

PROOF:

Suppose the ith item is the first placed in bin $M + 1$. We need to show that $s_i \leq \frac{1}{3}$. We will prove this by contradiction. Assume $s_i > \frac{1}{3}$.

It follows that $s_1, s_2, \ldots, s_{i-1} > \frac{1}{3}$, since the sizes are arranged in sorted order. From this it follows that all bins $B_1, B_2, \ldots, B_M$ have at most two items each.

Consider the state of the system after the $i - 1$st item is placed in a bin, but before the ith item is placed. We now want to show that (under the assumption that $s_i > \frac{1}{3}$) the first M bins are arranged as follows: First there are some bins with exactly one element, and then the remaining bins have two elements.

Suppose there were two bins B_x and B_y, such that $1 \leq x < y \leq M$, B_x has two items, and B_y has one item. Let x_1 and x_2 be the two items in B_x, and let y_1 be the item in B_y. $x_1 \geq y_1$, since x_1 was placed in the earlier bin. $x_2 \geq s_i$, by similar reasoning. Thus, $x_1 + x_2 \geq y_1 + s_i$. This implies that s_i could be placed in B_y. By our assumption this is not possible. Thus, if $s_i > \frac{1}{3}$, then, at the time that we try to process s_i, the first M bins are arranged such that the first j have one element and the next $M - j$ have two elements.

To prove the lemma we will show that there is no way to place all the items in M bins, which contradicts the premise of the lemma.

Clearly, no two items $s_1, s_2, \ldots, s_j$ can be placed in one bin, by any algorithm, since if they could, first fit would have done so too. We also know that first fit has not placed any of the items of size $s_{j+1}, s_{j+2}, \ldots, s_i$ into the first j bins, so none of them fit. Thus, in any packing, specifically the optimal packing, there must be j bins that do not contain these items. It follows that the items of size $s_{j+1}, s_{j+2}, \ldots, s_{i-1}$ must be contained in some set of $M - j$ bins, and from previous considerations, the total number of such items is $2(M - j)$.*

The proof is completed by noting that if $s_i > \frac{1}{3}$, there is no way for s_i to be placed in one of these M bins. Clearly, it cannot go in one of the j bins, since if it could, then first fit would have done so too. To place it in one of the remaining $M - j$ bins requires distributing $2(M - j) + 1$ items into the $M - j$ bins. Thus, some bin would have to have three items, each of which is larger than $\frac{1}{3}$, a clear impossibility.

This contradicts the fact that all the sizes can be placed in M bins, so the original assumption must be incorrect. Thus, $s_i \le \frac{1}{3}$.

LEMMA 10.2.

The number of objects placed in extra bins is at most $M - 1$.

PROOF:

Assume that there are at least M objects placed in extra bins. We know that $\sum_{i=1}^{N} s_i \le M$, since all the objects fit in M bins. Suppose that B_j is filled with W_j total weight for $1 \le j \le M$. Suppose the first M extra objects have sizes $x_1, x_2, \ldots, x_M$. Then, since the items in the first M bins plus the first M extra items are a subset of all the items, it follows that

$$\sum_{i=1}^{N} s_i \ge \sum_{j=1}^{M} W_j + \sum_{j=1}^{M} x_j \ge \sum_{j=1}^{M} (W_j + x_j)$$

Now $W_j + x_j > 1$, since otherwise the item corresponding to x_j would have been placed in B_j. Thus

$$\sum_{i=1}^{N} s_i > \sum_{j=1}^{M} 1 > M$$

But this is impossible if the N items can be packed in M bins. Thus, there can be at most $M - 1$ extra items.

THEOREM 10.4.

Let M be the optimal number of bins required to pack a list I of items. Then first fit decreasing never uses more than $(4M + 1)/3$ *bins.*

PROOF:

There are $M - 1$ extra items, of size at most $\frac{1}{3}$. Thus, there can be at most $\lceil (M - 1)/3 \rceil$ extra bins. The total number of bins used by first fit decreasing is thus at most $\lceil (4M - 1)/3 \rceil \le (4M + 1)/3$.

It is possible to prove a much tighter bound for both first fit decreasing and next fit decreasing.

---

*Recall that first fit packed these elements into $M - j$ bins and placed two items in each bin. Thus, there are $2(M - j)$ items.

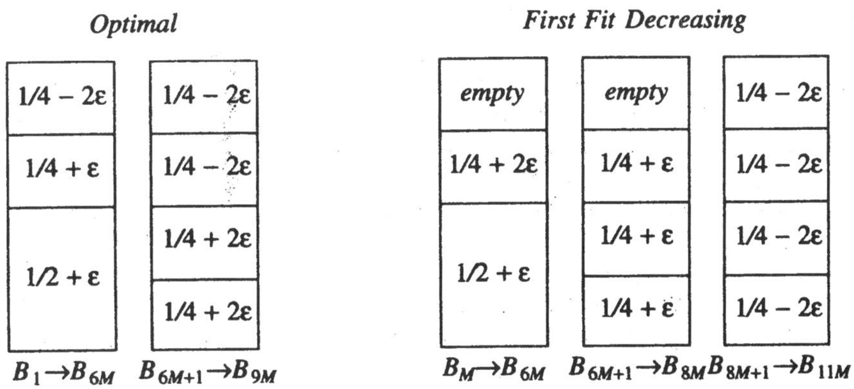

Figure 10.28 Example where first fit decreasing uses $11M$ bins, but only $9M$ bins are required

THEOREM 10.5.

Let M be the optimal number of bins required to pack a list I of items. Then first fit decreasing never uses more than $\frac{11}{9}M + 4$ bins. There exist sequences such that first fit decreasing uses $\frac{11}{9}M$ bins.

PROOF:

The upper bound requires a very complicated analysis. The lower bound is exhibited by a sequence consisting of $6M$ elements of size $\frac{1}{2} + \epsilon$, followed by $6M$ elements of size $\frac{1}{4} + 2\epsilon$, followed by $6M$ elements of size $\frac{1}{4} + \epsilon$, followed by $12M$ elements of size $\frac{1}{4} - 2\epsilon$. Figure 10.28 shows that the optimal packing requires $9M$ bins, but first fit decreasing uses $11M$ bins.

In practice, first fit decreasing performs extremely well. If sizes are chosen uniformly over the unit interval, then the expected number of extra bins is $\Theta(\sqrt{M})$. Bin packing is a fine example of how simple greedy heuristics can give good results.

10.2. Divide and Conquer

Another common technique used to design algorithms is *divide and conquer*. Divide and conquer algorithms consist of two parts:

Divide: Smaller problems are solved recursively (except, of course, base cases).

Conquer: The solution to the original problem is then formed from the solutions to the subproblems.

Traditionally, routines in which the text contains at least two recursive calls are called divide and conquer algorithms, while routines whose text contains only one recursive call are not. We generally insist that the subproblems be disjoint (that is, essentially nonoverlapping). Let us review some of the recursive algorithms that have been covered in this text.

We have already seen several divide and conquer algorithms. In Section 2.4.3, we saw an $O(N \log N)$ solution to the maximum subsequence sum problem. In Chapter 4, we saw linear-time tree traversal strategies. In Chapter 7, we saw the classic examples of divide and conquer, namely mergesort and quicksort, which have $O(N \log N)$ worst-case and average-case bounds, respectively.

We have also seen several examples of recursive algorithms that probably do not classify as divide and conquer, but merely reduce to a single simpler case. In Section 1.3, we saw a simple routine to print a number. In Chapter 2, we used recursion to perform efficient exponentiation. In Chapter 4, we examined simple search routines for binary search trees. In Section 6.6, we saw simple recursion used to merge leftist heaps. In Section 7.7, an algorithm was given for selection that takes linear average time. The disjoint set *Find* operation was written recursively in Chapter 8. Chapter 9 showed routines to recover the shortest path in Dijkstra's algorithm and other procedures to perform depth-first search in graphs. None of these algorithms are really divide and conquer algorithms, because only one recursive call is performed.

We have also seen, in Section 2.4, a very bad recursive routine to compute the Fibonacci numbers. This could be called a divide and conquer algorithm, but it is terribly inefficient, because the problem really is not divided at all.

In this section, we will see more examples of the divide and conquer paradigm. Our first application is a problem in *computational geometry*. Given N points in a plane, we will show that the closest pair of points can be found in $O(N \log N)$ time. The exercises describe some other problems in computational geometry which can be solved by divide and conquer. The remainder of the section shows some extremely interesting, but mostly theoretical, results. We provide an algorithm which solves the selection problem in $O(N)$ worst-case time. We also show that 2 N-bit numbers can be multiplied in $o(N^2)$ operations and that two $N \times N$ matrices can be multiplied in $o(N^3)$ operations. Unfortunately, even though these algorithms have better worst-case bounds than the conventional algorithms, none are practical except for very large inputs.

10.2.1. Running Time of Divide and Conquer Algorithms

All the efficient divide and conquer algorithms we will see divide the problems into subproblems, each of which is some fraction of the original problem, and then perform some additional work to compute the final answer. As an example, we have seen that mergesort operates on two problems, each of which is half the size of the original, and then uses $O(N)$ additional work. This yields the running time equation (with appropriate initial conditions)

$$T(N) = 2T(N/2) + O(N)$$

We saw in Chapter 7 that the solution to this equation is $O(N \log N)$. The following theorem can be used to determine the running time of most divide and conquer algorithms.

THEOREM 10.6.

The solution to the equation $T(N) = aT(N/b) + \Theta(N^k)$, *where* $a \geq 1$ *and* $b > 1$, *is*

$$T(N) = \begin{cases} O(N^{\log_b a}) & \text{if } a > b^k \\ O(N^k \log N) & \text{if } a = b^k \\ O(N^k) & \text{if } a < b^k \end{cases}$$

PROOF:

Following the analysis of mergesort in Chapter 7, we will assume that N is a power of b; thus, let $N = b^m$. Then $N/b = b^{m-1}$ and $N^k = (b^m)^k = b^{mk} = b^{km} = (b^k)^m$. Let us assume $T(1) = 1$, and ignore the constant factor in $\Theta(N^k)$. Then we have

$$T(b^m) = aT(b^{m-1}) + (b^k)^m$$

If we divide through by a^m, we obtain the equation

$$\frac{T(b^m)}{a^m} = \frac{T(b^{m-1})}{a^{m-1}} + \left\{\frac{b^k}{a}\right\}^m \tag{10.3}$$

We can apply this equation for other values of m, obtaining

$$\frac{T(b^{m-1})}{a^{m-1}} = \frac{T(b^{m-2})}{a^{m-2}} + \left\{\frac{b^k}{a}\right\}^{m-1} \tag{10.4}$$

$$\frac{T(b^{m-2})}{a^{m-2}} = \frac{T(b^{m-3})}{a^{m-3}} + \left\{\frac{b^k}{a}\right\}^{m-2} \tag{10.5}$$

$$\dots$$

$$\frac{T(b^1)}{a^1} = \frac{T(b^0)}{a^0} + \left\{\frac{b^k}{a}\right\}^1 \tag{10.6}$$

We use our standard trick of adding up the telescoping equations (10.3) through (10.6). Virtually all the terms on the left cancel the leading terms on the right, yielding

$$\frac{T(b^m)}{a^m} = 1 + \sum_{i=1}^{m}\left\{\frac{b^k}{a}\right\}^i \tag{10.7}$$

$$= \sum_{i=0}^{m}\left\{\frac{b^k}{a}\right\}^i \tag{10.8}$$

Thus

$$T(N) = T(b^m) = a^m \sum_{i=0}^{m}\left\{\frac{b^k}{a}\right\}^i \tag{10.9}$$

If $a > b^k$, then the sum is a geometric series with ratio smaller than 1. Since the sum of infinite series would converge to a constant, this finite sum is also bounded by a constant, and thus Equation (10.10) applies:

$$T(N) = O(a^m) = O(a^{\log_b N}) = O(N^{\log_b a}) \tag{10.10}$$

If $a = b^k$, then each term in the sum is 1. Since the sum contains $1 + \log_b N$ terms and $a = b^k$ implies that $\log_b a = k$,

$$\begin{aligned} T(N) &= O(a^m \log_b N) = O(N^{\log_b a} \log_b N) = O(N^k \log_b N) \\ &= O(N^k \log N) \end{aligned} \tag{10.11}$$

Finally, if $a < b^k$, then the terms in the geometric series are larger than 1, and the second formula in Section 1.2.3 applies. We obtain

$$T(N) = a^m \frac{(b^k/a)^{m+1} - 1}{(b^k/a) - 1} = O(a^m (b^k/a)^m) = O((b^k)^m) = O(N^k) \tag{10.12}$$

proving the last case of the theorem.

As an example, mergesort has $a = b = 2$ and $k = 1$. The second case applies, giving the answer $O(N \log N)$. If we solve three problems, each of which is half the original size, and combine the solutions with $O(N)$ additional work, then $a = 3$, $b = 2$ and $k = 1$. Case 1 applies here, giving a bound of $O(N^{\log_2 3}) = O(N^{1.59})$. An algorithm that solved three half-sized problems, but required $O(N^2)$ work to merge the solution, would have an $O(N^2)$ running time, since the third case would apply.

There are two important cases that are not covered by Theorem 10.6. We state two more theorems, leaving the proofs as exercises. Theorem 10.7 generalizes the previous theorem.

THEOREM 10.7.

The solution to the equation $T(N) = aT(N/b) + \Theta(N^k \log^p N)$, *where* $a \geq 1$, $b > 1$, *and* $p \geq 0$ *is*

$$T(N) = \begin{cases} O(N^{\log_b a}) & \text{if } a > b^k \\ O(N^k \log^{p+1} N) & \text{if } a = b^k \\ O(N^k \log^p N) & \text{if } a < b^k \end{cases}$$

THEOREM 10.8.

If $\sum_{i=1}^{k} \alpha_i < 1$, *then the solution to the equation* $T(N) = \sum_{i=1}^{k} T(\alpha_i N) + O(N)$ *is* $T(N) = O(N)$.

10.2.2. Closest-Points Problem

The input to our first problem is a list P of points in a plane. If $p_1 = (x_1, y_1)$ and $p_2 = (x_2, y_2)$, then the Euclidean distance between p_1 and p_2 is $[(x_1 - x_2)^2 + (y_1 - y_2)^2]^{1/2}$. We are required to find the closest pair of points. It is possible that two points have the same position; in that case that pair is the closest, with distance zero.

If there are N points, then there are $N(N-1)/2$ pairs of distances. We can check all of these, obtaining a very short program, but at the expense of an $O(N^2)$ algorithm. Since this approach is just an exhaustive search, we should expect to do better.

Let us assume that the points have been sorted by x coordinate. At worst, this adds $O(N \log N)$ to the final time bound. Since we will show an $O(N \log N)$ bound for the entire algorithm, this sort is essentially free, from a complexity standpoint.

Figure 10.29 shows a small sample point set P. Since the points are sorted by x coordinate, we can draw an imaginary vertical line that partitions the point set into two halves, P_L and P_R. This is certainly simple to do. Now we have almost exactly the same situation as we saw in the maximum subsequence sum problem in Section 2.4.3. Either the closest points are both in P_L, or they are both in P_R, or one is in P_L and the other is in P_R. Let us call these distances d_L, d_R, and d_C. Figure 10.30 shows the partition of the point set and these three distances.

We can compute d_L and d_R recursively. The problem, then, is to compute d_C. Since we would like an $O(N \log N)$ solution, we must be able to compute d_C with only $O(N)$ additional work. We have already seen that if a procedure consists of two half-sized recursive calls and $O(N)$ additional work, then the total time will be $O(N \log N)$.

Let $\delta = \min(d_L, d_R)$. The first observation is that we only need to compute d_C if d_C improves on δ. If d_C is such a distance, then the two points that define d_C must be within δ of the dividing line; we will refer to this area as a *strip*. As shown in Figure 10.31, this observation limits the number of points that need to be considered (in our case, $\delta = d_R$).

There are two strategies that can be tried to compute d_C. For large point sets that are uniformly distributed, the number of points that are expected to be in the strip is very small. Indeed, it is easy to argue that only $O(\sqrt{N})$ points are in the strip on average. Thus, we could perform a brute force calculation on these points in $O(N)$ time. The pseudocode in Figure 10.32 implements this strategy, assuming the C convention that the points are indexed starting at 0.

Figure 10.29 A small point set

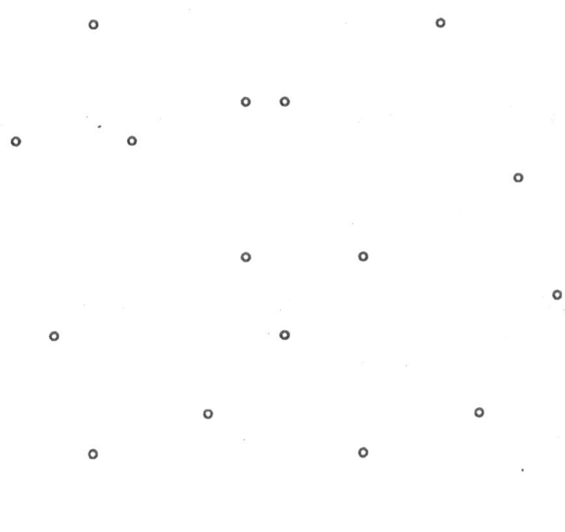

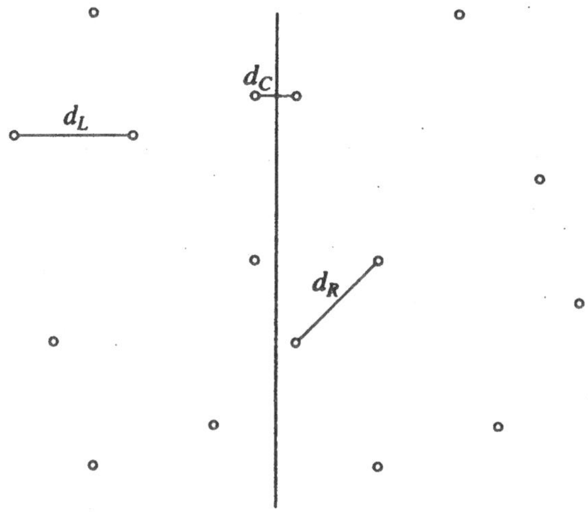

Figure 10.30 P partitioned into P_L and P_R; shortest distances are shown

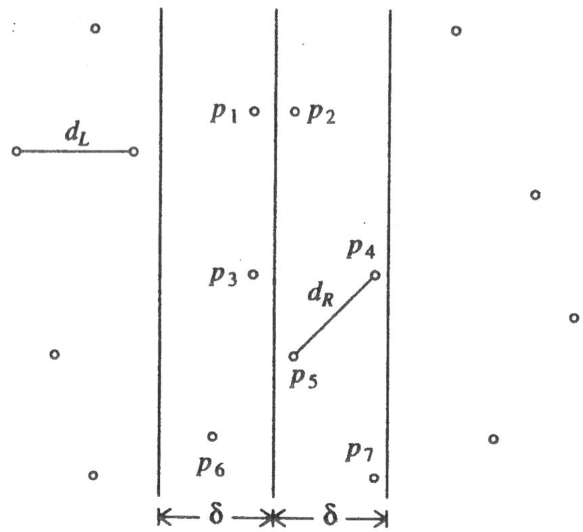

Figure 10.31 Two-lane strip, containing all points considered for d_C strip

```
/* Points are all in the strip */

for( i = 0; i < NumPointsInStrip; i++ )
    for( j = i + 1; j < NumPointsInStrip; j++ )
        if( Dist(P_i, P_j) < δ )
            δ = Dist(P_i, P_j);
```

Figure 10.32 Brute force calculation of $min(\delta, D_C)$

```
/* Points are all in the strip and sorted by y coordinate */

for( i = 0; i < NumPointsInStrip; i++ )
    for( j = i + 1; j < NumPointsInStrip; j++ )
        if( Pi and Pj 's coordinates differ by more than δ )
            break;      /* Go to next Pi . */
        else
        if( Dist(Pi,Pj) < δ )
            δ = Dist(Pi,Pj);
```

Figure 10.33 Refined calculation of $min(\delta, D_C)$

In the worst case, all the points could be in the strip, so this strategy does not always work in linear time. We can improve this algorithm with the following observation: The y coordinates of the two points that define d_C can differ by at most δ. Otherwise, $d_C > \delta$. Suppose that the points in the strip are sorted by their y coordinates. Therefore, if p_i and p_j's y coordinates differ by more than δ, then we can proceed to p_{i+1}. This simple modification is implemented in Figure 10.33.

This extra test has a significant effect on the running time, because for each p_i only a few points p_j are examined before p_i's and p_j's y coordinates differ by more than δ and force an exit from the inner *for* loop. Figure 10.34 shows, for instance, that for point p_3, only the two points p_4 and p_5 lie in the strip within δ vertical distance.

Figure 10.34 Only p_4 and p_5 are considered in the second *for* loop

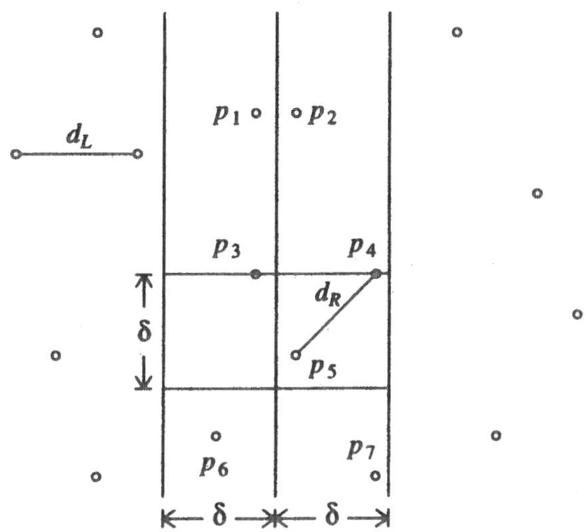

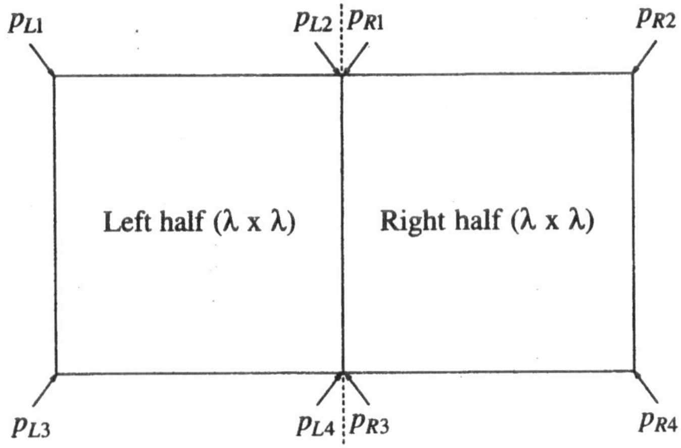

Figure 10.35 At most eight points fit in the rectangle; there are two coordinates shared by two points each

In the worst case, for any point p_i, at most 7 points p_j are considered. This is because these points must lie either in the δ by δ square in the left half of the strip or in the δ by δ square in the right half of the strip. On the other hand, all the points in each δ by δ square are separated by at least δ. In the worst case, each square contains four points, one at each corner. One of these points is p_i, leaving at most seven points to be considered. This worst-case situation is shown in Figure 10.35. Notice that even though p_{L2} and p_{R1} have the same coordinates, they could be different points. For the actual analysis, it is only important that the number of points in the λ by 2λ rectangle be $O(1)$, and this much is certainly clear.

Because at most seven points are considered for each p_i, the time to compute a d_C that is better than δ is $O(N)$. Thus, we appear to have an $O(N \log N)$ solution to the closest-points problem, based on the two half-sized recursive calls plus the linear extra work to combine the two results. However, we do not quite have an $O(N \log N)$ solution yet.

The problem is that we have assumed that a list of points sorted by y coordinate is available. If we perform this sort for each recursive call, then we have $O(N \log N)$ extra work: this gives an $O(N \log^2 N)$ algorithm. This is not all that bad, especially when compared to the brute force $O(N^2)$. However, it is not hard to reduce the work for each recursive call to $O(N)$, thus ensuring an $O(N \log N)$ algorithm.

We will maintain two lists. One is the point list sorted by x coordinate, and the other is the point list sorted by y coordinate. We will call these lists P and Q, respectively. These can be obtained by a preprocessing sorting step at cost $O(N \log N)$ and thus does not affect the time bound. P_L and Q_L are the lists passed to the left-half recursive call, and P_R and Q_R are the lists passed to the right-half recursive call. We have already seen that P is easily split in the middle. Once the dividing line is known, we step through Q sequentially, placing each element in Q_L or Q_R as appropriate. It is easy to see that Q_L and Q_R will be automatically sorted by y coordinate. When the recursive calls return, we scan through the Q list and discard all the points whose x coordinates are not within the strip. Then Q contains only points in the strip, and these points are guaranteed to be sorted by their y coordinates.

This strategy ensures that the entire algorithm is $O(N \log N)$, because only $O(N)$ extra work is performed.

10.2.3. The Selection Problem

The *selection problem* requires us to find the kth smallest element in a list S of N elements. Of particular interest is the special case of finding the median. This occurs when $k = \lceil N/2 \rceil$.

In Chapters 1, 6, and 7 we have seen several solutions to the selection problem. The solution in Chapter 7 uses a variation of quicksort and runs in $O(N)$ average time. Indeed, it is described in Hoare's original paper on quicksort.

Although this algorithm runs in linear average time, it has a worst case of $O(N^2)$. Selection can easily be solved in $O(N \log N)$ worst-case time by sorting the elements, but for a long time it was unknown whether or not selection could be accomplished in $O(N)$ worst-case time. The *quickselect* algorithm outlined in Section 7.7.6 is quite efficient in practice, so this was mostly a question of theoretical interest.

Recall that the basic algorithm is a simple recursive strategy. Assuming that N is larger than the cutoff point where elements are simply sorted, an element v, known as the pivot, is chosen. The remaining elements are placed into two sets, S_1 and S_2. S_1 contains elements that are guaranteed to be no larger than v, and S_2 contains elements that are no smaller than v. Finally, if $k \leq |S_1|$, then the kth smallest element in S can be found by recursively computing the kth smallest element in S_1. If $k = |S_1| + 1$, then the pivot is the kth smallest element. Otherwise, the kth smallest element in S is the $(k - |S_1| - 1)$st smallest element in S_2. The main difference between this algorithm and quicksort is that there is only one subproblem to solve instead of two.

In order to obtain a linear algorithm, we must ensure that the subproblem is only a fraction of the original and not merely only a few elements smaller than the original. Of course, we can always find such an element if we are willing to spend some time to do so. The difficult problem is that we cannot spend too much time finding the pivot.

For quicksort, we saw that a good choice for pivot was to pick three elements and use their median. This gives some expectation that the pivot is not too bad, but does not provide a guarantee. We could choose 21 elements at random, sort them in constant time, use the 11th largest as pivot, and get a pivot that is even more likely to be good. However, if these 21 elements were the 21 largest, then the pivot would still be poor. Extending this, we could use up to $O(N/\log N)$ elements, sort them using heapsort in $O(N)$ total time, and be almost certain, from a statistical point of view, of obtaining a good pivot. In the worst case, however, this does not work because we might select the $O(N/\log N)$ largest elements, and then the pivot would be the $[N - O(N/\log N)]$th largest element, which is not a constant fraction of N.

The basic idea is still useful. Indeed, we will see that we can use it to improve the expected number of comparisons that quickselect makes. To get a good worst case, however, the key idea is to use one more level of indirection. Instead of finding the median from a sample of random elements, we will find the median from a *sample of medians.*

The basic pivot selection algorithm is as follows:

1. Arrange the N elements into $\lfloor N/5 \rfloor$ groups of five elements, ignoring the (at most four) extra elements.
2. Find the median of each group. This gives a list M of $\lfloor N/5 \rfloor$ medians.
3. Find the median of M. Return this as the pivot, v.

We will use the term *median-of-median-of-five partitioning* to describe the quickselect algorithm that uses the pivot selection rule given above. We will now show that median-of-median-of-five partitioning guarantees that each recursive subproblem is at most roughly 70 percent as large as the original. We will also show that the pivot can be computed quickly enough to guarantee an $O(N)$ running time for the entire selection algorithm.

Let us assume for the moment that N is divisible by 5, so there are no extra elements. Suppose also that $N/5$ is odd, so that the set M contains an odd number of elements. This provides some symmetry, as we shall see. We are thus assuming, for convenience, that N is of the form $10k + 5$. We will also assume that all the elements are distinct. The actual algorithm must make sure to handle the case where this is not true. Figure 10.36 shows how the pivot might be chosen when $N = 45$.

In Figure 10.36, v represents the element which is selected by the algorithm as pivot. Since v is the median of nine elements, and we are assuming that all elements are distinct, there must be four medians that are larger than v and four that are smaller. We denote these by L and S, respectively. Consider a group of five elements with a large median (type L). The median of the group is smaller than two elements

Figure 10.36 How the pivot is chosen

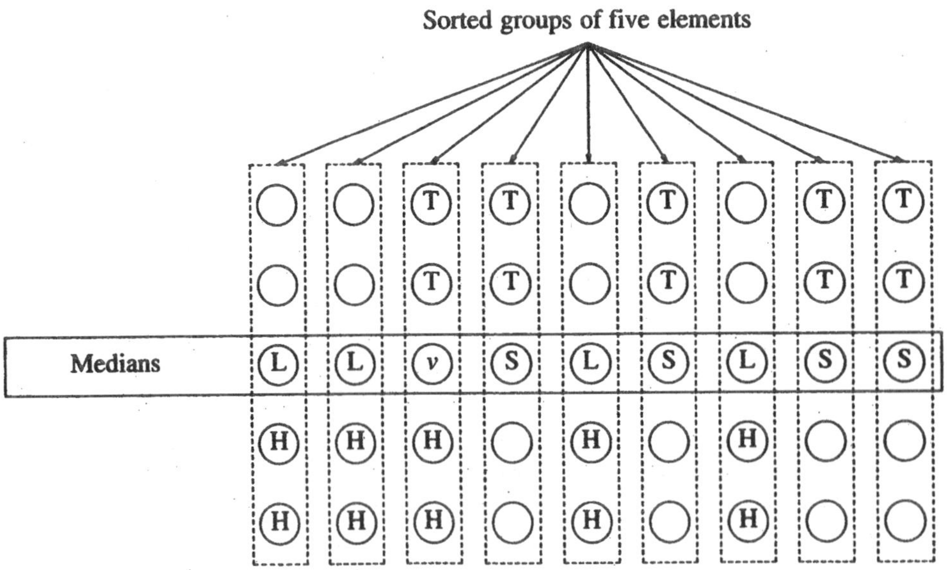

in the group and larger than two elements in the group. We will let H represent the *huge* elements. These are elements that are known to be larger than a large median. Similarly, T represents the *tiny* elements, which are smaller than a small median. There are 10 elements of type H: Two are in each of the groups with an L type median, and two elements are in the same group as v. Similarly, there are 10 elements of type T.

Elements of type L or H are guaranteed to be larger than v, and elements of type S or T are guaranteed to be smaller than v. There are thus guaranteed to be 14 large and 14 small elements in our problem. Therefore, a recursive call could be on at most $45 - 14 - 1 = 30$ elements.

Let us extend this analysis to general N of the form $10k + 5$. In this case, there are k elements of type L and k elements of type S. There are $2k + 2$ elements of type H, and also $2k + 2$ elements of type T. Thus, there are $3k + 2$ elements that are guaranteed to be larger than v and $3k + 2$ elements that are guaranteed to be smaller. Thus, in this case, the recursive call can contain at most $7k + 2 < 0.7N$ elements. If N is not of the form $10k + 5$, similar arguments can be made without affecting the basic result.

It remains to bound the running time to obtain the pivot element. There are two basic steps. We can find the median of five elements in constant time. For instance, it is not hard to sort five elements in eight comparisons. We must do this $\lfloor N/5 \rfloor$ times, so this step takes $O(N)$ time. We must then compute the median of a group of $\lfloor N/5 \rfloor$ elements. The obvious way to do this is to sort the group and return the element in the middle. But this takes $O(\lfloor N/5 \rfloor \log \lfloor N/5 \rfloor) = O(N \log N)$ time, so this does not work. The solution is to call the selection algorithm recursively on the $\lfloor N/5 \rfloor$ elements.

This completes the description of the basic algorithm. There are still some details that need to be filled in if an actual implementation is desired. For instance, duplicates must be handled correctly, and the algorithm needs a cutoff large enough to ensure that the recursive calls make progress. There is quite a large amount of overhead involved, and this algorithm is not practical at all, so we will not describe any more of the details that need to be considered. Even so, from a theoretical standpoint, the algorithm is a major breakthrough, because, as the following theorem shows, the running time is linear in the worst case.

THEOREM 10.9.

The running time of quickselect using median-of-median-of-five partitioning is $O(N)$.

PROOF:

The algorithm consists of two recursive calls of size $0.7N$ and $0.2N$, plus linear extra work. By Theorem 10.8, the running time is linear.

Reducing the Average Number of Comparisons

Divide and conquer can also be used to reduce the expected number of comparisons required by the selection algorithm. Let us look at a concrete example. Suppose we have a group S of 1,000 numbers and are looking for the 100th smallest number, which we will call X. We choose a subset S' of S consisting of 100 numbers. We would expect that the value of X is similar in size to the 10th smallest number in S'. More specifically, the fifth smallest number in S' is almost certainly less than X, and the 15th smallest number in S' is almost certainly greater than X.

More generally, a sample S' of s elements is chosen from the N elements. Let δ be some number, which we will choose later so as to minimize the average number of comparisons used by the procedure. We find the ($v_1 = ks/N - \delta$)th and ($v_2 = ks/N + \delta$)th smallest elements in S'. Almost certainly, the kth smallest element in S will fall between v_1 and v_2, so we are left with a selection problem on 2δ elements. With low probability, the kth smallest element does not fall in this range, and we have considerable work to do. However, with a good choice of s and δ, we can ensure, by the laws of probability, that the second case does not adversely affect the total work.

If an analysis is performed, we find that if $s = N^{2/3}\log^{1/3} N$ and $\delta = N^{1/3}\log^{2/3} N$, then the expected number of comparisons is $N + k + O(N^{2/3}\log^{1/3} N)$, which is optimal except for the low-order term. (If $k > N/2$, we can consider the symmetric problem of finding the $(N - k)$th largest element.)

Most of the analysis is easy to do. The last term represents the cost of performing the two selections to determine v_1 and v_2. The average cost of the partitioning, assuming a reasonably clever strategy, is equal to N plus the expected rank of v_2 in S, which is $N + k + O(N\delta/s)$. If the kth element winds up in S', the cost of finishing the algorithm is equal to the cost of selection on S', namely, $O(s)$. If the kth smallest element doesn't wind up in S', the cost is $O(N)$. However, s and δ have been chosen to guarantee that this happens with very low probability $o(1/N)$, so the expected cost of this possibility is $o(1)$, which is a term that goes to zero as N gets large. An exact calculation is left as Exercise 10.21.

This analysis shows that finding the median requires about $1.5N$ comparisons on average. Of course, this algorithm requires some floating-point arithmetic to compute s, which can slow down the algorithm on some machines. Even so, experiments have shown that if correctly implemented, this algorithm compares favorably with the quickselect implementation in Chapter 7.

10.2.4. *Theoretical Improvements for Arithmetic Problems*

In this section we describe a divide and conquer algorithm that multiplies two N-digit numbers. Our previous model of computation assumed that multiplication was done in constant time, because the numbers were small. For large numbers, this assumption is no longer valid. If we measure multiplication in terms of the size of numbers being multiplied, then the natural multiplication algorithm takes quadratic time. The divide and conquer algorithm runs in subquadratic time. We also present the classic divide and conquer algorithm that multiplies two N by N matrices in subcubic time.

Multiplying Integers

Suppose we want to multiply two N-digit numbers X and Y. If exactly one of X and Y is negative, then the answer is negative; otherwise it is positive. Thus, we can perform this check and then assume that $X, Y \geq 0$. The algorithm that almost everyone uses when multiplying by hand requires $\Theta(N^2)$ operations, because each digit in X is multiplied by each digit in Y.

If $X = 61{,}438{,}521$ and $Y = 94{,}736{,}407$, $XY = 5{,}820{,}464{,}730{,}934{,}047$. Let us break X and Y into two halves, consisting of the most significant and least significant digits, respectively. Then $X_L = 6{,}143$, $X_R = 8{,}521$, $Y_L = 9{,}473$, and $Y_R = 6{,}407$. We also have $X = X_L 10^4 + X_R$ and $Y = Y_L 10^4 + Y_R$. It follows that

$$XY = X_L Y_L 10^8 + (X_L Y_R + X_R Y_L)10^4 + X_R Y_R$$

Notice that this equation consists of four multiplications, $X_L Y_L$, $X_L Y_R$, $X_R Y_L$, and $X_R Y_R$, which are each half the size of the original problem ($N/2$ digits). The multiplications by 10^8 and 10^4 amount to the placing of zeros. This and the subsequent additions add only $O(N)$ additional work. If we perform these four multiplications recursively using this algorithm, stopping at an appropriate base case, then we obtain the recurrence

$$T(N) = 4T(N/2) + O(N)$$

From Theorem 10.6, we see that $T(N) = O(N^2)$, so, unfortunately, we have not improved the algorithm. To achieve a subquadratic algorithm, we must use less than four recursive calls. The key observation is that

$$X_L Y_R + X_R Y_L = (X_L - X_R)(Y_R - Y_L) + X_L Y_L + X_R Y_R$$

Figure 10.37 The divide and conquer algorithm in action

Function	Value	Computational Complexity
X_L	6,143	Given
X_R	8,521	Given
Y_L	9,473	Given
Y_R	6,407	Given
$D_1 = X_L - X_R$	−2,378	$O(N)$
$D_2 = Y_R - Y_L$	−3,066	$O(N)$
$X_L Y_L$	58,192,639	$T(N/2)$
$X_R Y_R$	54,594,047	$T(N/2)$
$D_1 D_2$	7,290,948	$T(N/2)$
$D_3 = D_1 D_2 + X_L Y_L + X_R Y_R$	120,077,634	$O(N)$
$X_R Y_R$	54,594,047	Computed above
$D_3 10^4$	1,200,776,340,000	$O(N)$
$X_L Y_L 10^8$	5,819,263,900,000,000	$O(N)$
$X_L Y_L 10^8 + D_3 10^4 + X_R Y_R$	5,820,464,730,934,047	$O(N)$

Thus, instead of using two multiplications to compute the coefficient of 10^4, we can use one multiplication, plus the result of two multiplications that have already been performed. Figure 10.37 shows how only three recursive subproblems need to be solved.

It is easy to see that now the recurrence equation satisfies

$$T(N) = 3T(N/2) + O(N)$$

and so we obtain $T(N) = O(N^{\log_2 3}) = O(N^{1.59})$. To complete the algorithm, we must have a base case, which can be solved without recursion.

When both numbers are one-digit, we can do the multiplication by table lookup. If one number has zero digits, then we return zero. In practice, if we were to use this algorithm, we would choose the base case to be that which is most convenient for the machine.

Although this algorithm has better asymptotic performance than the standard quadratic algorithm, it is rarely used, because for small N the overhead is significant, and for larger N there are even better algorithms. These algorithms also make extensive use of divide and conquer.

Matrix Multiplication

A fundamental numerical problem is the multiplication of two matrices. Figure 10.38 gives a simple $O(N^3)$ algorithm to compute $\mathbf{C} = \mathbf{AB}$, where $\mathbf{A}$, $\mathbf{B}$, and $\mathbf{C}$ are $N \times N$ matrices. The algorithm follows directly from the definition of matrix multiplication. To compute $C_{i,j}$, we compute the dot product of the ith row in $\mathbf{A}$ with the jth column in $\mathbf{B}$. As usual, arrays begin at index 0.

For a long time it was assumed that $\Omega(N^3)$ was required for matrix multiplication. However, in the late sixties Strassen showed how to break the $\Omega(N^3)$ barrier.

Figure 10.38 Simple $O(N^3)$ matrix multiplication

```
/* Standard matrix multiplication */
/* Arrays start at 0 */

void
MatrixMultiply( Matrix A, Matrix B, Matrix C, int N )
{
    int i, j, k;

    for( i = 0; i < N; i++ )  /* Initialization */
        for( j = 0; j < N; j++ )
            C[ i ][ j ] = 0.0;

    for( i = 0; i < N; i++ )
        for( j = 0; j < N; j++ )
            for( k = 0; k < N; k++ )
                C[ i ][ j ] += A[ i ][ k ] * B[ k ][ j ];
}
```

$$\begin{bmatrix} A_{1,1} & A_{1,2} \\ A_{2,1} & A_{2,2} \end{bmatrix} \begin{bmatrix} B_{1,1} & B_{1,2} \\ B_{2,1} & B_{2,2} \end{bmatrix} = \begin{bmatrix} C_{1,1} & C_{1,2} \\ C_{2,1} & C_{2,2} \end{bmatrix}$$

Figure 10.39 Decomposing **AB** = C into four quadrants

The basic idea of Strassen's algorithm is to divide each matrix into four quadrants, as shown in Figure 10.39. Then it is easy to show that

$$C_{1,1} = A_{1,1}B_{1,1} + A_{1,2}B_{2,1}$$

$$C_{1,2} = A_{1,1}B_{1,2} + A_{1,2}B_{2,2}$$

$$C_{2,1} = A_{2,1}B_{1,1} + A_{2,2}B_{2,1}$$

$$C_{2,2} = A_{2,1}B_{1,2} + A_{2,2}B_{2,2}$$

As an example, to perform the multiplication **AB**

$$\mathbf{AB} = \begin{bmatrix} 3 & 4 & 1 & 6 \\ 1 & 2 & 5 & 7 \\ 5 & 1 & 2 & 9 \\ 4 & 3 & 5 & 6 \end{bmatrix} \begin{bmatrix} 5 & 6 & 9 & 3 \\ 4 & 5 & 3 & 1 \\ 1 & 1 & 8 & 4 \\ 3 & 1 & 4 & 1 \end{bmatrix}$$

we define the following eight $N/2$ by $N/2$ matrices:

$$A_{1,1} = \begin{bmatrix} 3 & 4 \\ 1 & 2 \end{bmatrix} \quad A_{1,2} = \begin{bmatrix} 1 & 6 \\ 5 & 7 \end{bmatrix} \quad B_{1,1} = \begin{bmatrix} 5 & 6 \\ 4 & 5 \end{bmatrix} \quad B_{1,2} = \begin{bmatrix} 9 & 3 \\ 3 & 1 \end{bmatrix}$$

$$A_{2,1} = \begin{bmatrix} 5 & 1 \\ 4 & 3 \end{bmatrix} \quad A_{2,2} = \begin{bmatrix} 2 & 9 \\ 5 & 6 \end{bmatrix} \quad B_{2,1} = \begin{bmatrix} 1 & 1 \\ 3 & 1 \end{bmatrix} \quad B_{2,2} = \begin{bmatrix} 8 & 4 \\ 4 & 1 \end{bmatrix}$$

We could then perform eight $N/2$ by $N/2$ matrix multiplications and four $N/2$ by $N/2$ matrix additions. The matrix additions take $O(N^2)$ time. If the matrix multiplications are done recursively, then the running time satisfies

$$T(N) = 8T(N/2) + O(N^2)$$

From Theorem 10.6, we see that $T(N) = O(N^3)$, so we do not have an improvement. As we saw with integer multiplication, we must reduce the number of subproblems below 8. Strassen used a strategy similar to the integer multiplication divide and conquer algorithm and showed how to use only seven recursive calls by carefully arranging the computations. The seven multiplications are

$$M_1 = (A_{1,2} - A_{2,2})(B_{2,1} + B_{2,2})$$

$$M_2 = (A_{1,1} + A_{2,2})(B_{1,1} + B_{2,2})$$

$$M_3 = (A_{1,1} - A_{2,1})(B_{1,1} + B_{1,2})$$

$$M_4 = (A_{1,1} + A_{1,2})B_{2,2}$$

$$M_5 = A_{1,1}(B_{1,2} - B_{2,2})$$

$$M_6 = A_{2,2}(B_{2,1} - B_{1,1})$$

$$M_7 = (A_{2,1} + A_{2,2})B_{1,1}$$

Once the multiplications are performed, the final answer can be obtained with eight more additions.

$$C_{1,1} = M_1 + M_2 - M_4 + M_6$$
$$C_{1,2} = M_4 + M_5$$
$$C_{2,1} = M_6 + M_7$$
$$C_{2,2} = M_2 - M_3 + M_5 - M_7$$

It is straightforward to verify that this tricky ordering produces the desired values. The running time now satisfies the recurrence

$$T(N) = 7T(N/2) + O(N^2)$$

The solution of this recurrence is $T(N) = O(N^{\log_2 7}) = O(N^{2.81})$.

As usual, there are details to consider, such as the case when N is not a power of 2, but these are basically minor nuisances. Strassen's algorithm is worse than the straightforward algorithm until N is fairly large. It does not generalize for the case where the matrices are sparse (contain many zero entries), and it does not easily parallelize. When run with floating-point entries, it is less stable numerically than the classic algorithm. Thus, it has only limited applicability. Nevertheless, it represents an important theoretical milestone and certainly shows that in computer science, as in many other fields, even though a problem seems to have an intrinsic complexity, nothing is certain until proven.

10.3. Dynamic Programming

In the previous section, we saw that a problem that can be mathematically expressed recursively can also be expressed as a recursive algorithm, in many cases yielding a significant performance improvement over a more naïve exhaustive search.

Any recursive mathematical formula could be directly translated to a recursive algorithm, but the underlying reality is that often the compiler will not do justice to the recursive algorithm, and an inefficient program results. When we suspect that this is likely to be the case, we must provide a little more help to the compiler, by rewriting the recursive algorithm as a nonrecursive algorithm that systematically records the answers to the subproblems in a table. One technique that makes use of this approach is known as *dynamic programming*.

10.3.1. Using a Table Instead of Recursion

In Chapter 2, we saw that the natural recursive program to compute the Fibonacci numbers is very inefficient. Recall that the program shown in Figure 10.40 has a running time $T(N)$ that satisfies $T(N) \geq T(N-1) + T(N-2)$. Since $T(N)$ satisfies the same recurrence relation as the Fibonacci numbers and has the same initial conditions, $T(N)$ in fact grows at the same rate as the Fibonacci numbers, and is thus exponential.

On the other hand, since to compute F_N, all that is needed is F_{N-1} and F_{N-2}, we only need to record the two most recently computed Fibonacci numbers. This yields the $O(N)$ algorithm in Figure 10.41.

```
/* Compute Fibonacci numbers as discussed in Chapter 1 */

int
Fib( int N )
{
    if( N <= 1 )
        return 1;
    else
        return Fib( N - 1 ) + Fib( N - 2 );
}
```

Figure 10.40 Inefficient algorithm to compute Fibonacci numbers

```
int
Fibonacci( int N )
{
    int i, Last, NextToLast, Answer;

    if( N <= 1 )
        return 1;

    Last = NextToLast = 1;
    for( i = 2; i <= N; i++ )
    {
        Answer = Last + NextToLast;
        NextToLast = Last;
        Last = Answer;
    }

    return Answer;
}
```

Figure 10.41 Linear algorithm to compute Fibonacci numbers

The reason that the recursive algorithm is so slow is because of the algorithm used to simulate recursion. To compute F_N, there is one call to F_{N-1} and F_{N-2}. However, since F_{N-1} recursively makes a call to F_{N-2} and F_{N-3}, there are actually two separate calls to compute F_{N-2}. If one traces out the entire algorithm, then we can see that F_{N-3} is computed three times, F_{N-4} is computed five times, F_{N-5} is computed eight times, and so on. As Figure 10.42 shows, the growth of redundant calculations is explosive. If the compiler's recursion simulation algorithm were able to keep a list of all precomputed values and not make a recursive call for an already solved subproblem, then this exponential explosion would be avoided. This is why the program in Figure 10.41 is so much more efficient.

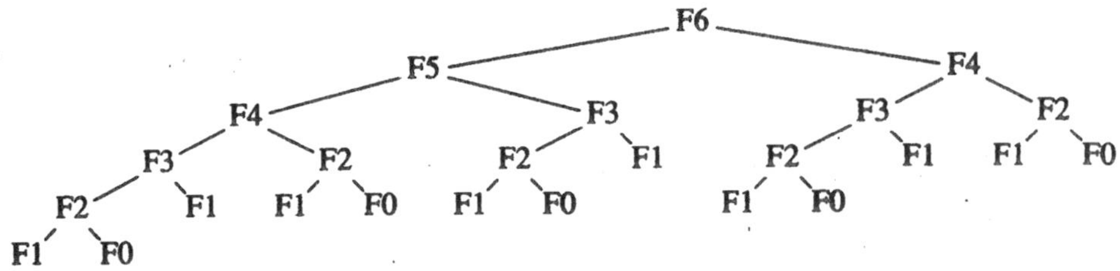

Figure 10.42 Trace of the recursive calculation of Fibonacci numbers

```
double
Eval( int N )
{
    int i;
    double Sum;

    if( N == 0 )
        return 1.0;
    else
    {
        Sum = 0.0;
        for( i = 0; i < N; i++ )
            Sum += Eval( i );
        return 2.0 * Sum / N + N;
    }
}
```

Figure 10.43 Recursive program to evaluate $C(N) = 2/N \sum_{i=0}^{N-1} C(i) + N$

Figure 10.44 Trace of the recursive calculation in *Eval*

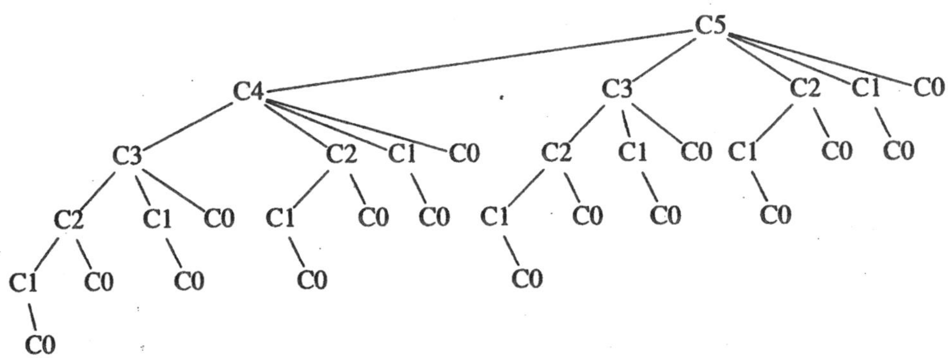

```
double
Eval( int N )
{
    int i, j;
    double Sum, Answer;
    double *C;

    C = malloc( sizeof( double ) * ( N + 1 ) );
    if( C == NULL )
        FatalError( "Out of space!!!" );

    C[ 0 ] = 1.0;
    for( i = 1; i <= N; i++ )
    {
        Sum = 0.0;
        for( j = 0; j < i; j++ )
            Sum += C[ j ];
        C[ i ] = 2.0 * Sum / i + i;
    }

    Answer = C[ N ];
    free( C );

    return Answer;
}
```

Figure 10.45 Evaluating $C(N) = 2/N \sum_{i=0}^{N-1} C(i) + N$ with a table

As a second example, we saw in Chapter 7 how to solve the recurrence $C(N) = (2/N)\sum_{i=0}^{N-1} C(i) + N$, with $C(0) = 1$. Suppose that we want to check, numerically, whether the solution we obtained is correct. We could then write the simple program in Figure 10.43 to evaluate the recursion.

Once again, the recursive calls duplicate work. In this case, the running time $T(N)$ satisfies $T(N) = \sum_{i=0}^{N-1} T(i) + N$, because, as shown in Figure 10.44, there is one (direct) recursive call of each size from 0 to $N - 1$, plus $O(N)$ additional work (where else have we seen the tree shown in Figure 10.44?). Solving for $T(N)$, we find that it grows exponentially. By using a table, we obtain the program in Figure 10.45. This program avoids the redundant recursive calls and runs in $O(N^2)$. It is not a perfect program; as an exercise, you should make the simple change that reduces its running time to $O(N)$.

10.3.2. Ordering Matrix Multiplications

Suppose we are given four matrices, **A**, **B**, **C**, and **D**, of dimensions $\mathbf{A} = 50 \times 10$, $\mathbf{B} = 10 \times 40$, $\mathbf{C} = 40 \times 30$, and $\mathbf{D} = 30 \times 5$. Although matrix multiplication is not commutative, it is associative, which means that the matrix product **ABCD** can be parenthesized, and thus evaluated, in any order. The obvious way to multiply two matrices of dimensions $p \times q$ and $q \times r$, respectively, uses pqr scalar multiplications. (Using a theoretically superior algorithm such as Strassen's algorithm does not significantly alter the problem we will consider, so we will assume this performance bound.) What is the best way to perform the three matrix multiplications required to compute **ABCD**?

In the case of four matrices, it is simple to solve the problem by exhaustive search, since there are only five ways to order the multiplications. We evaluate each case below:

- (**A**((**BC**)**D**)): Evaluating **BC** requires $10 \times 40 \times 30 = 12{,}000$ multiplications. Evaluating (**BC**)**D** requires the 12,000 multiplications to compute **BC**, plus an additional $10 \times 30 \times 5 = 1{,}500$ multiplications, for a total of 13,500. Evaluating (**A**((**BC**)**D**)) requires 13,500 multiplications for (**BC**)**D**, plus an additional $50 \times 10 \times 5 = 2{,}500$ multiplications, for a grand total of 16,000 multiplications.
- (**A**(**B**(**CD**))): Evaluating **CD** requires $40 \times 30 \times 5 = 6{,}000$ multiplications. Evaluating **B**(**CD**) requires the 6,000 multiplications to compute **CD**, plus an additional $10 \times 40 \times 5 = 2{,}000$ multiplications, for a total of 8,000. Evaluating (**A**(**B**(**CD**))) requires 8,000 multiplications for **B**(**CD**), plus an additional $50 \times 10 \times 5 = 2{,}500$ multiplications, for a grand total of 10,500 multiplications.
- ((**AB**)(**CD**)): Evaluating **CD** requires $40 \times 30 \times 5 = 6{,}000$ multiplications. Evaluating **AB** requires $50 \times 10 \times 40 = 20{,}000$ multiplications. Evaluating ((**AB**)(**CD**)) requires 6,000 multiplications for **CD**, 20,000 multiplications for **AB**, plus an additional $50 \times 40 \times 5 = 10{,}000$ multiplications for a grand total of 36,000 multiplications.
- (((**AB**)**C**)**D**): Evaluating **AB** requires $50 \times 10 \times 40 = 20{,}000$ multiplications. Evaluating (**AB**)**C** requires the 20,000 multiplications to compute **AB**, plus an additional $50 \times 40 \times 30 = 60{,}000$ multiplications, for a total of 80,000. Evaluating (((**AB**)**C**)**D**) requires 80,000 multiplications for (**AB**)**C**, plus an additional $50 \times 30 \times 5 = 7{,}500$ multiplications, for a grand total of 87,500 multiplications.
- ((**A**(**BC**))**D**): Evaluating **BC** requires $10 \times 40 \times 30 = 12{,}000$ multiplications. Evaluating **A**(**BC**) requires the 12,000 multiplications to compute **BC**, plus an additional $50 \times 10 \times 30 = 15{,}000$ multiplications, for a total of 27,000. Evaluating ((**A**(**BC**))**D**) requires 27,000 multiplications for **A**(**BC**), plus an additional $50 \times 30 \times 5 = 7{,}500$ multiplications, for a grand total of 34,500 multiplications.

The calculations show that the best ordering uses roughly one-ninth the number of multiplications as the worst ordering. Thus, it might be worthwhile to perform a few calculations to determine the optimal ordering. Unfortunately, none of the obvious greedy strategies seems to work. Moreover, the number of possible orderings grows quickly. Suppose we define $T(N)$ to be this number. Then $T(1) = T(2) = 1$,

$T(3) = 2$, and $T(4) = 5$, as we have seen. In general,

$$T(N) = \sum_{i=1}^{N-1} T(i)T(N-i)$$

To see this, suppose that the matrices are $\mathbf{A}_1, \mathbf{A}_2, \ldots, \mathbf{A}_N$, and the last multiplication performed is $(\mathbf{A}_1\mathbf{A}_2 \cdots \mathbf{A}_i)(\mathbf{A}_{i+1}\mathbf{A}_{i+2} \cdots \mathbf{A}_N)$. Then there are $T(i)$ ways to compute $(\mathbf{A}_1\mathbf{A}_2 \cdots \mathbf{A}_i)$ and $T(N-i)$ ways to compute $(\mathbf{A}_{i+1}\mathbf{A}_{i+2} \cdots \mathbf{A}_N)$. Thus, there are $T(i)T(N-i)$ ways to compute $(\mathbf{A}_1\mathbf{A}_2 \cdots \mathbf{A}_i)(\mathbf{A}_{i+1}\mathbf{A}_{i+2} \cdots \mathbf{A}_N)$ for each possible i.

The solution of this recurrence is the well-known Catalan numbers, which grow exponentially. Thus, for large N, an exhaustive search through all possible orderings is useless. Nevertheless, this counting argument provides a basis for a solution that is substantially better than exponential. Let c_i be the number of columns in matrix $\mathbf{A}_i$ for $1 \le i \le N$. Then $\mathbf{A}_i$ has c_{i-1} rows, since otherwise the multiplications are not valid. We will define c_0 to be the number of rows in the first matrix, $\mathbf{A}_1$.

Suppose $m_{Left,Right}$ is the number of multiplications required to multiply $\mathbf{A}_{Left}\mathbf{A}_{Left+1} \cdots \mathbf{A}_{Right-1}\mathbf{A}_{Right}$. For consistency, $m_{Left,Left} = 0$. Suppose the last multiplication is $(\mathbf{A}_{Left} \cdots \mathbf{A}_i)(\mathbf{A}_{i+1} \cdots \mathbf{A}_{Right})$, where $Left \le i < Right$. Then the number of multiplications used is $m_{Left,i} + m_{i+1,Right} + c_{Left-1}c_i c_{Right}$. These three terms represent the multiplications required to compute $(\mathbf{A}_{Left} \cdots \mathbf{A}_i)$, $(\mathbf{A}_{i+1} \cdots \mathbf{A}_{Right})$, and their product, respectively.

If we define $M_{Left,Right}$ to be the number of multiplications required in an *optimal* ordering, then, if $Left < Right$,

$$M_{Left,Right} = \min_{Left \le i < Right} \{M_{Left,i} + M_{i+1,Right} + c_{Left-1}c_i c_{Right}\}$$

This equation implies that if we have an optimal multiplication arrangement of $\mathbf{A}_{Left} \cdots \mathbf{A}_{Right}$, the subproblems $\mathbf{A}_{Left} \cdots \mathbf{A}_i$ and $\mathbf{A}_{i+1} \cdots \mathbf{A}_{Right}$ cannot be performed suboptimally. This should be clear, since otherwise we could improve the entire result by replacing the suboptimal computation by an optimal computation.

The formula translates directly to a recursive program, but, as we have seen in the last section, such a program would be blatantly inefficient. However, since there are only approximately $N^2/2$ values of $M_{Left,Right}$ that ever need to be computed, it is clear that a table can be used to store these values. Further examination shows that if $Right - Left = k$, then the only values $M_{x,y}$ that are needed in the computation of $M_{Left,Right}$ satisfy $y - x < k$. This tells us the order in which we need to compute the table.

If we want to print out the actual ordering of the multiplications in addition to the final answer $M_{1,N}$, then we can use the ideas from the shortest-path algorithms in Chapter 9. Whenever we make a change to $M_{Left,Right}$, we record the value of i that is responsible. This gives the simple program shown in Figure 10.46.

Although the emphasis of this chapter is not coding, it is worth noting that many programmers tend to shorten variable names to a single letter. c, i, and k are used as single-letter variables because this agrees with the names we have used in the description of the algorithm, which is very mathematical. However, it is generally best to avoid l as a variable name, because "l" looks too much like "1" and can make for very difficult debugging if you make a transcription error.

Returning to the algorithmic issues, this program contains a triply nested loop and is easily seen to run in $O(N^3)$ time. The references describe a faster algorithm, but since the time to perform the actual matrix multiplication is still likely to be much larger than the time to compute the optimal ordering, this algorithm is still quite practical.

```
/* Compute optimal ordering of matrix multiplication */
/* C contains number of columns for each of the N matrices */
/* C[ 0 ] is the number of rows in matrix 1 */
/* Minimum number of multiplications is left in M[ 1 ][ N ] */
/* Actual ordering is computed via */
/* another procedure using LastChange */
/* M and LastChange are indexed starting at 1, instead of 0 */
/* Note: Entries below main diagonals of M and LastChange */
/* are meaningless and uninitialized */

void
OptMatrix( const long C[ ], int N,
           TwoDimArray M, TwoDimArray LastChange )
{
    int i, k, Left, Right;
    long ThisM;

    for( Left = 1; Left <= N; Left++ )
        M[ Left ][ Left ] = 0;
    for( k = 1; k < N; k++ )   /* k is Right - Left */
        for( Left = 1; Left <= N - k; Left++ )
        {
            /* For each position */
            Right = Left + k;
            M[ Left ][ Right ] = Infinity;
            for( i = Left; i < Right; i++ )
            {
                ThisM = M[ Left ][ i ] + M[ i + 1 ][ Right ]
                        + C[ Left - 1 ] * C[ i ] * C[ Right ];
                if( ThisM < M[ Left ][ Right ] )
                {
                    /* Update min */
                    M[ Left ][ Right ] = ThisM;
                    LastChange[ Left ][ Right ] = i;
                }
            }
        }
}
```

Figure 10.46 Program to find optimal ordering of matrix multiplications

10.3.3. Optimal Binary Search Tree

Our second dynamic programming example considers the following input: We are given a list of words, $w_1, w_2, \ldots, w_N$, and *fixed* probabilities $p_1, p_2, \ldots, p_N$ of their occurrence. The problem is to arrange these words in a binary search tree in a way that minimizes the expected total access time. In a binary search tree, the number of comparisons needed to access an element at depth d is $d + 1$, so if w_i is placed at depth d_i, then we want to minimize $\sum_{i=1}^{N} p_i(1 + d_i)$.

As an example, Figure 10.47 shows seven words along with their probability of occurrence in some context. Figure 10.48 shows three possible binary search trees. Their searching costs are shown in Figure 10.49.

The first tree was formed using a greedy strategy. The word with the highest probability of being accessed was placed at the root. The left and right subtrees were then formed recursively. The second tree is the perfectly balanced search tree. Neither of these trees is optimal, as demonstrated by the existence of the third tree. From this we can see that neither of the obvious solutions works.

This is initially surprising, since the problem appears to be very similar to the construction of a Huffman encoding tree, which, as we have already seen, can be solved by a greedy algorithm. Construction of an optimal binary search tree is harder, because the data are not constrained to appear only at the leaves, and also because the tree must satisfy the binary search tree property.

A dynamic programming solution follows from two observations. Once again, suppose we are trying to place the (sorted) words $w_{Left}, w_{Left+1}, \ldots, w_{Right-1}, w_{Right}$ into a binary search tree. Suppose the optimal binary search tree has w_i as the root, where $Left \le i \le Right$. Then the left subtree must contain $w_{Left}, \ldots, w_{i-1}$, and the right subtree must contain $w_{i+1}, \ldots, w_{Right}$ (by the binary search tree property). Further, both of these subtrees must also be optimal, since otherwise they could be replaced by optimal subtrees, which would give a better solution for $w_{Left}, \ldots, w_{Right}$. Thus, we can write a formula for the cost $C_{Left,Right}$ of an optimal binary search tree. Figure 10.50 may be helpful.

If $Left > Right$, then the cost of the tree is 0; this is the *NULL* case, which we always have for binary search trees. Otherwise, the root costs p_i. The left subtree has a cost of $C_{Left,i-1}$, relative to its root, and the right subtree has a cost of $C_{i+1,Right}$ relative to its root. As Figure 10.50 shows, each node in these subtrees is one level deeper from w_i than from their respective roots, so we must add $\sum_{j=Left}^{i-1} p_j$ and $\sum_{j=i+1}^{Right} p_j$. This gives the formula

$$C_{Left,Right} = \min_{Left \le i \le Right} \left\{ p_i + C_{Left,i-1} + C_{i+1,Right} + \sum_{j=Left}^{i-1} p_j + \sum_{j=i+1}^{Right} p_j \right\}$$

$$= \min_{Left \le i \le Right} \left\{ C_{Left,i-1} + C_{i+1,Right} + \sum_{j=Left}^{Right} p_j \right\}$$

From this equation, it is straightforward to write a program to compute the cost of the optimal binary search tree. As usual, the actual search tree can be maintained by saving the value of i that minimizes $C_{Left,Right}$. The standard recursive routine can be used to print the actual tree.

Word	Probability
a	0.22
am	0.18
and	0.20
egg	0.05
if	0.25
the	0.02
two	0.08

Figure 10.47 Sample input for optimal binary search tree problem

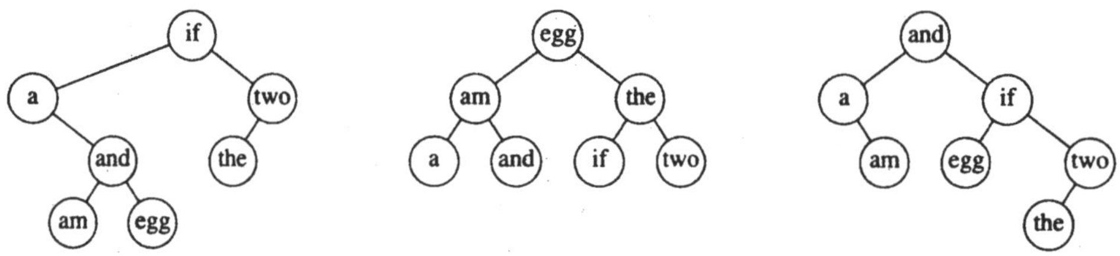

Figure 10.48 Three possible binary search trees for data in previous table

Input		Tree #1		Tree #2		Tree #3	
Word	Probability	Access Cost		Access Cost		Access Cost	
w_i	p_i	Once	Sequence	Once	Sequence	Once	Sequence
a	0.22	2	0.44	3	0.66	2	0.44
am	0.18	4	0.72	2	0.36	3	0.54
and	0.20	3	0.60	3	0.60	1	0.20
egg	0.05	4	0.20	1	0.05	3	0.15
if	0.25	1	0.25	3	0.75	2	0.50
the	0.02	3	0.06	2	0.04	4	0.08
two	0.08	2	0.16	3	0.24	3	0.24
Totals	1.00		2.43		2.70		2.15

Figure 10.49 Comparison of the three binary search trees

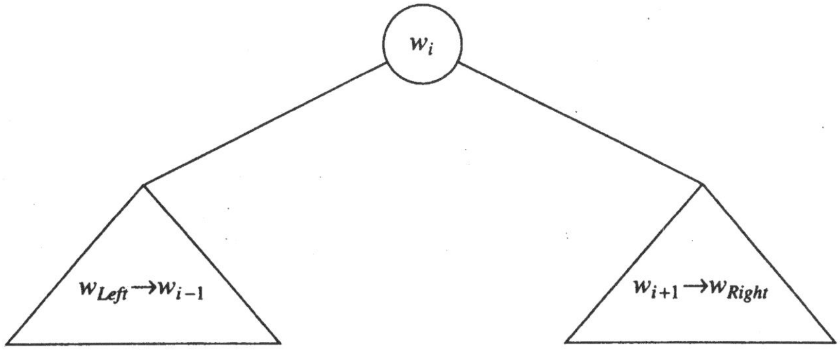

Figure 10.50 Structure of an optimal binary search tree

Figure 10.51 shows the table that will be produced by the algorithm. For each subrange of words, the cost and root of the optimal binary search tree are maintained. The bottommost entry computes the optimal binary search tree for the entire set of words in the input. The optimal tree is the third tree shown in Figure 10.48.

The precise computation for the optimal binary search tree for a particular subrange, namely, *am..if*, is shown in Figure 10.52. It is obtained by computing the minimum-cost tree obtained by placing *am, and, egg,* and *if* at the root. For instance, when *and* is placed at the root, the left subtree contains *am..am* (of cost 0.18, via previous calculation), the right subtree contains *egg..if* (of cost 0.35), and $p_{\text{am}} + p_{\text{and}} + p_{\text{egg}} + p_{\text{if}} = 0.68$, for a total cost of 1.21.

Figure 10.51 Computation of the optimal binary search tree for sample input

	Left=1	Left=2	Left=3	Left=4	Left=5	Left=6	Left=7
Iteration=1	a..a	am..am	and..and	egg..egg	if..if	the..the	two..two
	.22 a	.18 am	.20 and	.05 egg	.25 if	.02 the	.08 two
Iteration=2	a..am	am..and	and..egg	egg..if	if..the	the..two	
	.58 a	.56 and	.30 and	.35 if	.29 if	.12 two	
Iteration=3	a..and	am..egg	and..if	egg..the	if..two		
	1.02 am	.66 and	.80 if	.39 if	.47 if		
Iteration=4	a..egg	am..if	and..the	egg..two			
	1.17 am	1.21 and	.84 if	.57 if			
Iteration=5	a..if	am..the	and..two				
	1.83 and	1.27 and	1.02 if				
Iteration=6	a..the	am..two					
	1.89 and	1.53 and					
Iteration=7	a..two						
	2.15 and						

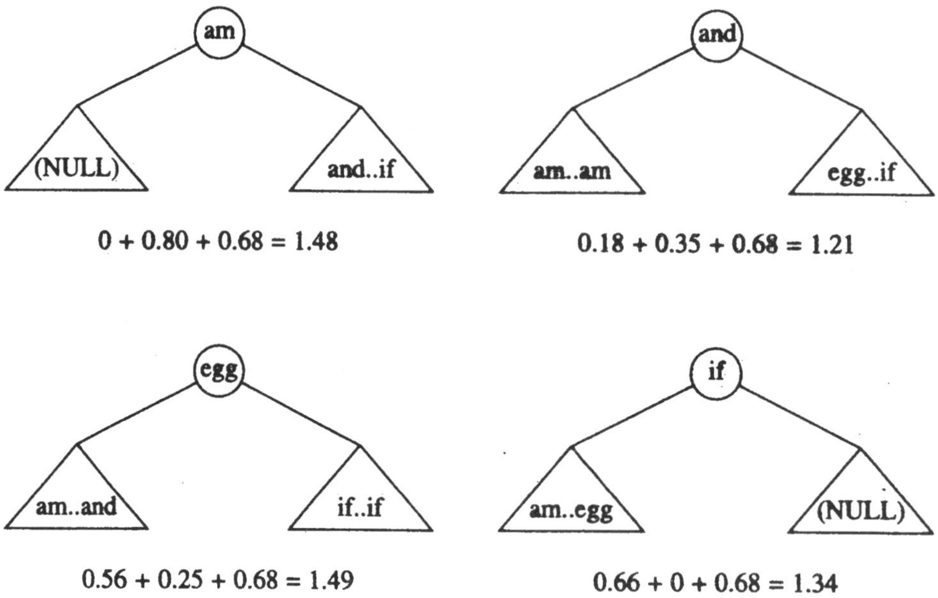

Figure 10.52 Computation of table entry (1.21, *and*) for *am..if*

The running time of this algorithm is $O(N^3)$, because when it is implemented, we obtain a triple loop. An $O(N^2)$ algorithm for the problem is sketched in the exercises.

10.3.4. All-Pairs Shortest Path

Our third and final dynamic programming application is an algorithm to compute shortest weighted paths between every pair of points in a directed graph $G = (V, E)$. In Chapter 9, we saw an algorithm for the *single-source shortest-path* problem, which finds the shortest path from some arbitrary vertex s to all others. That algorithm (Dijkstra's) runs in $O(|V|^2)$ time on dense graphs, but substantially faster on sparse graphs. We will give a short algorithm to solve the all-pairs problem for dense graphs. The running time of the algorithm is $O(|V|^3)$, which is not an asymptotic improvement over $|V|$ iterations of Dijkstra's algorithm but could be faster on a very dense graph, because its loops are tighter. The algorithm also performs correctly if there are negative edge costs, but no negative-cost cycles; Dijkstra's algorithm fails in this case.

Let us recall the important details of Dijkstra's algorithm (the reader may wish to review Section 9.3). Dijkstra's algorithm starts at a vertex s and works in stages. Each vertex in the graph is eventually selected as an intermediate vertex. If the current selected vertex is v, then for each $w \in V$, we set $d_w = \min(d_w, d_v + c_{v,w})$. This formula says that the best distance to w (from s) is either the previously known distance to w from s, or the result of going from s to v (optimally) and then directly from v to w.

Dijkstra's algorithm provides the idea for the dynamic programming algorithm: we select the vertices in sequential order. We will define $D_{k,i,j}$ to be the weight of

```
        /* Compute All-Shortest Paths */
        /* A[ ] contains the adjacency matrix */
        /* with A[ i ][ i ] presumed to be zero */
        /* D[ ] contains the values of the shortest path */
        /* N is the number of vertices */
        /* A negative cycle exists iff */
        /* D[ i ][ i ] is set to a negative value */
        /* Actual path can be computed using Path[ ] */
        /* All arrays are indexed starting at 0 */
        /* NotAVertex is -1 */

        void
        AllPairs( TwoDimArray A, TwoDimArray D,
                  TwoDimArray Path, int N )
        {
            int i, j, k;

            /* Initialize D and Path */
/* 1*/      for( i = 0; i < N; i++ )
/* 2*/          for( j = 0; j < N; j++ )
                {
/* 3*/              D[ i ][ j ] = A[ i ][ j ];
/* 4*/              Path[ i ][ j ] = NotAVertex;
                }

/* 5*/      for( k = 0; k < N; k++ )
                /* Consider each vertex as an intermediate */
/* 6*/          for( i = 0; i < N; i++ )
/* 7*/              for( j = 0; j < N; j++ )
/* 8*/                  if( D[ i ][ k ] + D[ k ][ j ] < D[ i ][ j ] )
                        {
                            /* Update shortest path */
/* 9*/                      D[ i ][ j ] = D[ i ][ k ] + D[ k ][ j ];
/*10*/                      Path[ i ][ k ] = k;
                        }
        }
```

Figure 10.53 All-pairs shortest path

the shortest path from v_i to v_j that uses only $v_1, v_2, \ldots, v_k$ as intermediates. By this definition, $D_{0,i,j} = c_{i,j}$, where $c_{i,j}$ is ∞ if (v_i, v_j) is not an edge in the graph. Also, by definition, $D_{|V|,i,j}$ is the shortest path from v_i to v_j in the graph.

As Figure 10.53 shows, when $k > 0$ we can write a simple formula for $D_{k,i,j}$. The shortest path from v_i to v_j that uses only $v_1, v_2, \ldots, v_k$ as intermediates is the shortest path that either does not use v_k as an intermediate at all, or consists of the merging of the two paths $v_i \rightarrow v_k$ and $v_k \rightarrow v_j$, each of which uses only the first $k - 1$ vertices as intermediates. This leads to the formula

$$D_{k,i,j} = \min\{D_{k-1,i,j}, D_{k-1,i,k} + D_{k-1,k,j}\}$$

The time requirement is once again $O(|V|^3)$. Unlike the two previous dynamic programming examples, this time bound has not been substantially lowered by another approach.

Because the kth stage depends only on the $(k-1)$th stage, it appears that only two $|V| \times |V|$ matrices need to be maintained. However, using k as an *intermediate* vertex on a path that starts or finishes with k does not improve the result unless there is a negative cycle. Thus, only one matrix is necessary, because $D_{k-1,i,k} = D_{k,i,k}$ and $D_{k-1,k,j} = D_{k,k,j}$, which implies that none of the terms on the right change values and need to be saved. This observation leads to the simple program in Figure 10.53, which numbers vertices starting at zero to conform with C's conventions.

On a complete graph, where every pair of vertices is connected (in both directions), this algorithm is almost certain to be faster than $|V|$ iterations of Dijkstra's algorithm, because the loops are so tight. Lines 1 through 4 can be executed in parallel, as can lines 6 through 10. Thus, this algorithm seems to be well suited for parallel computation.

Dynamic programming is a powerful algorithm design technique, which provides a starting point for a solution. It is essentially the divide and conquer paradigm of solving simpler problems first, with the important difference being that the simpler problems are not a clear division of the original. Because subproblems are repeatedly solved, it is important to record their solutions in a table rather than recompute them. In some cases, the solution can be improved (although it is certainly not always obvious and frequently difficult), and in other cases, the dynamic programming technique is the best approach known.

In some sense, if you have seen one dynamic programming problem, you have seen them all. More examples of dynamic programming can be found in the exercises and references.

10.4. Randomized Algorithms

Suppose you are a professor who is giving weekly programming assignments. You want to make sure that the students are doing their own programs or, at the very least, understand the code they are submitting. One solution is to give a quiz on the day that each program is due. On the other hand, these quizzes take time out of class, so it might only be practical to do this for roughly half of the programs. Your problem is to decide when to give the quizzes.

Of course, if the quizzes are announced in advance, that could be interpreted as an implicit license to cheat for the 50 percent of the programs that will not get a quiz. One could adopt the unannounced strategy of giving quizzes on alternate programs, but students would figure out the strategy before too long. Another possibility is to give quizzes on what seem like the important programs, but this would likely lead to similar quiz patterns from semester to semester. Student grapevines being what they are, this strategy would probably be worthless after a semester.

One method that seems to eliminate these problems is to use a coin. A quiz is made for every program (making quizzes is not nearly as time-consuming as grading them), and at the start of class, the professor will flip a coin to decide whether the quiz is to be given. This way, it is impossible to know before class whether or not the quiz will occur, and these patterns do not repeat from semester to semester. Thus, the students will have to expect that a quiz will occur with 50 percent probability, regardless of previous quiz patterns. The disadvantage is that it is possible that there is no quiz for an entire semester. This is not a likely occurrence, unless the coin is suspect. Each semester, the expected number of quizzes is half the number of programs, and with high probability, the number of quizzes will not deviate much from this.

This example illustrates what we call *randomized algorithms.* At least once during the algorithm, a random number is used to make a decision. The running time of the algorithm depends not only on the particular input, but also on the random numbers that occur.

The worst-case running time of a randomized algorithm is almost always the same as the worst-case running time of the nonrandomized algorithm. The important difference is that a good randomized algorithm has no bad inputs, but only bad random numbers (relative to the particular input). This may seem like only a philosophical difference, but actually it is quite important, as the following example shows.

Consider two variants of quicksort. Variant A uses the first element as pivot, while variant B uses a randomly chosen element as pivot. In both cases, the worst-case running time is $\Theta(N^2)$, because it is possible at each step that the largest element is chosen as pivot. The difference between these worst cases is that there is a particular input that can always be presented to variant A to cause the bad running time. Variant A will run in $\Theta(N^2)$ time every single time it is given an already-sorted list. If variant B is presented with the same input twice, it will have two different running times, depending on what random numbers occur.

Throughout the text, in our calculations of running times, we have assumed that all inputs are equally likely. This is not true, because nearly sorted input, for instance, occurs much more often than is statistically expected, and this causes problems, particularly for quicksort and binary search trees. By using a randomized algorithm, the particular input is no longer important. The random numbers are important, and we can get an *expected* running time, where we now average over all possible random numbers instead of over all possible inputs. Using quicksort with a random pivot gives an $O(N \log N)$-expected-time algorithm. This means that for any input, including already-sorted input, the running time is expected to be $O(N \log N)$, based on the statistics of random numbers. An expected running time bound is somewhat stronger than an average-case bound but, of course, is weaker than the corresponding worst-case bound. On the other hand, as we saw in the selection problem, solutions that obtain the worst-case bound are frequently not as practical as their average-case counterparts. Randomized algorithms usually are.

In this section we will examine two uses of randomization. First, we will see a novel scheme for supporting the binary search tree operations in $O(\log N)$ expected time. Once again, this means that there are no bad inputs, just bad random numbers. From a theoretical point of view, this is not terribly exciting, since balanced search trees achieve this bound in the worst case. Nevertheless, the use of randomization leads to relatively simple algorithms for searching, inserting, and especially deleting.

Our second application is a randomized algorithm to test the primality of large numbers. No efficient polynomial-time nonrandomized algorithms are known for this problem. The algorithm we present runs quickly but occasionally makes an error. The probability of error can, however, be made negligibly small.

10.4.1. Random Number Generators

Since our algorithms require random numbers, we must have a method to generate them. Actually, true randomness is virtually impossible to do on a computer, since these numbers will depend on the algorithm, and thus cannot possibly be random. Generally, it suffices to produce *pseudorandom* numbers, which are numbers that appear to be random. Random numbers have many known statistical properties; pseudorandom numbers satisfy most of these properties. Surprisingly, this too is much easier said than done.

Suppose we only need to flip a coin; thus, we must generate a 0 or 1 randomly. One way to do this is to examine the system clock. The clock might record time as an integer that counts the number of seconds since some starting time. We could then use the lowest bit. The problem is that this does not work well if a sequence of random numbers is needed. One second is a long time, and the clock might not change at all while the program is running. Even if the time were recorded in units of microseconds, if the program were running by itself the sequence of numbers that would be generated would be far from random, since the time between calls to the generator would be essentially identical on every program invocation. We see, then, that what is really needed is a *sequence* of random numbers.* These numbers should appear independent. If a coin is flipped and heads appears, the next coin flip should still be equally likely to come up heads or tails.

The simplest method to generate random numbers is the linear congruential generator, which was first described by Lehmer in 1951. Numbers $x_1, x_2, \ldots$ are generated satisfying

$$x_{i+1} = Ax_i \bmod M$$

To start the sequence, some value of x_0 must be given. This value is known as the *seed*. If $x_0 = 0$, then the sequence is far from random, but if A and M are correctly chosen, then any other $1 \le x_0 < M$ is equally valid. If M is prime, then x_i is never 0. As an example, if $M = 11$, $A = 7$, and $x_0 = 1$, then the numbers generated are

$$7, 5, 2, 3, 10, 4, 6, 9, 8, 1, 7, 5, 2, \ldots$$

Notice that after $M - 1 = 10$ numbers, the sequence repeats. Thus, this sequence has a period of $M - 1$, which is as large as possible (by the pigeonhole principle). If M is prime, there are always choices of A that give a full period of $M - 1$. Some choices of A do not; if $A = 5$ and $x_0 = 1$, the sequence has a short period of 5.

$$5, 3, 4, 9, 1, 5, 3, 4, \ldots$$

*We will use random in place of pseudorandom in the rest of this section.

If M is chosen to be a large, 31-bit prime, the period should be significantly large for most applications. Lehmer suggested the use of the 31-bit prime $M = 2^{31} - 1 = 2{,}147{,}483{,}647$. For this prime, $A = 48{,}271$ is one of the many values that gives a full-period generator. Its use has been well studied and is recommended by experts in the field. We will see later that with random number generators, tinkering usually means breaking, so one is well advised to stick with this formula until told otherwise.

This seems like a simple routine to implement. Generally, a global variable is used to hold the current value in the sequence of x's. This is the rare case where a global variable is useful. This global variable is initialized by some routine. When debugging a program that uses random numbers, it is probably best to set $x_0 = 1$, so that the same random sequence occurs all the time. When the program seems to work, either the system clock can be used or the user can be asked to input a value for the seed.

It is also common to return a random real number in the open interval $(0, 1)$ (0 and 1 are not possible values); this can be done by dividing by M. From this, a random number in any closed interval $[\alpha, \beta]$ can be computed by normalizing. This yields the "obvious" routine in Figure 10.54 which, unfortunately, works on few machines.

The problem with this routine is that the multiplication could overflow; although this is not an error, it affects the result and thus the pseudorandomness. Schrage gave a procedure in which all of the calculations can be done on a 32-bit machine without overflow. We compute the quotient and remainder of M/A and define these as Q and R, respectively. In our case, $Q = 44{,}488$, $R = 3{,}399$, and $R < Q$. We have

Figure 10.54 Random number generator that does not work

```
static unsigned long Seed = 1;

#define A 48271L
#define M 2147483647L

double
Random( void )
{
    Seed = ( A * Seed ) % M;
    return ( double ) Seed / M;
}

void
Initialize( unsigned long InitVal )
{
    Seed = InitVal;
}
```

$$\begin{aligned}x_{i+1} &= Ax_i \bmod M = Ax_i - M\left\lfloor \frac{Ax_i}{M} \right\rfloor \\ &= Ax_i - M\left\lfloor \frac{x_i}{Q} \right\rfloor + M\left\lfloor \frac{x_i}{Q} \right\rfloor - M\left\lfloor \frac{Ax_i}{M} \right\rfloor \\ &= Ax_i - M\left\lfloor \frac{x_i}{Q} \right\rfloor + M\left(\left\lfloor \frac{x_i}{Q} \right\rfloor - \left\lfloor \frac{Ax_i}{M} \right\rfloor\right)\end{aligned}$$

Since $x_i = Q\lfloor \frac{x_i}{Q} \rfloor + x_i \bmod Q$, we can replace the leading Ax_i and obtain

$$\begin{aligned}x_{i+1} &= A\left(Q\left\lfloor \frac{x_i}{Q} \right\rfloor + x_i \bmod Q\right) - M\left\lfloor \frac{x_i}{Q} \right\rfloor + M\left(\left\lfloor \frac{x_i}{Q} \right\rfloor - \left\lfloor \frac{Ax_i}{M} \right\rfloor\right) \\ &= (AQ - M)\left\lfloor \frac{x_i}{Q} \right\rfloor + A(x_i \bmod Q) + M\left(\left\lfloor \frac{x_i}{Q} \right\rfloor - \left\lfloor \frac{Ax_i}{M} \right\rfloor\right)\end{aligned}$$

Since $M = AQ + R$, it follows that $AQ - M = -R$. Thus, we obtain

$$x_{i+1} = A(x_i \bmod Q) - R\left\lfloor \frac{x_i}{Q} \right\rfloor + M\left(\left\lfloor \frac{x_i}{Q} \right\rfloor - \left\lfloor \frac{Ax_i}{M} \right\rfloor\right)$$

The term $\delta(x_i) = \lfloor \frac{x_i}{Q} \rfloor - \lfloor \frac{Ax_i}{M} \rfloor$ is either 0 or 1, because both terms are integers and their difference lies between 0 and 1. Thus, we have

$$x_{i+1} = A(x_i \bmod Q) - R\left\lfloor \frac{x_i}{Q} \right\rfloor + M\,\delta(x_i)$$

A quick check shows that because $R < Q$, all the remaining terms can be calculated without overflow (this is one of the reasons for choosing $A = 48{,}271$). Furthermore, $\delta(x_i) = 1$ only if the remaining terms evaluate to less than zero. Thus $\delta(x_i)$ does not need to be explicitly computed but can be determined by a simple test. This leads to the program in Figure 10.55.

This program works as long as INT_MAX $\geq 2^{31} - 1$. One might be tempted to assume that all machines have a random number generator at least as good as the one in Figure 10.55 in their standard library. Sadly, this is not true. Many libraries have generators based on the function

$$x_{i+1} = (Ax_i + C) \bmod 2^B$$

where B is chosen to match the number of bits in the machine's integer, and C is odd. These libraries also return x_i, instead of a value between 0 and 1. Unfortunately, these generators always produce values of x_i that alternate between even and odd—hardly a desirable property. Indeed, the lower k bits cycle with period 2^k (at best). Many other random number generators have much smaller cycles than the one provided in Figure 10.55. These are not suitable for the case where long sequences of random numbers are needed. Finally, it may seem that we can get a better random number generator by adding a constant to the equation. For instance, it seems that

$$x_{i+1} = (48{,}271x_i + 1) \bmod (2^{31} - 1)$$

would somehow be even more random. This illustrates how fragile these generators are.

```
static unsigned long Seed = 1;

#define A 48271L
#define M 2147483647L
#define Q ( M / A )
#define R ( M % A )

double
Random( void )
{
    long TmpSeed;

    TmpSeed = A * ( Seed % Q ) - R * ( Seed / Q );
    if( TmpSeed >= 0 )
        Seed = TmpSeed;
    else
        Seed = TmpSeed + M;

    return ( double ) Seed / M;
}

void
Initialize( unsigned long InitVal )
{
    Seed = InitVal;
}
```

Figure 10.55 Random number generator that works on 32-bit machines

$$[48{,}271(179{,}424{,}105) + 1] \bmod (2^{31} - 1) = 179{,}424{,}105$$

so if the seed is 179,424,105, the generator gets stuck in a cycle of period 1.

10.4.2. Skip Lists

Our first use of randomization is a data structure that supports both searching and insertion in $O(\log N)$ expected time. As mentioned in the introduction to this section, this means that the running time for each operation on *any input sequence* has expected value $O(\log N)$, where the expectation is based on the random number generator. It is possible to add deletion and all the operations that involve ordering and obtain expected time bounds that match the average time bounds of binary search trees.

The simplest possible data structure to support searching is the linked list. Figure 10.56 shows a simple linked list. The time to perform a search is proportional to the number of nodes that have to be examined, which is at most N.

Figure 10.56 Simple linked list

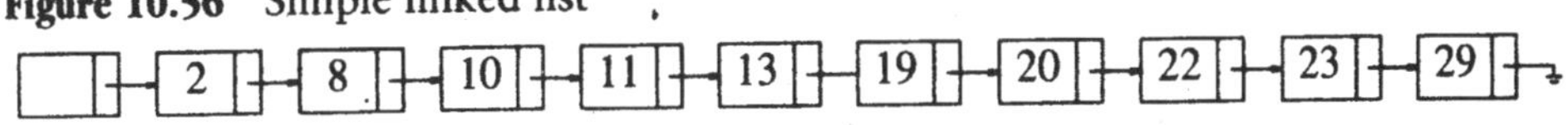

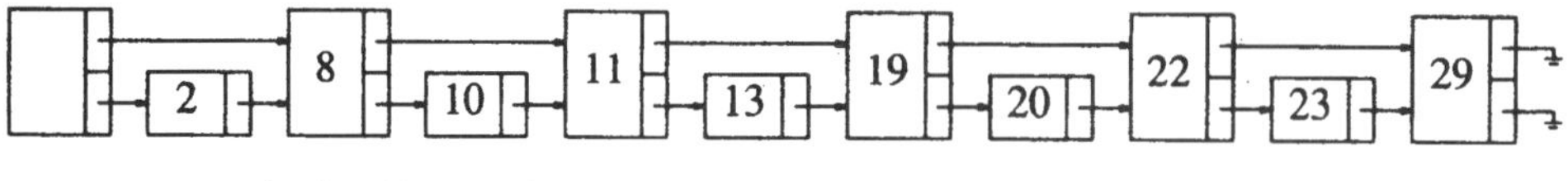

Figure 10.57 Linked list with pointers to two cells ahead

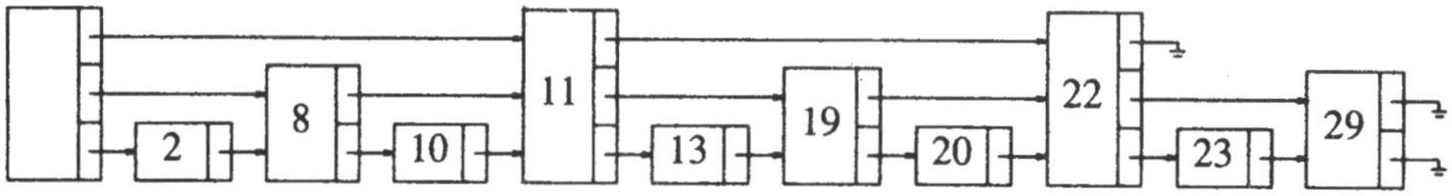

Figure 10.58 Linked list with pointers to four cells ahead

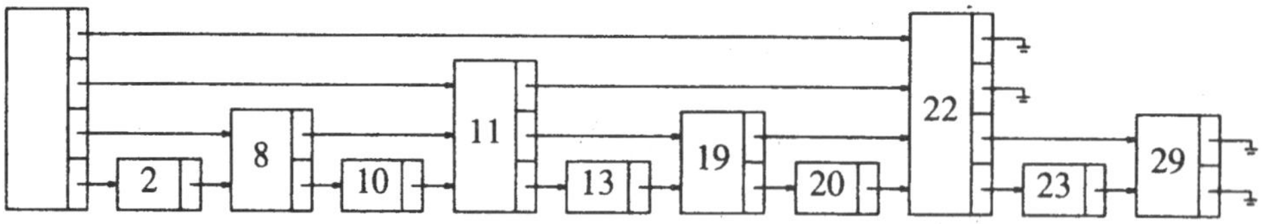

Figure 10.59 Linked list with pointers to 2^i cells ahead

Figure 10.57 shows a linked list in which every other node has an additional pointer to the node two ahead of it in the list. Because of this, at most $\lceil N/2 \rceil + 1$ nodes are examined in the worst case.

We can extend this idea and obtain Figure 10.58. Here, every fourth node has a pointer to the node four ahead. Only $\lceil N/4 \rceil + 2$ nodes are examined.

The limiting case of this argument is shown in Figure 10.59. Every 2^ith node has a pointer to the node 2^i ahead of it. The total number of pointers has only doubled, but now at most $\lceil \log N \rceil$ nodes are examined during a search. It is not hard to see that the total time spent for a search is $O(\log N)$, because the search consists of either advancing to a new node or dropping to a lower pointer in the same node. Each of these steps consumes at most $O(\log N)$ total time during a search. Notice that the search in this data structure is essentially a binary search.

The problem with this data structure is that it is much too rigid to allow efficient insertion. The key to making this data structure usable is to relax the structure conditions slightly. We define a *level k node* to be a node that has k pointers. As Figure 10.59 shows, the ith pointer in any level k node ($k \geq i$) points to the next node with at least i levels. This is an easy property to maintain; however, Figure 10.59 shows a more restrictive property than this. We thus drop the restriction that the ith pointer points to the node 2^i ahead, and we replace it with the less restrictive condition above.

When it comes time to insert a new element, we allocate a new node for it. We must at this point decide what level the node should be. Examining Figure 10.59, we find that roughly half the nodes are level 1 nodes, roughly a quarter are level 2, and, in general, approximately $1/2^i$ nodes are level i. We choose the level of the node randomly, in accordance with this probability distribution. The easiest way to do this is to flip a coin until a head occurs and use the total number of flips as the node level. Figure 10.60 shows a typical skip list.

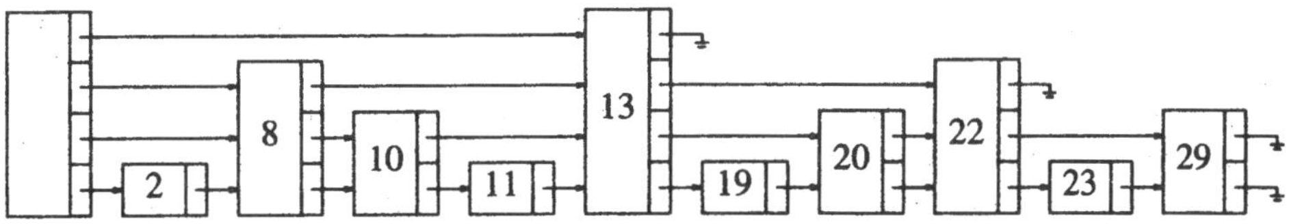

Figure 10.60 A skip list

Given this, the skip list algorithms are simple to describe. To perform a *Find*, we start at the highest pointer at the header. We traverse along this level until we find that the next node is larger than the one we are looking for (or *NULL*). When this occurs, we go to the next lower level and continue the strategy. When progress is stopped at level 1, either we are in front of the node we are looking for, or it is not in the list. To perform an *Insert*, we proceed as in a *Find*, and keep track of each point where we switch to a lower level. The new node, whose level is determined randomly, is then spliced into the list. This operation is shown in Figure 10.61.

A cursory analysis shows that since the expected number of nodes at each level is unchanged from the original (nonrandomized) algorithm, the total amount of work that is expected to be performed traversing to nodes on the same level is unchanged. This tells us that these operations have $O(\log N)$ *expected* costs. Of course, a more formal proof is required, but it is not much different from this.

Skip lists are similar to hash tables, in that they require an estimate of the number of elements that will be in the list (so that the number of levels can be determined). If an estimate is not available, we can assume a large number or use a technique similar to rehashing. Experiments have shown that skip lists are as efficient as many balanced search tree implementations and are certainly much simpler to implement in many languages.

10.4.3. Primality Testing

In this section we examine the problem of determining whether or not a large number is prime. As was mentioned at the end of Chapter 2, some cryptography schemes depend on the difficulty of factoring a large, 200-digit number into two

Figure 10.61 Before and after an insertion

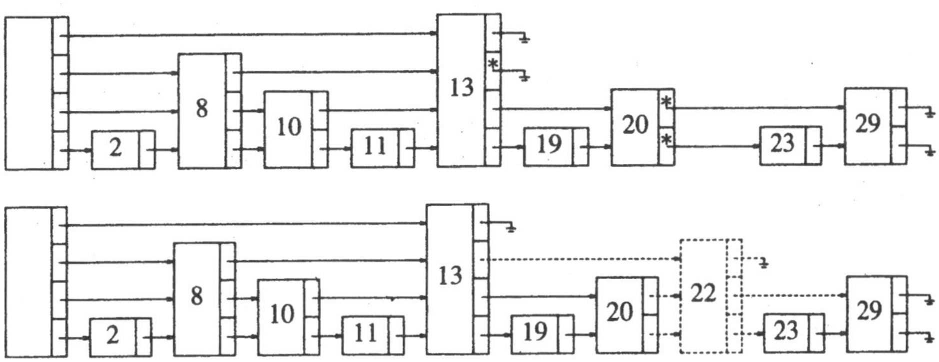

100-digit primes. In order to implement this scheme, we need a method of generating these two primes. The problem is of major theoretical interest, because nobody now knows how to test whether a d-digit number N is prime in time polynomial in d. For instance, the obvious method of testing for the divisibility by odd numbers from 3 to $\sqrt{N}$ requires roughly $\frac{1}{2}\sqrt{N}$ divisions, which is about $2^{d/2}$. On the other hand, this problem is not thought to be NP-complete; thus, it is one of the few problems on the fringe—its complexity is unknown at the time of this writing.

In this chapter, we will give a polynomial-time algorithm that can test for primality. If the algorithm declares that the number is not prime, we can be certain that the number is not prime. If the algorithm declares that the number is prime, then, with high probability but not 100 percent certainty, the number is prime. The error probability does not depend on the particular number that is being tested but instead depends on random choices made by the algorithm. Thus, this algorithm occasionally makes a mistake, but we will see that the error ratio can be made arbitrarily negligible.

The key to the algorithm is a well-known theorem due to Fermat.

THEOREM 10.10.

Fermat's Lesser Theorem: If P is prime, and $0 < A < P$, then $A^{P-1} \equiv 1 (\bmod P)$.

PROOF:

A proof of this theorem can be found in any textbook on number theory.

For instance, since 67 is prime, $2^{66} \equiv 1 (\bmod 67)$. This suggests an algorithm to test whether a number N is prime. Merely check whether $2^{N-1} \equiv 1 (\bmod N)$. If $2^{N-1} \not\equiv 1 (\bmod N)$, then we can be certain that N is not prime. On the other hand, if the equality holds, then N is probably prime. For instance, the smallest N that satisfies $2^{N-1} \equiv 1 (\bmod N)$ but is not prime is $N = 341$.

This algorithm will occasionally make errors, but the problem is that it will always make the same errors. Put another way, there is a fixed set of N for which it does not work. We can attempt to randomize the algorithm as follows: Pick $1 < A < N - 1$ at random. If $A^{N-1} \equiv 1 (\bmod N)$, declare that N is probably prime, otherwise declare that N is definitely not prime. If $N = 341$, and $A = 3$, we find that $3^{340} \equiv 56 (\bmod 341)$. Thus, if the algorithm happens to choose $A = 3$, it will get the correct answer for $N = 341$.

Although this seems to work, there are numbers that fool even this algorithm for most choices of A. One such set of numbers is known as the *Carmichael numbers*. These are not prime but satisfy $A^{N-1} \equiv 1 (\bmod N)$ for *all* $0 < A < N$ that are relatively prime to N. The smallest such number is 561. Thus, we need an additional test to improve the chances of not making an error.

In Chapter 7, we proved a theorem related to quadratic probing. A special case of this theorem is the following:

THEOREM 10.11.

If P is prime and $0 < X < P$, the only solutions to $X^2 \equiv 1 (\bmod P)$ are $X = 1, P - 1$.

PROOF:
$X^2 \equiv 1(\text{mod}\, P)$ implies that $X^2 - 1 \equiv 0(\text{mod}\, P)$. This implies $(X-1)(X+1) \equiv 0(\text{mod}\, P)$. Since P is prime, $0 < X < P$, and P must divide either $(X-1)$ or $(X+1)$, the theorem follows.

Therefore, if at any point in the computation of $A^{N-1}(\text{mod}\, N)$ we discover a violation of this theorem, we can conclude that N is definitely not prime. If we use *Power*, from Section 2.4.4, we see that there will be several opportunities to apply this test. We modify this routine to perform operations mod N, and apply the test of Theorem 10.11. This strategy is implemented in Figure 10.62.

Recall that if *Witness* returns anything but 1, it has *proven* that N cannot be prime. The proof is nonconstructive, because it gives no method of actually finding the factors. It has been shown that for any (sufficiently large) N, at most $(N-9)/4$ values of A fool this algorithm. Thus, if A is chosen at random, and the algorithm answers that N is (probably) prime, then the algorithm is correct at least 75 percent of the time. Suppose *Witness* is run 50 times. The probability that the algorithm is fooled once is at most $\frac{1}{4}$. Thus, the probability that 50 independent random trials fool the algorithm is never more than $1/4^{50} = 2^{-100}$. This is actually a very conservative estimate, which holds for only a few choices of N. Even so, one is more likely to see a hardware error than an incorrect claim of primality.

10.5. Backtracking Algorithms

The last algorithm design technique we will examine is *backtracking*. In many cases, a backtracking algorithm amounts to a clever implementation of exhaustive search, with generally unfavorable performance. This is not always the case, however, and even so, in some cases, the savings over a brute force exhaustive search can be significant. Performance is, of course, relative: an $O(N^2)$ algorithm for sorting is pretty bad, but an $O(N^5)$ algorithm for the traveling salesman (or any *NP*-complete) problem would be a landmark result.

A practical example of a backtracking algorithm is the problem of arranging furniture in a new house. There are many possibilities to try, but typically only a few are actually considered. Starting with no arrangement, each piece of furniture is placed in some part of the room. If all the furniture is placed and the owner is happy, then the algorithm terminates. If we reach a point where all subsequent placement of furniture is undesirable, we have to undo the last step and try an alternative. Of course, this might force another undo, and so forth. If we find that we undo all possible first steps, then there is no placement of furniture that is satisfactory. Otherwise, we eventually terminate with a satisfactory arrangement. Notice that although this algorithm is essentially brute force, it does not try all possibilities directly. For instance, arrangements that consider placing the sofa in the kitchen are never tried. Many other bad arrangements are discarded early, because an undesirable subset of the arrangement is detected. The elimination of a large group of possibilities in one step is known as *pruning*.

We will see two examples of backtracking algorithms. The first is a problem in computational geometry. Our second example shows how computers select moves in games, such as chess and checkers.

```
/* If Witness does not return 1, N is definitely */
/* composite. Do this by computing ( A ^ i ) mod N and */
/* looking for non-trivial square roots of 1 along the */
/* way. We are assuming very large numbers, so this */
/* is pseudocode */

HugeInt
Witness( HugeInt A, HugeInt i, HugeInt N )
{
    HugeInt X, Y;

    if( i == 0 )
        return 1;

    X = Witness( A, i / 2, N );
    if( X == 0 )  /* If N is recursively composite, stop */
        return 0;

    /* N is not prime if we find a non-trivial root of 1 */
    Y = ( X * X ) % N;
    if( Y == 1 && X != 1 && X != N - 1 )
        return 0;

    if( i % 2 != 0 )
        Y = ( A * Y ) % N;

    return Y;
}

/* IsPrime: Test if N >= 3 is prime using one value */
/* of A. Repeat this procedure as many times as needed */
/* for desired error rate */

int
IsPrime( HugeInt N )
{
    return Witness( RandInt( 2, N - 2 ), N - 1, N ) == 1;
}
```

Figure 10.62 A probabilistic primality testing algorithm

10.5.1. The Turnpike Reconstruction Problem

Suppose we are given N points, $p_1, p_2, \ldots, p_N$, located on the x-axis. x_i is the x coordinate of p_i. Let us further assume that $x_1 = 0$ and the points are given from left to right. These N points determine $N(N-1)/2$ (not necessarily unique) distances $d_1, d_2, \ldots, d_N$ between every pair of points of the form $|x_i - x_j|$ $(i \neq j)$. It is clear that if we are given the set of points, it is easy to *construct* the set of distances in $O(N^2)$ time. This set will not be sorted, but if we are willing to settle for an $O(N^2 \log N)$ time bound, the distances can be sorted, too. The ***turnpike reconstruction problem*** is to reconstruct a point set from the distances. This finds applications in physics and molecular biology (see the references for pointers to more specific information). The name derives from the analogy of points to turnpike exits on East Coast highways. Just as factoring seems harder than multiplication, the reconstruction problem seems harder than the construction problem. Nobody has been able to give an algorithm that is guaranteed to work in polynomial time. The algorithm that we will present generally runs in $O(N^2 \log N)$ but can take exponential time in the worst case.

Of course, given one solution to the problem, an infinite number of others can be constructed by adding an offset to all the points. This is why we insist that the first point is anchored at 0 and that the point set that constitutes a solution is output in nondecreasing order.

Let D be the set of distances, and assume that $|D| = M = N(N-1)/2$. As an example, suppose that

$$D = \{1, 2, 2, 2, 3, 3, 3, 4, 5, 5, 5, 6, 7, 8, 10\}$$

Since $|D| = 15$, we know that $N = 6$. We start the algorithm by setting $x_1 = 0$. Clearly, $x_6 = 10$, since 10 is the largest element in D. We remove 10 from D. The points that we have placed and the remaining distances are as shown in the following figure.

$$D = \{ 1, 2, 2, 2, 3, 3, 3, 4, 5, 5, 5, 6, 7, 8 \}$$

The largest remaining distance is 8, which means that either $x_2 = 2$ or $x_5 = 8$. By symmetry, we can conclude that the choice is unimportant, since either both choices lead to solutions (which are mirror images of each other), or neither do, so we can set $x_5 = 8$ without affecting the solution. We then remove the distances $x_6 - x_5 = 2$ and $x_5 - x_1 = 8$ from D, obtaining

$$D = \{ 1, 2, 2, 3, 3, 3, 4, 5, 5, 5, 6, 7 \}$$

The next step is not obvious. Since 7 is the largest value in D, either $x_4 = 7$ or $x_2 = 3$. If $x_4 = 7$, then the distances $x_6 - 7 = 3$ and $x_5 - 7 = 1$ must also be present in D. A quick check shows that indeed they are. On the other hand, if we set $x_2 = 3$, then $3 - x_1 = 3$ and $x_5 - 3 = 5$ must be present in D. These distances are also in D, so we have no guidance on which choice to make. Thus, we try one and see if it leads to a solution. If it turns out that it does not, we can come back and try the other. Trying the first choice, we set $x_4 = 7$, which leaves

$$D = \{2, 2, 3, 3, 4, 5, 5, 5, 6\}$$

At this point, we have $x_1 = 0$, $x_4 = 7$, $x_5 = 8$, and $x_6 = 10$. Now the largest distance is 6, so either $x_3 = 6$ or $x_2 = 4$. But if $x_3 = 6$, then $x_4 - x_3 = 1$, which is impossible, since 1 is no longer in D. On the other hand, if $x_2 = 4$ then $x_2 - x_0 = 4$, and $x_5 - x_2 = 4$. This is also impossible, since 4 only appears once in D. Thus, this line of reasoning leaves no solution, so we backtrack.

Since $x_4 = 7$ failed to produce a solution, we try $x_2 = 3$. If this also fails, we give up and report no solution. We now have

$$D = \{1, 2, 2, 3, 3, 4, 5, 5, 6\}$$

Once again, we have to choose between $x_4 = 6$ and $x_3 = 4$. $x_3 = 4$ is impossible, because D only has one occurrence of 4, and two would be implied by this choice. $x_4 = 6$ is possible, so we obtain

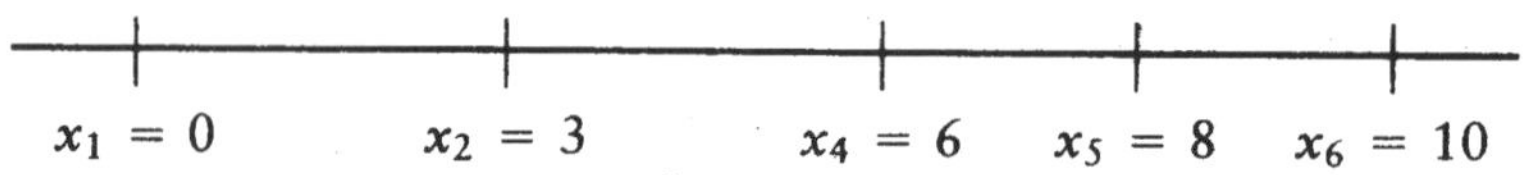

$$D = \{1, 2, 3, 5, 5\}$$

The only remaining choice is to assign $x_3 = 5$; this works because it leaves D empty, and so we have a solution.

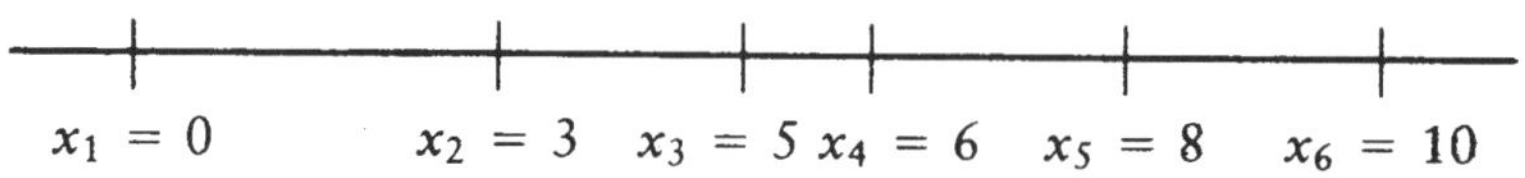

$$D = \{\}$$

Figure 10.63 shows a decision tree representing the actions taken to arrive at the solution. Instead of labeling the branches, we have placed the labels in the branches' destination nodes. A node with an asterisk indicates that the points chosen are inconsistent with the given distances; nodes with two asterisks have only impossible nodes as children, and thus represent an incorrect path.

The pseudocode to implement this algorithm is mostly straightforward. The driving routine, *Turnpike,* is shown in Figure 10.64. It receives the point array X (which need not be initialized), the distance array D, and N.* If a solution is discovered, then *true* will be returned, the answer will be placed in X, and D will be empty. Otherwise, *false* will be returned, X will be undefined, and the distance array D

*We have used one-letter variable names, which is generally poor style, for consistency with the worked example. We also, for simplicity, do not give the type of variables.

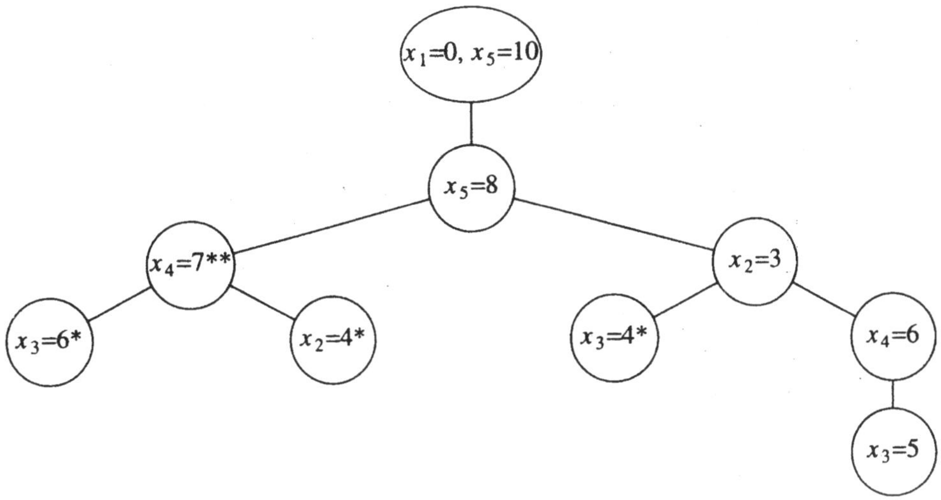

Figure 10.63 Decision tree for the worked turnpike reconstruction example

```
int
Turnpike( int X[ ], DistSet D, int N )
{
    X[ 1 ] = 0;
    X[ N ] = DeleteMax( D );
    X[ N - 1 ] = DeleteMax( D );
    if( X[ N ] - X[ N - 1 ] ∈ D )
    {
        Remove( X[ N ] - X[ N - 1 ], D );
        return Place( X, D, N, 2, N - 2 );
    }
    else
        return False;
}
```

Figure 10.64 Turnpike reconstruction algorithm: driver routine (pseudocode)

will be untouched. The routine sets x_1, x_{N-1}, and x_N, as described above, alters D, and calls the backtracking algorithm *Place* to place the other points. We presume that a check has already been made to ensure that $|D| = N(N-1)/2$.

The more difficult part is the backtracking algorithm, which is shown in Figure 10.65. Like most backtracking algorithms, the most convenient implementation is recursive. We pass the same arguments plus the boundaries *Left* and *Right*; $x_{Left}, \ldots, x_{Right}$ are the x coordinates of points that we are trying to place. If D is empty (or *Left* > *Right*), then a solution has been found, and we can return. Otherwise, we first try to place $x_{Right} = D_{\max}$. If all the appropriate distances are present (in

```
/* Backtracking algorithm to place the points */
/* X[ Left ... Right ] */
/* X[ 1 ... Left - 1 ] and X[ Right + 1 ... N ] */
/* are already tentatively placed */
/* If Place returns True, */
/* then X[ Left ... Right ] will have values */

int
Place( int X[ ], DistSet D, int N, int Left, int Right )
{
    int DMax, Found = False;

    if( D is empty )
        return True;
    DMax = FindMax( D );

    /* Check if setting X[ Right ] = DMax is feasible */
    if(| X[ j ] - DMax | ∈ D
                for all 1 ≤ j < Left and Right < j ≤ N )
    {
        X[ Right ] = DMax;  /* Try X[ Right ] = DMax */
        for( 1 ≤ j < Left, Right < j ≤ N )
            Delete( | X[ j ] - DMax |, D );
        Found = Place( X, D, N, Left, Right - 1 );

        if( !Found )  /* Backtrack */
        for( 1 ≤ j < Left, Right < j ≤ N )  /* Undo deletion */
            Insert( | X[ j ] - DMax |, D );
    }

    /* If first attempt failed, try to see if setting */
    /* X[ Left ] = X[ N ] - DMax is feasible */
    if( !Found && ( | X[ N ] - DMax - X[ j ] | ∈ D
                for all 1 ≤ j < Left and Right < j ≤ N ) )
    {
        X[ Left ] = X[ N ] - DMax;  /* Same logic as before */
        for( 1 ≤ j < Left, Right < j ≤ N )
            Delete( | X[ N ] - DMax - X[ j ] |, D );
        Found = Place( X, D, N, Left + 1, Right );

        if( !Found )  /* Backtrack */
            for( 1 ≤ j < Left, Right < j ≤ N )  /* Undo */
                Insert( | X[ N ] - DMax - X[ j ] |, D );
    }
    return Found;
}
```

Figure 10.65 Turnpike reconstruction algorithm: backtracking steps (pseudocode)

the correct quantity), then we tentatively place this point, remove these distances, and try to fill from *Left* to *Right* − 1. If the distances are not present, or the attempt to fill *Left* to *Right* − 1 fails, then we try setting $x_{Left} = x_N - d_{\max}$, using a similar strategy. If this does not work, then there is no solution; otherwise a solution has been found, and this information is eventually passed back to *Turnpike* by the *return* statement and X array.

The analysis of the algorithm involves two factors. Suppose lines 9 through 11 and 18 through 20 are never executed. We can maintain D as a balanced binary search (or splay) tree (this would require a code modification, of course). If we never backtrack, there are at most $O(N^2)$ operations involving D, such as deletion and the *Finds* implied at lines 4 and 12 to 13. This claim is obvious for deletions, since D has $O(N^2)$ elements and no element is ever reinserted. Each call to *Place* uses at most $2N$ *Finds*, and since *Place* never backtracks in this analysis, there can be at most $2N^2$ *Finds*. Thus, if there is no backtracking, the running time is $O(N^2 \log N)$.

Of course, backtracking happens, and if it happens repeatedly, then the performance of the algorithm is affected. This can be forced to happen by construction of a pathological case. Experiments have shown that if the points have integer coordinates distributed uniformly and randomly from $[0, D_{\max}]$, where $D_{\max} = \Theta(N^2)$, then, almost certainly, at most one backtrack is performed during the entire algorithm.

10.5.2. Games

As our last application, we will consider the strategy that a computer might use to play a strategic game, such as checkers or chess. We will use, as an example, the much simpler game of tic-tac-toe, because it makes the points easier to illustrate.

Tic-tac-toe is a draw if both sides play optimally. By performing a careful case-by-case analysis, it is not a difficult matter to construct an algorithm that never loses and always wins when presented the opportunity. This can be done, because certain positions are known traps and can be handled by a lookup table. Other strategies, such as taking the center square when it is available, make the analysis simpler. If this is done, then by using a table we can always choose a move based only on the current position. Of course, this strategy requires the programmer, and not the computer, to do most of the thinking.

Minimax Strategy

The more general strategy is to use an evaluation function to quantify the "goodness" of a position. A position that is a win for a computer might get the value of +1; a draw could get 0; and a position that the computer has lost would get a −1. A position for which this assignment can be determined by examining the board is known as a *terminal* position.

If a position is not terminal, the value of the position is determined by recursively assuming optimal play by both sides. This is known as a *minimax* strategy, because one player (the human) is trying to minimize the value of the position, while the other player (the computer) is trying to maximize it.

A *successor position* of P is any position P_s that is reachable from P by playing one move. If the computer is to move when in some position P, it recursively evaluates the value of all the successor positions. The computer chooses the move with the largest value; this is the value of P. To evaluate any successor position P_s, all of P_s's successors are recursively evaluated, and the smallest value is chosen. This smallest value represents the most favorable reply for the human player.

The code in Figure 10.66 makes the computer's strategy more clear. Lines 1 through 4 evaluate immediate wins or draws. If neither of these cases apply, then the position is nonterminal. Recalling that *Value* should contain the maximum of all possible successor positions, line 5 initializes it to the smallest possible value, and the loop in lines 6 through 13 searches for improvements. Each successor position is recursively evaluated in turn by lines 8 through 10. This is recursive, because, as we will see, the procedure *FindHumanMove* calls *FindCompMove*. If the human's response to a move leaves the computer with a more favorable position than that obtained with the previously best computer move, then the *Value* and *BestMove* are updated. Figure 10.67 shows the procedure for the human's move selection. The logic is virtually identical, except that the human player chooses the move that leads to the lowest-valued position. Indeed, it is not difficult to combine these two procedures into one by passing an extra variable, which indicates whose turn it is to move. This does make the code somewhat less readable, so we have stayed with separate routines.

Since these routines must pass back both the value of the position and the best move, we pass the address of two variables that will get this information, by using pointers. The last two parameters now answer the questions "WHERE?" instead of "WHAT?"

As an example, in Figure 10.66, *BestMove* contains the address where the best move can be placed. *FindCompMove* can examine or alter the data at that address by accessing **BestMove*. Line 9 shows how the calling routine should behave. Since the caller has two integers prepared to store the data, and *FindHumanMove* only wants the addresses of these two integers, the address operator (&) is used.

If the & operator is not used at line 9, and both *Dc* and *Response* are zero (which would be typical of uninitialized data), then *FindHumanMove* will try to place its best move and position value in memory location zero. Of course, this is not what was intended, and will almost certainly result in a program crash (try it!). This is the most common error when using the *scanf* family of library routines.

We leave supporting routines as an exercise. The most costly computation is the case where the computer is asked to pick the opening move. Since at this stage the game is a forced draw, the computer selects square 1.* A total of 97,162 positions were examined, and the calculation took a few seconds. No attempt was made to optimize the code. When the computer moves second, the number of positions examined is 5,185 if the human selects the center square, 9,761 when a corner square is selected, and 13,233 when a noncorner edge square is selected.

---

*We numbered the squares starting from the top left and moving right. However, this is only important for the supporting routines.

```
        /* Recursive procedure to find best move for computer */
        /* BestMove points to a number from 1 to 9 indicating square */
        /* Possible evaluations satisfy CompLoss < Draw < CompWin */
        /* Complementary procedure FindHumanMove is Figure 10.67 */
        /* Board is an array and thus can be changed by Place */

        void
        FindCompMove( BoardType Board, int *BestMove, int *Value )
        {
            int Dc, i, Response;  /* Dc means don't care */

/* 1*/      if( FullBoard( Board ) )
/* 2*/          *Value = Draw;
            else
/* 3*/      if( ImmediateCompWin( Board, BestMove ) )
/* 4*/          *Value = CompWin;
            else
            {
/* 5*/          *Value = CompLoss;
/* 6*/          for( i = 1; i <= 9; i++ )  /* Try each square */
                {
/* 7*/              if( IsEmpty( Board, i ) )
                    {
/* 8*/                  Place( Board, i, Comp );
/* 9*/                  FindHumanMove( Board, &Dc, &Response );
/*10*/                  Unplace( Board, i );  /* Restore Board */

/*11*/                  if( Response > *Value )
                        {
                            /* Update best move */
/*12*/                      *Value = Response;
/*13*/                      *BestMove = i;
                        }
                    }
                }
            }
        }
```

Figure 10.66 Minimax tic-tac-toe algorithm: computer selection

```
        void
        FindHumanMove( BoardType Board, int *BestMove, int *Value )
        {
            int Dc, i, Response; /* Dc means don't care */

/* 1*/      if( FullBoard( Board ) )
/* 2*/          *Value = Draw;
            else
/* 3*/      if( ImmediateHumanWin( Board, BestMove ) )
/* 4*/          *Value = CompLoss;
            else
            {
/* 5*/          *Value = CompWin;
/* 6*/          for( i = 1; i <= 9; i++ ) /* Try each square */
                {
/* 7*/              if( IsEmpty( Board, i ) )
                    {
/* 8*/                  Place( Board, i, Human );
/* 9*/                  FindCompMove( Board, &Dc, &Response );
/*10*/                  Unplace( Board, i ); /* Restore board */

/*11*/                  if( Response < *Value )
                        {
                            /* Update best move */
/*12*/                      *Value = Response;
/*13*/                      *BestMove = i;
                        }
                    }
                }
            }
        }
```

Figure 10.67 Minimax tic-tac-toe algorithm: human selection

For more complex games, such as checkers and chess, it is obviously infeasible to search all the way to the terminal nodes.* In this case, we have to stop the search after a certain depth of recursion is reached. The nodes where the recursion is stopped become terminal nodes. These terminal nodes are evaluated with a function that estimates the value of the position. For instance, in a chess program, the evaluation function measures such variables as the relative amount and strength of pieces and positional factors. The evaluation function is crucial for success, because the computer's move selection is based on maximizing this function. The best computer chess programs have surprisingly sophisticated evaluation functions.

*It is estimated that if this search were conducted for chess, at least 10^{100} positions would be examined for the first move. Even if the improvements described later in this section were incorporated, this number could not be reduced to a practical level.

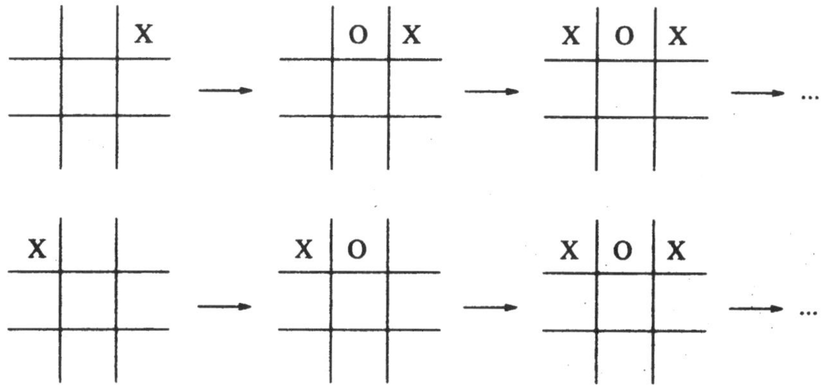

Figure 10.68 Two searches that arrive at identical position

Nevertheless, for computer chess, the single most important factor seems to be number of moves of look-ahead the program is capable of. This is sometimes known as *ply;* it is equal to the depth of the recursion. To implement this, an extra parameter is given to the search routines.

The basic method to increase the look-ahead factor in game programs is to come up with methods that evaluate fewer nodes without losing any information. One method which we have already seen is to use a table to keep track of all positions that have been evaluated. For instance, in the course of searching for the first move, the program will examine the positions in Figure 10.68. If the values of the positions are saved, the second occurrence of a position need not be recomputed; it essentially becomes a terminal position. The data structure that records this is known as a *transposition table;* it is almost always implemented by hashing. In many cases, this can save considerable computation. For instance, in a chess endgame, where there are relatively few pieces, the time savings can allow a search to go several levels deeper.

α–β Pruning

Probably the most significant improvement one can obtain in general is known as α–β pruning. Figure 10.69 shows the trace of the recursive calls used to evaluate some hypothetical position in a hypothetical game. This is commonly referred to as a *game tree.* (We have avoided the use of this term until now, because it is somewhat misleading: no tree is actually constructed by the algorithm. The game tree is just an abstract concept.) The value of the game tree is 44.

Figure 10.70 shows the evaluation of the same game tree, with several unevaluated nodes. Almost half of the terminal nodes have not been checked. We show that evaluating them would not change the value at the root.

First, consider node *D*. Figure 10.71 shows the information that has been gathered when it is time to evaluate *D*. At this point, we are still in *FindHumanMove* and are contemplating a call to *FindCompMove* on *D*. However, we already know that *FindHumanMove* will return at most 40, since it is a *min* node. On the other hand, its *max* node parent has already found a sequence that guarantees 44.

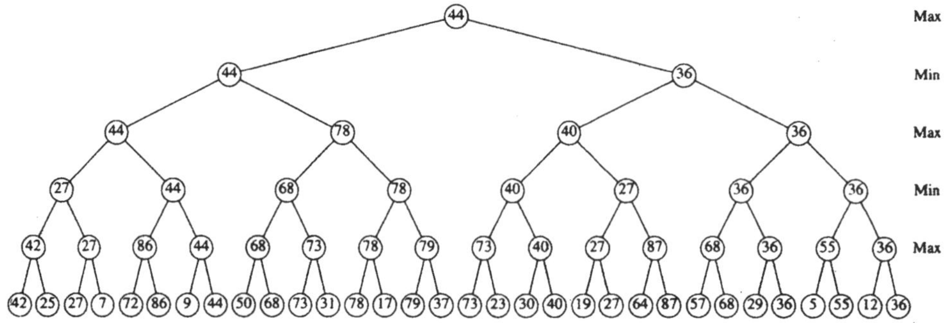

Figure 10.69 A hypothetical game tree

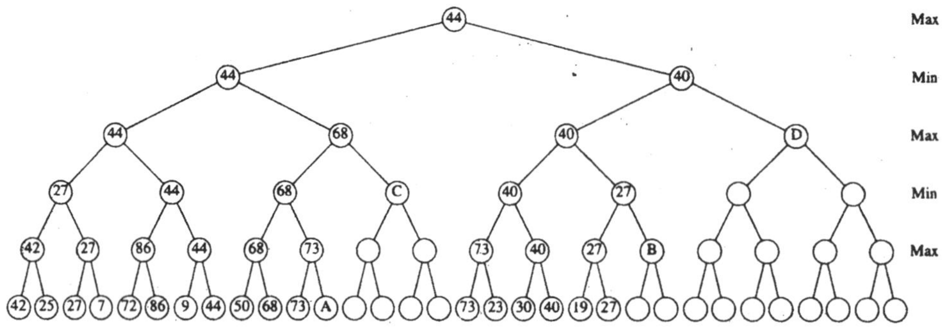

Figure 10.70 A pruned game tree

Nothing that *D* does can possibly increase this value. Therefore, *D* does not need to be evaluated. This pruning of the tree is known as α pruning. An identical situation occurs at node *B*. To implement α pruning, *FindCompMove* passes its tentative maximum (α) to *FindHumanMove*. If the tentative minimum of *FindHumanMove* falls below this value, then *FindHumanMove* returns immediately.

A similar thing happens at nodes *A* and *C*. This time, we are in the middle of a *FindCompMove* and are about to make a call to *FindHumanMove* to evaluate *C*.

Figure 10.71 The node marked ? is unimportant

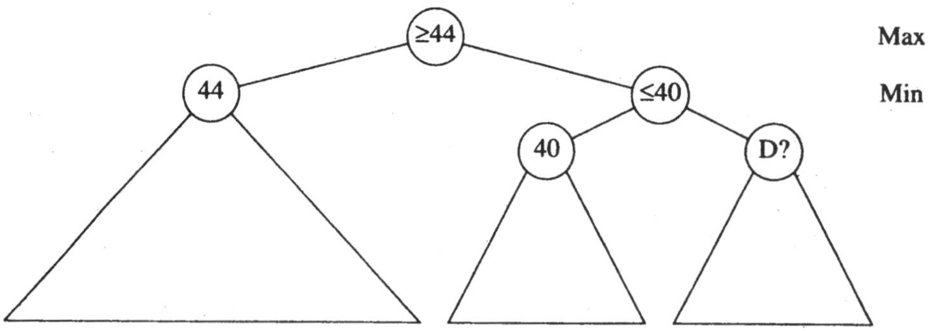

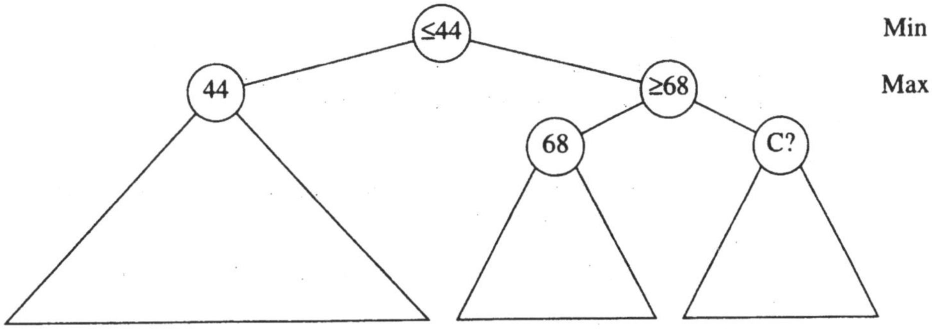

Figure 10.72 The node marked ? is unimportant

Figure 10.72 shows the situation that is encountered at node C. However, the *FindHumanMove,* at the *min* level, which has called *FindCompMove,* has already determined that it can force a value of at most 44 (recall that low values are good for the human side). Since *FindCompMove* has a tentative maximum of 68, nothing that C does will affect the result at the *min* level. Therefore, C should not be evaluated. This type of pruning is known as β pruning; it is the symmetric version of α pruning. When both techniques are combined, we have α–β pruning.

Implementing α–β pruning requires surprisingly little code. Figure 10.73 shows half of the α–β pruning scheme (minus type declarations); you should have no trouble coding the other half.

To take full advantage of α–β pruning, game programs usually try to apply the evaluation function to nonterminal nodes in an attempt to place the best moves early in the search. The result is even more pruning than one would expect from a random ordering of the nodes. Other techniques, such as searching deeper in more active lines of play, are also employed.

In practice, α–β pruning limits the searching to only $O(\sqrt{N})$ nodes, where N is the size of the full game tree. This is a huge saving and means that searches using α–β pruning can go twice as deep as compared to an unpruned tree. Our tic-tac-toe example is not ideal, because there are so many identical values, but even so, the initial search of 97,162 nodes is reduced to 4,493 nodes. (These counts include nonterminal nodes).

In many games, computers are among the best players in the world. The techniques used are very interesting, and can be applied to more serious problems. More details can be found in the references.

Summary

This chapter illustrates five of the most common techniques found in algorithm design. When confronted with a problem, it is worthwhile to see if any of these methods apply. A proper choice of algorithm, combined with judicious use of data structures, can often lead quickly to efficient solutions.

```
        /* Same as before, but perform alpha-beta pruning */
        /* The main routine should make the call with */
        /* Alpha = CompLoss and Beta = CompWin */

        void
        FindCompMove( BoardType Board, int *BestMove, int *Value,
                            int Alpha, int Beta )
        {
            int Dc, i, Response; /* Dc means don't care */

/* 1*/      if( FullBoard( Board ) )
/* 2*/          *Value = Draw;
            else
/* 3*/      if( ImmediateCompWin( Board, BestMove ) )
/* 4*/          *Value = CompWin;
            else
            {
/* 5*/          *Value = Alpha;
/* 6*/          for( i = 1; i <= 9 && *Value < Beta; i++ )
                {
/* 7*/              if( IsEmpty( Board, i ) )
                    {
/* 8*/                  Place( Board, i, Comp );
/* 9*/                  FindHumanMove( Board, &Dc, &Response,
                                                *Value, Beta );
/*10*/                  Unplace( Board, i ); /* Restore board */

/*11*/                  if( Response > *Value )
                        {
                            /* Update best move */
/*12*/                      *Value = Response;
/*13*/                      *BestMove = i;
                        }
                    }
                }
            }
        }
```

Figure 10.73 Minimax tic-tac-toe algorithm with $\alpha - \beta$ pruning: computer selection

Exercises

10.1 Show that the greedy algorithm to minimize the mean completion time for multiprocessor job scheduling works.

10.2 The input is a set of jobs $j_1, j_2, \ldots, j_N$, each of which takes one time unit to complete. Each job j_i earns d_i dollars if it is completed by the time limit t_i, but no money if completed after the time limit.

a. Give an $O(N^2)$ greedy algorithm to solve the problem.

**b. Modify your algorithm to obtain an $O(N \log N)$ time bound. *Hint: The time bound is due entirely to sorting the jobs by money. The rest of the algorithm can be implemented, using the disjoint set data structure, in* $o(N \log N)$.

10.3 A file contains only colons, spaces, newlines, commas, and digits in the following frequency: colon (100), space (605), newline (100), comma (705), 0 (431), 1 (242), 2 (176), 3 (59), 4 (185), 5 (250), 6 (174), 7 (199), 8 (205), 9 (217). Construct the Huffman code.

10.4 Part of the encoded file must be a header indicating the Huffman code. Give a method for constructing the header of size at most $O(N)$ (in addition to the symbols), where N is the number of symbols.

10.5 Complete the proof that Huffman's algorithm generates an optimal prefix code.

10.6 Show that if the symbols are sorted by frequency, Huffman's algorithm can be implemented in linear time.

10.7 Write a program to implement file compression (and uncompression) using Huffman's algorithm.

*10.8 Show that any on-line bin-packing algorithm can be forced to use at least $\frac{3}{2}$ the optimal number of bins, by considering the following sequence of items: N items of size $\frac{1}{6} - 2\epsilon$, N items of size $\frac{1}{3} + \epsilon$, N items of size $\frac{1}{2} + \epsilon$.

10.9 Explain how to implement first fit and best fit in $O(N \log N)$ time.

10.10 Show the operation of all of the bin-packing strategies discussed in Section 10.1.3 on the input 0.42, 0.25, 0.27, 0.07, 0.72, 0.86, 0.09, 0.44, 0.50, 0.68, 0.73, 0.31, 0.78, 0.17, 0.79, 0.37, 0.73, 0.23, 0.30.

10.11 Write a program that compares the performance (both in time and number of bins used) of the various bin packing heuristics.

10.12 Prove Theorem 10.7.

10.13 Prove Theorem 10.8.

*10.14 N points are placed in a unit square. Show that the distance between the closest pair is $O(N^{-1/2})$.

*10.15 Argue that for the closest-points algorithm, the average number of points in the strip is $O(\sqrt{N})$. *Hint: Use the result of the previous exercise.*

10.16 Write a program to implement the closest-pair algorithm.

10.17 What is the asymptotic running time of quickselect, using a median-of-median-of-three partitioning strategy?

10.18 Show that quickselect with median-of-median-of-seven partitioning is linear. Why is median-of-median-of-seven partitioning not used in the proof?

10.19 Implement the quickselect algorithm in Chapter 7, quickselect using median-of-median-of-five partitioning, and the sampling algorithm at the end of Section 10.2.3. Compare the running times.

10.20 Much of the information used to compute the median-of-median-of-five is thrown away. Show how the number of comparisons can be reduced by more careful use of the information.

*10.21 Complete the analysis of the sampling algorithm described at the end of Section 10.2.3, and explain how the values of δ and s are chosen.

10.22 Show how the recursive multiplication algorithm computes XY, where $X = 1234$ and $Y = 4321$. Include all recursive computations.

10.23 Show how to multiply two complex numbers $X = a + bi$ and $Y = c + di$ using only three multiplications.

10.24 a. Show that

$$X_L Y_R + X_R Y_L = (X_L + X_R)(Y_L + Y_R) - X_L Y_L - X_R Y_R$$

b. This gives an $O(N^{1.59})$ algorithm to multiply N-bit numbers. Compare this method to the solution in the text.

10.25 *a. Show how to multiply two numbers by solving five problems that are roughly one-third of the original size.

**b. Generalize this problem to obtain an $O(N^{1+\epsilon})$ algorithm for any constant $\epsilon > 0$.

c. Is the algorithm in part (b) better than $O(N \log N)$?

10.26 Why is it important that Strassen's algorithm does not use commutativity in the multiplication of 2×2 matrices?

10.27 Two 70×70 matrices can be multiplied using 143,640 multiplications. Show how this can be used to improve the bound given by Strassen's algorithm.

10.28 What is the optimal way to compute $\mathbf{A}_1\mathbf{A}_2\mathbf{A}_3\mathbf{A}_4\mathbf{A}_5\mathbf{A}_6$, where the dimensions of the matrices are: $\mathbf{A}_1 : 10 \times 20$, $\mathbf{A}_2 : 20 \times 1$, $\mathbf{A}_3 : 1 \times 40$, $\mathbf{A}_4 : 40 \times 5$, $\mathbf{A}_5 : 5 \times 30$, $\mathbf{A}_6 : 30 \times 15$?

10.29 Show that none of the following greedy algorithms for chained matrix multiplication work. At each step

a. Compute the cheapest multiplication.

b. Compute the most expensive multiplication.

c. Compute the multiplication between the two matrices M_i and M_{i+1}, such that the number of columns in M_i is minimized (breaking ties by one of the rules above).

10.30 Write a program to compute the best ordering of matrix multiplication. Include the routine to print out the actual ordering.

10.31 Show the optimal binary search tree for the following words, where the frequency of occurrence is in parentheses: ***a*** (0.18), ***and*** (0.19), ***I*** (0.23), ***it*** (0.21), *or* (0.19).

*10.32 Extend the optimal binary search tree algorithm to allow for unsuccessful searches. In this case, q_j, for $1 \le j < N$, is the probability that a search is performed for any word W satisfying $w_j < W < w_{j+1}$. q_0 is the probability of performing a search for $W < w_1$, and q_N is the probability of performing a search for $W > w_N$. Notice that $\sum_{i=1}^{N} p_i + \sum_{j=0}^{N} q_j = 1$.

*10.33 Suppose $C_{i,i} = 0$ and that otherwise

$$C_{i,j} = W_{i,j} + \min_{i<k\le j} (C_{i,k-1} + C_{k,j})$$

Suppose that **W** satisfies the *quadrangle inequality*, namely, for all $i \le i' \le j \le j'$,

$$W_{i,j} + W_{i',j'} \le W_{i',j} + W_{i,j'}$$

Suppose further, that **W** is *monotone:* If $i \le i'$ and $j \le j'$, then $W_{i,j} \le W_{i',j'}$.

a. Prove that **C** satisfies the quadrangle inequality.

b. Let $R_{i,j}$ be the largest k that achieves the minimum $C_{i,k-1} + C_{k,j}$. (That is, in case of ties, choose the largest k). Prove that

$$R_{i,j} \le R_{i,j+1} \le R_{i+1,j+1}$$

c. Show that **R** is nondecreasing along each row and column.

d. Use this to show that all entries in **C** can be computed in $O(N^2)$ time.

e. Which of the dynamic programming algorithms can be solved in $O(N^2)$ using these techniques?

10.34 Write a routine to reconstruct the shortest paths from the algorithm in Section 10.3.4.

10.35 Examine the random number generator on your system. How random is it?

10.36 Write the routines to perform insertion, deletion, and searching in skip lists.

10.37 Give a formal proof that the expected time for the skip list operations is $O(\log N)$.

10.38 Figure 10.74 shows a routine to flip a coin, assuming that *random* returns an integer (which is prevalent in many systems). What is the expected performance of the skip list algorithms if the random number generator uses a modulus of the form $M = 2^B$ (which is unfortunately prevalent on many systems)?

10.39 a. Use the exponentiation algorithm to prove that $2^{340} \equiv 1 (\bmod 341)$.

b. Show how the randomized primality test works for $N = 561$ with several choices of A.

10.40 Implement the turnpike reconstruction algorithm.

```
enum CoinSide { Heads, Tails };
typedef enum CoinSide CoinSide;

CoinSide
Flip( void )
{
    if( ( rand( ) % 2 ) == 0 )
        return Heads;
    else
        return Tails;
}
```

Figure 10.74 Questionable coin flipper

10.41 Two point sets are *homometric* if they yield the same distance set and are not rotations of each other. The following distance set gives two distinct point sets: $\{ 1, 2, 3, 4, 5, 6, 7, 8, 9, 10, 11, 12, 13, 16, 17 \}$. Find the two point sets.

10.42 Extend the reconstruction algorithm to find *all* homometric point sets given a distance set.

10.43 Show the result of α–β pruning of the tree in Figure 10.75.

10.44 a. Does the code in Figure 10.73 implement α pruning or β pruning?

b. Implement the complementary routine.

10.45 Write the remaining procedures for tic-tac-toe.

10.46 The *one-dimensional circle packing problem* is as follows: You have N circles of radii $r_1, r_2, \ldots, r_N$. These circles are packed in a box such that each circle is tangent to the bottom of the box, and are arranged in the original order. The problem is to find the width of the minimum-sized box. Figure 10.76 shows an example with circles of radii 2, 1, 2 respectively. The minimum-sized box has width $4 + 4\sqrt{2}$.

*10.47 Suppose that the edges in an undirected graph G satisfy the triangle inequality: $c_{u,v} + c_{v,w} \geq c_{u,w}$. Show how to compute a traveling salesman tour of cost at most twice optimal. *Hint: Construct a minimum spanning tree.*

Figure 10.75 Game tree, which can be pruned

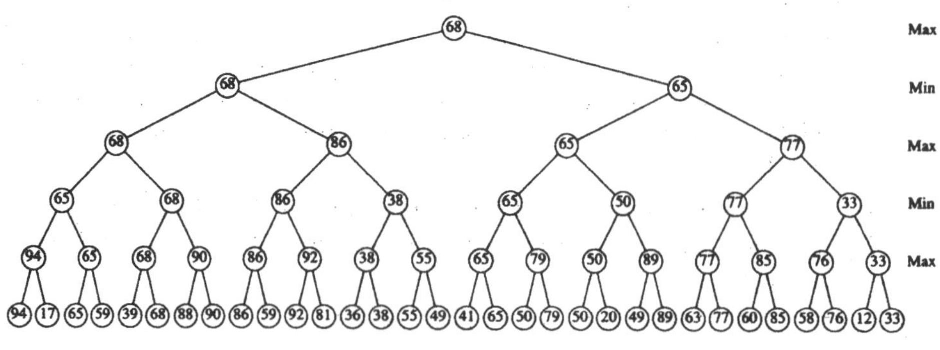

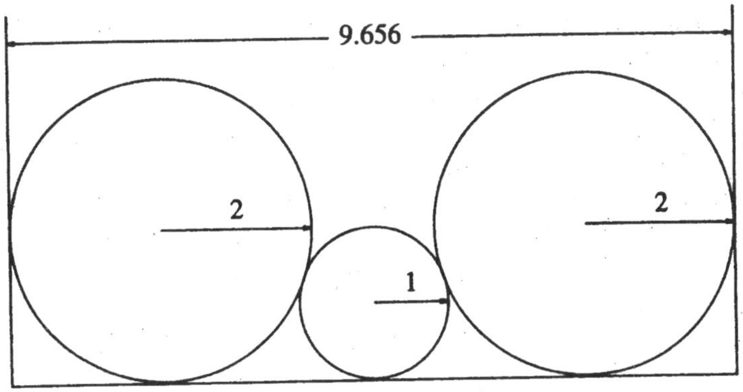

Figure 10.76 Sample for circle packing problem

*10.48 You are a tournament director and need to arrange a round robin tournament among $N = 2^k$ players. In this tournament, everyone plays exactly one game each day; after $N - 1$ days, a match has occurred between every pair of players. Give a recursive algorithm to do this.

10.49 *a. Prove that in a round robin tournament it is always possible to arrange the players in an order $p_{i_1}, p_{i_2}, \ldots, p_{i_N}$ such that for all $1 \leq j < N$, p_{i_j} has won the match against $p_{i_{j+1}}$.

b. Give an $O(N \log N)$ algorithm to find one such arrangement. Your algorithm may serve as a proof for part (a).

*10.50 We are given a set $P = p_1, p_2, \ldots, p_N$ of N points in a plane. A *Voronoi diagram* is a partition of the plane into N regions R_i such that all points in R_i are closer to p_i than any other point in P. Figure 10.77 shows a sample

Figure 10.77 Voronoi diagram

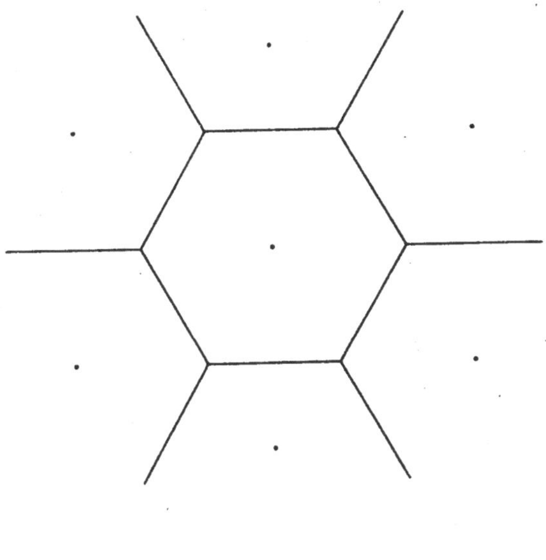

Voronoi diagram for seven (nicely arranged) points. Give an $O(N \log N)$ algorithm to construct the Voronoi diagram.

*10.51 A *convex polygon* is a polygon with the property that any line segment whose endpoints are on the polygon lies entirely within the polygon. The *convex hull* problem consists of finding the smallest (area) convex polygon which encloses a set of points in the plane. Figure 10.78 shows the convex hull for a set of 40 points. Give an $O(N \log N)$ algorithm to find the convex hull.

*10.52 Consider the problem of right-justifying a paragraph. The paragraph contains a sequence of words $w_1, w_2, \ldots, w_N$ of length $a_1, a_2, \ldots, a_N$, which we wish to break into lines of length L. Words are separated by blanks whose ideal length is b (millimeters), but blanks can stretch or shrink as necessary (but must be >0), so that a line $w_i w_{i+1} \ldots w_j$ has length exactly L. However, for each blank b' we charge $|b' - b|$ *ugliness points*. The exception to this is the last line, for which we charge only if $b' < b$ (in other words, we charge only for shrinking), since the last line does not need to be justified. Thus, if b_i is the length of the blank between a_i and a_{i+1}, then the *ugliness* of setting any line (but the last) $w_i w_{i+1} \ldots w_j$ for $j > i$ is $\sum_{k=i}^{j-1} |b_k - b| = (j - i)|b' - b|$, where b' is the average size of a blank on this line. This is true of the last line only if $b' < b$, otherwise the last line is not ugly at all.

a. Give a dynamic programming algorithm to find the least ugly setting of $w_1, w_2, \ldots, w_N$ into lines of length L. *Hint: For* $i = N, N-1, \ldots, 1$, *compute the best way to set* $w_i, w_{i+1}, \ldots, w_N$.

b. Give the time and space complexities for your algorithm (as a function of the number of words, N).

c. Consider the special case where we are using a line printer instead of a laser printer, and assume the optimal value of b is 1 (space). In this case, no shrinking of blanks is allowed, since the next smallest blank space would be 0. Give a linear-time algorithm to generate the least ugly setting on a line printer.

*10.53 The *longest increasing subsequence* problem is as follows: Given numbers $a_1, a_2, \ldots, a_N$, find the maximum value of k such that $a_{i_1} < a_{i_2} < \cdots < a_{i_k}$,

Figure 10.78 Example of a convex hull

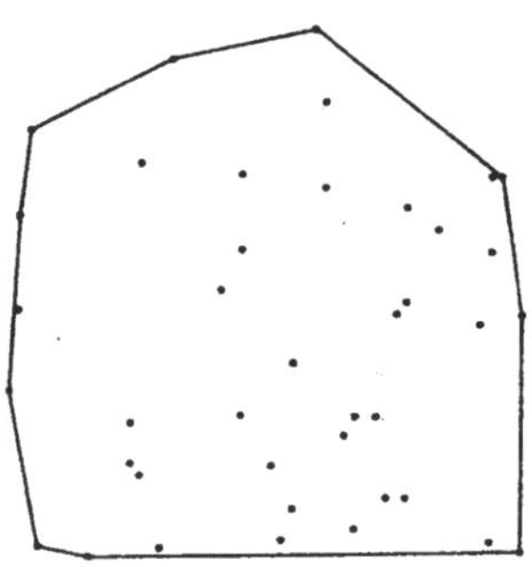

and $i_1 < i_2 < \cdots < i_k$. As an example, if the input is 3, 1, 4, 1, 5, 9, 2, 6, 5, the maximum increasing subsequence has length four (1, 4, 5, 9 among others). Give an $O(N^2)$ algorithm to solve the longest increasing subsequence problem.

*10.54 The *longest common subsequence* problem is as follows: Given two sequences $A = a_1, a_2, \ldots, a_M$, and $B = b_1, b_2, \ldots, b_N$, find the length, k, of the longest sequence $C = c_1, c_2, \ldots, c_k$ such that C is a subsequence of both A and B. As an example, if

$$A = \text{d, y, n, a, m, i, c}$$

and

$$B = \text{p, r, o, g, r, a, m, m, i, n, g,}$$

then the longest common subsequence is a,m and has length 2. Give an algorithm to solve the longest common subsequence problem. Your algorithm should run in $O(MN)$ time.

*10.55 The *pattern matching problem* is as follows: Given a string S of text, and a pattern P, find the first occurrence of P in S. *Approximate pattern matching* allows k mismatches of three types:

1. A character can be in S that is not in P.
2. A character can be in P that is not in S.
3. P and S can differ in a position.

As an example, if we are searching for the pattern "textbook" with at most three mismatches in the string "data structures txtborkk", we find a match (insert an *e*, change an *r* to an *o*, delete a *k*). Give an $O(MN)$ algorithm to solve the approximate string matching problem, where $M = |P|$ and $N = |S|$.

*10.56 One form of the *knapsack problem* is as follows: We are given a set of integers $A = a_1, a_2, \ldots, a_N$ and an integer K. Is there a subset of A whose sum is exactly K?

a. Give an algorithm that solves the knapsack problem in $O(NK)$ time.
b. Why does this not show that $P = NP$?

*10.57 You are given a currency system with coins of (decreasing) value $c_1, c_2, \ldots, c_N$ cents.

a. Give an algorithm that computes the minimum number of coins required to give K cents in change.
b. Give an algorithm that computes the number of different ways to give K cents in change.

*10.58 Consider the problem of placing eight queens on an (eight by eight) chess board. Two queens are said to attack each other if they are on the same row, column, or (not necessarily main) diagonal.

a. Give a randomized algorithm to place eight nonattacking queens on the board.

```
Distance
Shortest( S, T, G )
{
    Distance d_T, Tmp;

    if( S == T )
        return 0;

    d_T = ∞;
    for each Vertex V adjacent to S
    {
        Tmp = Shortest( V, T, G );
        if( c_S,V + Tmp < D_T )
            d_T = c_S,V + Tmp;
    }
    return d_T
}
```

Figure 10.79 Recursive shortest path algorithm

b. Give a backtracking algorithm to solve the same problem.

c. Implement both algorithms and compare the running time.

*10.59 In the game of chess, a knight in row R and column C may move to row $1 \le R' \le B$ and column $1 \le C' \le B$ (where B is the size of the board) provided that either

$$|R - R'| = 2 \text{ and } |C - C'| = 1$$

or

$$|R - R'| = 1 \text{ and } |C - C'| = 2$$

A *knight's tour* is a sequence of moves that visits all squares exactly once before returning to the starting point.

a. If B is odd, show that a knight's tour cannot exist.

b. Give a backtracking algorithm to find a knight's tour.

10.60 Consider the recursive algorithm in Figure 10.79 for finding the shortest weighted path in an acyclic graph, from S to T.

a. Why does this algorithm not work for general graphs?

b. Prove that this algorithm terminates for acyclic graphs.

c. What is the worst-case running time of the algorithm?

References

The original paper on Huffman codes is [22]. Variations on the algorithm are discussed in [31], [33], and [34]. Another popular compression scheme is Ziv–

Lempel encoding [60], [61]. Here the codes have a fixed length but represent strings instead of characters. [8] and [36] are good surveys of the common compression schemes.

The analysis of bin-packing heuristics first appeared in Johnson's Ph.D. thesis and was published in [23]. The improved lower bound for on-line bin packing given in Exercise 10.8 is from [57]; this result has been improved further in [37] and [55]. [49] describes another approach to on-line bin packing.

Theorem 10.7 is from [7]. The closest points algorithm appeared in [50]. [52] describes the turnpike reconstruction problem and its applications. The exponential worst case input was given by [59]. Two older books on the relatively new field of computational geometry are [14] and [45]. [41] and [42] contain more recent results. [2] contains the lecture notes for a computational geometry course taught at MIT; it includes an extensive bibliography.

The linear-time selection algorithm appeared in [9]. [17] discusses the sampling approach that finds the median in $1.5N$ expected comparisons. The $O(N^{1.59})$ multiplication is from [24]. Generalizations are discussed in [10] and [26]. Strassen's algorithm appears in the short paper [53]. The paper states the results and not much else. Pan [43] gives several divide and conquer algorithms, including the one in Exercise 10.27. The best known bound is $O(N^{2.376})$, which is due to Coppersmith and Winograd [13].

The classic references on dynamic programming are the books [5] and [6]. The matrix ordering problem was first studied in [19]. It was shown in [21] that the problem can be solved in $O(N \log N)$ time.

An $O(N^2)$ algorithm was provided for the construction of optimal binary search trees by Knuth [27]. The all-pairs shortest-path algorithm is from Floyd [16]. A theoretically better $O(N^3(\log\log N/\log N)^{1/3})$ algorithm is given by Fredman [18], but not surprisingly, it is not practical. A slightly improved bound (with 1/2 instead of 1/3) is given in [54]; see also [3] for related results. Under certain conditions, the running time of dynamic programs can automatically be improved by a factor of N or more. This is discussed in Exercise 10.33, [15], and [58].

The discussion of random number generators is based on [44]. Park and Miller attribute the portable implementation to Schrage [51]. Skip lists are discussed by Pugh in [46]. An alternative, namely the *treap*, is discussed in Chapter 12. The randomized primality-testing algorithm is due to Miller [38] and Rabin [48]. The theorem that at most $(N - 9)/4$ values of A fool the algorithm is from Monier [39]. Other randomized algorithms are discussed in [47]. More examples of randomization techniques can be found in [21], [25], and [40].

More information on α–β pruning can be found in [1], [28], and [29]. The top programs that play chess, checkers, Othello, and backgammon have all achieved world class status. [35] describes an Othello program. The paper appears in a special issue on computer games (mostly chess); this issue is a gold mine of ideas. One of the papers describes the use of dynamic programming to solve chess endgames completely when only a few pieces are left on the board. Related research has resulted in the change of the 50-move rule in certain cases.

Exercise 10.41 is solved in [8]. Determining whether a homometric point set with no duplicate distances exists for $N > 6$ is open. Christofides [12] gives a solution to Exercise 10.47, and also an algorithm which generates a tour at most $\frac{3}{2}$ optimal. Exercise 10.52 is discussed in [30]. Exercise 10.55 is solved in [56]. An $O(kN)$ algorithm is given in [32]. Exercise 10.57 is discussed in [11], but do not be misled by the title of the paper.

1. B. Abramson, "Control Strategies for Two-Player Games," *ACM Computing Surveys*, 21 (1989), 137–161.
2. A. Aggarwal and J. Wein, *Computational Geometry: Lecture Notes for 18.409*, MIT Laboratory for Computer Science, 1988.
3. N. Alon, Z. Galil, and O. Margalit, "On the Exponent of the All-Pairs Shortest Path Problem," *Proceedings of the Thirty-Second Annual Symposium on the Foundations of Computer Science*, (1991), 569–575.
4. T. Bell, I. H. Witten, and J. G. Cleary, "Modeling for Text Compression," *ACM Computing Surveys*, 21 (1989), 557–591.
5. R. E. Bellman, *Dynamic Programming*, Princeton University Press, Princeton, N.J., 1957.
6. R. E. Bellman and S. E. Dreyfus, *Applied Dynamic Programming*, Princeton University Press, Princeton, N.J., 1962.
7. J. L. Bentley, D. Haken, and J. B. Saxe, "A General Method for Solving Divide-and-Conquer Recurrences," *SIGACT News*, 12 (1980), 36–44.
8. G. S. Bloom, "A Counterexample to the Theorem of Piccard," *Journal of Combinatorial Theory A* (1977), 378–379.
9. M. Blum, R. W. Floyd, V. R. Pratt, R. L. Rivest, and R. E. Tarjan, "Time Bounds for Selection," *Journal of Computer and System Sciences*, 7 (1973), 448–461.
10. A. Borodin and J. I. Munro, *The Computational Complexity of Algebraic and Numerical Problems*, American Elsevier, New York, 1975.
11. L. Chang and J. Korsh, "Canonical Coin Changing and Greedy Solutions," *Journal of the ACM*, 23 (1976), 418–422.
12. N. Christofides, "Worst-case Analysis of a New Heuristic for the Traveling Salesman Problem," *Management Science Research Report #388*, Carnegie-Mellon University, Pittsburgh, PA, 1976.
13. D. Coppersmith and S. Winograd, "Matrix Multiplication via Arithmetic Progressions," *Proceedings of the Nineteenth Annual ACM Symposium on the Theory of Computing* (1987), 1–6.
14. H. Edelsbrunner, *Algorithms in Combinatorial Geometry*, Springer-Verlag, Berlin, 1987.
15. D. Eppstein, Z. Galil, and R. Giancarlo, "Speeding up Dynamic Programming," *Proceedings of the Twenty-ninth Annual IEEE Symposium on the Foundations of Computer Science*, (1988), 488–495.
16. R. W. Floyd, "Algorithm 97: Shortest Path," *Communications of the ACM*, 5 (1962), 345.
17. R. W. Floyd and R. L. Rivest, "Expected Time Bounds for Selection," *Communications of the ACM*, 18 (1975), 165–172.
18. M. L. Fredman, "New Bounds on the Complexity of the Shortest Path Problem," *SIAM Journal on Computing*, 5 (1976), 83–89.
19. S. Godbole, "On Efficient Computation of Matrix Chain Products," *IEEE Transactions on Computers*, 9 (1973), 864–866.
20. R. Gupta, S. A. Smolka, and S. Bhaskar, "On Randomization in Sequential and Distributed Algorithms," *ACM Computing Surveys*, 26 (1994), 7–86.
21. T. C. Hu and M. R. Shing, "Computations of Matrix Chain Products, Part I," *SIAM Journal on Computing*, 11 (1982), 362–373.

22. D. A. Huffman, "A Method for the Construction of Minimum Redundancy Codes," *Proceedings of the IRE,* 40 (1952), 1098–1101.
23. D. S. Johnson, A. Demers, J. D. Ullman, M. R. Garey, and R. L. Graham, "Worst-case Performance Bounds for Simple One-Dimensional Packing Algorithms," *SIAM Journal on Computing,* 3 (1974), 299–325.
24. A. Karatsuba and Y. Ofman, "Multiplication of Multi-digit Numbers on Automata," *Doklady Akademii Nauk SSSR,* 145 (1962), 293–294.
25. D. R. Karger, "Random Sampling in Graph Optimization Problems," Ph. D. thesis, Stanford University, 1995.
26. D. E. Knuth, *The Art of Computer Programming, Vol 2: Seminumerical Algorithms,* second edition, Addison-Wesley, Reading, Mass., 1981.
27. D. E. Knuth, "Optimum Binary Search Trees," *Acta Informatica,* 1 (1971), 14–25.
28. D. E. Knuth and R. W. Moore, "Estimating the Efficiency of Backtrack Programs," *Mathematics of Computation,* 29 (1975), 121–136.
29. D. E. Knuth, "An Analysis of Alpha-Beta Cutoffs," *Artificial Intelligence,* 6 (1975), 293–326.
30. D. E. Knuth, *TEX and Metafont, New Directions in Typesetting,* Digital Press, Bedford, Mass., 1981.
31. D. E. Knuth, "Dynamic Huffman Coding," *Journal of Algorithms,* 6 (1985), 163–180.
32. G. M. Landau and U. Vishkin, "Introducing Efficient Parallelism into Approximate String Matching and a New Serial Algorithm," *Proceedings of the Eighteenth Annual ACM Symposium on Theory of Computing* (1986), 220–230.
33. L. L. Larmore, "Height-Restricted Optimal Binary Trees," *SIAM Journal on Computing,* 16 (1987), 1115–1123.
34. L. L. Larmore and D. S. Hirschberg, "A Fast Algorithm for Optimal Length-Limited Huffman Codes," *Journal of the ACM,* 37 (1990), 464–473.
35. K. Lee and S. Mahajan, "The Development of a World Class Othello Program," *Artificial Intelligence,* 43 (1990), 21–36.
36. D. A. Lelewer and D. S. Hirschberg, "Data Compression," *ACM Computing Surveys,* 19 (1987), 261–296.
37. F. M. Liang, "A Lower Bound for On-line Bin Packing," *Information Processing Letters,* 10 (1980), 76–79.
38. G. L. Miller, "Riemann's Hypothesis and Tests for Primality," *Journal of Computer and System Sciences,* 13 (1976), 300–317.
39. L. Monier, "Evaluation and Comparison of Two Efficient Probabilistic Primality Testing Algorithms," *Theoretical Computer Science,* 12 (1980), 97–108.
40. R. Motwani and P. Raghavan, *Randomized Algorithms,* Cambridge University Press, New York (1995).
41. K. Mulmuley, *Computational Geometry: An Introduction Through Randomized Algorithms,* Prentice–Hall, Englewood Cliffs, N.J. (1994).
42. J. O'Rourke, *Computational Geometry in C,* Cambridge University Press, New York (1994).
43. V. Pan, "Strassen's Algorithm is Not Optimal," *Proceedings of the Nineteenth Annual IEEE Symposium on the Foundations of Computer Science* (1978), 166–176.
44. S. K. Park and K. W. Miller, "Random Number Generators: Good Ones are Hard To Find," *Communications of the ACM,* 31 (1988), 1192–1201. (See also *Technical Correspondence,* in 36 (1993) 105–110).
45. F. P. Preparata and M. I. Shamos, *Computational Geometry: An Introduction,* Springer-Verlag, New York, 1985.
46. W. Pugh, "Skip Lists: A Probabilistic Alternative to Balanced Trees," *Communications of the ACM,* 33 (1990), 668–676.

47. M. O. Rabin, "Probabilistic Algorithms," in *Algorithms and Complexity, Recent Results and New Directions* (J. F. Traub, ed.), Academic Press, New York, 1976, 21–39.
48. M. O. Rabin, "Probabilistic Algorithms for Testing Primality," *Journal of Number Theory,* 12 (1980), 128–138.
49. P. Ramanan, D. J. Brown, C. C. Lee, and D. T. Lee, "On-line Bin Packing in Linear Time," *Journal of Algorithms,* 10 (1989), 305–326.
50. M. I. Shamos and D. Hoey, "Closest-Point Problems," *Proceedings of the Sixteenth Annual IEEE Symposium on the Foundations of Computer Science* (1975), 151–162.
51. L. Schrage, "A More Portable FORTRAN Random Number Generator," *ACM Transactions on Mathematics Software,* 5 (1979), 132–138.
52. S. S. Skiena, W. D. Smith, and P. Lemke, "Reconstructing Sets From Interpoint Distances," *Proceedings of the Sixth Annual ACM Symposium on Computational Geometry* (1990), 332–339.
53. V. Strassen, "Gaussian Elimination is Not Optimal," *Numerische Mathematik,* 13 (1969), 354–356.
54. T. Takaoka, "A New Upper Bound on the Complexity of the All-Pairs Shortest Path Problem," *Information Processing Letters,* 43 (1992), 195–199.
55. A. van Vleit, "An Improved Lower Bound for On-Line Bin Packing Algorithms," *Information Processing Letters,* 43 (1992), 277–284.
56. R. A. Wagner and M. J. Fischer, "The String-to-String Correction Problem," *Journal of the ACM,* 21 (1974), 168–173.
57. A. C. Yao, "New Algorithms for Bin Packing," *Journal of the ACM,* 27 (1980), 207–227.
58. F. F. Yao, "Efficient Dynamic Programming Using Quadrangle Inequalities," *Proceedings of the Twelfth Annual ACM Symposium on the Theory of Computing* (1980), 429–435.
59. Z. Zhang, "An Exponential Example for a Partial Digest Mapping Algorithm," *Journal of Computational Molecular Biology,* 1 (1994), 235–239.
60. J. Ziv and A. Lempel, "A Universal Algorithm for Sequential Data Compression," *IEEE Transactions on Information Theory* IT23 (1977), 337–343.
61. J. Ziv and A. Lempel, "Compression of Individual Sequences via Variable-rate Coding," *IEEE Transactions on Information Theory* IT24 (1978), 530–536.

CHAPTER 11

Amortized Analysis

In this chapter, we will analyze the running times for several of the advanced data structures that have been presented in Chapters 4 and 6. In particular, we will consider the worst-case running time for any sequence of M operations. This contrasts with the more typical analysis, in which a worst-case bound is given for any *single* operation.

As an example, we have seen that AVL trees support the standard tree operations in $O(\log N)$ worst-case time per operation. AVL trees are somewhat complicated to implement, not only because there are a host of cases, but also because height balance information must be maintained and updated correctly. The reason that AVL trees are used is that a sequence of $\Theta(N)$ operations on an unbalanced search tree could require $\Theta(N^2)$ time, which would be expensive. For search trees, the $O(N)$ worst-case running time of an operation is not the real problem. The major problem is that this could happen repeatedly. Splay trees offer a pleasant alternative. Although any operation can still require $\Theta(N)$ time, this degenerate behavior cannot occur repeatedly, and we can prove that any sequence of M operations takes $O(M \log N)$ worst-case time (total). Thus, in the long run this data structure behaves as though each operation takes $O(\log N)$. We call this an *amortized time bound.*

Amortized bounds are weaker than the corresponding worst-case bounds, because there is no guarantee for any single operation. Since this is generally not important, we are willing to sacrifice the bound on a single operation, if we can retain the same bound for the sequence of operations and at the same time simplify the data structure. Amortized bounds are stronger than the equivalent average-case bound. For instance, binary search trees have $O(\log N)$ average time per operation, but it is still possible for a sequence of M operations to take $O(MN)$ time.

Because deriving an amortized bound requires us to look at an entire sequence of operations instead of just one, we expect that the analysis will be more tricky. We will see that this expectation is generally realized.

In this chapter we shall

- Analyze the binomial queue operations.
- Analyze skew heaps.
- Introduce and analyze the Fibonacci heap.
- Analyze splay trees.

11.1. An Unrelated Puzzle

Consider the following puzzle: Two kittens are placed on opposite ends of a football field, 100 yards apart. They walk toward each other at the speed of 10 yards per minute. At the same time, their mother is at one end of the field. She can run at 100 yards per minute. The mother runs from one kitten to the other, making turns with no loss of speed, until the kittens (and thus the mother) meet at midfield. How far does the mother run?

It is not hard to solve this puzzle with a brute-force calculation. We leave the details to you, but one expects that this calculation will involve computing the sum of an infinite geometric series. Although this straightforward calculation will lead to an answer, it turns out that a much simpler solution can be arrived at by introducing an extra variable, namely, time.

Because the kittens are 100 yards apart and approach each other at a combined velocity of 20 yards per minute, it takes them five minutes to get to midfield. Since the mother runs 100 yards per minute, her total is 500 yards.

This puzzle illustrates the point that sometimes it is easier to solve a problem indirectly than directly. The amortized analyses that we will perform will use this idea. We will introduce an extra variable, known as the *potential,* to allow us to prove results that seem very difficult to establish otherwise.

11.2. Binomial Queues

The first data structure we will look at is the binomial queue of Chapter 6, which we now review briefly. Recall that a *binomial tree* B_0 is a one-node tree, and for $k > 0$, the binomial tree B_k is built by melding two binomial trees B_{k-1} together. Binomial trees B_0 through B_4 are shown in Figure 11.1.

Figure 11.1 Binomial trees B_0, B_1, B_2, B_3, and B_4

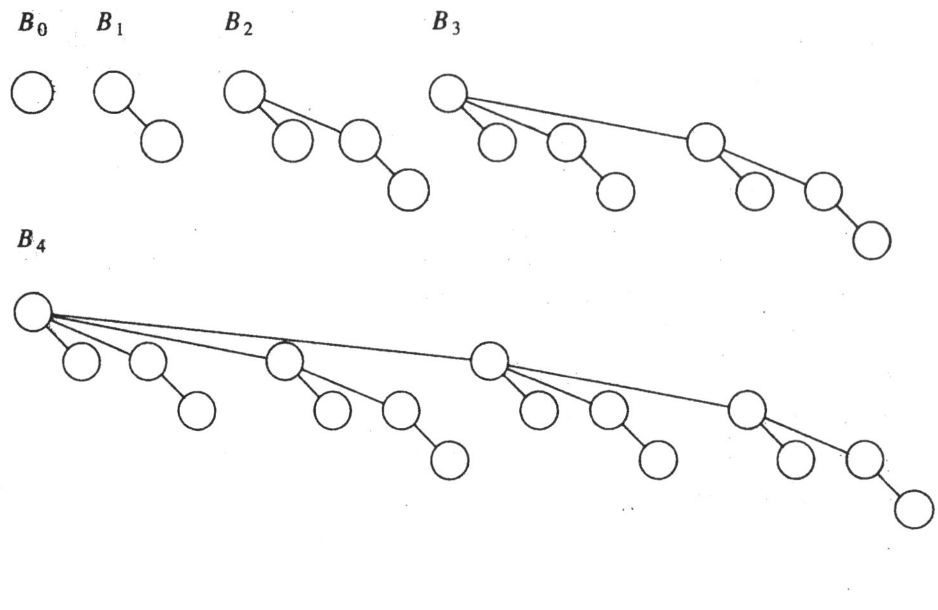

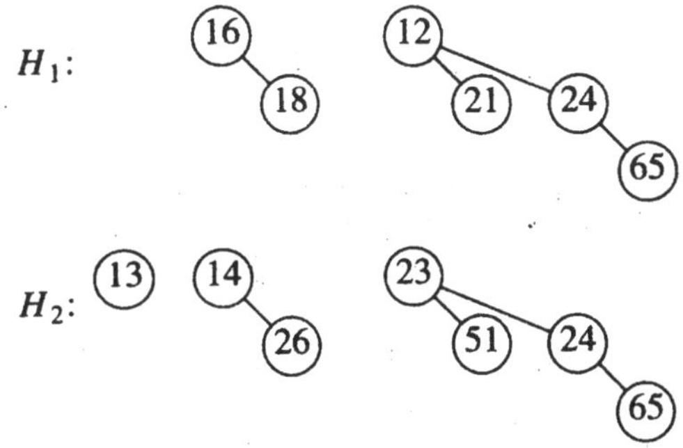

Figure 11.2 Two binomial queues H_1 and H_2

The *rank* of a node in a binomial tree is equal to the number of children; in particular, the rank of the root of B_k is k. A *binomial queue* is a collection of heap-ordered binomial trees, in which there can be at most one binomial tree B_k for any k. Two binomial queues, H_1 and H_2, are shown in Figure 11.2.

The most important operation is *Merge*. To merge two binomial queues, an operation similar to addition of binary integers is performed: At any stage we may have zero, one, two, or possibly three B_k trees, depending on whether or not the two priority queues contain a B_k tree and whether or not a B_k tree is carried over from the previous step. If there is zero or one B_k tree, it is placed as a tree in the resultant binomial queue. If there are two B_k trees, they are melded into a B_{k+1} tree and carried over; if there are three B_k trees, one is placed as a tree in the binomial queue and the other two are melded and carried over. The result of merging H_1 and H_2 is shown in Figure 11.3.

Insertion is performed by creating a one-node binomial queue and performing a *Merge*. The time to do this is $M + 1$, where M represents the smallest type of binomial tree B_M not present in the binomial queue. Thus, insertion into a binomial queue that has a B_0 tree but no B_1 tree requires two steps. Deletion of the minimum is accomplished by removing the minimum and splitting the original binomial queue into two binomial queues, which are then merged. A less terse explanation of these operations is given in Chapter 6.

We consider a very simple problem first. Suppose we want to build a binomial queue of N elements. We know that building a binary heap of N elements can be done in $O(N)$, so we expect a similar bound for binomial queues.

Figure 11.3 Binomial queue H_3: the result of merging H_1 and H_2

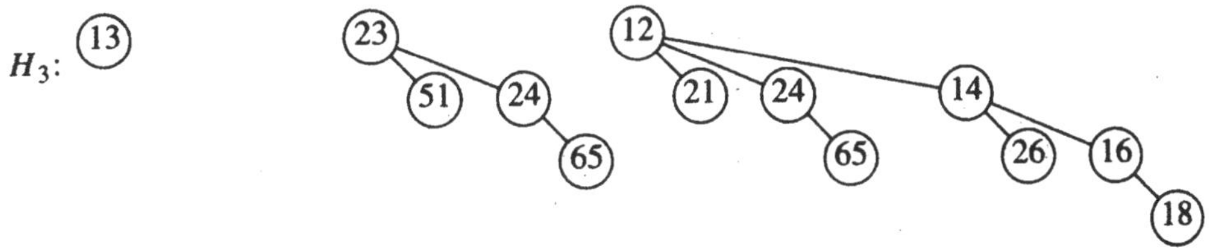

CLAIM:

A binomial queue of N elements can be built by N successive insertions in $O(N)$ time.

The claim, if true, would give an extremely simple algorithm. Since the worst-case time for each insertion is $O(\log N)$, it is not obvious that the claim is true. Recall that if this algorithm were applied to binary heaps, the running time would be $O(N \log N)$.

To prove the claim, we could do a direct calculation. To measure the running time, we define the cost of each insertion to be one time unit plus an extra unit for each linking step. Summing this cost over all insertions gives the total running time. This total is N units plus the total number of linking steps. The 1st, 3rd, 5th, and all odd-numbered steps require no linking steps, since there is no B_0 present at the time of insertion. Thus, half of the insertions require no linking steps. A quarter of the insertions require only one linking step (2nd, 6th, 10th, and so on). An eighth require two, and so on. We could add this all up and bound the number of linking steps by N, proving the claim. This brute-force calculation will not help when we try to analyze a sequence of operations that include more than just insertions, so we will use another approach to prove this result.

Consider the result of an insertion. If there is no B_0 tree present at the time of the insertion, then the insertion costs a total of one unit, using the same accounting as above. The result of the insertion is that there is now a B_0 tree, and thus we have added one tree to the forest of binomial trees. If there is a B_0 tree but no B_1 tree, then the insertion costs two units. The new forest will have a B_1 tree but will no longer have a B_0 tree, so the number of trees in the forest is unchanged. An insertion that costs three units will create a B_2 tree but destroy a B_0 and B_1 tree, yielding a net loss of one tree in the forest. In fact, it is easy to see that, in general, an insertion that costs c units results in a net increase of $2 - c$ trees in the forest, because a B_{c-1} tree is created but all B_i trees $0 \le i < c - 1$ are removed. Thus, expensive insertions remove trees, while cheap insertions create trees.

Let C_i be the cost of the ith insertion. Let T_i be the number of trees *after* the ith insertion. $T_0 = 0$ is the number of trees initially. Then we have the invariant

$$C_i + (T_i - T_{i-1}) = 2 \tag{11.1}$$

We then have

$$\begin{aligned} C_1 + (T_1 - T_0) &= 2 \\ C_2 + (T_2 - T_1) &= 2 \\ &\vdots \\ C_{N-1} + (T_{N-1} - T_{N-2}) &= 2 \\ C_N + (T_N - T_{N-1}) &= 2 \end{aligned}$$

If we add all these equations, most of the T_i terms cancel, leaving

$$\sum_{i=1}^{N} C_i + T_N - T_0 = 2N$$

or equivalently,

$$\sum_{i=1}^{N} C_i = 2N - (T_N - T_0)$$

Recall that $T_0 = 0$ and T_N, the number of trees after the N insertions, is certainly not negative, so $(T_N - T_0)$ is not negative. Thus

$$\sum_{i=1}^{N} C_i \leq 2N$$

which proves the claim.

During the *BuildBinomialQueue* routine, each insertion had a worst-case time of $O(\log N)$, but since the entire routine used at most $2N$ units of time, the insertions behaved as though each used no more than two units each.

This example illustrates the general technique we will use. The state of the data structure at any time is given by a function known as the *potential.* The potential function is not maintained by the program, but rather is an accounting device that will help with the analysis. When operations take less time than we have allocated for them, the unused time is "saved" in the form of a higher potential. In our example, the potential of the data structure is simply the number of trees. In the analysis above, when we have insertions that use only one unit instead of the two units that are allocated, the extra unit is saved for later by an increase in potential. When operations occur that exceed the allotted time, then the excess time is accounted for by a decrease in potential. One may view the potential as representing a savings account. If an operation uses less than its allotted time, the difference is saved for use later on by more expensive operations. Figure 11.4 shows

Figure 11.4 A sequence of N *Inserts*

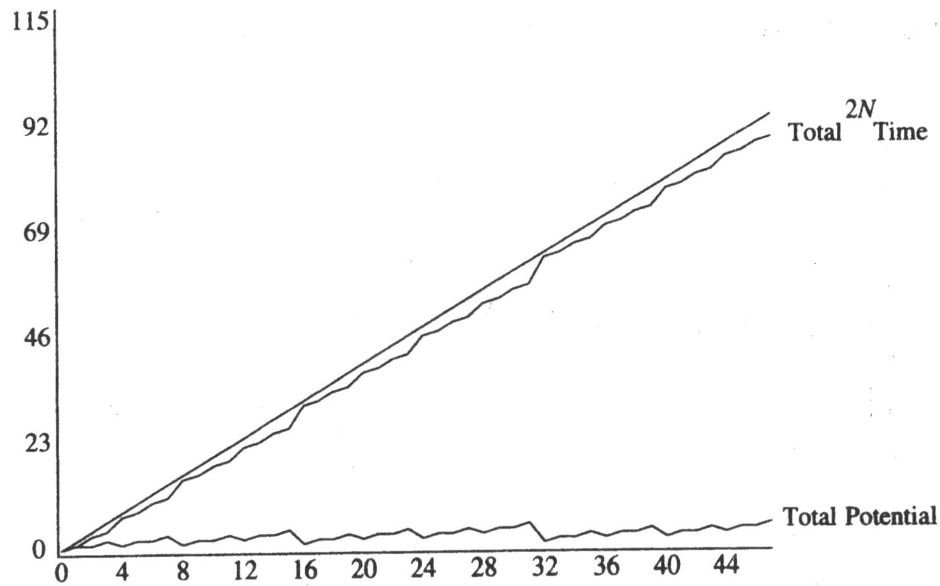

the cumulative running time used by *BuildBinomialQueue* over a sequence of insertions. Observe that the running time never exceeds $2N$ and that the potential in the binomial queue after any insertion measures the amount of savings.

Once a potential function is chosen, we write the main equation:

$$T_{\text{actual}} + \Delta Potential = T_{\text{amortized}} \tag{11.2}$$

T_{actual}, the *actual time* of an operation, represents the exact (observed) amount of time required to execute a particular operation. In a binary search tree, for example, the actual time to perform a *Find(X)* is 1 plus the depth of the node containing X. If we sum the basic equation over the entire sequence, and if the final potential is at least as large as the initial potential, then the amortized time is an upper bound on the actual time used during the execution of the sequence. Notice that while T_{actual} varies from operation to operation, $T_{\text{amortized}}$ is stable.

Picking a potential function that proves a meaningful bound is a very tricky task; there is no one method that is used. Generally, many potential functions are tried before the one that works is found. Nevertheless, the discussion above suggests a few rules, which tell us the properties that good potential functions have. The potential function should

- Always assume its minimum at the start of the sequence. A popular method of choosing potential functions is to ensure that the potential function is initially 0, and always nonnegative. All of the examples that we will encounter use this strategy.
- Cancel a term in the actual time. In our case, if the actual cost was c, then the potential change was $2 - c$. When these are added, an amortized cost of 2 is obtained. This is shown in Figure 11.5.

We can now perform a complete analysis of binomial queue operations.

Figure 11.5 The insertion cost and potential change for each operation in a sequence

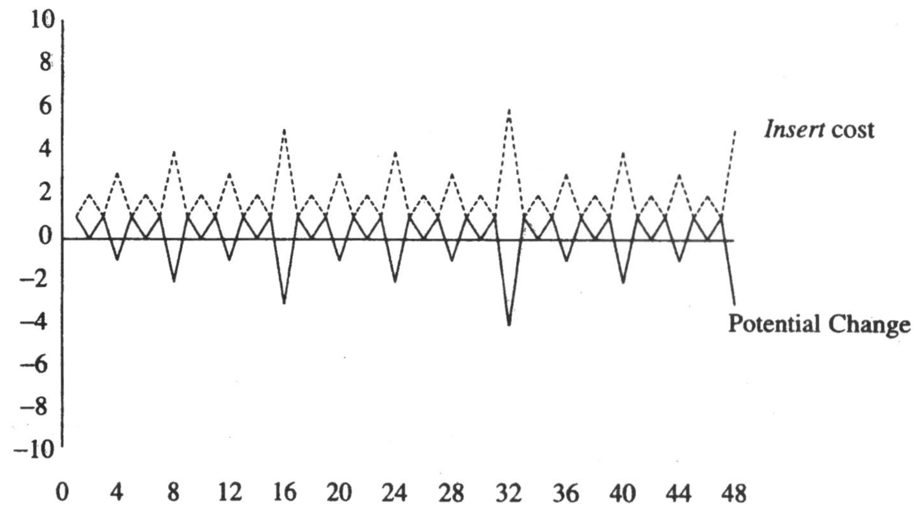

THEOREM 11.1.

The amortized running times of Insert, DeleteMin, and Merge are $O(1)$, $O(\log N)$, *and* $O(\log N)$, *respectively, for binomial queues.*

PROOF:

The potential function is the number of trees. The initial potential is 0, and the potential is always nonnegative, so the amortized time is an upper bound on the actual time. The analysis for *Insert* follows from the argument above. For *Merge*, assume the two trees have N_1 and N_2 nodes with T_1 and T_2 trees, respectively. Let $N = N_1 + N_2$. The actual time to perform the merge is $O(\log(N_1) + \log(N_2)) = O(\log N)$. After the merge, there can be at most $\log N$ trees, so the potential can increase by at most $O(\log N)$. This gives an amortized bound of $O(\log N)$. The *DeleteMin* bound follows in a similar manner.

11.3. Skew Heaps

The analysis of binomial queues is a fairly easy example of an amortized analysis. We now look at skew heaps. As is common with many of our examples, once the right potential function is found, the analysis is easy. The difficult part is choosing a meaningful potential function.

Recall that for skew heaps, the key operation is merging. To merge two skew heaps, we merge their right paths and make this the new left path. For each node on the new path, except the last, the old left subtree is attached as the right subtree. The last node on the new left path is known to not have a right subtree, so it is silly to give it one. The bound does not depend on this exception, and if the routine is coded recursively, this is what will happen naturally. Figure 11.6 shows the result of merging two skew heaps.

Suppose we have two heaps, H_1 and H_2, and there are r_1 and r_2 nodes on their respective right paths. Then the actual time to perform the merge is proportional to $r_1 + r_2$, so we will drop the Big-Oh notation and charge one unit of time for each node on the paths. Since the heaps have no structure, it is possible that all the nodes in both heaps lie on the right path, and this would give a $\Theta(N)$ worst-case bound to merge the heaps (Exercise 11.3 asks you to construct an example). We will show that the amortized time to merge two skew heaps is $O(\log N)$.

Figure 11.6 Merging of two skew heaps

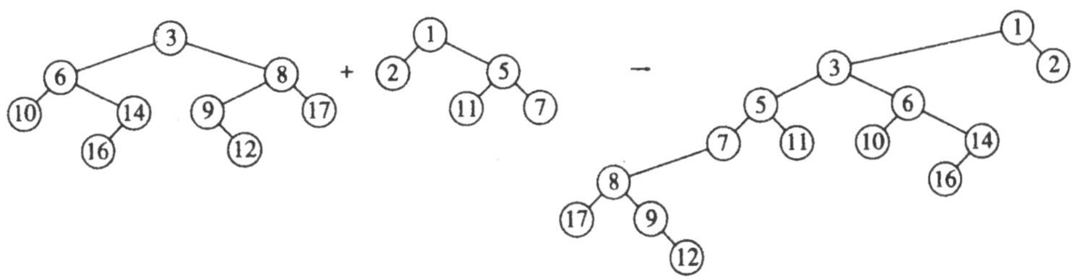

What is needed is some sort of a potential function that captures the effect of skew heap operations. Recall that the effect of a *Merge* is that every node on the right path is moved to the left path, and its old left child becomes the new right child. One idea might be to classify each node as a right node or left node, depending on whether or not it is a right child, and use the number of right nodes as a potential function. Although the potential is initially 0 and always nonnegative, the problem is that the potential does not decrease after a merge and thus does not adequately reflect the savings in the data structure. The result is that this potential function cannot be used to prove the desired bound.

A similar idea is to classify nodes as either heavy or light, depending on whether or not the right subtree of any node has more nodes than the left subtree.

DEFINITION: A node p is *heavy* if the number of descendants of p's right subtree is at least half of the number of descendants of p, and *light* otherwise. Note that the number of descendants of a node includes the node itself.

As an example, Figure 11.7 shows a skew heap. The nodes with keys 15, 3, 6, 12, and 7 are heavy, and all other nodes are light.

The potential function we will use is the number of heavy nodes in the (collection) of heaps. This seems like a good choice, because a long right path will contain an inordinate number of heavy nodes. Because nodes on this path have their children swapped, these nodes will be converted to light nodes as a result of the merge.

THEOREM 11.2.

The amortized time to merge two skew heaps is $O(\log N)$.

PROOF:

Let H_1 and H_2 be the two heaps, with N_1 and N_2 nodes respectively. Suppose the right path of H_1 has l_1 light nodes and h_1 heavy nodes, for a total of $l_1 + h_1$.

Figure 11.7 Skew heap—heavy nodes are 3, 6, 7, 12, and 15

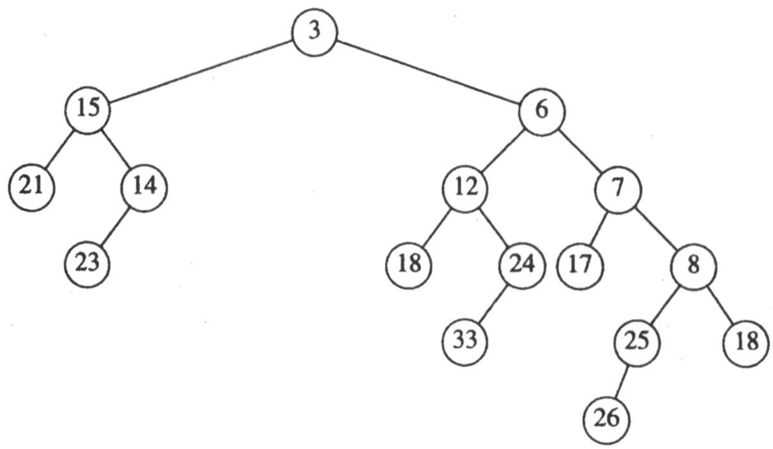

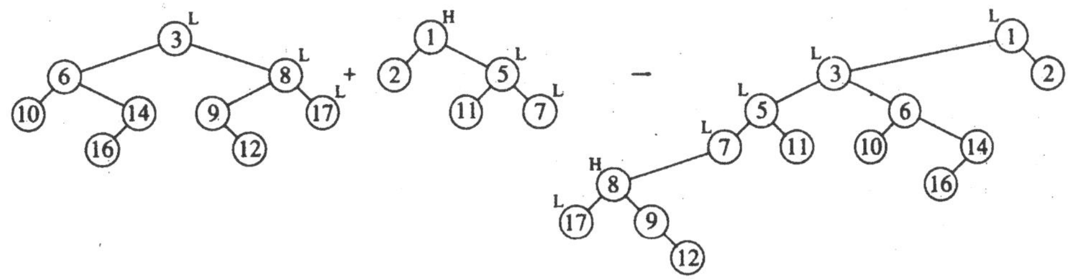

Figure 11.8 Change in heavy/light status after a merge

Likewise, H_2 has l_2 light and h_2 heavy nodes on its right path, for a total of $l_2 + h_2$ nodes.

If we adopt the convention that the cost of merging two skew heaps is the total number of nodes on their right paths, then the actual time to perform the merge is $l_1 + l_2 + h_1 + h_2$. Now the only nodes whose heavy/light status can change are nodes that are initially on the right path (and wind up on the left path), since no other nodes have their subtrees altered. This is shown by the example in Figure 11.8.

If a heavy node is initially on the right path, then after the merge it must become a light node. The other nodes that were on the right path were light and may or may not become heavy, but since we are proving an upper bound, we will have to assume the worst, which is that they become heavy and increase the potential. Then the net change in the number of heavy nodes is at most $l_1 + l_2 - h_1 - h_2$. Adding the actual time and the potential change (Equation (11.2)) gives an amortized bound of $2(l_1 + l_2)$.

Now we must show that $l_1 + l_2 = O(\log N)$. Since l_1 and l_2 are the number of light nodes on the original right paths, and the right subtree of a light node is less than half the size of the tree rooted at the light node, it follows directly that the number of light nodes on the right path is at most $\log N_1 + \log N_2$, which is $O(\log N)$.

The proof is completed by noting that the initial potential is 0 and that the potential is always nonnegative. It is important to verify this, since otherwise the amortized time does not bound the actual time and is meaningless.

Since the *Insert* and *DeleteMin* operations are basically just *Merges*, they also have $O(\log N)$ amortized bounds.

11.4. Fibonacci Heaps

In Section 9.3.2, we showed how to use priority queues to improve on the näive $O(|V|^2)$ running time of Dijkstra's shortest-path algorithm. The important observation was that the running time was dominated by $|E|$ *DecreaseKey* operations and $|V|$ *Insert* and *DeleteMin* operations. These operations take place on a set of size at most $|V|$. By using a binary heap, all these operations take $O(\log|V|)$ time, so the resulting bound for Dijkstra's algorithm can be reduced to $O(|E| \log|V|)$.

In order to lower this time bound, the time required to perform the *DecreaseKey* operation must be improved. d-heaps, which were described in Section 6.5, give an $O(\log_d |V|)$ time bound for the *DecreaseKey* operation as well as for *Insert,* but an $O(d \log_d |V|)$ bound for *DeleteMin*. By choosing d to balance the costs of $|E|$ *DecreaseKey* operations with $|V|$ *DeleteMin* operations, and remembering that d must always be at least 2, we see that a good choice for d is

$$d = \max(2, \lfloor |E|/|V| \rfloor)$$

This improves the time bound for Dijkstra's algorithm to

$$O(|E| \log_{(2+\lfloor |E|/|V| \rfloor)} |V|)$$

The *Fibonacci heap* is a data structure that supports all the basic heap operations in $O(1)$ amortized time, with the exception of *DeleteMin* and *Delete,* which take $O(\log N)$ amortized time. It immediately follows that the heap operations in Dijkstra's algorithm will require a total of $O(|E| + |V| \log |V|)$ time.

Fibonacci heaps* generalize binomial queues by adding two new concepts:

A different implementation of DecreaseKey: The method we have seen before is to percolate the element up toward the root. It does not seem reasonable to expect an $O(1)$ amortized bound for this strategy, so a new method is needed.

Lazy merging: Two heaps are merged only when it is required to do so. This is similar to lazy deletion. For lazy merging, *Merge*s are cheap, but because lazy merging does not actually combine trees, the *DeleteMin* operation could encounter lots of trees, making that operation expensive. Any one *DeleteMin* could take linear time, but it is always possible to charge the time to previous *Merge* operations. In particular, an expensive *DeleteMin* must have been preceded by a large number of unduly cheap *Merges*, which were able to store up extra potential.

11.4.1. Cutting Nodes in Leftist Heaps

In binary heaps, the *DecreaseKey* operation is implemented by lowering the value at a node and then percolating it up toward the root until heap order is established. In the worst case, this can take $O(\log N)$ time, which is the length of the longest path toward the root in a balanced tree.

This strategy does not work if the tree that represents the priority queue does not have $O(\log N)$ depth. As an example, if this strategy is applied to leftist heaps, then the *DecreaseKey* operation could take $\Theta(N)$ time, as the example in Figure 11.9 shows.

We see that for leftist heaps, another strategy is needed for the *DecreaseKey* operation. Our example will be the leftist heap in Figure 11.10. Suppose we want to decrease the key with value 9 down to 0. If we make the change, we find that we have created a violation of heap order, which is indicated by a dashed line in Figure 11.11.

*The name comes from a property of this data structure, which we will prove later in the section.

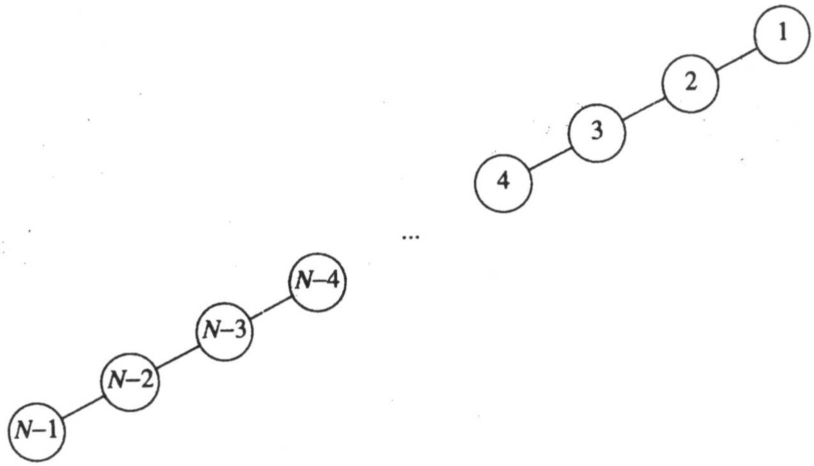

Figure 11.9 Decreasing $N - 1$ to 0 via percolate up would take $\Theta(N)$ time

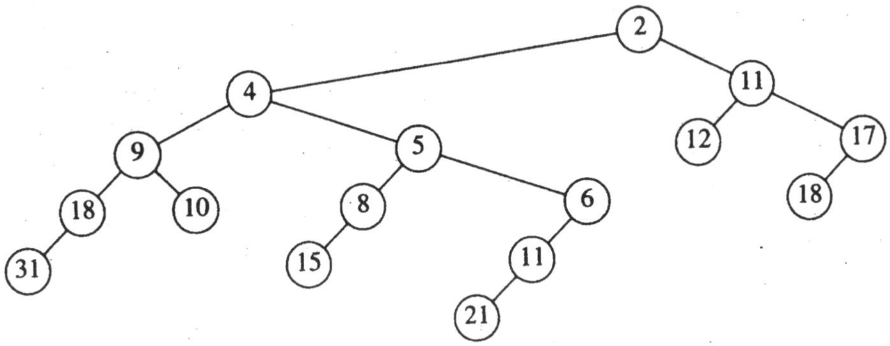

Figure 11.10 Sample leftist heap H

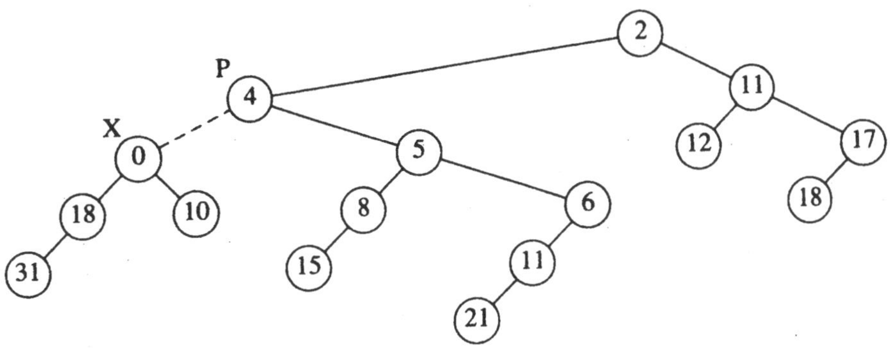

Figure 11.11 Decreasing 9 to 0 creates a heap order violation

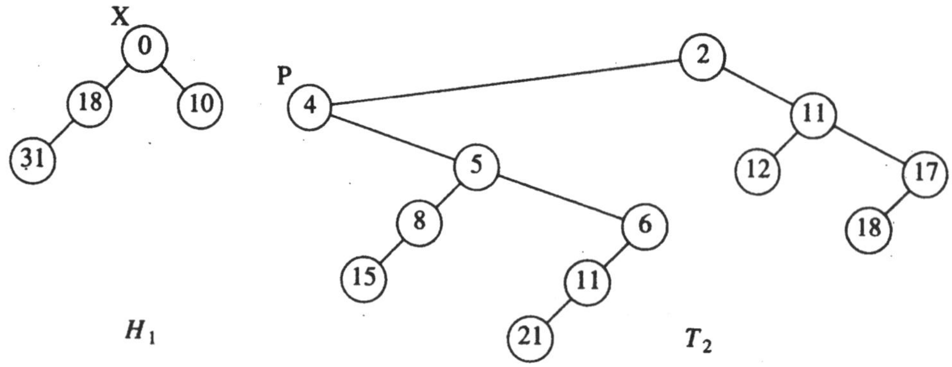

Figure 11.12 The two trees after the cut

We do not want to percolate the 0 to the root, because, as we have seen, there are cases where this could be expensive. The solution is to *cut* the heap along the dashed line, thus creating two trees, and then merge the two trees back into one. Let X be the node to which the *DecreaseKey* operation is being applied, and let P be its parent. After the cut, we have two trees, namely, H_1 with root X, and T_2, which is the original tree with H_1 removed. The situation is shown in Figure 11.12.

If these two trees were both leftist heaps, then they could be merged in $O(\log N)$ time, and we would be done. It is easy to see that H_1 is a leftist heap, since none of its nodes have had any changes in their descendants. Thus, since all of its nodes originally satisfied the leftist property, they still must.

Nevertheless, it seems that this scheme will not work, because T_2 is not necessarily leftist. However, it is easy to reinstate the leftist heap property by using two observations:

- Only nodes on the path from P to the root of T_2 can be in violation of the leftist heap property; these can be fixed by swapping children.
- Since the maximum right path length has at most $\lfloor \log(N + 1) \rfloor$ nodes, we only need to check the first $\lfloor \log(N + 1) \rfloor$ nodes on the path from P to the root of T_2. Figure 11.13 shows H_1 and T_2 after T_2 is converted to a leftist heap.

Figure 11.13 T_2 converted to the leftist heap H_2

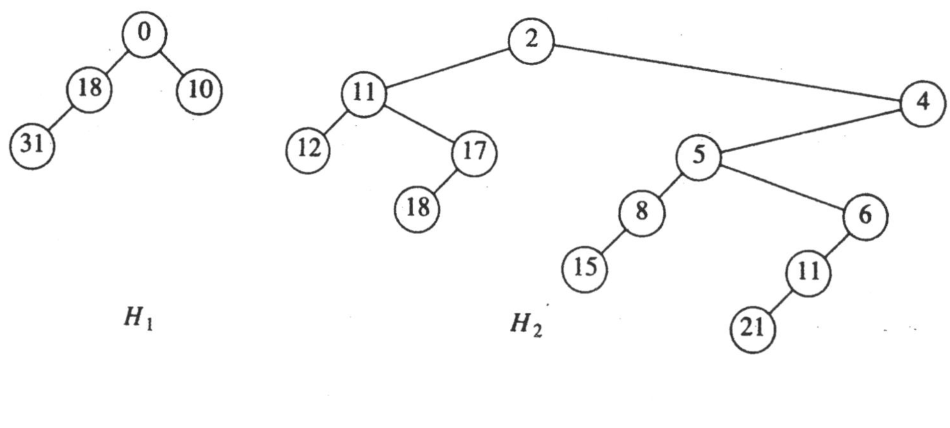

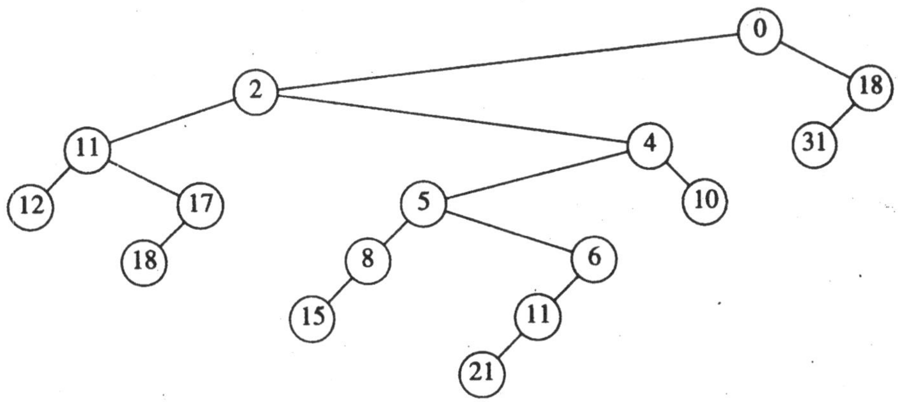

Figure 11.14 *DecreaseKey*$(H, X, 9)$ completed by merging H_1 and H_2

Because we can convert T_2 to the leftist heap H_2 in $O(\log N)$ steps, and then merge H_1 and H_2, we have an $O(\log N)$ algorithm for performing the *DecreaseKey* operation in leftist heaps. The heap that results in our example is shown in Figure 11.14.

11.4.2. Lazy Merging for Binomial Queues

The second idea that is used by Fibonacci heaps is *lazy merging*. We will apply this idea to binomial queues and show that the amortized time to perform a *Merge* operation (as well as insertion, which is a special case) is $O(1)$. The amortized time for *DeleteMin* will still be $O(\log N)$.

The idea is as follows: To merge two binomial queues, merely concatenate the two lists of binomial trees, creating a new binomial queue. This new queue may have several trees of the same size, so it violates the binomial queue property. We will call this a *lazy binomial queue* in order to maintain consistency. This is a fast operation that always takes constant (worst-case) time. As before, an insertion is done by creating a one-node binomial queue and merging. The difference is that the *Merge* is lazy.

The *DeleteMin* operation is much more painful, because it is where we finally convert the lazy binomial queue back into a standard binomial queue, but, as we will show, it is still $O(\log N)$ amortized time—but not $O(\log N)$ worst-case time, as before. To perform a *DeleteMin*, we find (and eventually return) the minimum element. As before, we delete it from the queue, making each of its children new trees. We then merge all the trees into a binomial queue by merging two equal-sized trees until it is no longer possible.

As an example, Figure 11.15 shows a lazy binomial queue. In a lazy binomial queue, there can be more than one tree of the same size. To perform the *DeleteMin*, we remove the smallest element, as before, and obtain the tree in Figure 11.16.

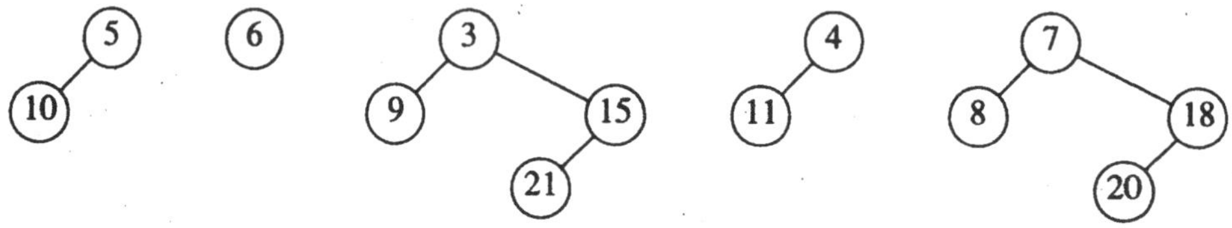

Figure 11.15 Lazy binomial queue

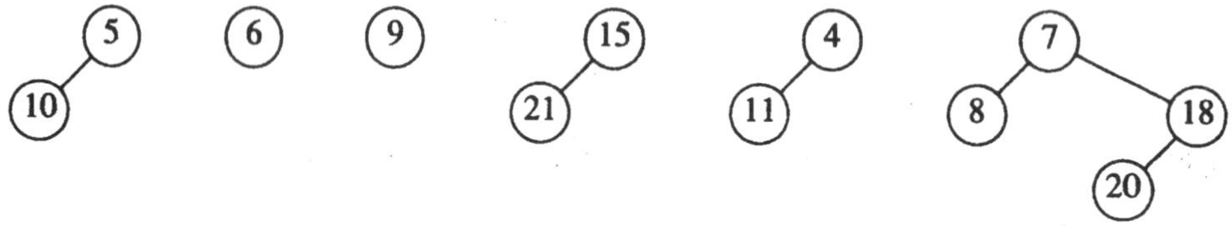

Figure 11.16 Lazy binomial queue after removing the smallest element (3)

We now have to merge all the trees and obtain a standard binomial queue. A standard binomial queue has at most one tree of each rank. In order to do this efficiently, we must be able to perform the *Merge* in time proportional to the number of trees present (T) (or $\log N$, whichever is larger). To do this, we form an array of lists, $L_0, L_1, \ldots, L_{R_{\max}+1}$, where $R_{\max}$ is the rank of the largest tree. Each list L_R contains all of the trees of rank R. The procedure in Figure 11.17 is then applied.

Each time through the loop, at lines 3 through 5, the total number of trees is reduced by 1. This means that this part of the code, which takes constant time per execution, can only be performed $T - 1$ times, where T is the number of trees. The *for* loop counters and tests at the end of the *while* loop take $O(\log N)$ time, so the running time is $O(T + \log N)$, as required. Figure 11.18 shows the execution of this algorithm on the previous collection of binomial trees.

Amortized Analysis of Lazy Binomial Queues

To carry out the amortized analysis of lazy binomial queues, we will use the same potential function that was used for standard binomial queues. Thus, the potential of a lazy binomial queue is the number of trees.

THEOREM 11.3.

The amortized running times of Merge *and* Insert *are both* $O(1)$ *for lazy binomial queues. The amortized running time of* DeleteMin *is* $O(\log N)$.

PROOF:

The potential function is the number of trees in the collection of binomial queues. The initial potential is 0, and the potential is always nonnegative. Thus,

```
for( R = 0; R <= ⌊log N⌋; R++ )
    while |L_R| ≥ 2 do
    {
        Remove two trees from L_R;
        Merge the two trees into a new tree;
        Add the new tree to L_{R+1};
    }
```

Figure 11.17 Procedure to reinstate a binomial queue

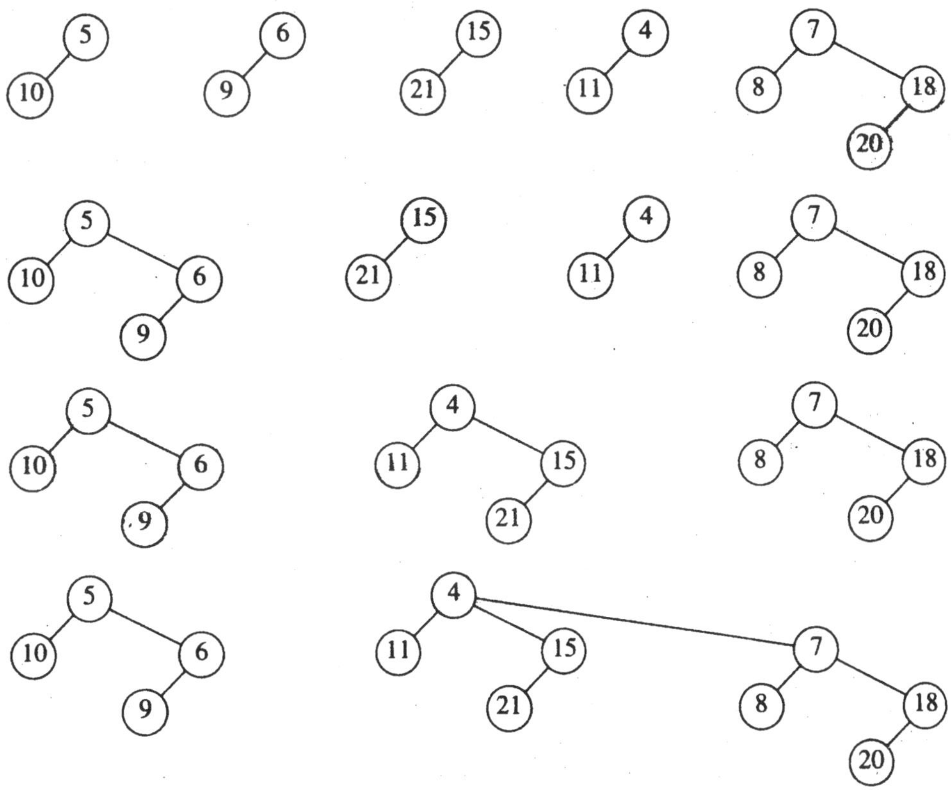

Figure 11.18 Combining the binomial trees into a binomial queue

over a sequence of operations, the total amortized time is an upper bound on the total actual time.

For the *Merge* operation, the actual time is constant, and the number of trees in the collection of binomial queues is unchanged, so, by Equation (11.2), the amortized time is $O(1)$.

For the *Insert* operation, the actual time is constant, and the number of trees can increase by at most 1, so the amortized time is $O(1)$.

The *DeleteMin* operation is more complicated. Let R be the rank of the tree that contains the minimum element, and let T be the number of trees. Thus, the potential at the start of the *DeleteMin* operation is T. To perform a *DeleteMin*, the children of the smallest node are split off into separate trees. This creates $T + R$ trees, which must be merged into a standard binomial queue. The actual time to perform this is $T + R + \log N$, if we ignore the constant in the Big-Oh notation, by the argument above.* On the other hand, once this is done, there can be at most $\log N$ trees remaining, so the potential function can increase by at most $(\log N) - T$. Adding the actual time and the change in potential gives an amortized bound of $2 \log N + R$. Since all the trees are binomial trees, we know that $R \leq \log N$. Thus we arrive at an $O(\log N)$ amortized time bound for the *DeleteMin* operation.

11.4.3. The Fibonacci Heap Operations

As we mentioned before, the Fibonacci heap combines the leftist heap *DecreaseKey* operation with the lazy binomial queue *Merge* operation. Unfortunately, we cannot use both operations without a slight modification. The problem is that if arbitrary cuts are made in the binomial trees, the resulting forest will no longer be a collection of binomial trees. Because of this, it will no longer be true that the rank of every tree is at most $\lfloor \log N \rfloor$. Since the amortized bound for *DeleteMin* in lazy binomial queues was shown to be $2 \log N + R$, we need $R = O(\log N)$ for the *DeleteMin* bound to hold.

In order to ensure that $R = O(\log N)$, we apply the following rules to all nonroot nodes:

- Mark a (nonroot) node the first time that it loses a child (because of a cut).
- If a marked node loses another child, then cut it from its parent. This node now becomes the root of a separate tree and is no longer marked. This is called a *cascading cut,* because several of these could occur in one *DecreaseKey* operation.

Figure 11.19 shows one tree in a Fibonacci heap prior to a *DecreaseKey* operation. When the node with key 39 is changed to 12, the heap order is violated. Therefore, the node is cut from its parent, becoming the root of a new tree. Since the node containing 33 is marked, this is its second lost child, and thus it is cut from its parent (10). Now 10 has lost its second child, so it is cut from 5. The process stops here, since 5 was unmarked. The node 5 is now marked. The result is shown in Figure 11.20.

Notice that 10 and 33, which used to be marked nodes, are no longer marked, because they are now root nodes. This will be a crucial observation in our proof of the time bound.

*We can do this because we can place the constant implied by the Big-Oh notation in the potential function and still get the cancellation of terms, which is needed in the proof.

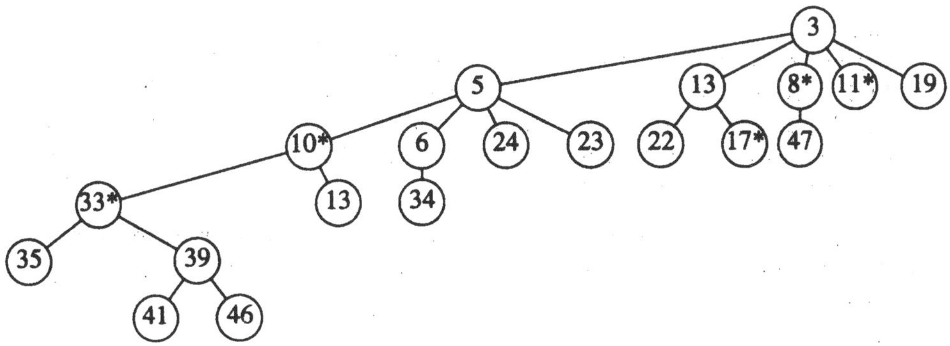

Figure 11.19 A tree in the Fibonacci heap prior to decreasing 39 to 12

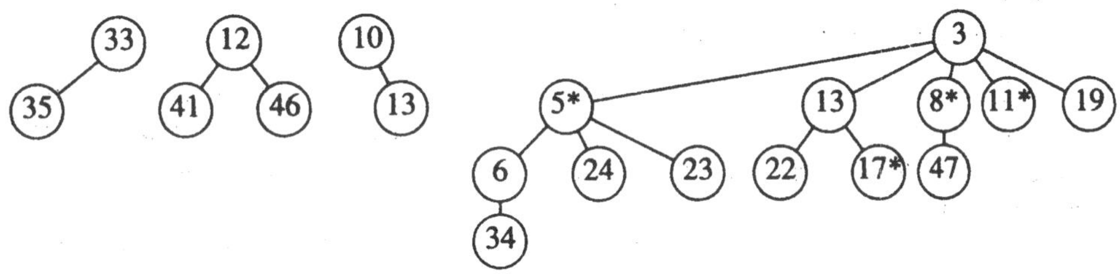

Figure 11.20 The resulting segment of the Fibonacci heap after the *DecreaseKey* operation

11.4.4. *Proof of the Time Bound*

Recall that the reason for marking nodes is that we needed to bound the rank (number of children) R of any node. We will now show that any node with N descendants has rank $O(\log N)$.

LEMMA 11.1.

Let X be any node in a Fibonacci heap. Let c_i be the ith youngest child of X. Then the rank of c_i is at least $i - 2$.

PROOF:

At the time when c_i was linked to X, X already had (older) children $c_1, c_2, \ldots, c_{i-1}$. Thus, X had at least $i - 1$ children when it linked to c_i. Since nodes are linked only if they have the same rank, it follows that at the time that c_i was linked to X, c_i had at least $i - 1$ children. Since that time, it could have lost at most one child, or else it would have been cut from X. Thus, c_i has at least $i - 2$ children.

From Lemma 11.1, it is easy to show that any node of rank R must have a lot of descendants.

LEMMA 11.2.

Let F_k be the Fibonacci numbers defined (in Section 1.2) by $F_0 = 1$, $F_1 = 1$, and $F_k = F_{k-1} + F_{k-2}$. Any node of rank $R \geq 1$ has at least F_{R+1} descendants (including itself).

PROOF:

Let S_R be the smallest tree of rank R. Clearly, $S_0 = 1$ and $S_1 = 2$. By Lemma 11.1, a tree of rank R must have subtrees of rank at least $R-2, R-3, \ldots, 1$, and 0, plus another subtree, which has at least one node. Along with the root of S_R itself, this gives a minimum value for $S_{R>1}$ of $S_R = 2 + \sum_{i=0}^{R-2} S_i$. It is easy to show that $S_R = F_{R+1}$ (Exercise 1.9a).

Because it is well known that the Fibonacci numbers grow exponentially, it immediately follows that any node with s descendants has rank at most $O(\log s)$. Thus, we have

LEMMA 11.3.

The rank of any node in a Fibonacci heap is $O(\log N)$.

PROOF:

Immediate from the discussion above.

If all we were concerned about were the time bounds for the *Merge, Insert,* and *DeleteMin* operations, then we could stop here and prove the desired amortized time bounds. Of course, the whole point of Fibonacci heaps is to obtain an $O(1)$ time bound for *DecreaseKey* as well.

The actual time required for a *DecreaseKey* operation is 1 plus the number of cascading cuts that are performed during the operation. Since the number of cascading cuts could be much more than $O(1)$, we will need to pay for this with a loss in potential. If we look at Figure 11.20, we see that the number of trees actually increases with each cascading cut, so we will have to enhance the potential function to include something that decreases during cascading cuts. Notice that we cannot just throw out the number of trees from the potential function, since then we will not be able to prove the time bound for the *Merge* operation. Looking at Figure 11.20 again, we see that a cascading cut causes a decrease in the number of marked nodes, because each node that is the victim of a cascading cut becomes an unmarked root. Since each cascading cut costs 1 unit of actual time and increases the tree potential by 1, we will count each marked node as two units of potential. This way, we have a chance of canceling out the number of cascading cuts.

THEOREM 11.4

The amortized time bounds for Fibonacci heaps are $O(1)$ *for Insert, Merge, and DecreaseKey and* $O(\log N)$ *for DeleteMin.*

PROOF:

The potential is the number of trees in the collection of Fibonacci heaps plus twice the number of marked nodes. As usual, the initial potential is 0 and is always nonnegative. Thus, over a sequence of operations, the total amortized time is an upper bound on the total actual time.

For the *Merge* operation, the actual time is constant, and the number of trees and marked nodes is unchanged, so, by Equation (11.2), the amortized time is $O(1)$.

For the *Insert* operation, the actual time is constant, the number of trees increases by 1, and the number of marked nodes is unchanged. Thus, the potential increases by at most 1, so the amortized time is $O(1)$.

For the *DeleteMin* operation, let R be the rank of the tree that contains the minimum element, and let T be the number of trees before the operation. To perform a *DeleteMin,* we once again split the children of a tree, creating an additional R new trees. Notice that, although this can remove marked nodes (by making them unmarked roots), this cannot create any additional marked nodes. These R new trees, along with the other T trees, must now be merged, at a cost of $T + R + \log N = T + O(\log N)$, by Lemma 11.3. Since there can be at most $O(\log N)$ trees, and the number of marked nodes cannot increase, the potential change is at most $O(\log N) - T$. Adding the actual time and potential change gives the $O(\log N)$ amortized bound for *DeleteMin.*

Finally, for the *DecreaseKey* operation, let C be the number of cascading cuts. The actual cost of a *DecreaseKey* is $C + 1$, which is the total number of cuts performed. The first (noncascading) cut creates a new tree and thus increases the potential by 1. Each cascading cut creates a new tree, but converts a marked node to an unmarked (root) node, for a net loss of one unit per cascading cut. The last cut also can convert an unmarked node (in Fig. 11.20 it is node 5) into a marked node, thus increasing the potential by 2. The total change in potential is thus at most $3 - C$. Adding the actual time and the potential change gives a total of 4, which is $O(1)$.

11.5. Splay Trees

As a final example, we analyze the running time of splay trees. Recall, from Chapter 4, that after an access of some item X is performed, a splaying step moves X to the root by a series of three operations: *zig, zig-zag,* and *zig-zig.* These tree rotations are shown in Figure 11.21. We adopt the convention that if a tree rotation is being performed at node X, then prior to the rotation P is its parent and G is its grandparent (if X is not a child of the root).

Recall that the time required for any tree operation on node X is proportional to the number of nodes on the path from the root to X. If we count each *zig* operation as one rotation and each *zig-zig* or *zig-zag* as two rotations, then the cost of any access is equal to 1 plus the number of rotations.

In order to show an $O(\log N)$ amortized bound for the splaying step, we need a potential function that can increase by at most $O(\log N)$ over the entire splaying step but that will also cancel out the number of rotations performed during the step. It is not at all easy to find a potential function that satisfies these criteria. A simple first guess at a potential function might be the sum of the depths of all the nodes in the tree. This does not work, because the potential can increase by $\Theta(N)$ during an access. A canonical example of this occurs when elements are inserted in sequential order.

A potential function Φ that does work is defined as

$$\Phi(T) = \sum_{i \in T} \log S(i)$$

where $S(i)$ represents the number of descendants of i (including i itself). The potential function is the sum, over all nodes i in the tree T, of the logarithm of $S(i)$.

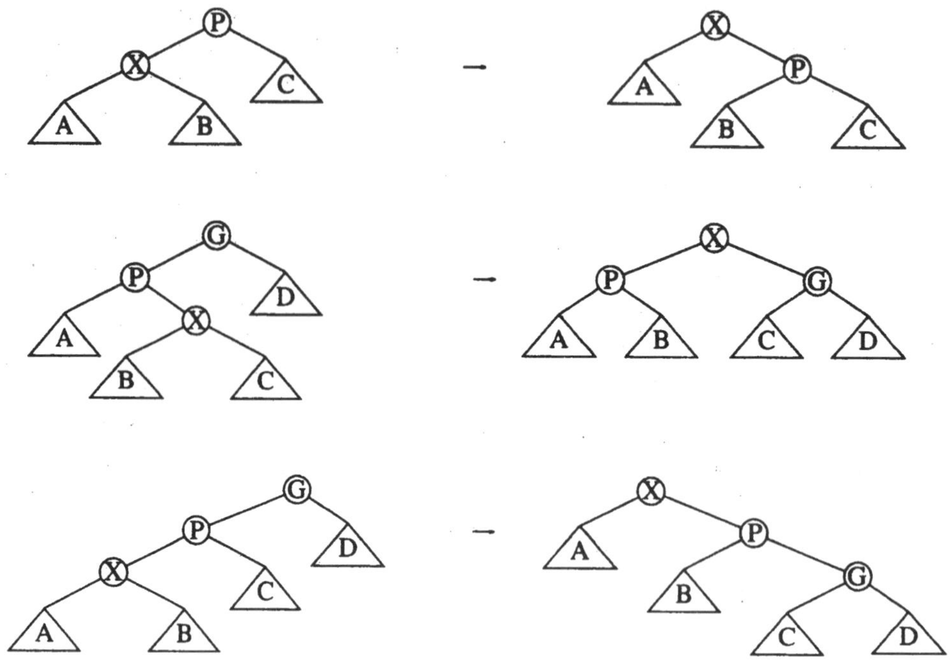

Figure 11.21 *zig, zig-zag,* and *zig-zig* operations; each has a symmetric case (not shown)

To simplify the notation, we will define

$$R(i) = \log S(i)$$

This makes

$$\Phi(T) = \sum_{i \in T} R(i)$$

$R(i)$ represents the *rank* of node i. The terminology is similar to what we used in the analysis of the disjoint set algorithm, binomial queues, and Fibonacci heaps. In all these data structures, the meaning of *rank* is somewhat different, but the rank is generally meant to be on the order (magnitude) of the logarithm of the size of the tree. For a tree T with N nodes, the rank of the root is simply $R(T) = \log N$. Using the sum of ranks as a potential function is similar to using the sum of heights as a potential function. The important difference is that while a rotation can change the heights of many nodes in the tree, only X, P, and G can have their ranks changed.

Before proving the main theorem, we need the following lemma.

LEMMA 11.4.

If $a + b \leq c$, and a and b are both positive integers, then

$$\log a + \log b \leq 2 \log c - 2$$

PROOF:

By the arithmetic-geometric mean inequality,

$$\sqrt{ab} \le (a+b)/2$$

Thus

$$\sqrt{ab} \le c/2$$

Squaring both sides gives

$$ab \le c^2/4$$

Taking logarithms of both sides proves the lemma.

With the preliminaries taken care of, we are ready to prove the main theorem.

THEOREM 11.5

The amortized time to splay a tree with root T at node X is at most $3(R(T) - R(X)) + 1 = O(\log N)$.

PROOF:

The potential function is the sum of the ranks of the nodes in T.

If X is the root of T, then there are no rotations, so there is no potential change. The actual time is 1 to access the node; thus, the amortized time is 1 and the theorem is true. Thus, we may assume that there is at least one rotation.

For any splaying step, let $R_i(X)$ and $S_i(X)$ be the rank and size of X before the step, and let $R_f(X)$ and $S_f(X)$ be the rank and size of X immediately after the splaying step. We will show that the amortized time required for a *zig* is at most $3(R_f(X) - R_i(X)) + 1$ and that the amortized time for either a *zig-zag* or *zig-zig* is at most $3(R_f(X) - R_i(X))$. We will show that when we add over all steps, the sum telescopes to the desired time bound.

Zig step: For the *zig* step, the actual time is 1 (for the single rotation), and the potential change is $R_f(X) + R_f(P) - R_i(X) - R_i(P)$. Notice that the potential change is easy to compute, because only X's and P's trees change size. Thus

$$AT_{\text{zig}} = 1 + R_f(X) + R_f(P) - R_i(X) - R_i(P)$$

From Figure 11.21 we see that $S_i(P) \ge S_f(P)$; thus, it follows that $R_i(P) \ge R_f(P)$. Thus,

$$AT_{\text{zig}} \le 1 + R_f(X) - R_i(X)$$

Since $S_f(X) \ge S_i(X)$, it follows that $R_f(X) - R_i(X) \ge 0$, so we may increase the right side, obtaining

$$AT_{\text{zig}} \le 1 + 3(R_f(X) - R_i(X))$$

Zig-zag step: For the *zig-zag* case, the actual cost is 2, and the potential change is $R_f(X) + R_f(P) + R_f(G) - R_i(X) - R_i(P) - R_i(G)$. This gives an amortized time bound of

$$AT_{\text{zig-zag}} = 2 + R_f(X) + R_f(P) + R_f(G) - R_i(X) - R_i(P) - R_i(G)$$

From Figure 11.21 we see that $S_f(X) = S_i(G)$, so their ranks must be equal. Thus, we obtain

$$AT_{\text{zig-zag}} = 2 + R_f(P) + R_f(G) - R_i(X) - R_i(P)$$

We also see that $S_i(P) \geq S_i(X)$. Consequently, $R_i(X) \leq R_i(P)$. Making this substitution gives

$$AT_{\text{zig-zag}} \leq 2 + R_f(P) + R_f(G) - 2R_i(X)$$

From Figure 11.21 we see that $S_f(P) + S_f(G) \leq S_f(X)$. If we apply Lemma 11.4, we obtain

$$\log S_f(P) + \log S_f(G) \leq 2 \log S_f(X) - 2$$

By the definition of *rank*, this becomes

$$R_f(P) + R_f(G) \leq 2R_f(X) - 2$$

Substituting this, we obtain

$$\begin{aligned} AT_{\text{zig-zag}} &\leq 2R_f(X) - 2R_i(X) \\ &\leq 2(R_f(X) - R_i(X)) \end{aligned}$$

Since $R_f(X) \geq R_i(X)$, we obtain

$$AT_{\text{zig-zag}} \leq 3(R_f(X) - R_i(X))$$

Zig-zig step: The third case is the *zig-zig*. The proof of this case is very similar to the *zig-zag* case. The important inequalities are $R_f(X) = R_i(G)$, $R_f(X) \geq R_f(P)$, $R_i(X) \leq R_i(P)$, and $S_i(X) + S_f(G) \leq S_f(X)$. We leave the details as Exercise 11.8.

Figure 11.22 The splaying steps involved in splaying at node 2

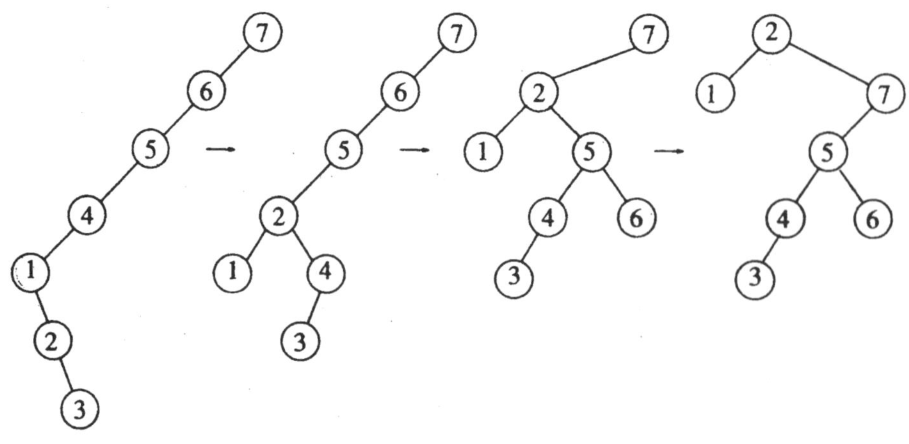

The amortized cost of an entire splay is the sum of the amortized costs of each splay step. Figure 11.22 shows the steps that are performed in a splay at node 2. Let $R_1(2)$, $R_2(2)$, $R_3(2)$, and $R_4(2)$ be the rank of node 2 in each of the four trees. The cost of the first step, which is a *zig-zag*, is at most $3(R_2(2)-R_1(2))$. The cost of the second step, which is a *zig-zig*, is $3(R_3(2) - R_2(2))$. The last step is a *zig* and has cost no larger than $3(R_4(2) - R_3(2)) + 1$. The total cost thus telescopes to $3(R_4(2) - R_1(2)) + 1$.

In general, by adding up the amortized costs of all the rotations, of which at most one can be a *zig*, we see that the total amortized cost to splay at node X is at most $3(R_f(X) - R_i(X)) + 1$, where $R_i(X)$ is the rank of X before the first splaying step and $R_f(X)$ is the rank of X after the last splaying step. Since the last splaying step leaves X at the root, we obtain an amortized bound of $3(R_f(T) - R_i(X)) + 1$, which is $O(\log N)$.

Because every operation on a splay tree requires a splay, the amortized cost of any operation is within a constant factor of the amortized cost of a splay. Thus, all splay tree operations take $O(\log N)$ amortized time. By using a more general potential function, it is possible to show that splay trees have several remarkable properties. This is discussed in more detail in the exercises.

Summary

In this chapter, we have seen how an amortized analysis can be used to apportion charges among operations. To perform the analysis, we invent a fictitious potential function. The potential function measures the state of the system. A high-potential data structure is volatile, having been built on relatively cheap operations. When the expensive bill comes for an operation, it is paid for by the savings of previous operations. One can view potential as standing for *potential for disaster*, in that very expensive operations can occur only when the data structure has a high potential and has used considerably less time than has been allocated.

Low potential in a data structure means that the cost of each operation has been roughly equal to the amount allocated for it. Negative potential means debt; more time has been spent than has been allocated, so the allocated (or amortized) time is not a meaningful bound.

As expressed by Equation (11.2), the amortized time for an operation is equal to the sum of the actual time and potential change. Taken over an entire sequence of operations, the amortized time for the sequence is equal to the total sequence time plus the net change in potential. As long as this net change is positive, then the amortized bound provides an *upper bound* for the actual time spent and is meaningful.

The keys to choosing a potential function are to guarantee that the minimum potential occurs at the beginning of the algorithm, and to have the potential increase for cheap operations and decrease for expensive operations. It is important that the excess or saved time be measured by an opposite change in potential. Unfortunately, this is sometimes easier said than done.

Exercises

11.1 When do M consecutive insertions into a binomial queue take less than $2M$ time units?

11.2 Suppose a binomial queue of $N = 2^k - 1$ elements is built. Alternately perform M *Insert* and *DeleteMin* pairs. Clearly, each operation takes $O(\log N)$ time. Why does this not contradict the amortized bound of $O(1)$ for insertion?

*11.3 Show that the amortized bound of $O(\log N)$ for the skew heap operations described in the text cannot be converted to a worst-case bound, by giving a sequence of operations that lead to a *Merge* requiring $\Theta(N)$ time.

*11.4 Show how to merge two skew heaps with one top-down pass and reduce the *Merge* cost to $O(1)$ amortized time.

11.5 Extend skew heaps to support the *DecreaseKey* operation in $O(\log N)$ amortized time.

11.6 Implement Fibonacci heaps and compare their performance with that of binary heaps when used in Dijkstra's algorithm.

11.7 A standard implementation of Fibonacci heaps requires four pointers per node (parent, child, and two siblings). Show how to reduce the number of pointers, at the cost of at most a constant factor in the running time.

11.8 Show that the amortized time of a *zig-zig* splay is at most $3(R_f(X) - R_i(X))$.

11.9 By changing the potential function, it is possible to prove different bounds for splaying. Let the *weight function* $W(i)$ be some function assigned to each node in the tree, and let $S(i)$ be the sum of the weights of all the nodes in the subtree rooted at i, including i itself. The special case $W(i) = 1$ for all nodes corresponds to the function used in the proof of the splaying bound. Let N be the number of nodes in the tree, and let M be the number of accesses. Prove the following two theorems:

a. The total access time is $O(M + (M + N)\log N)$.

*b. If q_i is the number of times that item i is accessed, and $q_i > 0$ for all i, then the total access time is

$$O\left(M + \sum_{i=1}^{N} q_i \log(M/q_i)\right)$$

11.10 a. Show how to implement the *Merge* operation on splay trees so that any sequence of $N - 1$ *Merges* starting from N single-element trees takes $O(N \log^2 N)$ time.

*b. Improve the bound to $O(N \log N)$.

11.11 In Chapter 5, we described *rehashing*: When a table becomes more than half full, a new table twice as large is constructed, and the entire old table is rehashed. Give a formal amortized analysis, with potential function, to show that the amortized cost of an insertion is still $O(1)$.

11.12 Show that if deletions are not allowed, then any sequence of M insertions into an N-node 2–3 tree produces $O(M + N)$ node splits.

11.13 A *deque* with *heap order* is a data structure consisting of a list of items, on which the following operations are possible:

Push(X,D): Insert item *X* on the front end of deque *D*.

Pop(D): Remove the front item from deque *D* and return it.

Inject(X,D): Insert item *X* on the rear end of deque *D*.

Eject(D): Remove the rear item from deque *D* and return it.

FindMin(D): Return the smallest item from deque *D* (breaking ties arbitrarily).

a. Describe how to support these operations in constant amortized time per operation.

**b. Describe how to support these operations in constant worst-case time per operation.

11.14 Show that the binomial queues actually support merging in $O(1)$ amortized time. Define the potential of a binomial queue to be the number of trees plus the rank of the largest tree.

References

An excellent survey of amortized analysis is provided in [10].

Most of the references below duplicate citations in earlier chapters. We cite them again for convenience and completeness. Binomial queues were first described in [11] and analyzed in [1]. Solutions to Exercises 11.3 and 11.4 appear in [9]. Fibonacci heaps are described in [3]. Exercise 11.9(a) shows that splay trees are optimal, to within a constant factor of the the best static search trees. Exercise 11.9(b) shows that splay trees are optimal, to within a constant factor of the best optimal search trees. These, as well as two other strong results, are proved in the original splay tree paper [7].

The *Merge* operation for splay trees is described in [6]. Exercise 11.12 is solved, with an implicit use of amortization, in [2]. The paper also shows how to merge 2–3 trees efficiently. A solution to Exercise 11.13 can be found in [4]. Exercise 11.14 is from [5].

Amortized analysis is used in [8] to design an on-line algorithm that processes a series of queries in time only a constant factor larger than any off-line algorithm in its class.

1. M. R. Brown, "Implementation and Analysis of Binomial Queue Algorithms," *SIAM Journal on Computing,* 7 (1978), 298–319.
2. M. R. Brown and R. E. Tarjan, "Design and Analysis of a Data Structure for Representing Sorted Lists," *SIAM Journal on Computing,* 9 (1980), 594–614.
3. M. L. Fredman and R. E. Tarjan, "Fibonacci Heaps and Their Uses in Improved Network Optimization Algorithms," *Journal of the ACM,* 34 (1987), 596–615.
4. H. Gajewska and R. E. Tarjan, "Deques with Heap Order," *Information Processing Letters,* 22 (1986), 197–200.
5. C. M. Khoong and H. W. Leong, "Double-Ended Binomial Queues," *Proceedings of the Fourth Annual International Symposium on Algorithms and Computation* (1993), 128–137.

6. G. Port and A. Moffat, "A Fast Algorithm for Melding Splay Trees," *Proceedings of First Workshop on Algorithms and Data Structures* (1989), 450–459.
7. D. D. Sleator and R. E. Tarjan, "Self-adjusting Binary Search Trees," *Journal of the ACM*, 32 (1985), 652–686.
8. D. D. Sleator and R. E. Tarjan, "Amortized Efficiency of List Update and Paging Rules," *Communications of the ACM*, 28 (1985), 202–208.
9. D. D. Sleator and R. E. Tarjan, "Self-adjusting Heaps," *SIAM Journal on Computing*, 15 (1986), 52–69.
10. R. E. Tarjan, "Amortized Computational Complexity," *SIAM Journal on Algebraic and Discrete Methods*, 6 (1985), 306–318.
11. J. Vuillemin, "A Data Structure for Manipulating Priority Queues," *Communications of the ACM*, 21 (1978), 309–314.

CHAPTER 12

Advanced Data Structures and Implementation

In this chapter, we discuss seven data structures with an emphasis on practicality. We begin by examining alternatives to the AVL tree discussed in Chapter 4. These include an optimized version of the splay tree, the red black tree, a deterministic form of the skip list (previously discussed in Chapter 10), the AA-tree, and the treap.

We then examine a data structure that can be used for multidimensional data. In this case, each item may have several keys. The *k*-d tree allows searching relative to any key.

Finally, we examine the pairing heap, which, despite a lack of analytical results, seems to be the most practical alternative to the Fibonacci heap.

Recurring themes include

- Nonrecursive, top-down (instead of bottom-up) search tree implementations when appropriate.
- Detailed, optimized implementations that make use of, among other things, sentinel nodes.

12.1. Top-Down Splay Trees

In Chapter 4, we discussed the basic splay tree operation. When an item X is inserted as a leaf, a series of tree rotations, known as a *splay*, makes X the new root of the tree. A splay is also performed during searches, and if an item is not found, a splay is performed on the last node on the access path. In Chapter 11, we showed that the *amortized* cost of a splay tree operation is $O(\log N)$.

A direct implementation of this strategy requires a traversal from the root down the tree, and then a bottom-up traversal to implement the splaying step. This can be done either by maintaining parent pointers (as shown in Chapter 4), or by storing the access path on a stack. Unfortunately, both methods require a substantial amount of overhead, and both must handle many special cases. In this section, we show how to perform rotations on the initial access path. The result is a procedure that is faster in practice, uses only $O(1)$ extra space, but retains the $O(\log N)$ amortized time bound.

Figure 12.1 shows the rotations for the *zig*, *zig-zig*, and *zig-zag* cases. (As is customary, three symmetric rotations are omitted.) At any point in the access, we have a current node X that is the root of its subtree; this is represented in our diagrams as the "middle" tree.* Tree L stores nodes in the tree T that are less than X, but not in X's subtree; similarly tree R stores nodes in the tree T that are larger than X, but not in X's subtree. Initially, X is the root of T, and L and R are empty.

If the rotation should be a *zig*, then the tree rooted at Y becomes the new root of the middle tree. X and subtree B are attached as a left child of the smallest item in R; X's left child is logically made *NULL*.† As a result, X is the new smallest item in R. Note carefully that Y does not have to be a leaf for the *zig* case to apply. If we are searching for an item that is smaller than Y, and Y has no left child (but does have a right child), then the *zig* case will apply.

Figure 12.1 Top-down splay rotations: *zig*, *zig-zig*, and *zig-zag*

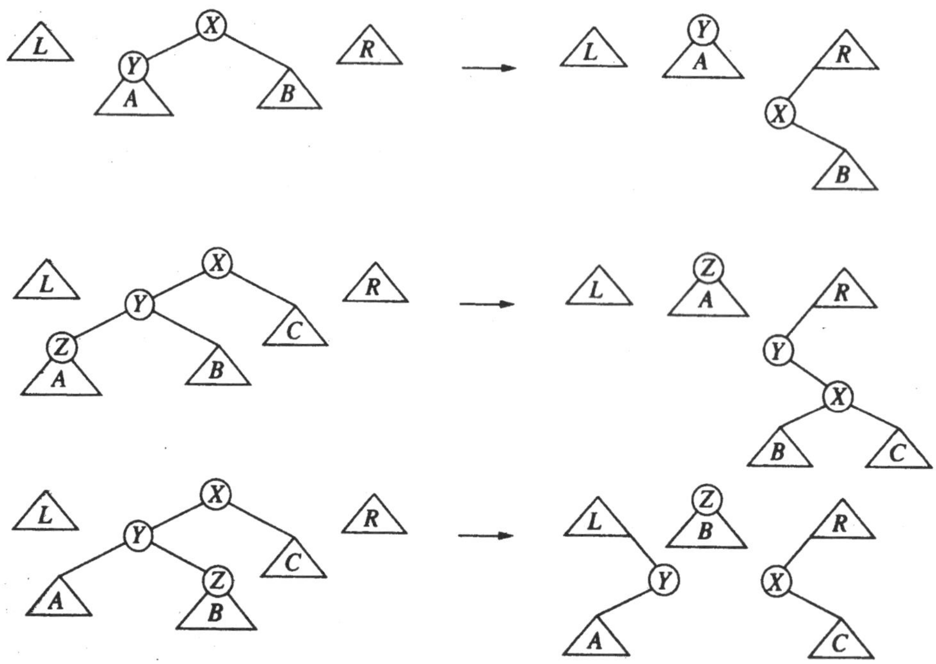

*For simplicity we don't distinguish between a "node" and the item in the node.

†In the code, the smallest node in R does not have a *NULL* left pointer because there is no need for it. This means that *PrintTree*(R) will include some items that logically are not in R.

For the *zig-zig* case, we have a similar dissection. The crucial point is that a rotation between X and Y is performed. The *zig-zag* case brings the bottom node Z to the top in the middle tree, and attaches subtrees X and Y to R and L, respectively. Note that Y is attached to, and then becomes, the largest item in L.

The *zig-zag* step can be simplified somewhat because no rotations are performed. Instead of making Z the root of the middle tree, we make Y the root. This is shown in Figure 12.2. This simplifies the coding because the action for the *zig-zag* case becomes identical to the *zig* case. This would seem advantageous because testing for a host of cases is time-consuming. The disadvantage is that by descending only one level, we have more iterations in the splaying procedure.

Once we have performed the final splaying step, Figure 12.3 shows how L, R, and the middle tree are arranged to form a single tree. Note carefully that the result is different from bottom-up splaying. The crucial fact is that the $O(\log N)$ amortized bound is preserved (Exercise 12.1).

An example of the top-down splaying algorithm is shown in Figure 12.4. We attempt to access 19 in the tree. The first step is a *zig-zag*. In accordance with (a symmetric version of) Figure 12.2, we bring the subtree rooted at 25 to the root of the middle tree, and attach 12 and its left subtree to L.

Next we have a *zig-zig*: 15 is elevated to the root of the middle tree, and a rotation between 20 and 25 is performed, with the resulting subtree being attached to R. The search for 19 then results in a terminal *zig*. The middle tree's new root is 18, and 15 and its left subtree are attached as a right child of L's largest node. The reassembly, in accordance with Figure 12.3, terminates the splay step.

We will use a header with left and right pointers to eventually contain the roots of the left and right trees. Since these trees are initially empty, a header is used to correspond to the min or max node of the right or left tree, respectively, in this initial state. This way the code can avoid checking for empty trees. The first time the left tree becomes nonempty, the right pointer will get initialized and will not change

Figure 12.2 Simplified top-down *zig-zag*

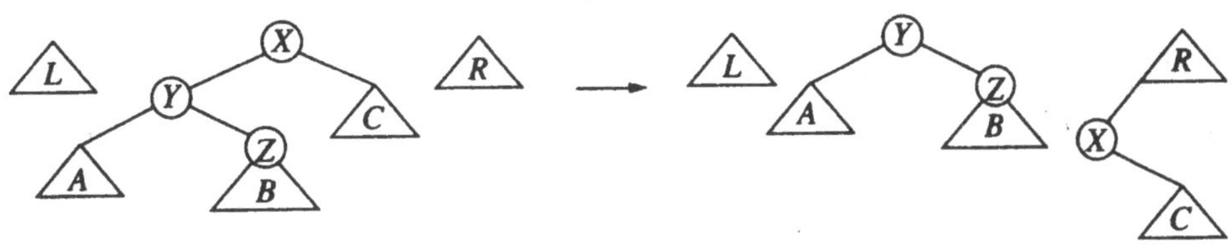

Figure 12.3 Final arrangement for top-down splaying

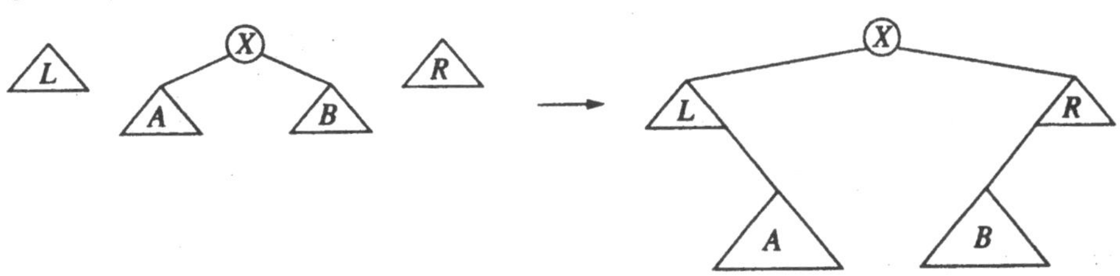

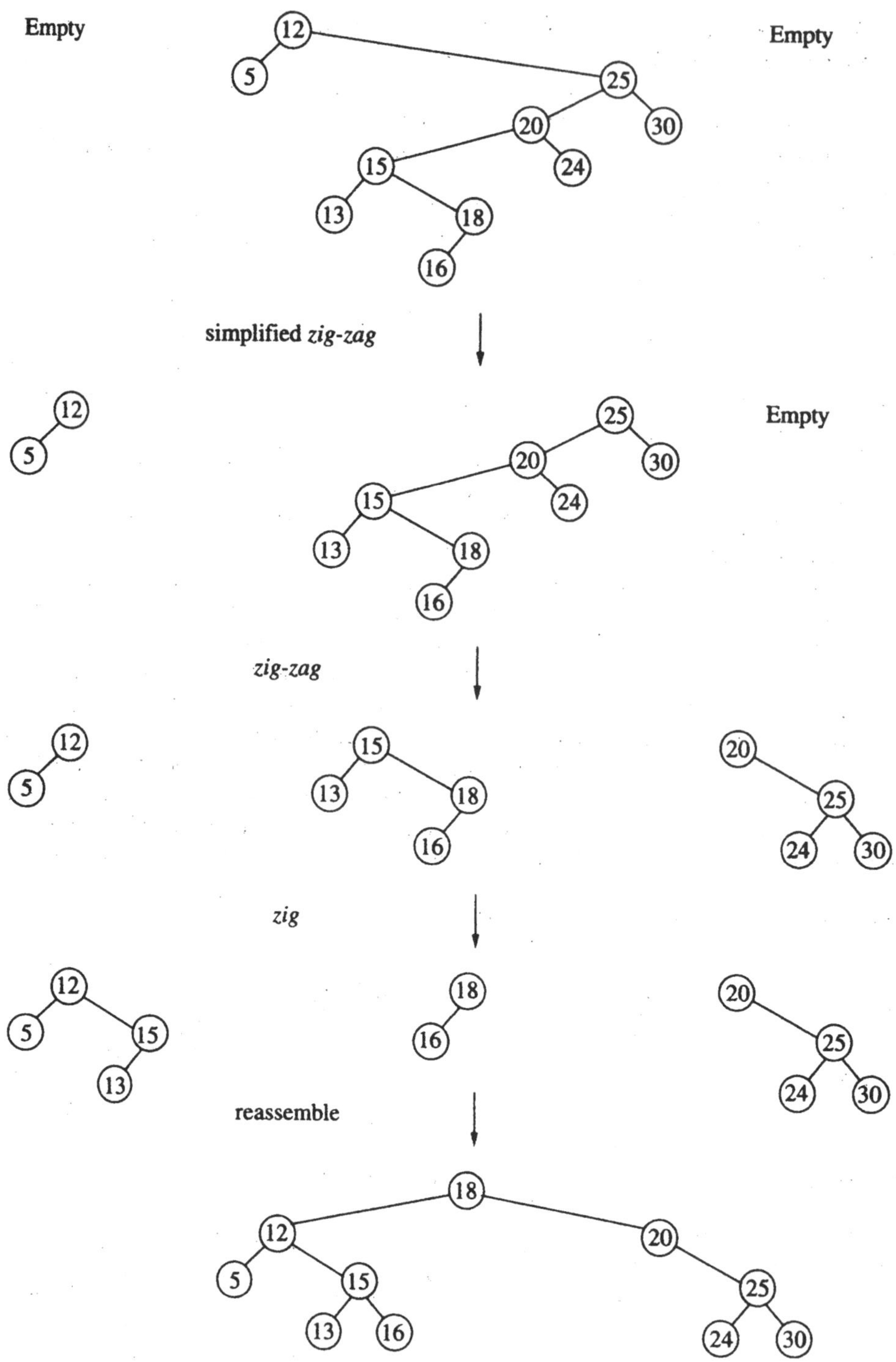

Figure 12.4 Steps in top-down splay (access 19 in top tree)

in the future; thus it will contain the root of the right tree at the end of the top-down search. Similarly, the left pointer will eventually contain the root of the right tree.

The procedure *Initialize*, shown in Figure 12.5, is used to allocate the *NullNode* sentinel. We use the sentinel *NullNode* to represent a *NULL* pointer. We will repeatedly use this technique to simplify the code (and consequently make the code somewhat faster). Figure 12.6 gives the code for the splaying procedure. The *Header* node allows us to be certain that we can attach X to the largest node in R without having to worry that R might be empty (and similarly for the symmetric case dealing with L).

As we mentioned above, before the reassembly at the end of the splay, *Header.Left* and *Header.Right* point at R and L, respectively (this is not a typo—follow the pointers). Except for this detail, the code is relatively straightforward.

Figure 12.7 shows the procedure to insert an item into a tree T. A new node is allocated (if necessary), and if T is empty, a one-node tree is created. Otherwise we splay T around *Item*. If the data in T's new root is equal to *Item*, we have a duplicate; instead of reinserting *Item*, we preserve *NewNode* for a future insertion and return immediately. If T's new root contains a value larger than *Item*, then T's new root and its right subtree become a right subtree of *NewNode*, and T's left subtree becomes the left subtree of *NewNode*. Similar logic applies if T's new root contains a value smaller than *Item*. In either case, *NewNode* becomes the new root.

In Chapter 4, we showed that deletion in splay trees is easy, because a splay will place the target of the deletion at the root. We close by showing the deletion routine in Figure 12.8. It is indeed rare that a deletion procedure is shorter than the corresponding insertion procedure.

12.2. Red Black Trees

A historically popular alternative to the AVL tree is the *red black* tree. Operations on red black trees take $O(\log N)$ time in the worst case, and, as we will see, a careful nonrecursive implementation (for insertion) can be done relatively effortlessly (compared with AVL trees).

A red black tree is a binary search tree with the following coloring properties:

1. Every node is colored either red or black.
2. The root is black.
3. If a node is red, its children must be black.
4. Every path from a node to a *NULL* pointer must contain the same number of black nodes.

A consequence of the coloring rules is that the height of a red black tree is at most $2\log(N+1)$. Consequently, searching is guaranteed to be a logarithmic operation. Figure 12.9 shows a red black tree. Red nodes are shown with double circles.

```
#ifndef _Splay_H

struct SplayNode;
typedef struct SplayNode *SplayTree;

SplayTree MakeEmpty( SplayTree T );
SplayTree Find( ElementType X, SplayTree T );
SplayTree FindMin( SplayTree T );
SplayTree FindMax( SplayTree T );
SplayTree Initialize( void );
SplayTree Insert( ElementType X, SplayTree T );
SplayTree Remove( ElementType X, SplayTree T );
ElementType Retrieve( SplayTree T ); /* Gets root item */

#endif /* _Splay_H */


/* Place in the implementation file */
struct SplayNode
{
    ElementType Element;
    SplayTree   Left;
    SplayTree   Right;
};

typedef struct SplayNode *Position;
static Position NullNode = NULL; /* Needs initialization */

SplayTree
Initialize( void )
{
    if( NullNode == NULL )
    {
        NullNode = malloc( sizeof( struct SplayNode ) );
        if( NullNode == NULL )
            FatalError( "Out of space!!!" );
        NullNode->Left = NullNode->Right = NullNode;
    }
    return NullNode;
}
```

Figure 12.5 Splay trees: declarations and initialization

```
/* Top-down splay procedure, */
/* not requiring Item to be in the tree */

SplayTree
Splay( ElementType Item, Position X )
{
    static struct SplayNode Header;
    Position LeftTreeMax, RightTreeMin;

    Header.Left = Header.Right = NullNode;
    LeftTreeMax = RightTreeMin = &Header;
    NullNode->Element = Item;

    while( Item != X->Element )
    {
        if( Item < X->Element )
        {
            if( Item < X->Left->Element )
                X = SingleRotateWithLeft( X );
            if( X->Left == NullNode )
                break;
            /* Link right */
            RightTreeMin->Left = X;
            RightTreeMin = X;
            X = X->Left;
        }
        else
        {
            if( Item > X->Right->Element )
                X = SingleRotateWithRight( X );
            if( X->Right == NullNode )
                break;
            /* Link Left */
            LeftTreeMax->Right = X;
            LeftTreeMax = X;
            X = X->Right;
        }
    }   /* while Item != X->Element */

    /* Reassemble */
    LeftTreeMax->Right = X->Left;
    RightTreeMin->Left = X->Right;
    X->Left = Header.Right;
    X->Right = Header.Left;

    return X;
}
```

Figure 12.6 Top-down splaying procedure

```
SplayTree
Insert( ElementType Item, SplayTree T )
{
    static Position NewNode = NULL;

    if( NewNode == NULL )
    {
        NewNode = malloc( sizeof( struct SplayNode ) );
        if( NewNode == NULL )
            FatalError( "Out of space!!!" );
    }
    NewNode->Element = Item;

    if( T == NullNode )
    {
        NewNode->Left = NewNode->Right = NullNode;
        T = NewNode;
    }
    else
    {
        T = Splay( Item, T );
        if( Item < T->Element )
        {
            NewNode->Left = T->Left;
            NewNode->Right = T;
            T->Left = NullNode;
            T = NewNode;
        }
        else
        if( T->Element < Item )
        {
            NewNode->Right = T->Right;
            NewNode->Left = T;
            T->Right = NullNode;
            T = NewNode;
        }
        else
            return T; /* Already in the tree */
    }

    NewNode = NULL; /* So next insert will call malloc */
    return T;
}
```

Figure 12.7 Top-down splay tree insert

```
SplayTree
Remove( ElementType Item, SplayTree T )
{
    Position NewTree

    if( T != NullNode )
    {
        T = Splay( Item, T );
        if( Item == T->Element )
        {
            /* Found it! */
            if( T->Left == NullNode )
                NewTree = T->Right;
            else
            {
                NewTree = T->Left
                NewTree = Splay( Item, NewTree );
                NewTree->Right = T->Right;
            }
            free( T );
            T = NewTree;
        }
    }

    return T;
}
```

Figure 12.8 Top-down deletion procedure

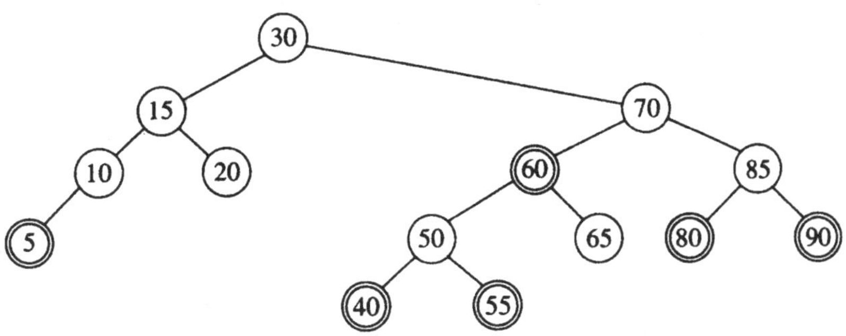

Figure 12.9 Example of a red black tree (insertion sequence is: 10, 85, 15, 70, 20, 60, 30, 50, 65, 80, 90, 40, 5, 55)

The difficulty, as usual, is inserting a new item into the tree. The new item, as usual is placed as a leaf in the tree. If we color this item black, then we are certain to violate condition 4, because we will create a longer path of black nodes. Thus the item must be colored red. If the parent is black, we are done. If the parent is already red, then we will violate condition 3 by having consecutive red nodes. In this case, we have to adjust the tree to ensure that condition 3 is enforced (without introducing a violation of condition 4). The basic operations that are used to do this are color changes and tree rotations.

12.2.1. Bottom-Up Insertion

As we have already mentioned, if the parent of the newly inserted item is black, we are done. Thus insertion of 25 into the tree in Figure 12.9 is trivial.

There are several cases (each with a mirror image symmetry) to consider if the parent is red. First, suppose that the sibling of the parent is black (we adopt the convention that *NULL* nodes are black). This would apply for an insertion of 3 or 8, but not for the insertion of 99. Let *X* be the newly added leaf, *P* be its parent, *S* be the sibling of the parent (if it exists), and *G* be the grandparent. Only *X* and *P* are red in this case; *G* is black, because otherwise there would be two consecutive red nodes *prior* to the insertion, in violation of red black rules. Adopting the splay tree terminology, *X*, *P*, and *G* can form either a *zig-zig* chain or a *zig-zag* chain (in either of two directions). Figure 12.10 shows how we can rotate the tree for the case where *P* is a left child (note there is a symmetric case). Even though *X* is a leaf, we have drawn a more general case that allows *X* to be in the middle of the tree. We will use this more general rotation later.

The first case corresponds to a single rotation between *P* and *G*, and the second case corresponds to a double rotation, first between *X* and *P* and then between *X*

Figure 12.10 *Zig* rotation and *zig-zag* rotation work if *S* is black

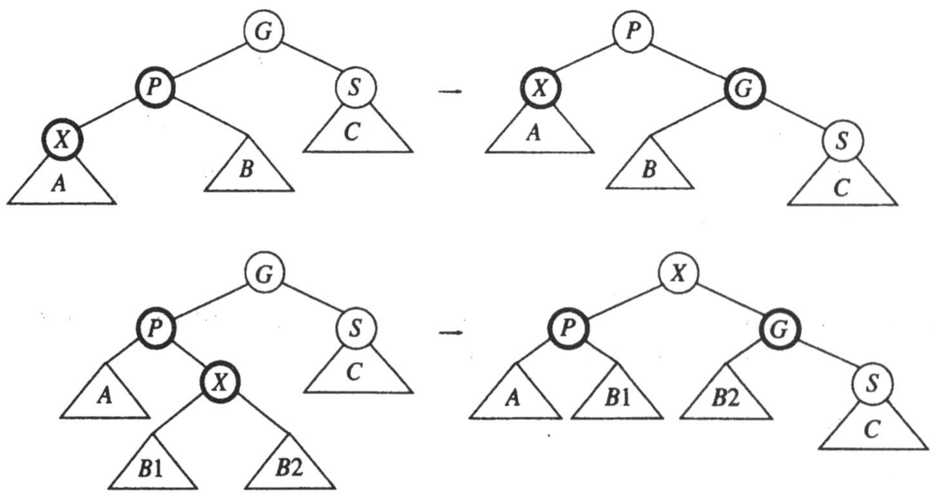

and G. When we write the code, we have to keep track of the parent, the grandparent, and, for reattachment purposes, the great-grandparent.

In both cases, the subtree's new root is colored black, and so even if the original great-grandparent was red, we removed the possibility of two consecutive red nodes. Equally important, the number of black nodes on the paths into A, B, and C has remained unchanged as a result of the rotations.

So far so good. But what happens if S is red, as is the case when we attempt to insert 79 in the tree in Figure 12.9? In that case, initially there is one black node on the path from the subtree's root to C. After the rotation, there must still be only one black node. But in both cases, there are three nodes (the new root, G, and S) on the path to C. Since only one may be black, and since we cannot have consecutive red nodes, it follows that we'd have to color both S and the subtree's new root red, and G (and our fourth node) black. That's great, but what happens if the great-grandparent is also red? In that case, we could percolate this procedure up toward the root as is done for B-trees and binary heaps, until we no longer have two consecutive red nodes, or we reach the root (which will be recolored black).

12.2.2. *Top-Down Red Black Trees*

Implementing the percolation would require maintaining the path using a stack or parent pointers. We saw that splay trees are more efficient if we use a top-down procedure, and it turns out that we can apply a top-down procedure to red black trees that guarantees that S won't be red.

The procedure is conceptually easy. On the way down, when we see a node X that has two red children, we make X red and the two children black. Figure 12.11 shows this color flip. This will induce a red black violation only if X's parent P is also red. But in that case, we can apply the appropriate rotations in Figure 12.10. What if X's parent's sibling is red? This possibility has been removed by our actions on the way down, and so X's parent's sibling can't be red! Specifically, if on the way down the tree we see a node Y that has two red children, we know that Y's grandchildren must be black, and that since Y's children are made black too, even after the rotation that may occur, we won't see another red node for two levels. Thus when we see X, if X's parent is red, it is not possible for X's parent's sibling to be red also.

As an example, suppose we want to insert 45 into the tree in Figure 12.9. On the way down the tree, we see node 50, which has two red children. Thus, we perform a color flip, making 50 red, and 40 and 55 black. Now 50 and 60 are both red. We perform the single rotation between 60 and 70, making 60 the black root of 30's right subtree, and 70 and 50 both red. We then continue, performing an identical action if we see other nodes on the path that contain two red children. When we get to the leaf, we insert 45 as a red node, and since the parent is black, we are done. The resulting tree is shown in Figure 12.12.

As Figure 12.12 shows, the red black tree that results is frequently very well balanced. Experiments suggest that the average red black tree is about as deep as an average AVL tree and that, consequently, the searching times are typically near

Figure 12.11 Color flip: only if X's parent is red do we continue with a rotation

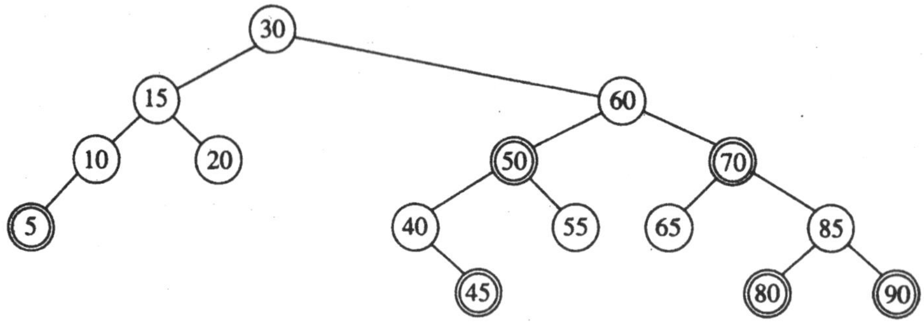

Figure 12.12 Insertion of 45 into Figure 12.9

optimal. The advantage of red black trees is the relatively low overhead required to perform insertion, and the fact that in practice rotations occur relatively infrequently.

An actual implementation is complicated not only by the host of possible rotations, but also by the possibility that some subtrees (such as 10's right subtree) might be empty, and the special case of dealing with the root (which among other things, has no parent). Thus, we use two sentinel nodes: one for the root, and *NullNode*, which indicates a *NULL* pointer as it did for splay trees. The root sentinel will store the key $-\infty$ and a right pointer to the real root. Because of this, the searching and printing procedures need to be adjusted. The recursive routines are trickiest. Rather than compel the user to pass $T \rightarrow Right$ we use a hidden recursive procedure. Consequently, the user need not be concerned with the header node. Figure 12.13 shows how the inorder traversal is rewritten.

We also need to require the user to call the routine *Initialize* to allocate the header node. *Initialize* should also allocate *NullNode* if this is the first tree constructed (subsequent trees can share *NullNode*). This, along with a sketch of the type declarations, is shown in Figure 12.14.

Next, Figure 12.15 shows the routine to perform a single rotation. Because the resultant tree must be attached to a parent, *Rotate* takes the parent node as a parameter. Rather than keeping track of the type of rotation as we descend the tree, we pass *Item* as a parameter. Since we expect very few rotations during the insertion procedure, it turns out that it is not only simpler, but actually faster, to do it this way. *Rotate* simply returns the result of performing an appropriate single rotation.

Finally, we provide the insertion procedure in Figure 12.16. The routine *HandleReorient* is called when we encounter a node with two red children, and also when we insert a leaf. The only tricky part is the observation that a double rotation is really two single rotations, and is done only when branching to X takes opposite directions. As we mentioned in the earlier discussion, *Insert* must keep track of the parent, grandparent, and great-grandparent as the tree is descended. Note that after a rotation, the values stored in the grandparent and great-grandparent are no longer correct. However, we are assured that they will be restored by the time they are next needed.

12.2.3. Top-Down Deletion

Deletion in red black trees can also be performed top-down. Everything boils down to being able to delete a leaf. This is because to delete a node that has two children, we replace it with the smallest node in the right subtree; that node, which must have at most one child, is then deleted. Nodes with only a right child can be deleted in the same manner, while nodes with only a left child can be deleted by replacement with the largest node in the left subtree, and subsequent deletion of that node. Note that for red black trees, we don't want to use the strategy of bypassing for the case of a node with one child because that may connect two red nodes in the middle of the tree, making enforcement of the red black condition difficult.

Deletion of a red leaf is, of course, trivial. If a leaf is black, however, the deletion is more complicated because removal of a black node will violate condition 4. The solution is to ensure during the top-down pass that the leaf is red.

Figure 12.13 Inorder traversal for tree with two sentinels

```
/* Print the tree, watch out for NullNode, */
/* and skip header */

static void
DoPrint( RedBlackTree T )
{
    if( T != NullNode )
    {
        DoPrint( T->Left );
        Output( T->Element );
        DoPrint( T->Right );
    }
}

void
PrintTree( RedBlackTree T )
{
    DoPrint( T->Right );
}
```

```
typedef enum ColorType { Red, Black } ColorType;

struct RedBlackNode
{
    ElementType  Element
    RedBlackTree Left;
    RedBlackTree Right;
    ColorType    Color;
};

Position NullNode = NULL; /* Needs initialization */

/* Initialization procedure */
RedBlackTree
Initialize( void )
{
    RedBlackTree T;

    if( NullNode = NULL )
    {
        NullNode == malloc( sizeof( struct RedBlackNode ) );
        if( NullNode == NULL )
            FatalError( "Out of space!!!" );
        NullNode->Left = NullNode->Right = NullNode;
        NullNode->Color = Black;
        NullNode->Element = Infinity;
    }

    /* Create the header node */
    T = malloc( sizeof( struct RedBlackNode ) );
    if( T == NULL )
        FatalError( "Out of space!!!" );
    T->Element = NegInfinity;
    T->Left = T->Right = NullNode;
    T->Color = Black;

    return T;
}
```

Figure 12.14 Type declarations and initialization

Throughout this discussion, let X be the current node, T be its sibling, and P be their parent. We begin by coloring the root red. As we traverse down the tree, we attempt to ensure that X is red. When we arrive at a new node, we are certain that P is red (inductively, by the invariant we are trying to maintain), and that X and T are black (because we can't have two consecutive red nodes). There are two main cases.

```
/* Perform a rotation at node X */
/* (whose parent is passed as a parameter) */
/* The child is deduced by examining Item */

static Position
Rotate( ElementType Item, Position Parent )
{
    if( Item < Parent->Element )
        return Parent->Left = Item < Parent->Left->Element ?
            SingleRotateWithLeft( Parent->Left ) :
            SingleRotateWithRight( Parent->Left );
    else
        return Parent->Right = Item < Parent->Right->Element ?
            SingleRotateWithLeft( Parent->Right ) :
            SingleRotateWithRight( Parent->Right );
}
```

Figure 12.15 Rotate procedure

```
static Position X, P, GP, GGP;

static
void HandleReorient( ElementType Item, RedBlackTree T )
{
    X->Color = Red; /* Do the color flip */
    X->Left->Color = Black;
    X->Right->Color = Black;

    if( P->Color == Red ) /* Have to rotate */
    {
        GP->Color = Red;
        if( (Item < GP->Element) != (Item < P->Element) )
            P = Rotate( Item, GP ); /* Start double rotation */
        X = Rotate( Item, GGP );
        X->Color = Black;
    }
    T->Right->Color = Black; /* Make root black */
}

RedBlackTree
Insert( ElementType Item, RedBlackTree T )
{
    X = P = GP = T;
    NullNode->Element = Item;
    while( X->Element != Item ) /* Descend down the tree */
```

Figure 12.16 Insertion procedure

```
        {
            GGP = GP; GP = P; P = X;
            if( Item < X->Element )
                X = X->Left;
            else
                X = X->Right;
            if( X->Left->Color == Red && X->Right->Color == Red )
                HandleReorient( Item, T );
        }

        if( X != NullNode )
            return NullNode; /* Duplicate */

        X = malloc( sizeof( struct RedBlackNode ) );
        if( X == NULL )
            FatalError( "Out of space!!!" );
        X->Element = Item;
        X->Left = X->Right = NullNode;

        if( Item < P->Element ) /* Attach to its parent */
            P->Left = X;
        else
            P->Right = X;
        HandleReorient( Item, T ); /* Color red; maybe rotate */

        return T;
    }
```

Figure 12.16 Insertion procedure *(continued)*

First, suppose X has two black children. Then there are three subcases, which are shown in Figure 12.17. If T also has two black children, we can flip the colors of X, T, and P to maintain the invariant. Otherwise, one of T's children is red. Depending on which one it is,* we can apply the rotation shown in the second and third cases of Figure 12.17. Note carefully that this case will apply for the leaf, because *NullNode* is considered to be black.

Otherwise one of X's children is red. In this case, we fall through to the next level, obtaining new X, T, and P. If we're lucky, X will land on the red child, and we can continue onward. If not, we know that T will be red, and X and P will be black. We can rotate T and P, making X's new parent red; X and its grandparent will, of course, be black. At this point we can go back to the first main case.

*If both children are red, we can apply either rotation. As usual, there are symmetric rotations for the case when X is a right child that are not shown.

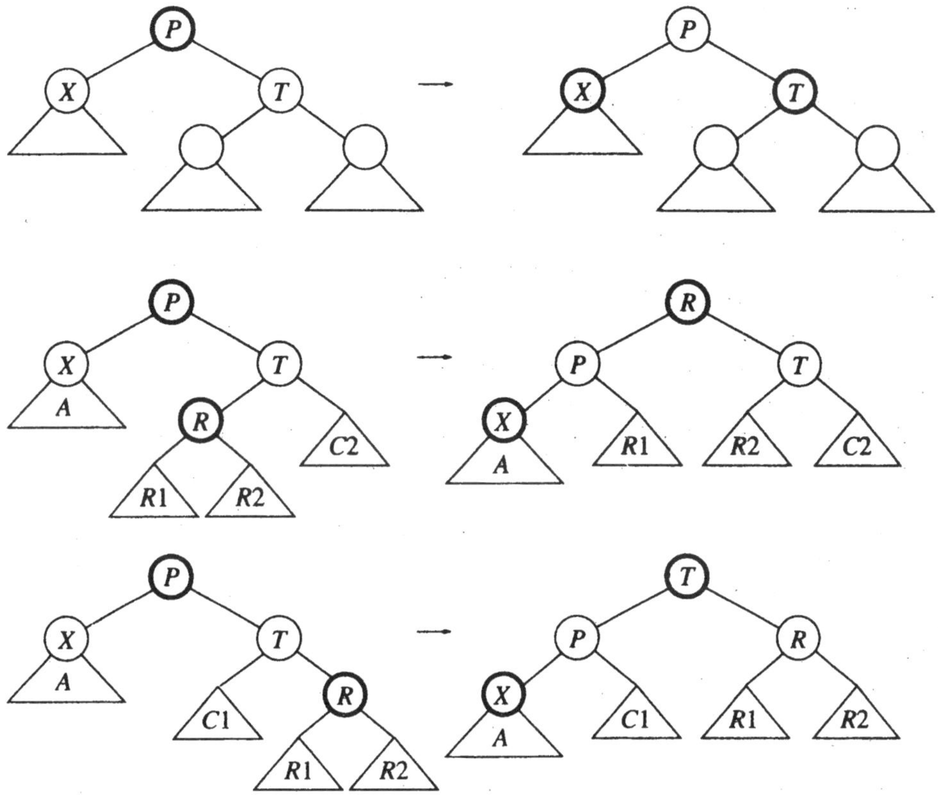

Figure 12.17 Three cases when X is a left child and has two black children

12.3. Deterministic Skip Lists

The ideas that we saw used for red black trees can be applied to skip lists to ensure logarithmic worst-case operation. In this section, we describe the simplest implementation of the resulting data structure, the *1-2-3 deterministic skip list*.

Recall from Chapter 10 that nodes in a skip list have randomly assigned heights. A node of height h contains h forward pointers $p_1, p_2, \ldots, p_h$; p_i points to the next node of height i or larger. The probability that a node has height h is 0.5^h (0.5 can be replaced by any number between 0 and 1.0, to implement time/space trade-offs). As a consequence, we expect to process only a few forward pointers until we drop down a level; since we will have roughly $\log N$ levels, we obtain $O(\log N)$ expected running time per operation.

To make this bound a worst-case bound, we need to guarantee that only a constant number of forward pointers need to be examined until we can drop down to a lower level. To do this, we add a balancing condition. First we need two definitions.

DEFINITION: Two elements are *linked* if there exists at least one pointer going from one to another.

DEFINITION: The *gap size* between two elements linked at height h is equal to the number of elements of height $h - 1$ between them.

A 1-2-3 deterministic skip list satisfies the property that every gap (except possibly the zero gap between the header and tail) is of size 1, 2, or 3. As an example, Figure 12.18 shows a 1-2-3 deterministic skip list. There are two gaps of size 3: The first is the three elements of height 1 that are between 25 and 45, and the second is the three elements of height 2 between the list header and tail. The tail node contains ∞; its presence simplifies the algorithms and also makes it easier to define the notion of a gap at the end of the list.

Clearly, only a constant number of pointers are traversed along any level before we drop to a lower level. Consequently, the searching time is $O(\log N)$ in the worst case.

To perform insertion, we must make sure that when a new node of height h is added, it doesn't create a gap of four height h nodes. This turns out to be simple. We adopt a top-down strategy that is similar to what was done with red black trees.

Let us suppose we are on level L and are about to drop one level. If the gap we are about to drop into has size 3, then we raise the middle item in the gap to have height L, thereby forming two gaps of size 1. Since this eliminates gaps of size 3 on the way toward the insertion, we know that the insertion is safe, as is any increase of middle item heights.

As an example, Figure 12.19 shows the insertion of item 27 into the deterministic skip list in Figure 12.18. At the header node, we are about to drop from level 3 to level 2. Since the drop would be into a 3 gap, the middle item (25) is raised to height 3 and spliced in. The search at level 2 takes us to 25, at which point we need to drop to level 1. Again we see a 3 gap, so 35 is raised to height 2. The result is shown in Figure 12.20. When it is time to insert 27, it is spliced into the list, as shown in Figure 12.21.

Difficulty in deletion occurs with gaps of size 1. When we see that we are about to drop into a 1 gap, we enlarge it, either by borrowing from a neighbor (if it is not a 1 gap) or by lowering the height of the node that separates the gap from the neighbor. Since both of these are gaps of size 1, the result is a 3 gap. The code is a bit more complex than this description because there are several cases to deal with.

Figure 12.18 A 1-2-3 deterministic skip list

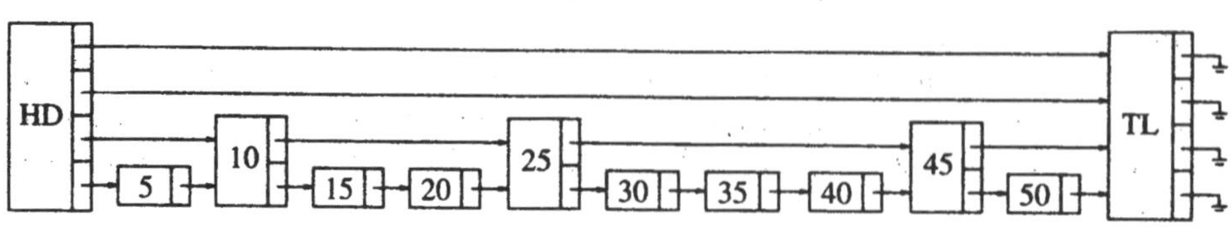

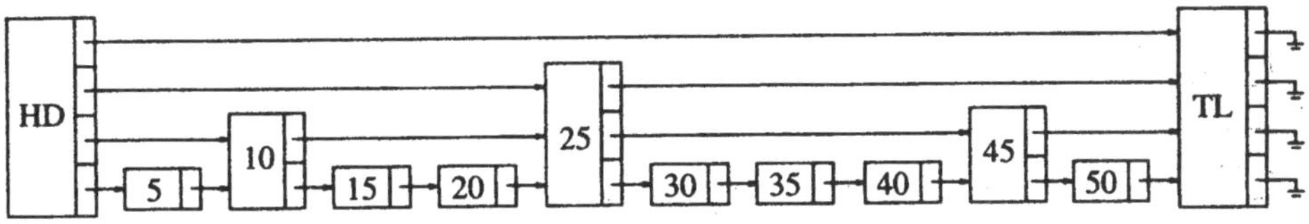

Figure 12.19 Insertion of 27: first, a gap of three height 2 nodes is split by raising 25

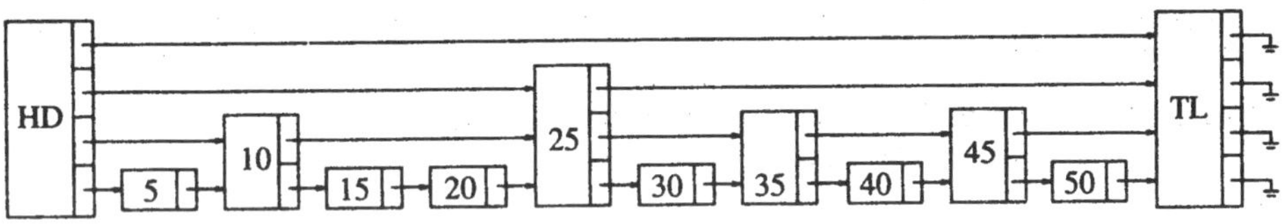

Figure 12.20 Insertion of 27: next, a gap of three height 1 nodes is split by raising 35

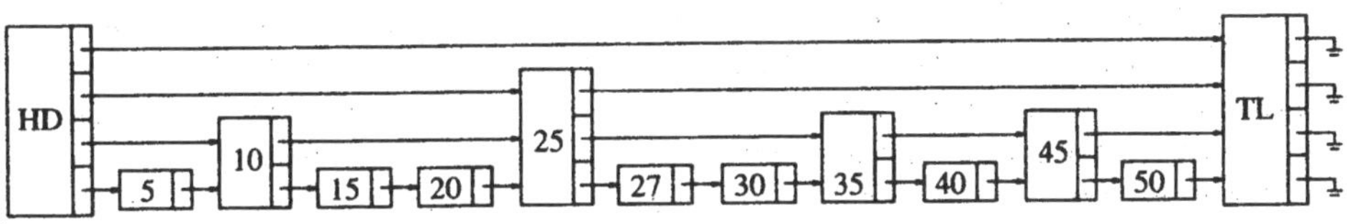

Figure 12.21 Insertion of 27: finally, 27 is inserted as height 1 node

How is all of this implemented? After we describe all the details, we will see that the amount of code is actually quite small.

The first important detail is that when we promote a height h node to height $h + 1$, we can't spend the $O(h)$ time that would be used to copy h pointers to a new array. Otherwise, the time bound would be $O(\log^2 N)$ for insertion. A reasonable method is to represent the h forward pointers in a height h node by a linked list. Since we go down levels, the linked list for a node would start with the level h forward pointer and end with the level 1 forward pointer.

The second optimization is more tricky, and could cost space. Instead of storing a node as an item and a linked list of forward pointers, we store a linked list of forward pointer, forward item pairs. The easiest way to see what this means is to look at Figure 12.22, which is another representation of Figure 12.21. We'll use the term *abstract* or *logical* representation to describe Figure 12.21 and refer to Figure 12.22 as the (actual) implementation.

First, notice that the skyline (i.e., heights as we scan from left to right) of both the abstract representation and actual implementation are identical except that the tail node has been removed. In our implementation, each node maintains a pointer that allows us to descend a level, a pointer to the next node on the same level, and the item that is logically stored in that next item (as shown in our original abstract description).

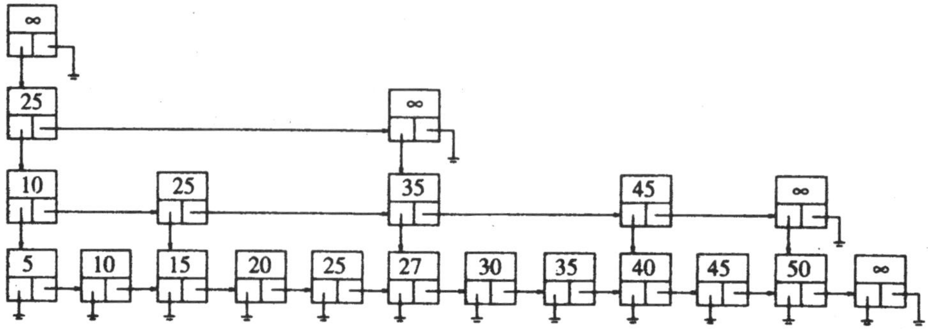

Figure 12.22 Linked list implementation of 1-2-3 deterministic skip list in Figure 12.21

Notice that some items appear more than once: for instance, 25 appears in three places. Indeed, if a node has height h in the abstract representation, its item will appear in h places in our actual implementation. There are important consequences and surprising results that we will explain after we provide an implementation.

The basic node consists of a key and two pointers. To make the coding faster and more simple, we have used the tail node; if it is impossible or undesirable to assign ∞, then some other mechanism must be used. We'll also have a sentinel for the header, and a sentinel for the bottom, to replace the *NULL* pointers. The declarations and an initialization routine are shown in Figure 12.23.

The searching function is the same as for randomized skip lists. Figure 12.24 shows that if we don't have a match, then we either go down or right, depending on the result of a comparison. Insertion, shown in Figure 12.25, is greatly simplified by the sentinels. As we can see by some of the outrageous pointer trails, if we had to test each pointer against *NULL*, we would easily triple the size of the code.

As Figure 12.25 indicates, the code for deterministic skip list insertion is somewhat shorter, with many fewer cases than for the red black tree. The price we pay seems to be space: In the worst case we have $2N$ nodes that contain two pointers and an item. For a red black tree, we have N nodes that contain two pointers, an item, and a color bit. So we might be using twice as much space. However, things aren't necessarily that bad. First, experiments suggest that on average the deterministic skip list has about $1.57N$ nodes. Second, in some cases, the deterministic skip list actually uses less space than the red black tree.

Here's a real-life example. On a 32-bit machine, pointers and integers are 4 bytes. On some systems, including some versions of UNIX, memory is allocated in chunks that are powers of 2, but 4 bytes of that chunk are used by the memory management routines. Thus a request for 12 bytes is filled by a 16-byte chunk: 12 bytes for the user and 4 bytes overhead. A request for 13 bytes, however, must be filled by a 32-byte chunk. So, in this case, a deterministic skip list will use 16 bytes per node, and on average there are $1.57N$ nodes, so the total is typically about $25N$ bytes. The red black tree uses $32N$ bytes! This illustrates that on some machines an extra bit is very expensive; this is one of the attractions of self-organizing structures.

```
struct SkipNode
{
    ElementType Element;
    SkipList    Right;
    SkipList    Down;
};

static Position Bottom  = NULL; /* Needs initialization */
static Position Tail    = NULL; /* Needs initialization */

/* Initialization procedure */

SkipList
Initialize( void )
{
    SkipList L;

    if( Bottom == NULL )
    {
        Bottom = malloc( sizeof( struct SkipNode ) );
        if( Bottom == NULL )
            FatalError( "Out of space!!!" );
        Bottom->Right = Bottom->Down = Bottom;

        Tail = malloc( sizeof( struct SkipNode ) );
        if( Tail == NULL )
            FatalError( "Out of space!!!" );
        Tail->Element = Infinity;
        Tail->Right = Tail;
    }

    /* Create the header node */
    L = malloc( sizeof( struct SkipNode ) );
    if( L == NULL )
        FatalError( "Out of space!!!" );
    L->Element = Infinity;
    L->Right = Tail;
    L->Down = Bottom;

    return L;
}
```

Figure 12.23 Deterministic skip list: types and initialization (none of this is part of the header file)

```
/* Return position of node containing Item, */
/* or Bottom if not found */

Position
Find( ElementType Item, SkipList L )
{
    Position Current = L;

    Bottom->Element = Item;
    while( Item != Current->Element )
        if( Item < Current->Element )
            Current = Current->Down;
        else
            Current = Current->Right;

    return Current;
}
```

Figure 12.24 Deterministic skip list: *Find* routine

```
SkipList
Insert( ElementType Item, SkipList L )
{
    Position Current = L;
    Position NewNode;

    Bottom->Element = Item;
    while( Current != Bottom )
    {
        while( Item > Current->Element )
            Current = Current->Right;

        /* If gap size is 3 or at bottom level */
        /* and must insert, then promote the middle element */
        if( Current->Element >
                Current->Down->Right->Right->Element )
        {
            NewNode = malloc( sizeof( struct SkipNode ) );
            if( NewNode == NULL )
                FatalError( "Out of space!!!" );
            NewNode->Right = Current->Right;
            NewNode->Down = Current->Down->Right->Right;
            Current->Right = NewNode;
            NewNode->Element = Current->Element;
            Current->Element = Current->Down->Right->Element;
        }
        else
            Current = Current->Down;
    }
```

Figure 12.25 Deterministic skip list: insertion procedure

```
        /* Raise height of DSL if necessary */
        if( L->Right != Tail )
        {
            NewNode = malloc( sizeof( struct SkipNode ) );
            if( NewNode == NULL )
                FatalError( "Out of space!!!" );
            NewNode->Down = L;
            NewNode->Right = Tail;
            NewNode->Element = Infinity;
            L = NewNode;
        }

        return L;
    }
```

Figure 12.25 Deterministic skip list: insertion procedure *(continued)*

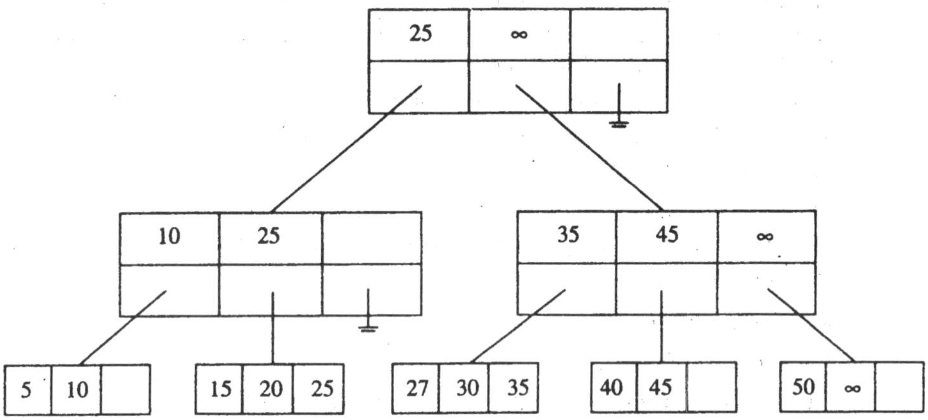

Figure 12.26 Horizontal array implementation of Figure 12.22

The performance of deterministic skip lists seems to compare favorably with red black trees. When looking for improvement in the insertion time, the line of code

```
if( Current->Element > Current->Down->Right->Right->Element )
```

seems to stand out;* if we store items in an array of up to three elements, the access of the third item could be direct, rather than through two *Right* pointers. Figure 12.26 shows the resulting structure, which, ironically, bears strong resemblance to the B-tree discussed in Chapter 4. This is known as the *horizontal array implementation*

*Indeed, the more "obvious" test

```
Current->Element == Current->Down->Right->Right->Right->Element
```

takes 20 percent longer on some systems!

of the 1-2-3 deterministic skip list. Just as there are higher-order B-trees, we can have higher-order deterministic skip lists, in both linked list and horizontal array forms. Which of these methods is best remains to be studied, and may well depend on the particular system and application.

12.4. AA-Trees

Because of a host of possible rotations, the red black tree is fairly tricky to code, especially for deletion. The deterministic skip list requires somewhat less code, but it is still quite tricky, as indicated by the three sentinels that are required. Deletion in a deterministic skip list is certainly a nontrivial task. In this section we describe a simple but competitive implementation of the *binary B-tree*, known as a BB-tree. A BB-tree is a red black tree with one extra condition: A node may have at most one red child. To make coding easier, we adopt a few rules.

1. First, we add the condition that only right children can be red. This eliminates about half of the possible restructuring cases. It also eliminates an annoying case in the deletion algorithm: if an internal node has only one child, the child must be a right child (that happens to be red), because a black left child would violate condition 4 for red black trees. Thus, we can always replace an internal node with the smallest node in its right subtree.
2. We code our procedures recursively.
3. Instead of storing a color bit with each node, we store information in a short integer (for instance, eight bits). This information is the *level* of a node. The level of a node is
    - One if the node is a leaf.
    - The level of its parent, if the node is red.
    - One less than the level of its parent, if the node is black.

The result is an AA-tree. Figure 12.27 shows the type declarations that are used for an AA-tree. Once again, we use a sentinel to represent *NULL*.

If we translate the AA structure requirements from colors to levels, we see that the left child must be exactly one level lower than its parent, and the right child may be zero or one level lower than its parent (but not more).

A *horizontal link* is a connection between a node and a child of equal levels; the structure requirements mandate that horizontal links are right pointers, and that there may not be two consecutive horizontal links. Figure 12.28 shows a sample AA-tree. Searching is done using the usual algorithm. Insertion of a new item is always done at the bottom level. However, two problems can result: insertion of a 2 would generate a left horizontal link, while insertion of 45 would create two consecutive right horizontal links.

In both cases a single rotation fixes the problem: we remove left horizontal links by right rotations, and consecutive right horizontal links by a left rotation. These

```
    /* Returned for failures */
Position NullNode = NULL;    /* Needs more initialization */

struct AANode
{
    ElementType Element;
    AATree      Left;
    AATree      Right;
    int         Level;
};

AATree
Initialize( void )
{
    if( NullNode == NULL )
    {
        NullNode = malloc( sizeof( struct AANode ) );
        if( NullNode == NULL )
            FatalError( "Out of space!!!" );
        NullNode->Left = NullNode->Right = NullNode;
        NullNode->Level = 0;
    }
    return NullNode;
}
```

Figure 12.27 AA-trees: some type declarations and initialization

procedures are called *Skew* and *Split*, respectively. Figure 12.29 shows the code for these primitives. A *Skew* removes a left horizontal link, but may create consecutive right horizontal links; thus we process *Skew* first, and then *Split*. After a *Split*, the middle node *R* increases in level. This may cause problems for the original parent of *X* by creating either a left horizontal node or consecutive right horizontal nodes; both problems can be fixed by percolating up the *Skew/Split* strategy. This is done automatically if we use recursion. Figure 12.30 depicts both procedures.

The actions taken to insert 45 in the AA-tree in Figure 12.28 are shown in Figures 12.31 through 12.35. The insertion procedure is then only two lines longer than an unbalanced implementation, as shown in Figure 12.36.

Figure 12.28 AA-tree resulting from insertion of 10, 85, 15, 70, 20, 60, 30, 50, 65, 80, 90, 40, 5, 55, 35

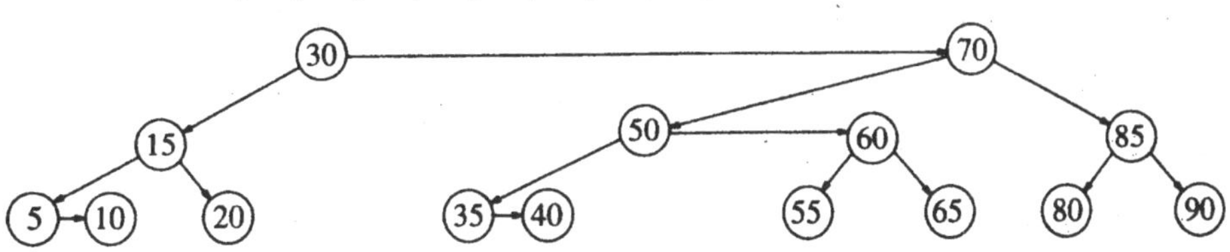

```
/* If T's left child is on the same level as T, */
/* perform a rotation */

AATree
Skew( AATree T )
{
    if( T->Left->Level == T->Level )
        T = SingleRotateWithLeft( T );
    return T;
}

/* If T's rightmost grandchild is on the same level, */
/* rotate right child up */

AATree
Split( AATree T )
{
    if( T->Right->Right->Level == T->Level )
    {
        T = SingleRotateWithRight( T );
        T->Level++;
    }
    return T;
}
```

Figure 12.29 AA-trees: *Skew* and *Split* procedures

Deletion is, of course, more complex, but since we removed many of the special cases, the code is actually pretty reasonable. Recall, first of all, that if a node is not a leaf, then it must have a right child. This means that when deleting a node, we can always replace the node with the smallest child in the right subtree, which is guaranteed to be at level 1.

To help us out, we keep two static local variables *DeletePtr* and *LastPtr*. These must be static because *Remove* is a recursive procedure. When we traverse a right pointer, we adjust *DeletePtr*; because we call *Remove* recursively until we reach the bottom (we don't test for equality on the way down), we are guaranteed that if the item to be removed is in the tree, *DeletePtr* will be pointing at the node that contains it.* *LastPtr* points at the leaf at which the search terminates. Because we don't stop until we reach the bottom, if the item is in the tree, *LastPtr* will point at the level 1 node that contains the replacement value, and must be removed from the tree.

When we reach the bottom of the tree, we perform step 2, which copies the level 1 node value into the internal node and then calls *free* to remove the level 1 node.

Nonleaf nodes check to see if their levels have been destroyed by a recursive call. Let *T* be the current node. If the deletion has lowered the level of one of *T*'s

*This technique can be used in the *Find* procedure to replace the three-way comparisons done at each node with two-way comparisons at each node, plus one extra equality test at the bottom.

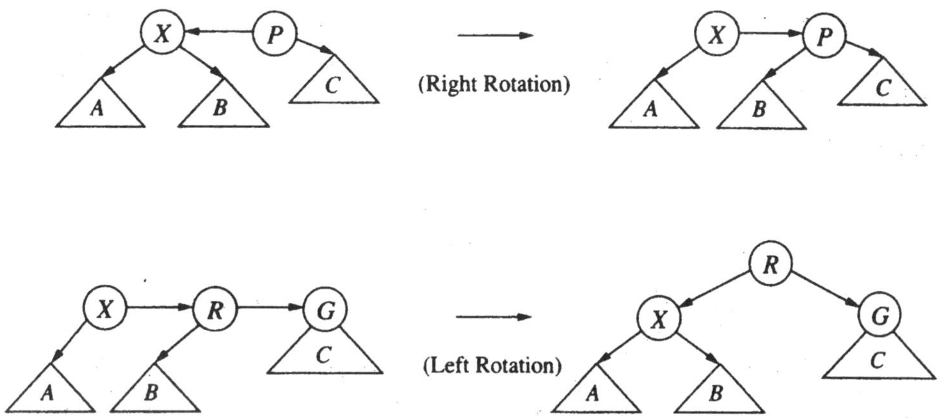

Figure 12.30 *Skew* and *Split*. Note that *R*'s level increases in a *Split*

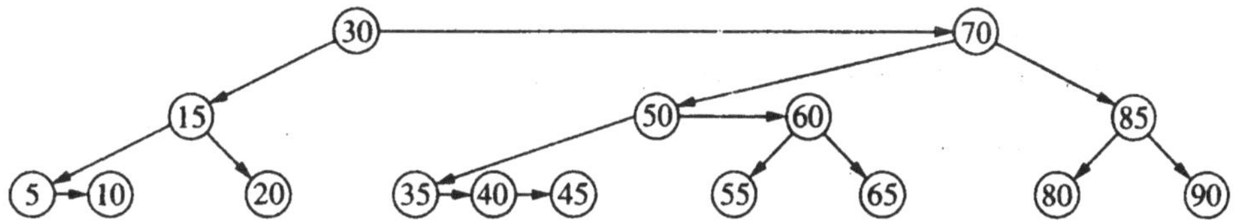

Figure 12.31 After inserting 45 into sample tree

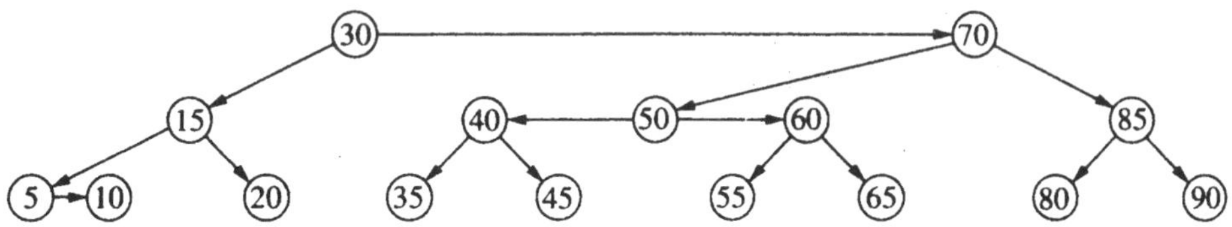

Figure 12.32 After *Split* at 35

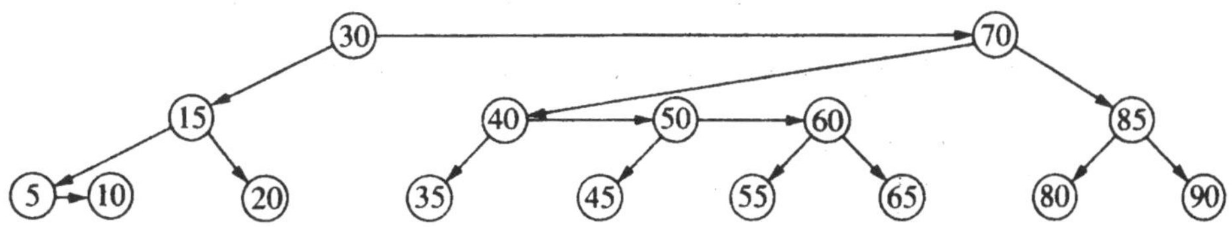

Figure 12.33 After *Skew* at 50

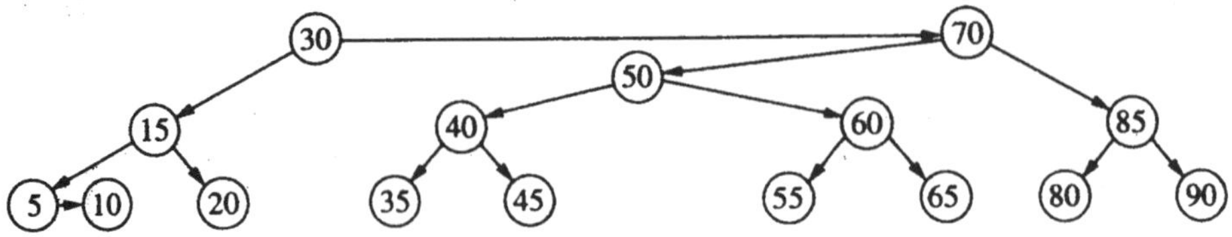

Figure 12.34 After *Split* at 40

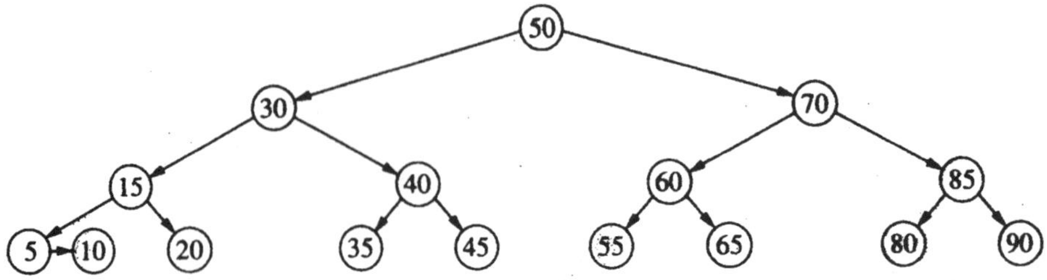

Figure 12.35 Final tree after *Skew* 70 and *Split* 30

```
AATree
Insert( ElementType Item, AATree T )
{
    if( T == NullNode )
    {
        /* Create and return a one-node tree */
        T = malloc( sizeof( struct AANode ) );
        If( T == NULL )
            FatalError( "Out of space!!!" );
        else
        {
            T->Element = Item; T->Level = 1;
            T->Left = T->Right = NullNode;
        }
        return T;
    }
    else
    if( Item < T->Element )
        T->Left = Insert( Item, T->Left );
    else
    if( Item > T->Element )
        T->Right = Insert( Item, T->Right );

    /* Otherwise it's a duplicate; do nothing */

    T = Skew( T );
    T = Split( T );
    return T;
}
```

Figure 12.36 AA-trees: insertion procedure

children (only the child entered by the recursive call could actually be affected, but for simplicity, we don't keep track of it) to two less than *T*'s level, then *T*'s level needs to be lowered also. Furthermore, if *T* has a right red child, *T*'s right child must also have its level lowered. At this point, we could have six nodes on the same level: *T*, *T*'s right red child *R*, *R*'s two children, and those children's right red children. Figure 12.37 shows the simplest possible scenario.

After node 1 is removed, node 2 and thus node 5 become level 1 nodes. First we must fix the left horizontal link that is now introduced between nodes 5 and 3. This essentially requires two rotations (one between nodes 5 and 3, and then one between nodes 5 and 4). In this case, the current node, *T*, is not involved. On the other hand, if a deletion came from the right side, then *T*'s left node could suddenly become horizontal; that would also require a similar double rotation (starting at *T*). To avoid testing all these cases, we just call *Skew* three times. Once we've done that, two calls to *Split* suffice to rearrange the horizontal edges. The entire deletion routine is shown in Figure 12.38. All in all, this is a relatively simple data structure to code.

Figure 12.37 When 1 is deleted, all nodes become level 1, introducing horizontal left links. Getting links to point right is accomplished by three calls to *Skew*. Two calls to *Split* remove consecutive horizontal links.

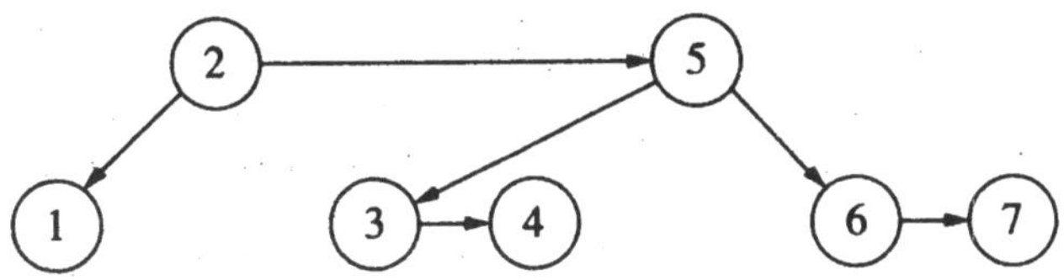

Figure 12.38 AA-Trees: deletion procedure

```
AATree
Remove( ElementType Item, AATree T )
{
    static Position DeletePtr, LastPtr;

    if( T != NullNode )
    {
        /* Step 1: Search down tree */
        /*         set LastPtr and DeletePtr */
        LastPtr = T;
        if( Item < T->Element )
            T->Left = Remove( Item, T->Left );
        else
        {
            DeletePtr = T;
            T->Right = Remove( Item, T->Right );
        }
```

```
                /* Step 2: If at the bottom of the tree and */
                /*         item is present, we remove it */
                if( T == LastPtr )
                {
                    if( DeletePtr != NullNode &&
                            Item == DeletePtr->Element )
                    {
                        DeletePtr->Element = T->Element;
                        DeletePtr = NullNode;
                        T = T->Right;
                        free( LastPtr );
                    }
                }

                /* Step 3: Otherwise, we are not at the bottom; */
                /*         rebalance */
                else
                    if( T->Left->Level < T->Level - 1 ||
                        T->Right->Level < T->Level - 1 )
                    {
                        if( T->Right->Level > --T->Level )
                            T->Right->Level = T->Level;
                        T = Skew( T );
                        T->Right = Skew( T->Right );
                        T->Right->Right = Skew( T->Right->Right );
                        T = Split( T );
                        T->Right = Split( T->Right );
                    }
            }
            return T;
        }
```

Figure 12.38 AA-Trees: deletion procedure *(continued)*

12.5. Treaps

Our last type of binary search tree, known as the *treap*, is probably the simplest of all. Like the skip list, it uses random numbers and gives $O(\log N)$ expected time behavior for any input. Searching time is identical to an unbalanced binary search tree (and thus slower than balanced search trees), while insertion time is only slightly slower than a recursive unbalanced binary search tree implementation. Although deletion is much slower, it is still $O(\log N)$ expected time.

The treap is so simple that we can describe it without a picture. Each node in the tree stores an item, a left and right pointer, and a priority that is randomly assigned when the node is created. A treap is a binary search tree with the property that the node priorities satisfy heap order: any node's priority must be at least as large as its parent's.

A collection of distinct items each of which has a distinct priority can only be represented by one treap. This is easily deduced by induction, since the node with the lowest priority must be the root. Consequently, the tree is formed on the basis of the $N!$ possible arrangements of priority instead of the $N!$ item orderings. The type declarations are straightforward, requiring only the addition of the *Priority* field. The sentinel *NullNode* will have priority of ∞, as shown in Figure 12.39.

Insertion into the treap is simple: After an item is added as a leaf, we rotate it up the treap until its priority satisfies heap order. It can be shown that the expected number of rotations is less than 2. After the item to be deleted has been found, it can be deleted by increasing its priority to ∞ and rotating it down through the path of low-priority children. Once it is a leaf, it can be removed. The routines in Figure 12.40 and Figure 12.41 implement these strategies using recursion. A nonrecursive implementation is left for the reader (Exercise 12.17). For deletion, note that when the node is logically a leaf, it still has *NullNode* as both its left and right children. Consequently, it is rotated with the left child. After the rotation, T is *NullNode*, and the left child can be freed. Note also that our implementation assumes that there are no duplicates; if this is not true, then the *Remove* could fail (why?).

The treap is particularly easy to implement because we never have to worry about adjusting the priority field. One of the difficulties of the balanced tree approaches is that it is difficult to track down errors that result from failing to update balance information in the course of an operation. In terms of total lines for a reasonable insertion and deletion package, the treap, especially a nonrecursive implementation, seems like the hands-down winner.

Figure 12.39 Initialization for treaps

```
Treap
Initialize( void )
{
    if( NullNode == NULL )
    {
        NullNode = malloc( sizeof( struct TreapNode ) );
        if( NullNode == NULL )
            FatalError( "Out of space!!!" );
        NullNode->Left = NullNode->Right = NullNode;
        NullNode->Priority = Infinity;
    }
    return NullNode;
}
```

```
Treap
Insert( ElementType Item, Treap T )
{
    if( T == NullNode )
    {
        /* Create and return a one-node tree */
        T = malloc( sizeof( struct TreapNode ) );
        if( T == NULL )
            FatalError( "Out of space!!!" );
        else
        {
            T->Element = Item; T->Priority = Random( );
            T->Left = T->Right = NullNode;
        }
    }
    else
    if( Item < T->Element )
    {
        T->Left = Insert( Item, T->Left );
        if( T->Left->Priority < T->Priority )
            T = SingleRotateWithLeft( T );
    }
    else
    if( Item > T->Element )
    {
        T->Right = Insert( Item, T->Right );
        if( T->Right->Priority < T->Priority )
            T = SingleRotateWithRight( T );
    }

    /* Otherwise it's a duplicate; do nothing */

    return T;
}
```

Figure 12.40 Treaps: insertion routine

```
Treap
Remove( ElementType Item, Treap T )
{
    if( T != NullNode )
    {
        if( Item < T->Element )
            T->Left = Remove( Item, T->Left );
        else
        if( Item > T->Element )
            T->Right = Remove( Item, T->Right );
        else
        {
            /* Match found */
            if( T->Left->Priority < T->Right->Priority )
                T = SingleRotateWithLeft( T );
            else
                T = SingleRotateWithRight( T );

            if( T != NullNode )    /* Continue on down */
                T = Remove( Item, T );
            else
            {
                /* At a leaf */
                free( T->Left );
                T->Left = NullNode;
            }
        }
    }
    return T;
}
```

Figure 12.41 Treaps: deletion procedure

12.6. *k*-d Trees

Suppose that an advertising company maintains a database and needs to generate mailing labels for certain constituencies. A typical request might require sending out a mailing to people who are between the ages of 34 and 49 and whose annual income is between \$100,000 and \$150,000. This problem is known as a two-dimensional range query. In one dimension, the problem can be solved by a simple recursive algorithm in $O(M + \log N)$ average time, by traversing a preconstructed binary search tree. Here M is the number of matches reported by the query. We would like to obtain a similar bound for two or more dimensions.

The two-dimensional search tree has the simple property that branching on odd levels is done with respect to the first key, and branching on even levels is done with respect to the second key. The root is arbitrarily chosen to be an odd

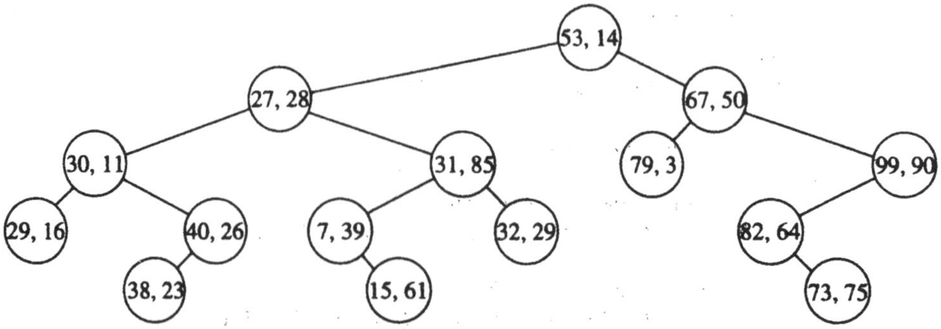

Figure 12.42 Sample 2-d tree

level. Figure 12.42 shows a 2-d tree. Insertion into a 2-d tree is a trivial extension of insertion into a binary search tree: as we go down the tree, we need to maintain the current level. To keep our code simple, we assume that a basic item is an array of two elements. We then need to toggle the level between 0 and 1. Figure 12.43 shows the code to perform an insertion. We use recursion in this section; a nonrecursive

Figure 12.43 Insertion into 2-d trees

```
static KdTree
RecursiveInsert( ItemType Item, KdTree T, int Level )
{
    if( T == NULL )
    {
        T = malloc( sizeof( struct KdNode ) );
        if( T == NULL )
            FatalError( "Out of space!!!" );
        T->Left = T->Right = NULL;
        T->Data[ 0 ] = Item[ 0 ];
        T->Data[ 1 ] = Item[ 1 ];
    }
    else
    if( Item[ Level ] < T->Data[ Level ] )
        T->Left = RecursiveInsert( Item, T->Left, !Level );
    else
        T->Right = RecursiveInsert( Item, T->Right, !Level );
    return T;
}

KdTree
Insert( ItemType Item, KdTree T )
{
    return RecursiveInsert( Item, T, 0 );
}
```

implementation that would be used in practice is straightforward and left as Exercise 12.23. One difficulty is duplicates, particularly since several items can agree in one field. Our code allows duplicates, and always places them in right branches; clearly this can be a problem if there are too many duplicates.

A moment's thought will convince you that a randomly constructed 2-d tree has the same structural properties as a random binary search tree: the height is $O(\log N)$ on average, but $O(N)$ in the worst case.

Unlike binary search trees, for which clever $O(\log N)$ worst-case variants exist, there are no schemes that are known to guarantee a balanced 2-d tree. The problem is that such a scheme would likely be based on tree rotations, and tree rotations don't work in 2-d trees. The best one can do is to periodically rebalance the tree by reconstructing a subtree, as described in the exercises. Similarly, there are no deletion algorithms beyond the obvious lazy deletion strategy. If all of the items arrive before we need to process queries, then we can construct a perfectly balanced 2-d tree in $O(N \log N)$ time; we leave this as Exercise 12.21c.

Several kinds of queries are possible on a 2-d tree. We can ask for an exact match, or a match based on one of the two keys; the latter type of request is a *partial match query*. Both of these are special cases of an (orthogonal) *range query*.

An orthogonal range query gives all items whose first key is between a specified set of values and whose second key is between another specified set of values. This is exactly the problem that was described in the introduction to this section. A range query is easily solved by a recursive tree traversal, as shown in Figure 12.44. By testing before making a recursive call, we can avoid unnecessarily visiting all nodes.

To find a specific item, we can set *Low* equal to *High* equal to the item we are searching for. To perform a partial match query, we set the range for the key not involved in the match to $-\infty$ to ∞. The other range is set with the low and high point equal to the value of the key involved in the match.

An insertion or exact match search in a 2-d tree takes time that is proportional to the depth of the tree, namely, $O(\log N)$ on average and $O(N)$ in the worst case. The running time of a range search depends on how balanced the tree is, whether or not a partial match is requested, and how many items are actually found. We mention three results that have been shown.

For a perfectly balanced tree, a range query could take $O(M + \sqrt{N})$ time in the worst case, to report M matches. At any node, we may have to visit two of the four grandchildren, leading to the equation $T(N) = 2T(N/4) + O(1)$. In practice, however, these searches tend to be very efficient, and even the worst case is not poor because for typical N, the difference between $\sqrt{N}$ and $\log N$ is compensated by the smaller constant that is hidden in the Big-Oh notation.

For a randomly constructed tree, the average running time of a partial match query is $O(M + N^{\alpha})$, where $\alpha = (-3 + \sqrt{17})/2$ (see below). A recent, and somewhat surprising, result is that this essentially describes the average running time of a range search of a random 2-d tree.

For k dimensions, the same algorithm works; we just cycle through the keys at each level. However, in practice the balance starts getting worse because typically the effect of duplicates and nonrandom inputs becomes more pronounced. We leave the coding details as an exercise for the reader, and mention the analytical results: For a perfectly balanced tree, the worst-case running time of a range query

```
/* Print items satisfying */
/* Low[ 0 ] <= Item[ 0 ] <= High[ 0 ] and */
/* Low[ 1 ] <= Item[ 1 ] <= High[ 1 ] */

static void
RecPrintRange( ItemType Low, ItemType High,
               KdTree T, int Level )
{
    if( T != NULL )
    {
        if( Low[ 0 ] <= T->Data[ 0 ] &&
                        T->Data[ 0 ] <= High[ 0 ] &&
                        Low[ 1 ] <= T->Data[ 1 ] &&
                        T->Data[ 1 ] <= High[ 1 ] )
            PrintItem( T->Data );

        if( Low[ Level ] <= T->Data[ Level ] )
            RecPrintRange( Low, High, T->Left, !Level );
        if( High[ Level ] >= T->Data[ Level ] )
            RecPrintRange( Low, High, T->Right, !Level );
    }
}

void
PrintRange( ItemType Low, ItemType High, KdTree T )
{
    RecPrintRange( Low, High, T, 0 );
}
```

Figure 12.44 2-d trees: range search

is $O(M + kN^{1-1/k})$. In a randomly constructed k-d tree, a partial match query that involves p of the k keys takes $O(M + N^{\alpha})$, where α is the (only) positive root of

$$(2 + \alpha)^p(1 + \alpha)^{k-p} = 2^k$$

Computation of α for various p and k is left as an exercise; the value for $k = 2$ and $p = 1$ is reflected in the result stated above for partial matching in random 2-d trees.

Although there are several exotic structures that support range searching, the k-d tree is probably the simplest such structure that achieves respectable running times.

12.7. Pairing Heaps

The last data structure we examine is the *pairing heap*. The analysis of the pairing heap is still open, but when *DecreaseKey* operations are needed, it seems to outperform other heap structures. The most likely reason for its efficiency is its simplicity. The pairing heap is represented as a heap ordered tree. Figure 12.45 shows a sample pairing heap.

The actual pairing heap implementation uses a left child, right sibling representation as discussed in Chapter 4. The *DecreaseKey* operation, as we will see, requires that each node contain an additional pointer. A node that is a leftmost child contains a pointer to its parent; otherwise the node is a right sibling, and contains a pointer to its left sibling. We'll refer to this field as the *Prev* field. The type declarations are omitted for brevity; they are completely straightforward. Figure 12.46 shows the actual representation of the pairing heap in Figure 12.45.

We begin by sketching the basic operations. To merge two pairing heaps, we make the heap with the larger root a left child of the heap with the smaller root. Insertion is, of course, a special case of merging. To perform a *DecreaseKey*, we lower the value in the requested node. Because we are not maintaining parent pointers for all nodes, we don't know if this violates the heap order. Thus we cut the adjusted node from its parent and complete the *DecreaseKey* by merging the two heaps that result. To perform a *DeleteMin*, we remove the root, creating a collection of heaps. If there are c children of the root, then $c - 1$ calls to the merge procedure will reassemble the heap. The most important detail is the method used to perform the merge and how the $c - 1$ merges are applied.

Figure 12.47 shows how two subheaps are combined. The procedure is generalized to allow the second subheap to have siblings. As we mentioned earlier, the subheap with the larger root is made a leftmost child of the other subheap. The

Figure 12.45 Sample pairing heap: abstract representation

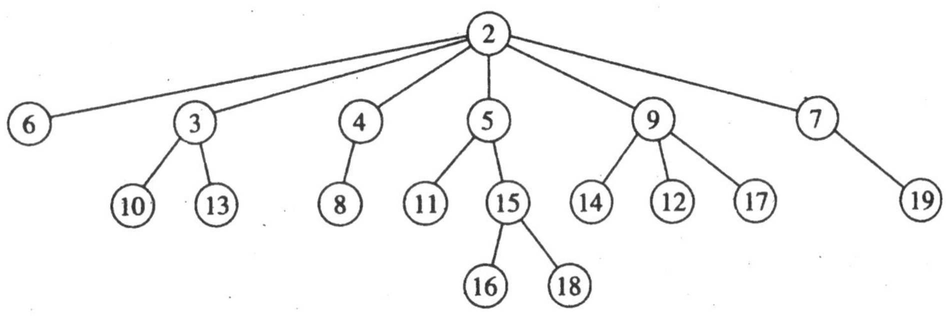

Figure 12.46 Actual representation of previous pairing heap

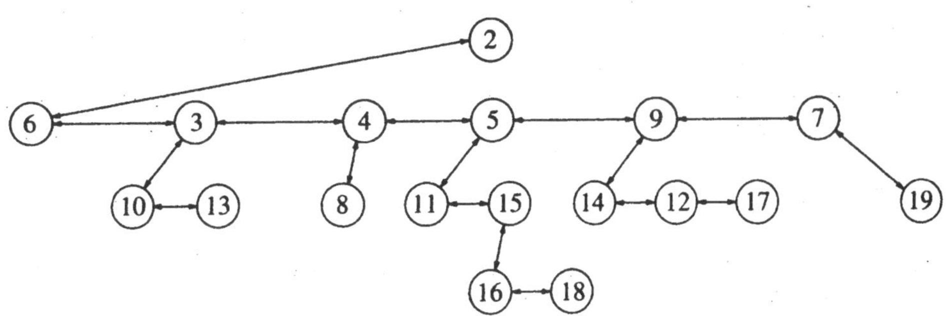

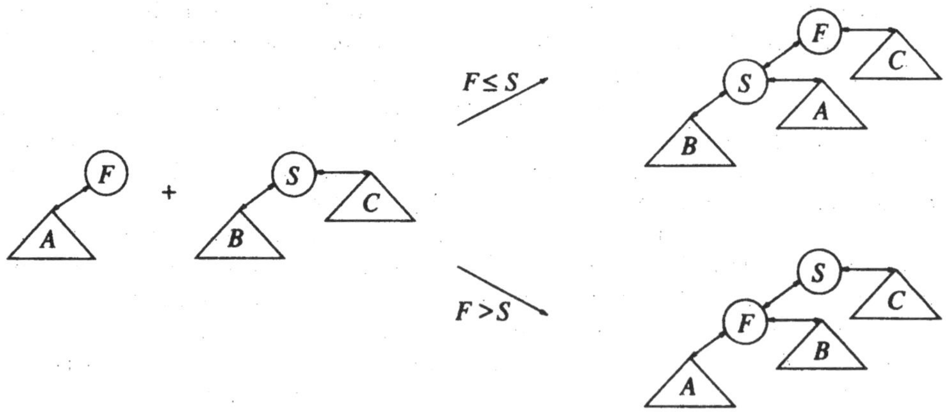

Figure 12.47 *CompareAndLink* merges two subheaps

code is straightforward and shown in Figure 12.48. Notice that we have several instances in which a pointer is tested against *NULL* before assigning its *Prev* field; this suggests that perhaps it would be useful to have a *NullNode* sentinel, which was customary in this chapter's search tree implementations.

The *Insert* and *DecreaseKey* operations are, then, simple implementations of the abstract description. *DecreaseKey* requires a *Position* object. Since an item's *Position* is determined (irrevocably) when it is first inserted, *Insert* sends the *Position* back to the caller through the third parameter, *Loc*, which is passed by reference. The code is shown in Figure 12.49. Our routine for *DecreaseKey* prints a warning message if the new key value is not smaller than the old. If this is the case, the resulting structure might not obey heap order. The basic *DeleteMin* procedure follows directly from the abstract description and is shown in Figure 12.50.

The devil, of course, is in the details: How is *CombineSiblings* implemented? Several variants have been proposed, but none has been shown to provide the same amortized bounds as the Fibonacci heap. Even so, the method coded in Figure 12.51 always seems to perform as well as or better than other heap structures, including the binary heap, for the typical graph theory uses that involve a host of *DecreaseKey* operations.

This method, known as *two-pass merging*, is the simplest and most practical of the many variants that have been suggested. We first scan left to right, merging pairs of children.* After the first scan, we have half as many trees to merge. A second scan is then performed, right to left. At each step we merge the rightmost tree remaining from the first scan with the current merged result. As an example, if we have eight children, c_1 through c_8, the first scan performs the merges c_1 and c_2, c_3 and c_4, c_5 and c_6, and c_7 and c_8. As a result we obtain d_1, d_2, d_3, and d_4. We perform the

*We must be careful if there is an odd number of children. When that happens, we merge the last child with the result of the rightmost merge to complete the first scan.

```
/* This is the basic operation to maintain order */
/* Links First and Second together to satisfy heap order */
/* Returns the resulting tree */
/* First is assumed NOT NULL */
/* First->NextSibling MUST be NULL on entry */

Position
CompareAndLink( Position First, Position Second )
{
    if( Second == NULL )
        return First;
    else
    if( First->Element <= Second->Element )
    {
        /* Attach Second as the leftmost child of First */
        Second->Prev = First;
        First->NextSibling = Second->NextSibling;
        if( First->NextSibling != NULL )
            First->NextSibling->Prev = First;
        Second->NextSibling = First->LeftChild;
        if( Second->NextSibling != NULL )
            Second->NextSibling->Prev = Second;
        First->LeftChild = Second;
        return First;
    }
    else
    {
        /* Attach First as the leftmost child of Second */
        Second->Prev = First->Prev;
        First->Prev = Second;
        First->NextSibling = Second->LeftChild;
        if( First->NextSibling != NULL )
            First->NextSibling->Prev = First;
        Second->LeftChild = First;
        return Second;
    }
}
```

Figure 12.48 Pairing heaps: routine to merge two subheaps

```
/* Insert Item into pairing heap H */
/* Return resulting pairing heap */
/* A pointer to the newly allocated node */
/* is passed back by reference and accessed as *Loc */

PairHeap
Insert( ElementType Item, PairHeap H, Position *Loc )
{
    Position NewNode;

    NewNode = malloc( sizeof( struct PairNode ) );
    if( NewNode == NULL )
        FatalError( "Out of space!!!" );
    NewNode->Element = Item;
    NewNode->LeftChild = NewNode->NextSibling = NULL;
    NewNode->Prev = NULL;

    *Loc = NewNode;
    if( H == NULL )
        return NewNode;
    else
        return CompareAndLink( H, NewNode );
}


/* Lower item in Position P by Delta */

PairHeap
DecreaseKey( Position P, ElementType Delta, PairHeap H )
{
    if( Delta < 0 )
        Error( "DecreaseKey called with negative Delta" );

    P->Element -= Delta;
    if( P == H )
        return H;

    if( P->NextSibling != NULL )
        P->NextSibling->Prev = P->Prev;
    if( P->Prev->LeftChild == P )
        P->Prev->LeftChild = P->NextSibling;
    else
        P->Prev->NextSibling = P->NextSibling;

    P->NextSibling = NULL;
    return CompareAndLink( H, P );
}
```

Figure 12.49 Pairing heaps: *Insert* and *DecreaseKey*

```
PairHeap
DeleteMin( ElementType *MinItem, PairHeap H )
{
    Position NewRoot = NULL;

    if( IsEmpty( H ) )
        Error( "Pairing heap is empty!" );
    else
    {
        *MinItem = H->Element;
        if( H->LeftChild != NULL )
            NewRoot = CombineSiblings( H->LeftChild );
        free( H );
    }
    return NewRoot;
}
```

Figure 12.50 Pairing heap *DeleteMin*

```
/* Assumes FirstSibling is NOT NULL */

PairHeap
CombineSiblings( Position FirstSibling )
{
    static Position TreeArray[ MaxSiblings ];
    int i, j, NumSiblings;

    /* If only one tree, return it */
    if( FirstSibling->NextSibling == NULL )
        return FirstSibling;

    /* Place each subtree in TreeArray */
    for( NumSiblings = 0; FirstSibling != NULL; NumSiblings++ )
    {
        TreeArray[ NumSiblings ] = FirstSibling;
        FirstSibling->Prev->NextSibling = NULL; /* Break links */
        FirstSibling = FirstSibling->NextSibling;
    }
    TreeArray[ NumSiblings ] = NULL;

    /* Combine the subtrees two at a time, */
    /* going left to right */
    for( i = 0; i + 1 < NumSiblings; i += 2 )
        TreeArray[ i ] = CompareAndLink(
                TreeArray[ i ], TreeArray[ i + 1 ] );
```

Figure 12.51 Pairing heaps: two-pass merging

```
        /* j has the result of the last CompareAndLink */
        /* If an odd number of trees, get the last one */
        j = i - 2;
        if( j == NumSiblings - 3 )
            TreeArray[ j ] = CompareAndLink(
                    TreeArray[ j ], TreeArray[ j + 2 ] );

        /* Now go right to left, merging last tree with */
        /* next to last. The result becomes the new last */
        for( ; j >= 2; j -= 2 )
            TreeArray[ j - 2 ] = CompareAndLink(
                    TreeArray[ j - 2 ], TreeArray[ j ] );

        return TreeArray[ 0 ];
    }
```

Figure 12.51 Pairing heaps: two-pass merging *(continued)*

second pass by merging d_3 and d_4; d_2 is then merged with that result, and then d_1 is merged with the result of the previous merge.

Our implementation requires an array to store the subtrees. In the worst case, $N - 1$ items could be children of the root, so this array must be large.

Other merging strategies are discussed in the exercises. The only simple merging strategy that is easily seen to be poor is a left-to-right single-pass merge (Exercise 12.35). The pairing heap is a good example of "simple is better" and seems to be the method of choice for serious applications requiring the *DecreaseKey* or *Merge* operation.

Summary

In this chapter, we've seen several efficient variations of the binary search tree. The top-down splay tree provides $O(\log N)$ amortized performance, the treap gives $O(\log N)$ randomized performance, and the red black tree, deterministic skip list, and AA tree all give $O(\log N)$ worst-case performance for the basic operations. The trade-offs between the various structures involve code complexity, ease of deletion, and differing searching and insertion costs. It is difficult to say that any one structure is a clear winner. Recurring themes include tree rotations and the use of sentinel nodes to eliminate many of the annoying tests for *NULL* pointers that would otherwise be necessary. The *k*-d tree provides a practical method for performing range searches, even though the theoretical bounds are not optimal.

Finally, we described and coded the pairing heap, which seems to be the most practical mergeable priority queue, especially when *DecreaseKey* operations are required. Unfortunately, the empirical results have not yet been confirmed analytically.

Exercises

12.1 Prove that the amortized cost of a top-down splay is $O(\log N)$.

**12.2 Prove that there exist access sequences that require $2\log N$ rotations per access for bottom-up splaying. Show that a similar result holds for top-down splaying.

12.3 Modify the splay tree to support queries for the kth smallest item. How would this be done in a deterministic skip list?

12.4 Compare, empirically, the simplified top-down splay with the originally described top-down splay.

12.5 Write the deletion procedure for red black trees.

12.6 Prove that the height of a red black tree is at most $2\log N$, and that this bound cannot be substantially lowered.

12.7 Show that every AVL tree can be colored as a red black tree. Are all red black trees AVL?

12.8 Show that a 1-2-3 deterministic skip list can be represented as a 2-3-4 tree, with items at internal nodes as well as leaves.

12.9 What happens if we try to insert an item that is already in the deterministic skip list?

12.10 Show that at most $2N$ nodes are used in a 1-2-3 deterministic skip list.

*12.11 In C, we can represent each abstract node as a dynamically allocated array of forward pointers, instead of a linked list of pointers. Show how to implement a 1-2-3 deterministic skip list in this scheme and maintain the $O(\log N)$ bound for each operation.

12.12 Write the deletion procedure for a 1-2-3 deterministic skip list.

12.13 Prove that the algorithm for deletion in AA-trees is correct.

12.14 Give a nonrecursive top-down implementation of AA-trees. Compare the implementation with the text's for simplicity and efficiency.

12.15 Write the *Skew* and *Split* procedures recursively, so that only one call of each is need for deletion.

12.16 How many fewer lines of code than the BB-tree does the AA-tree use? Does this make AA-trees faster?

12.17 Implement the insertion routine for treaps nonrecursively by maintaining a stack. Is it worth the effort?

12.18 We can make treaps self-adjusting by using the number of accesses as a priority and performing rotations as needed after each access. Compare this method with the randomized strategy. Alternatively, generate a random number each time an item X is accessed. If this number is smaller than X's current priority, use it as X's new priority (performing the appropriate rotation).

**12.19 Show that if the items are sorted, then a treap can be constructed in linear time, even if the priorities are not sorted.

12.20 Implement some of the tree structures without using the *NullNode* sentinel. How much coding effort is saved by using the sentinel?

12.21 Suppose we store, for each node, the number of *NULL* pointers in its subtree; call this the node's *weight*. Adopt the following strategy: If the left and right subtrees have weights that are not within a factor of 2 of each other, then completely rebuild the subtree rooted at the node. Show the following:

a. We can rebuild a node in $O(S)$, where S is the weight of the node.

b. The algorithm has amortized cost of $O(\log N)$ per insertion.

c. We can rebuild a node in a k-d tree in $O(S \log S)$ time, where S is the weight of the node.

d. We can apply the algorithm to k-d trees, at a cost of $O(\log^2 N)$ per insertion.

12.22 Suppose we call *SingleRotateWithLeft* on an arbitrary 2-d tree. Explain in detail all the reasons that the result is no longer a usable 2-d tree.

12.23 Implement the insertion and range search for the k-d tree. Do not use recursion.

12.24 Determine the time for partial match query for values of P corresponding to $k = 3$, 4, and 5.

12.25 For a perfectly balanced k-d tree, derive the worst-case running time of a range query that is quoted in the text (see pp. 487–488).

12.26 The *2-d heap* is a data structure that allows each item to have two individual keys. *DeleteMin* can be performed with respect to either of these keys. The 2-d heap is a complete binary tree with the following order property: For any node X at even depth, the item stored at X has the smallest key #1 in its subtree, while for any node X at odd depth, the item stored at X has the smallest key #2 in its subtree.

a. Draw a possible 2-d heap for the items (1,10), (2,9), (3,8), (4,7), (5,6).

b. How do we find the item with minimum key #1?

c. How do we find the item with minimum key #2?

d. Give an algorithm to insert a new item into the 2-d heap.

e. Give an algorithm to perform *DeleteMin* with respect to either key.

f. Give an algorithm to perform *FixHeap* in linear time.

12.27 Generalize the preceding exercise to obtain a *k-d heap,* in which each item can have k individual keys. You should be able to obtain the following bounds: *Insert* in $O(\log N)$, *DeleteMin* in $O(2^k \log N)$, and *FixHeap* in $O(kN)$.

12.28 Show that the k-d heap can be used to implement a double-ended priority queue.

12.29 Abstractly, generalize the k-d heap so that only levels that branch on key 1 have two children (all others have one).

a. Do we need pointers?

b. Clearly, the basic algorithms still work; what are the new time bounds?

12.30 Use a *k*-d tree to implement *DeleteMin*. What would you expect the average running time to be for a random tree?

12.31 Use a *k*-d heap to implement a *deque* (Exercise 3.26) that also supports *DeleteMin*.

12.32 Implement the pairing heap with a *NullNode* sentinel.

**12.33 Show that the amortized cost of each operation is $O(\log N)$ for the pairing heap algorithm in the text.

12.34 An alternative method for *CombineSiblings* is to place all of the siblings on a queue, and repeatedly *Dequeue* and merge the first two items on the queue, placing the result at the end of the queue. Implement this variation.

12.35 Show that using a stack instead of a queue in the previous exercise is bad, by giving a sequence that leads to $\Omega(N)$ cost per operation. This is the left-to-right single-pass merge.

12.36 Without *DecreaseKey*, we can remove parent pointers. How competitive is the result with the skew heap?

References

Top-down splay trees were described in the original splay tree paper [27]. A similar strategy, but without the crucial rotation, was described in [29]. The top-down red black tree algorithm is from [16]; a more accessible description can be found in [26]. An implementation of top-down red black trees without sentinel nodes is given in [14]; this provides a convincing demonstration of the usefulness of *NullNode*. Deterministic skip lists and their variants are discussed in [22] and [25]. Symmetric binary B-trees are from [6]; the AA-tree implementation shown in the text is adapted from the description in [1] and [3]. Treaps [4] are based on the *Cartesian tree* described in [30]. A related data structure is the *priority search tree* [20].

The *k*-d tree was first presented in [7]. Other range-searching algorithms are described in [8]. The worst case for range searching in a balanced *k*-d tree was obtained in [18], and the average-case results cited in the text are from [13] and [10].

The pairing heap and the alternatives suggested in the exercises were described in [15]. The study [17] suggests that the splay tree is the priority queue of choice when the *DecreaseKey* operation is not required. Another study [28] suggests that the pairing heap achieves the same asymptotic bounds as the Fibonacci heap, with better performance in practice. However, a related study [21] using priority queues to implement minimum spanning tree algorithms suggests that the amortized cost of *DecreaseKey* is not $O(1)$.

The solutions to most of the exercises can be found in the primary references. Exercise 12.21 represents a "lazy" balancing strategy that has become somewhat popular. [19], [5], [11], and [9] describe specific strategies; [2] shows how to implement all of these strategies in one framework. A tree that satisfies the property in Exercise 12.21 is *weight-balanced*. These trees can also be maintained by rotations [23]. Part (d) is from [24]. A solution to Exercises 12.26 to 12.28 can be found in [12].

1. A. Andersson, "A Note on Searching a Binary Search Tree," *Software—Practice and Experience*, 21 (1991), 1125–1128.
2. A. Andersson, "General Balanced Trees," *Journal of Algorithms*, to appear.
3. A. Andersson, "Balanced Search Trees Made Simple," *Proceedings on the Third Workshop on Algorithms and Data Structures* (1993), 61–71.
4. C. Aragon and R. Seidel, "Randomized Search Trees," *Proceedings of the Thirtieth Annual Symposium on Foundations of Computer Science* (1989), 540–545.
5. J. L. Baer and B. Schwab, "A Comparison of Tree-Balancing Algorithms," *Communications of the ACM*, 20 (1977), 322–330.
6. R. Bayer, "Symmetric Binary B-Trees: Data Structure and Maintenance Algorithms," *Acta Informatica*, 1 (1972), 290–306.
7. J. L. Bentley, "Multidimensional Binary Search Trees Used for Associative Searching," *Communications of the ACM*, 18 (1975), 509–517.
8. J. L. Bentley and J. H. Friedman, "Data Structures for Range Searching," *Computing Surveys*, 11 (1979), 397–409.
9. H. Chang and S. S. Iyengar, "Efficient Algorithms to Globally Balance a Binary Search Tree," *Communications of the ACM*, 27 (1984), 695–702.
10. P. Chanzy, "Range Search and Nearest Neighbor Search," *Master's Thesis*, McGill University (1993).
11. A. C. Day, "Balancing a Binary Tree," *Computer Journal*, 19 (1976), 360–361.
12. Y. Ding and M. A. Weiss, "The k-d Heap: An Efficient Multi-Dimensional Priority Queue," *Proceedings of the Third Workshop on Algorithms and Data Structures* (1993), 302–313.
13. P. Flajolet and C. Puech, "Partial Match Retrieval of Multidimensional Data," *Journal of the ACM*, 33 (1986), 371–407.
14. B. Flamig, *Practical Data Structures in C++*, John Wiley, New York (1994).
15. M. L Fredman, R. Sedgewick, D. D. Sleator, and R. E. Tarjan, "The Pairing Heap: A New Form of Self-Adjusting Heap," *Algorithmica*, 1 (1986), 111–129.
16. L. J. Guibas and R. Sedgewick, "A Dichromatic Framework for Balanced Trees," *Proceedings of the Nineteenth Annual Symposium on Foundations of Computer Science* (1978), 8–21.
17. D. W. Jones, "An Empirical Comparison of Priority-Queue and Event-Set Implementations," *Communications of the ACM*, 29 (1986), 300–311.
18. D. T. Lee and C. K. Wong, "Worst-Case Analysis for Region and Partial Region Searches in Multidimensional Binary Search Trees and Balanced Quad Trees," *Acta Informatica*, 9 (1977), 23–29.
19. W. A. Martin and D. N. Ness, "Optimizing Binary Trees Grown with a Sorting Algorithm," *Communications of the ACM*, 15 (1972), 88–93.
20. E. McCreight, "Priority Search Trees," *SIAM Journal of Computing*, 14 (1985), 257–276.
21. B. M. E. Moret and H. D. Shapiro, "An Empirical Analysis of Algorithms for Constructing a Minimum Spanning Tree," *Proceedings of the Second Workshop on Algorithms and Data Structures*, (1991), 400–411.
22. J. I. Munro, T. Papadakis, and R. Sedgewick, "Deterministic Skip Lists," *Proceedings of the Third Annual Symposium of Discrete Algorithms* (1992), 367–375.
23. J. Nievergelt and E. M. Reingold, "Binary Search Trees of Bounded Balance," *SIAM Journal on Computing*, 2 (1973), 33–43.

24. M. H. Overmars and J. van Leeuwen, "Dynamic Multidimensional Data Structures Based on Quad and K-D Trees," *Acta Informatica*, 17 (1982), 267–285.
25. T. Papadakis, *Skip Lists and Probabilistic Analysis of Algorithms*, Ph.D. Dissertation, University of Waterloo (1993).
26. R. Sedgewick, *Algorithms in* C, Addison-Wesley, Reading, Mass. (1990).
27. D. D. Sleator and R. E. Tarjan, "Self Adjusting Binary Search Trees," *Journal of the ACM*, 32 (1985), 652–686.
28. J. T. Stasko and J. S. Vitter, "Pairing Heaps: Experiments and Analysis," *Communications of the ACM*, 30 (1987), 234–249.
29. C. J. Stephenson, "A Method for Constructing Binary Search Trees by Making Insertions at the Root," *International Journal of Computer and Information Science*, 9 (1980), 15–29.
30. J. Vuillemin, "A Unifying Look at Data Structures," *Communications of the ACM*, 23 (1980), 229–239.

Index

Q

R

S

T